To my wife Ilona

Preface

"Evolving: to develop gradually; to evolve a scheme."
Dictionary.com

In today's industrial systems, economic markets, life and health-care sciences fuzzy systems play an important role in many application scenarios such as system identification, fault detection and diagnosis, quality control, instrumentation and control, decision support systems, visual inspection, psychological modelling, biomedical and bioinformatic systems, financial domains. Fuzzy systems are specific mathematical models aimed at imitating the real natural dependencies, procedures as accurately and as transparent as possible, building upon the concept of fuzzy logic, which was first introduced in 1965 by Lotfi A. Zadeh. The great attraction of fuzzy systems is because they are providing a reliable tradeoff between precise and linguistic modelling and offer a tool for solving complex problems in the real world. Precise modelling aims to achieve high accuracy in model outputs and is supported by the fact that fuzzy systems are universal approximators, i.e. having the potential to identify a model of the system with sufficient accuracy in order to guarantee process-save and correct model feedbacks. On the other hand, in linguistic modelling the main objective is to obtain models with a good interpretable capability, offering insights into the system processes and behaviors in an understandable way. Furthermore, fuzzy systems are able to express 1.) any form of uncertainty implicitly contained in the system as natural fact and 2.) any form of vagueness contained in the knowledge of human beings by a possibilistic point of view.

While during the stone ages in the field of fuzzy systems research (70s and 80s) most of the fuzzy systems relied on (vague) expert knowledge, during the 90s a big shift in the design of these systems towards data-driven aspects could be observed. This development went hand in hand with the technological progress of automated information processing in computers and the increasing amount of data in industrial systems. The extraction of fuzzy systems automatically from data was a cornerstone for omitting time-intensive design phases and discussions with the experts and bringing in more automatization capability. Another major advantage of data-driven fuzzy systems is that they rely on objective data rather than subjective experiences, which may be affected by different moods of the experts or be contradictory. Data-driven fuzzy systems are extracting basic trends and patterns out of the data to gain a deeper understanding of 'what the data says'. As such, they can be seen as an important contribution to both, the *fuzzy (systems)* as well as the *machine learning* and *data mining* community.

In order to account for changing systems dynamics and behaviors as well as new operating conditions and environmental influences, during the last decade (2000 to 2010) a new topic in the field of data-driven design of fuzzy systems emerged, the so-called *evolving fuzzy systems* (EFS). An evolving fuzzy system updates its structural components and parameters on demand based on new process characteristic, system behavior and operating conditions; it also expands its range of influence and evolves new model structures in order to integrate new knowledge (reflected by new incoming data). In this sense, evolving (fuzzy) systems can be seen as an important step towards *computational intelligence*, as the models permanently and automatically learn from changing system states and environmental conditions. *Evolving* should be here not confused with *evolutionary*. An evolutionary approach learns parameters and structures based on genetic operators, but does this by using all the data in an iterative optimization procedure rather than integrating new knowledge permanently on-the-fly. Nowadays, the emerging field of evolving fuzzy systems is reflected by many publications in international conferences and journals such as Fuzzy Sets and Systems, IEEE Transactions on Fuzzy Systems, Evolving Systems, IEEE Transactions on Systems, Man and Cybernetics part B, International Journal of Approximate Reasoning, Applied Soft Computing and many others. In the second half of the last decade, a lot of workshops (EFS '06, GEFS '08, ESDIS '09, EIS '10 and others) and special sessions at different international conferences (FUZZ-IEEE, EUSFLAT, IFSA, IPMU) were organized, mainly by the 'three musketeers' Plamen Angelov, Dimitar Filev and Nik Kasabov, who, along with other researchers, carried out pioneering research work at the beginning of the last decade, resulting in the evolving fuzzy systems approaches such as (alphabetically) *DENFIS* (Kasabov), *eTS* (Angelov and Filev), *FLEXFIS* (author of this book), *SAFIS* (Rong, Sundararajan et al.), *SOFNN* (Leng, McGinnity, Prasad) and *SONFIN* (Juang and Lin). These were completed during the second half of the last decade by approaches such as *EFP* (Wang and Vrbanek), *ePL* (Lima, Gomide, Ballini et al.), *SEIT2FNN* (Juang and Tsao) and others.

The aim of this book is to provide a round picture of the whole emerging field of evolving fuzzy systems within a consistent and comprehensive monograph. The first part will demonstrate the most important evolving fuzzy systems approaches developed during the last decade (as mentioned above), including a description of the most essential learning and parameter optimization steps used in these. Applications in the third part of the book will underline the necessity of evolving fuzzy systems in today's industrial systems and life sciences, including *on-line system identification,*

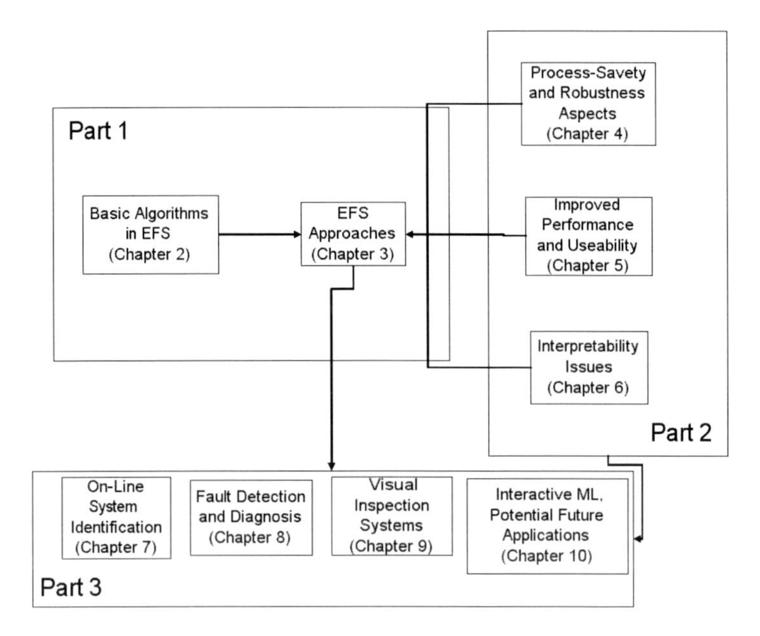

Fig. 0.1 Overview of the content in this book

fault detection in multi-channel measurement systems, visual inspection scenarios, perception-based texture analysis, bioinformatics, enhanced human-machine interaction and many more. The second part of the book deals with advanced concepts and novel aspects which were in large parts just partially handled in EFS approaches and can be seen as promising directions for improving *process safety, predictive quality, user-friendliness* and *interpretability* as well as *understandability* of evolving fuzzy systems. The epilogue concludes the book with *achieved issues, open problems* and *future challenges*, which may serve as inspiration and motivation for all (evolving) fuzzy systems researchers to enrich this newly emerging field with their ideas, point of views, further developments etc. A global view on the content of the book and the connections between the different parts and chapter are shown in Figure 0.1.

The book has not particularly been written in a mathematical theorem/proof style, but more in a way where ideas, concepts and algorithms are highlighted by numerous figures, tables, examples and applications together with their explanations. The book looks to be read not only from the field of fuzzy sets and systems research, but also from the machine learning, data mining, system identification and control community as well as attracting technicians from engineering and industrial practice.

Linz, October 2010 Edwin Lughofer

Acknowledgements

First of all, special thanks are dedicated to my wife Ilona, who showed great patience and tolerance during the writing period of this book, which lasted about 2 and 1/2 years and was proceeded in large parts during weekends and free time. Special thanks are also dedicated to my whole family and friends for supporting me morally and to all of my colleagues in our research department for providing me with inspirations, hints and discussions about the book. Further special thanks go to Günther Frizberg who brought me to the topic of on-line adaptation of data-driven models in 2000/2001 within the scope of an European project, which can be seen as my research starting point, Erich Peter Klement who acted as supervisor during my PhD on Evolving Fuzzy Models (2001-2005), supported by Eyke Hüllermeier who provided me with many useful inspirations and hints during the preparation of the thesis and beyond. Special thanks are also dedicated to Anne Margery for proof-reading and improving the English language of the whole monography, Janusz Kacprzyk for establishing the contract with Springer Verlag Heidelberg and Christian Eitzinger, the project coordinator of two European research projects (2005-2010) where I was involved as key researcher of our university and who supported the concept of on-line learning of fuzzy systems in various work-packages and application scenarios.

Furthermore, I would like to acknowledge the following persons (alphabetically according to family names) with whom I had intensive contacts and collaborations in several joint publications, projects (proposals), organizations of special sessions/issues and workshops, research visits as well as discussions and exchanging of ideas during the last years:

Plamen Angelov, Frank Bauer, Ulrich Bodenhofer, Janos Botzheim, Abdelhamid Bouchachia, Hendrik van Brussel, Oscar Cordon, Antonio Dourado, Mario Drobics, Hajrudin Efendic, Dimitar Filev, Veniero Giglio, Fernando Gomide, Carlos Guardiola, Christoph Hametner, Wolfgang Heidl, Frank Hoffmann, Eyke Hüllermeier, Richard Jacobs, Stefan Jakubek, Nik Kasabov, Uzay Kaymak, Stefan Kindermann, Frank Klawonn, Thomas Krieger, Wolfgang Märzinger, Bernhard Moser, Robert Murphy, Thomas Natschläger, Kurt Niehl, Marnix Nuttin, Witold Pedrycz,

Daniel Sanchez, Davy Sannen, Moamar Sayed-Mouchaweh, Michael Scherrer, Holger Schöner, Rudolf Seising, James Edward Smith, Muhammad Atif Tahir, Stefan Thumfart, Bogdan Trawinski, Krzysztof Trawinski, Gancho Vachkov, Karin Wiesauer, Xiaowei Zhou.

Contents

Part III: Applications

Abbreviations

A/D	Analog/Digital
ANFIS	Adaptive Neuro-Fuzzy Inference System
APE	Average Percent Error
AR	Auto-Regressive
ARMA	Auto-Regressive Moving Average
CART	Classification and Regression Trees
CBS	Center- and Border-Based Selection
CD	Compact Disc
COG	Center of Gravity
COM	Center of Maximum
CV	Cross-Validation
DBSCAN	Density Based Spatial Clustering of Applications with Noise
DDM	Dynamic Data Mining
DENFIS	Dynamic Evolving Neural Fuzzy-Inference System
D-FNN	Dynamic Fuzzy Neural Network
DNA	Deoxyribonucleic Acid
DVD	Digital Versatile Disc
EBF	Ellipsoidal Basis Function
ECM	Evolving Clustering Method
ECMc	Evolving Clustering Method with constraints
ECU	Engine Control Unit
EEG	Electroencephalography
EFC	Evolving Fuzzy Classifier
EFP	Evolving Fuzzy Predictor
EFS	Evolving Fuzzy System
EFuNN	Evolving Fuzzy Neural Network
ELS	Extended Least Squares
EM	Expectation-Maximization (algorithm)
ENFRN	Evolving Neuro-Fuzzy Recurrent Network
EGR	Exhaust Gas Recirculation

ePCA	Evolving Principal Component Analysis
ePL	Evolving Participatory Learning
eTS	Evolving Takagi-Sugeno fuzzy systems
eTS+	Evolving Takagi-Sugeno fuzzy systems from Data Streams
eVQ	Evolving Vector Quantization
eVQ-Class	Evolving Vector Quantization for Classification
FC	Fault Correction
FCM	Fuzzy C-Means
FD	Fault Detection
FI	Fault Isolation
FLEXFIS	Flexible Fuzzy Inference Systems
FLEXFIS-MOD	Flexible Fuzzy Inference Systems using Modified Version of eVQ
FLEXFIS-Class	Flexible Fuzzy Inference Systems for Classification
FLL	Fuzzy Lazy Learning
FMCLUST	Fuzzy Model Clustering
fMRI	functional Magnetic Resonance Imaging
FOU	Footprint of Uncertainty
FW	Feature Weighting
FWLS	Fuzzily Weighted Least Squares (solution)
GAP-RBF	Growing And Pruning Radial Basis Function Network
GCV	Generalized Cross-Validation
GD-FNN	Generalized Dynamic Fuzzy Neural Network
GLVQ	Generalized Learning Vector Quantization
GSVD	Generalized Singular Value Decomposition
GTLS	Generalized Total Least Squares
GUI	Graphical User Interface
HAL	Hybrid Active Learning
HIEM	Human-Inspired Evolving Machines/Models
ICF	Incremental Classifier Fusion
IDC	Incremental direct cluster-based fusion
IFAC	International Federation of Automatic Control
ILVQ	Incremental Learning Vector Quantization
IML	Interactive Machine Learning
iPCA	Incremental Principal Component Analysis
KDD	Knowledge Discovery in Databases
LM	Levenberg-Marquardt (optimization method)
LOESS	Locally Weighted Scatterplot Smoothing
LOFO	Leave-One-Feature-Out (weighting strategy)
LS	Least Squares (solution)
LVQ	Learning Vector Quantization
MAE	Mean Absolute Error
MF	Membership Function
MIMO	Multiple Input Multiple Output
MISO	Multiple Input Single Output

ML	Machine Learning
MM	Multi Model (architecture)
MRAN	Minimal Resource Allocation Network
MSE	Mean Squared Error
MTS	Mamdani-Takagi-Sugeno (fuzzy system)
MV	Machine Vision
MVEG	Motor Vehicle Emission Group
ND	Noise Detection
NDEI	Non-Dimensional Error Index
NEFCLASS	Neuro-Fuzzy Classification
NF	Neuro-Fuzzy
NIR	Near-InfraRed
NOx	Nitrogen Oxides
NP-hard	Non-deterministic Polynomial-time hard
OFW	On-line Feature Weighting
PAC	Process Analytical Chemistry
PCA	Principal Component Analysis
PCFR	Principal Component Fuzzy Regression
PCR	Principal Component Regression
PI	ProportionalIntegral (Controller)
PID	ProportionalIntegralDerivative (Controller)
PL	Participatory Learning
PLS	Partial Least Squares
PRESS	Predicted REsidual Sums of Squares
QCL	Quantum Cascade Laser
RAN	Resource Allocation Network
RBCLS	Rule-Based Constraint Least Squares
RBF	Radial Basis Function
RBFN	Radial Basis Function Network
RBS	Ribosome Binding Site
RENO	Regularized Numerical Optimization (of Fuzzy Systems)
RFWLS	Recursive Fuzzily Weighted Least Squares
RLM	Recursive Levenberg-Marquardt
RLS	Recursive Least Squares
RMSE	Root Mean Squared Error
ROFWLS	Recursive Orthogonal Fuzzily Weighted Least Squares
ROI	Range of Influence (clusters, rules)
ROI	Region of Interest (objects)
ROLS	Recursive Orthogonal Least Squares
RS	Random Selection
RWLS	Recursive Weighted Least Squares
SAFIS	Sequential Adaptive Fuzzy Inference System
SEIT2FNN	Self-Evolving Interval Type-2 Fuzzy Neural Network
SISO	Single Input Single Output
SM	Single Model (architecture)

SOFNN	Self-Organizing Fuzzy Neural Network
SONFIN	Self-constructing Neural Fuzzy Inference Network
SparseFIS	Sparse Fuzzy Inference Systems
SD	Steepest Descent
SVD	Singular Value Decomposition
SVM(s)	Support Vector Machines
S^3VM	Semi-Supervised Support Vector Machines
SVR	Support Vector Regression
TLS	Total Least Squares
TS	Takagi-Sugeno
TSK	Takagi-Sugeno-Kang
TSVM	Transductive Support Vector Machines
UCI	University of California Irvine (Machine Learning Repository)
VQ	Vector Quantization
WEKA	Waikato Environment for Knowledge Analysis
WLS	Weighted Least Squares (solution)
WTLS	Weighted Total Least Squares

Mathematical Symbols

Here we give an overview of mathematical symbols used repetitively in this book and having the same meaning in all the chapters. The definition of further used symbols, dedicated to and only appearing in combination with specific approaches, methods and algorithms, are given in the text.

N	Number of training samples (seen so far)
p	Dimensionality of the learning problem = the number of input features/variables
$\mathbf{x}$	One (current) data sample, containing p input variables
$\mathbf{x}(k)$ or $\mathbf{x}_k$	The kth data sample in a sequence of samples
$\{x_1, ..., x_p\}$	Input variables/features
y	Measured output/target value for regression problems
$\hat{y}$	Estimated output/target value for regression problems
$\hat{y}_i$	Estimated output/target value from the ith rule in a fuzzy system
m	Polynomial degree of the consequent function, regression model
r, reg	Regressors
X or R	Regression matrix
P	Inverse Hessian matrix
Jac	Jacobian matrix
C	Number of rules, clusters
Φ	Set of parameters (in fuzzy systems)
Φ_{lin}	Set of linear parameters
Φ_{nonlin}	Set of non-linear parameters
$\mathbf{c}_i$	Center vector of the ith rule, cluster
$\boldsymbol{\sigma}_i$	Ranges of influence/widths vector of the ith rule, cluster
cov_{ij}	Covariance between features x_i and x_j
Σ	Covariance matrix

win	Indicates the index of the winning rule, cluster (nearest to sample $\mathbf{x}$)
$\{\mu_{i1}, ..., \mu_{ip}\}$	Antecedent fuzzy sets of the ith rule (for p inputs)
$\mu_i(\mathbf{x})$	Membership/Activation degree of the ith rule for the current sample $\mathbf{x}$
$\Psi_i(\mathbf{x})$	Normalized membership degree of the ith rule for the current sample $\mathbf{x}$
l_i	Consequent function, singleton (class label) in the ith rule
$\{w_{i0}, ..., w_{ip}\}$	Linear weights in the consequent function of the ith rule
L_r	Real class label of a data sample in classification problems
L	Output class label for classification problems (from a fuzzy classifier)
L_i	Output class label of the ith rule (cluster)
K	Number of classes in classification problems
$conf_{ij}$	Confidence of the ith rule in the jth class
$conf_i$	Confidence of the ith rule to its output class label
$conf$	Overall confidence of the fuzzy classifier to its output class label
Small Greek letters	Various thresholds, parameters in the learning algorithms

Chapter 1
Introduction

Abstract. At the beginning of this chapter a small introduction regarding mathematical modelling in general will be given, including first principle and knowledge-based models. Then, we outline a motivation for the usage of a data-driven model design in industrial systems and the necessity for an evolving component therein, especially in on-line applications and in case of huge data bases. Hereby, we will describe concrete requirements and mention application examples, where on-line evolving models serve as key components for increasing predictive accuracy and product quality. After explaining why fuzzy systems provide a potential model architecture for data-driven modelling tasks, we define several types of fuzzy systems mathematically and formulate the goals of an incremental learning algorithm for evolving fuzzy systems from a mathematical and machine learning point of view. Hereby, we also highlight which components in fuzzy systems have to be adapted and evolved in order to guarantee stable models with high predictive power. The chapter is concluded by outlining a summary of the content in this book.

E. Lughofer: Evolving Fuzzy Systems – Method., Adv. Conc. and Appl., STUDFUZZ 266, pp. 1–42.
springerlink.com

1.1 Motivation

1.1.1 Mathematical Modelling in General

In today's industrial installations and applications, mathematical process models play a central role in order to cope with permanently increasing environmental regulations, process complexities and customer demands.

Eykhoff defined a *mathematical model* as 'a representation of the essential aspects of an existing system (or a system to be constructed) which presents knowledge of that system in usable form' [124].

While in former times (50s, 60s, 70s) a high percentage of industrial processes were manually conducted, controlled and/or supervised by several operators employed in the companies, during the last 30 years more and more process models have been developed in order to increase the automatization capabilities of the systems. Process models are often used in full-online production lines, sending instructions to manipulators or robots without the necessity of any form of interaction to human operators. In other cases, process models are also used to support the operators in decision processes, quality control applications, (bio)medical and health care supervision systems, financial domains etc., usually within a semi-online interaction mode. Both types of usage usually result in higher production through-put or savings in production times, hence finally in higher sales for the companies. Clearly, the development, implementation and on-line installation of system models requires significant investments at the early development phase, however in the long term view, rising profits can be expected, especially because of rationalizing working time for operators or even reducing the number of operators and employees.

Process models are a simplification of the real nature of the process, mimicking the real natural (physical, chemical) dependencies and procedures as accurately as possible by compressed knowledge representation.

This is usually achieved in form of input-output relations (supervised case) or characterizations of the system variable domains, e.g. grouping effects (unsupervised case). From this point of view, it is obvious that modelling can become a bottleneck for the development of a whole system, as the quality of a model usually determines an upper bound on the quality of the final problem solution and the whole output of the system. Therefore, it is very important to develop models with high precision within the scope of advanced knowledge investigation and identification procedures.

Mathematical process models can be useful for system analysis in order to get a better insight and furthermore understanding of the complete system, some specific system behaviors or process properties through analyzing the model structures and parameters. Such knowledge gains from process models may be taken as cornerstones for future design decisions or planned developments for extending components or changing principal strategies at the system(s) [263].

A wide and important field of application for process models in industrial systems is the design of controllers [105]. Process models in form of input-output relations are necessary in order to automatically guide certain objects to pre-defined desired positions or to achieve desired systems states, ideally in a smooth convergent way (without over-shooting the target). Some system variables together with the current deviation to the desired value are used as inputs, the output is usually a control signal sent to engines, motors or other devices responsible for steering the whole process. For instance consider an elevator, moving from one floor to the other: the target floor can be associated with the desired position and the control signal is a voltage load to the elevator engine; in this application it is very important to perform a smooth moving behavior in order to guarantee a comfortable feeling for the passengers. The more accurate the model is, the better a smooth control behavior can be achieved.

Quality control is another important application field for process models, where plausibility checks of the system components and device are usually required. Such plausibility checks are necessary in order to detect system failures in an early stage or even to prevent failures by interpreting and reacting on upcoming fault indicators with predictive process models (the reaction is often done manually by system operators through switching off some components or directly interacting with the hardware). This is important for 1.) increasing the quality of the sold products and 2.), even more important, to avoid broken pieces/items and therefore to prevent severe breakdowns of system components, in extreme cases avoiding real danger for the operators working at the system (for instance consider a waste gas gushing out of a broken pipe) or users of the products (e.g. consider defects on aeroplane wings). Currently, in quality control there are two main lines of approaches for plausibility analysis:

- One is based on the analysis of process measurements (temperatures, pressures, etc.) directly during production or online operation modes and to detect any deviation from the normal situation in these measurements (ideally in on-line mode) and to immediately react on it — to identify any such anormal behavior, often mathematical reference or fault models are used [229] [399] [82]. This approach is called *direct plausibility analysis*.
- The other is based on the analysis of produced items in a post-processing stage through surface inspection: for this task, often images are recorded showing the surface of the produced items [99]. These images can then be classified into different quality levels by specific mathematical models, often called (image) classifiers [382]. In case of images showing bad surfaces of the production items, a fault reasoning process can be initiated to trace back the origin of the fault(s) in the process. Therefore, this approach is also called *indirect plausibility analysis*.

Mathematical models are also applied in decision support systems [430], where the models usually provide suggestions about the state of objects or the system to the expert, hence serving as supporting role for experts: for instance consider the supervision of the health conditions of patients: due to specific form of measurement taken from the patients (EEG signals, fMRi images), the subconscious behavior and condition of a patient can be supervised and interpreted by mathematical decision models.

In a more general context, mathematical models are frequently used in prediction, simulation and optimization, which are often sub-tasks in above described applications. Prediction and simulation refers to predicting a model behavior, usually represented by one or more output variables, based on input signals such as measurement variables, extracted features from visual and audible source context, data sets stored in data-bases etc. The prediction horizon, which is defined by the time span how long the predictions are made into the future, serves as key quantity: the larger this horizon is, the lower the certainty of the models in their predictions are (think of forecasting stock markets or the weather — longer time horizons leads to higher uncertainties in the predictions). The difference between prediction and simulation is that in simulation tasks past output variable values are not used as inputs, so input and output variable set are disjoint. Both techniques can be applied in the field of soft sensors, substituting noisy, non-robust with respect to environment conditions and often too expensive hardware sensors. Mathematical models in optimization are often applied for finding the optimal operating point or optimal input profile (e.g. consider optimization of costs according to some constraints implicitly contained in the system).

Some of these applications will be handled in more detail in Part 3 of this book within the context of applying evolving fuzzy systems (as specific mathematical models) to industrial systems.

1.1.2 First Principle and Knowledge-Based Models

Mathematical models can be loosely divided into the following four classes:

- First principle models
- Knowledge-based models
- Data-driven models
- Any form of hybrid models

First principle models are models which are analytically deduced from an established theory about well-known physical, chemical laws and/or medical, biological process behaviors implicitly contained in the system.

In this regard, first principle models are also often called analytical models. They may appear in different forms, such as exact formulas in the case of describing static non-changing dependencies, as differential and integral equations in the case of describing continuous dynamic and time-varying behaviors of the system [291] (e.g. in combustion engines, elevators, burning procedures in blast furnaces), or as analytical wave equations (e.g. as used in the fields of audio signal processing [473] or structural health monitoring [143]). Usually, they are deduced once and kept fixed for the whole application process, meaning that no model parameters are updated and hence the model itself is never changed over time. When the complete process behavior, various operating conditions and underlying phenomena in the system are fully understood, first principle models can be accurate representations of these and therefore provide a save process behavior and an exact insight into system dependencies and relations. In this sense, in literature they are also often called *white box models*. On the other hand, the main disadvantage of first principle models is their high development time. Sometimes, setting up reliable and accurate models, covering all possible system states and operating conditions (especially the extreme cases) may take several months or even years. This is because a full understanding of systems with high complexity has to be investigated in several rounds during research projects or programmes. In some cases, the complexity of the system and especially of the environmental influences might be that high, that a deduction of analytical, physical-oriented models is simply not possible at all. For instance, the development of a feasible control strategy for the water level at a water power plant requires many environmental factors such as current and forecasted rainfalls, water levels in the feeder rivers or the current water level situation and reaction in the preliminary plant (anticipatory control). It is hardly possible to fulfill these requirements with classical analytical controllers (PID) deduced from physical laws prevailing at the water power plant. Moreover, the evaluation of first principle models may cause high computational effort (e.g. in complex differential equation systems) to such an extent that they are not applicable in an on-line scenario.

Such cases require the usage of knowledge-based or data-driven models.

Knowledge-based models are models which are extracted from the long-term experience of experts, operators and users working on the underlying system.

This long-term experience results in a wide knowledge about the process in various operating conditions which can be collected and formulated, mostly in form of linguistic rules, as the operators/experts expresses their experiences with spoken words. This wide knowledge, depending on the experience/skill level, may even cover extreme cases (e.g. floodwaters at water power plants) and very complex dependencies including environmental influences, which cannot be captured by first principle models deduced from physical, chemical etc. laws. The collection of rules is usually coded into specific mathematical models, so-called knowledge-based systems or expert systems [71]. Fuzzy systems [235] are a specific case of such rule-based expert systems and have the advantage over crisp rules that they are able to achieve a kind of interpolation effect between different system states and therefore providing a transition from one rule to the other, covering cases which were not explicitly formulated by the expert(s). During the 80ties and the first half of the 90ties, knowledge-based (hard-coded) fuzzy systems were used in many applications such as the the control of cement kilns [181], performance prediction of rock-cutting trenchers [172], waster water treatments [425] or steering of robots [333]. The performance of knowledge-based models often suffer from their vague components (linguistic rules), which may even be contradictory among different operators of the system (for instance performing different control strategies at a control system) — therefore, knowledge-based models are also called *weak white box models*. As operators are human beings, the performance of the coded knowledge may also suffer from various moods, levels of frustration and distraction of the operators when working at the system. This means that their experience also depends sometimes on quite subjective influences and may even affect the knowledge gaining process during several meetings and discussions. Hence, over the last 15 years, knowledge-based models have been replaced more and more by data-driven models.

1.1.3 Data-Driven Models

Data-driven models are understood as any form of mathematical models which are fully designed, extracted or learned from data.

Opposed to knowledge-based models, where the input contains subjective experiences and impressions, data usually represent objective observations of the underlying mechanisms in the systems, characterizing various system states, behaviors and operating conditions [1]. Depending on the type of application, data can be gathered from different sources. Based on the sources, data can be loosely divided into the following types:

- Measurement data: data is recorded by so-called measurement channels, which usually consist of all the hardware and data acquisition settings utilized to make measurements [4]: specimen, transducers, preamplifier, transmission path, analog and digital filters and A/D converters; the measurements obtained from measurement channels are also called raw data and have to be pre-processed (down-sampled, compressed, de-noised) to guarantee a proper application of data-driven models to this data. Measuring different components in a system, results in measurement variables.
- Context-based data: data is represented in any context-based form, such as audio signals, images, textures or video sequences; segmenting this context based information is often necessary in order to extract regions of interest or objects. Usually, the context-based data is not used in its native form (signal amplitudes or pixels as single samples), but pre-processed by feature extraction algorithms. A feature can be seen as a representative characterization of single objects, of

[1] An exception is data gathered in association or communication with human beings such as interview data or any form of subconscious data.

group of objects, of regions of interest or of whole audio segments, textures or images; examples of features are the roundness of an object in an image or the energy in the high frequency domain in audio signals. In this sense, a measurement variable can be also seen as a feature, as characterizing one single physical, chemical property at a system.
- Structured data: structured data are usually data blocks contained in data bases or in web domains, not necessarily represented by just numerical values (as is the case in measurement data and features extracted from context-based data), but also by categorical attributes or descriptive strings. A simple example of structured data are customer addresses (one data block consisting of strings and numbers).
- Data streams: Opposed to data sets, which are containing a whole collection of data samples/blocks (usually stored onto hard-discs), data streams are characterized by their dynamic spontaneous behavior and appearance, usually within an on-line processing context. As such, data streams are consisting of sample- or block-wise on-line recordings and may contain any form of above mentioned types of data (measurements, context-based and structured data).

Due to the objective property of data, contradictory input can be omitted, as for the same system states always the same or at least very similar data samples will be obtained (in some cases, exactly the same data is not possible due to uncertainties in the measurement process or errors in the measurement channel afflicting the data samples with noise). This property often makes a data-driven model more trustworthy than a knowledge-based model. On the other hand, data-driven models are learned from a concrete portion of data samples, collected over a time frame in the past and used as so-called training data in the learning process. This means that the generalization capability, i.e. the predictive accuracy of a data-driven model on completely new samples, strongly depends on the selection of training data samples. This can severely constrain a data-driven model in case of a too restricted spread of samples. Furthermore, significant noise levels (usually caused within the data recording process) may spoil the models, even though nowadays there are a lot of training tools available which can deal with quite high noise levels and still extract the correct trend of the underlying input/output relations/dependencies (preventing over-fitting).

A major advantage of data-driven models among both, first principle and knowledge-based models, is the generic applicability of their training phases: in fact data-driven models can be generically built up without knowing any underlying physical, chemical etc. laws, just the data has to be fed into the learning and evaluation algorithms, which can be automatically run through on computers and for which the meaning of the system variables and the origin of the data source usually play no role at all (in fact, for the learning algorithms, it does not matter whether a vector of numerical values stems from image features or from physical measurement variables). In this sense, data-driven models are also often referred to as black box models. It would be exaggerating to say that data-driven models can be used in a plug-and-play manner, but, assuming that the data recording process is also often

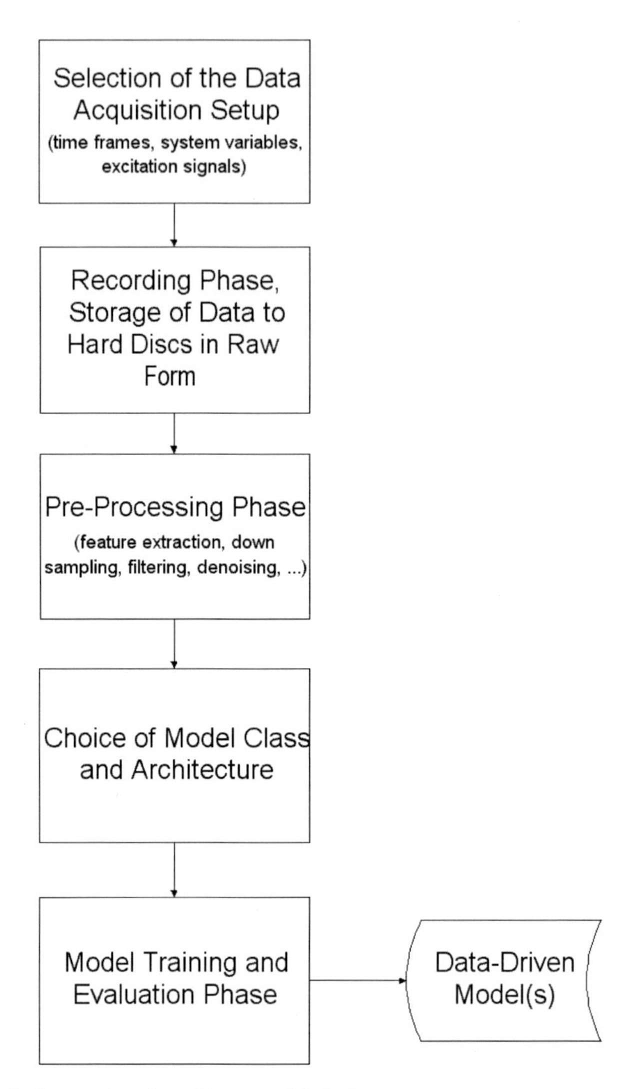

Fig. 1.1 Typical steps in a data-driven model design

automatized, the development effort for experts can be expected as quite low, at least much lower than in the case of first principle and knowledge-based models.

Figure 1.1 summarizes the basic components in a typical data-driven model design. The first phase clarifies the setup for the whole data acquisition setup and/or measurement process to be carried out at the system. This includes an appropriate choice of time frames when the data is recorded/acquired as well as the design of excitation signals (triggering data transfer) and the selection of important system influences and variables — the answer to the question which process data is really relevant in the modelling phase, in the case of context-based data it is more the

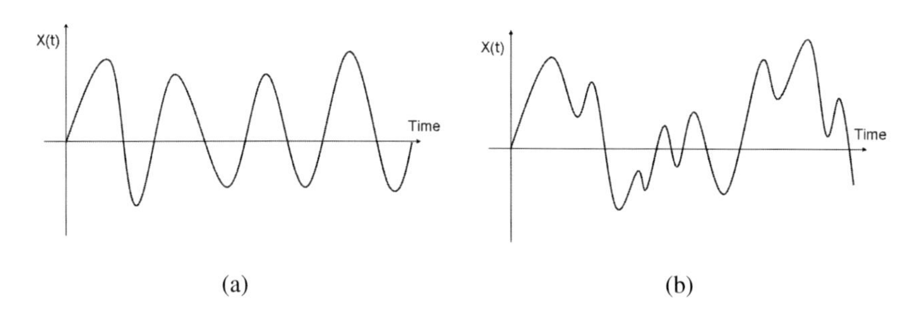

(a) (b)

Fig. 1.2 (a): Stationary time series signal showing only a few deflection patterns (nearly sinusoidal); (b) non-stationary time series signal showing many deflection patterns and an aperiodic behavior

selection of the types of audio signals, images, textures. This phase is extremely important for providing a good coverage of various (especially the most common) process and system states. The next phase concerns the concrete recordings and acquisition of data, including the storage onto hard disc(s). The data pre-processing phase is an essential component in order to extract features from context-based data, performing cleaning operations such as de-noising and data filtering and down-sampling procedures, especially in case of high-frequent data (data recorded with a high frequency, e.g. producing several 100 (very similar) samples per second). The choice of an appropriate model class depends on the formulation of the learning problem, which may be an un-supervised problem (yielding a pure characterization of the data), a supervised classification (yielding classifiers being able to identify and discriminate between groups of data) or a supervised regression/approximation problem (yielding models describing functional input/output relations). The choice of the model architecture depends on the complexity and the dynamics of the problem at hand. If the complexity is low, often pure linear models can explain sufficiently the nature of the data and the relations it implicitly carries. In case of higher complexities, non-linear models are needed, as linear models will cause significant bias due to their low flexibility regarding tracking the trends of non-linearities. In case of stationary data, the dynamics of the system is negligible, hence static models will do their job with sufficient accuracy. In case of non-stationary data, mostly dynamic models are needed which are able to incorporate time delays of the output and certain input variables to achieve reliable performance. For instance, for the classification of single images static models (in form of classifiers) are a feasible choice, whereas the tracking of dependencies among a series of images in a video sequence requires dynamic models. Figure 1.2 (a) demonstrates a data stream (time series) which is stationary as following a simple (nearly sinusoidal) periodic pattern along its time line, whereas in (b) a non-stationary signal is visualized, which is aperiodic, exciting many deflection patterns.

After the choice of the model class and architecture is clear, the model training and evaluation process can start. Figure 1.3 shows the essential component of

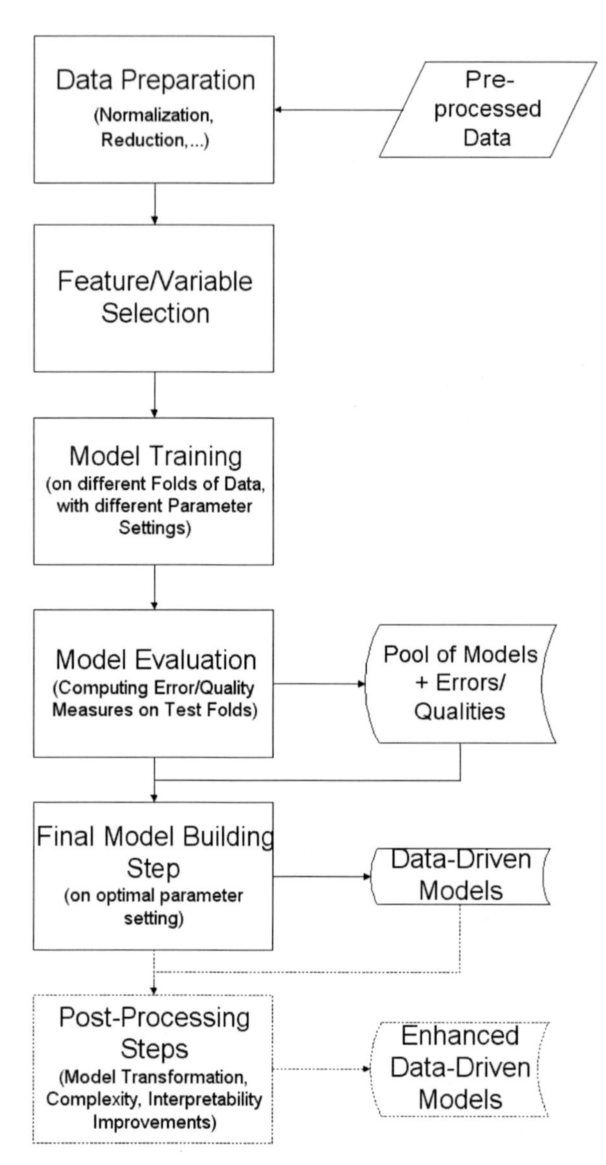

Fig. 1.3 Components in the model training framework, optional blocks highlighted with dotted lines

such a model building framework. Initial preparation of the data by normalizing and compressing the information is sometimes required depending on the chosen model training algorithm (e.g. some use distance measures for cluster extraction, hence normalization is necessary in order to omit range dominations by certain variables). Feature/variable selection is important for reducing dimensionality of the input feature space and therefore preventing over-fitting, especially when the number of training samples is low (see also Section 2.8 for further clarification). After choosing the most important variables/features for the modelling process, the actual model training phase begins. For this task, several methodologies exist, which, independently from the learning problem (un-supervised, supervised classification, supervised regression/approximation), stem from different motivations in the past. While statistical approaches try to build models based on statistical information criteria and uses the concept of statistical distribution theory [174], machine learning techniques are motivated by model induction and iterative knowledge extraction [311], also exploiting optimization techniques such as backward propagation (as used in neural networks [175]), least squares optimization and variants (as used in data-driven fuzzy modelling [28]) or quadratic programming (as applied in support vector machines [434] and kernel-based learning techniques [389]). Another big group of learning algorithms are pattern recognition approaches, motivated by the assumption of available data patterns among different samples providing a unique representation of different classes or local regions in the feature space [111]. During the last years, the clear boundaries between statistical, pattern recognition and machine learning techniques have washed more and more up, to such an extent that now a clear separation is often no longer possible. Once the model training technique is fixed, usually several training and evaluation runs are performed in order to find an optimal parameter setting (each method usually has a couple of essential parameters to tune which are sensitive to the performance of the final model). Afterwards, a final model is trained based on the complete training data set. Quality as well as error measures are calculated indicating the expected future performance of the models (see also Section 2.1.4). In an optional post-processing step, the model is reduced to a model with lower complexity or transformed to a model with higher interpretability, opening the possibility for experts to provide additional insights into the system process etc. (see also Chapter 6).

1.1.4 Evolving Models

> *Evolving models* are data-driven models, which are automatically adapted, extended and evolved dynamically on-the-fly based on new incoming data samples (new information they contain).

As such, evolving models are able to support any modelling scenarios for data streams, on-line measurements and dynamic data which changes its nature and characteristics over time and space. Thereby, it plays no role, whether the data is stationary or non-stationary: in the first case static (often not too complex) models are step-wise updated, in the second case dynamic models (including time delays of the variables/features) are continuously extended on-the-fly. Model adaptation usually refers to the update of some model parameters in order to shift, move some decision boundaries, cluster partitions or approximation surfaces, while the evolving concept

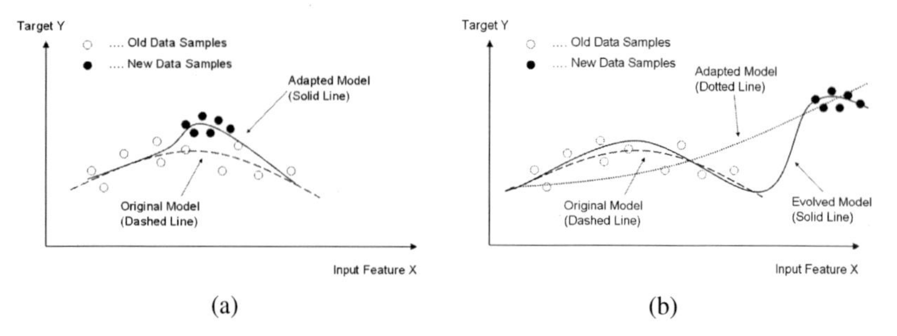

Fig. 1.4 (a): Situation where parameter adaptation is sufficient for obtaining an updated reliable model; (b) situation where the evolution of a structural component is required in order to extend the model appropriately to the extended feature space, increasing its flexibility (compare solid with dotted line, the latter achieved by only updating the parameters and keeping the structure of the model)

also takes the extension of models into account by generating (evolving) new structural components on demand, i.e. based on the characteristics of the new data. In some cases, adaptation of parameters is sufficient to include new information from the data source(s) into the (previously generated) models. For instance, consider the left case in Figure 1.4: the adapted model (solid line) is able to follow the trend of the new samples. In other cases, like for instance shown in Figure 1.4 (b), where the new samples are falling into a new region of the feature space, the structure of the model has to be extended (adding rules, neuron, leaves etc.) in order to represent a model with higher non-linearity and therefore better accuracy — compare solid line with dotted line, the latter achieved by updating only the parameters.

The necessity of evolving models in today's industrial systems becomes clear when considering that often models, which are originally generated in a batch offline modelling scenario by pre-collected data sets, are not fully covering all possible operating conditions, systems states or some extreme occurrences. This is simply because the available data does not reflect all these possible system variations. In some cases new occurrences may happen which were not even known before (consider a century flood at water power plant as was the case in August 2002 in middle Europe). Hence, it is impossible to collect data for all possible variations. In classification systems, the problem of collecting representative data with universal coverage of system states a priori is even less likely, as in these cases usually experts or operators working at the system have to annotate the training samples. This requires a high number of man hours or even man months and therefore significant money investments for the employers. Previously unseen system states or operating conditions in fact usually cause data samples lying in the extrapolation space of the previously collected data (as shown in Figure 1.4 (b)). In these regions, the reliability of the predictions obtained from data-driven models suffer significantly. An automatic range extension of data-driven models by evolving new structural components in the new (extended) feature space helps to prevent extrapolation cases and therefore increasing reliability and accuracy of the models, leading to more process-safety.

This can be achieved by refining, extending and evolving the models with newly recorded data. Often, it is a requirement to perform this in an on-line learning scenario, omitting the application of time-intensive re-training and annotation phases and guaranteeing that the models provide high flexibility by always being up-to-date as early as possible. In these cases, it is indispensable to perform the model update and evolution by *incremental learning steps, ideally in a single-pass manner*. Incrementality prevents time-intensive re-training phases, which would not terminate in real-time. In fact, in order to achieve life-long learning scenarios [167], where the models include all relations/dependencies/behaviors in and of the system seen so far, the re-training steps have to be carried out with all samples seen so far, which over time would result in severely decreasing performance. Life-long learning is the most common practice in incremental learning scenarios, as usually forgetting of previously learned relations is not desired (as long as there are no drift cases, which denote specific situations - see Section 4.2), meeting the *plasticity-stability dilemma* [2]. Single-pass means that the incremental learning steps do not require any past data, in fact a single sample is loaded, sent into the incremental update/evolution algorithm and immediately discarded, afterwards. Hence, virtual memory demands can be expected as very low. Single-pass incremental learning will be also used as engine for evolving fuzzy system approaches discussed in this book, therefore defined and explained in more detail in Section 1.2.2 below. A principal framework of the application of evolving models in industrial systems is visualized in Figure 1.5, dotted boxes denote alternative options which are not absolutely necessary. The feedback from operators on the model predictions is needed in case of classification system, but not necessarily in case of on-line evolving identification frameworks. If initial models built up in former off-line batch or on-line training cycles are not available, evolving modelling is carried out from scratch (applied from first data samples on).

Summarizing, essential requirements for evolving models in industrial systems are:

- Updating and/or extending already available models with new samples, in order to obtain a better performance in terms of accuracy of decision, prediction, control statements, a good coverage of the data space and to prevent extrapolation when operating on new samples. In this regard, three different cases can be distinguished:
  - Improving the accuracy and process safety of data-driven models due to a dynamic incorporation of new system states and refinement of parameters.
  - Refinement of (vague) knowledge-based models with data by enriching the experts' knowledge with objective information (*gray box models*).
  - Updating of parameters in analytical models, which usually need to be tuned manually for different system setups (*light-gray box models*).
- Building up models from huge data bases, where the whole data set cannot be loaded at once into the virtual memory as causing a memory overflow (e.g. consider 100s of Gigabytes or Terabytes of data sent into a training algorithm) $\rightarrow$ this requires a sample-wise or at least block-wise loading of the data and evolving the models by incremental learning steps.

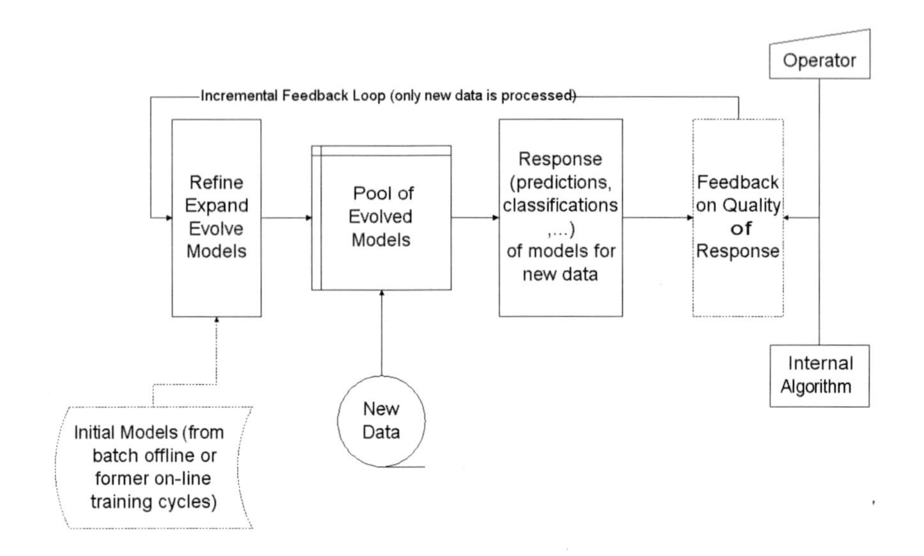

Fig. 1.5 On-line application framework for evolving modelling scenarios using incremental learning as engine for evolving/updating models

- Incorporating operator's feedback/knowledge during on-line mode for half-automatic systems, e.g. an operator may overrule a classifier's decision or even give enhanced comments on the model structure (enriched interaction scenarios).
- Model adaptation to a similar test object: for instance, at an engine test bench models are trained for a specific type of engine. The engine is exchanged and the model should be updated quickly to a quite similar type of engine. In this case, a complete re-building with a large collection of new samples can be prevented, when doing adaptation steps based on just a few samples recorded from the new engine.
- Forgetting of older learned model components, parameters in a gradual and smooth way over time in order to react on changes in the underlying data characteristics.

Apart from the requirements and demands in industrial (production and control) systems, another important aspect about evolving models is that they provide the opportunity for self-learning computer systems and machines. In fact, evolving models are permanently updating their knowledge and understanding about diverse complex relationships and dependencies in real-world application scenarios by integrating new system behaviors and environmental influences (which are manifested in the captured data). Their learning follows a life-long learning context and is never really terminated, but lasts as long as new information arrives. Therefore, they can be seen as a valuable contribution within the field of computational and artificial intelligence [14]. Combined with modelling aspects in the field of neural networks, where the human brain is roughly modelled by means of neurons and synaptic connectors between these [175], they are in fact mimicking somehow the learning capabilities of human beings (in fact, within the scope of small ideal circumstances, basically reduced to information in form of objectively measured data). Therefore, it is no surprise that evolving models and evolving intelligent systems were already investigated or at least considered to be investigated in the following real-world processes [209]:

- Modelling of cognitive tasks of the human brain: needs to be adaptive as all cognitive processes are evolving by nature.
- Gene expression modelling in bio-informatics: needs to adapt to new information.
- Intelligent robots should be able to learn from exploring and unknown environment.
- Stock market prediction models should adjust their decision criteria according to variations of the stock index, which can be caused by newly arising political, economical or financial situations.
- Evolution of very complex real-world models such as prediction models about the global climate change on earth, simulation models for forecasting the weather or diagnostic models for heart diseases by incorporating newly observed/recorded data automatically and dynamically on demand.

Finally, due to all these aspects discussed in this section, evolving models or evolving systems can be seen as a promising new emerging field for providing more intelligence, more automatization, more process safety and performance in various fields of real-world applications and engineering studies.

1.1.5 Evolving Fuzzy Systems

Fuzzy systems are specific mathematical models which build upon the concept of fuzzy logic, which was first introduced in 1965 by Lotfi A. Zadeh [464]. Opposed to Boolean logic, where only the full truth and the full falseness exist (associated with the truth values of 0 and 1), in fuzzy logic the truth values are 'fuzzified' as assigned a value in the range [0,1]. For instance when a statement such as 'Gerald is tall' is assigned a truth value of 0.2, the fact is just partially fulfilled, so in this case he is just a bit (to 20%) tall. This means that fuzzy logic is able to represent vague statements, expressing uncertainties and/or incomplete knowledge of humans. The basic functional elements in a fuzzy system are the fuzzy sets, used as building blocks for partitioning the input/output space into several granules. Each of these granules represents a local region associated with a linguistic terms [465] plus a membership function. Within the concept of fuzzy sets the truth values are related to membership function degrees, lying in [0,1] [339]. The membership functions are combined by specific conjunction operators, the so-called t-norms [220], to form the antecedent parts of the rules, whose collection forms the complete rule base of a fuzzy system. The consequents of the rules are either fuzzy sets, singleton crisp values or more complex functions (a detailed mathematical formulation of different types of fuzzy systems follows in the subsequent section).

In this sense, fuzzy systems belong to the group of knowledge-based models [397], which are able to express vague human experience and knowledge in a linguistic manner and in a rule-based form. In this sense, they are mimicking the human way of thinking and storing knowledge in the brain better than conventional models through vague concepts. Another advantage over conventional rule-based models are that, through providing fuzzy state descriptions, they are able to provide overlapping between different pre-conditions (in the rule antecedent space), hence achieving an interpolation effect, finally resulting in continuous model surfaces [122]. A conventional rule has crisp antecedent conditions and conclusions, usually represented in the following form:

IF x1 is Value1 AND x2 is Value2 THEN y=ValueOut

where Value1, Value2 and ValueOut are crisp values, x1, x2 the input variables and y the output variable. In case of categorical variables, the rules have some generalization capabilities, but in case of continuous numerical measurements, one rule will represent just one very specific case of the system. This means that inequations are needed, which however still provide crisp switches from one condition to the other. A fuzzy rule has fuzzy antecedent conditions in the following form:

IF x1 is HIGH AND x2 is LOW THEN y=VLOW

where HIGH, LOW and VLOW are fuzzy sets represented by membership functions which cover a certain part of the input/output space. Hence, a fuzzy rule covers a certain range of system cases (all those which are falling into the HIGH-times-LOW area spanned by the membership functions, at least to a certain amount of degree), and smooth transitions between different ranges are possible due to the fuzzy boundaries.

During the last two decades, a new branch within the field of fuzzy system design has emerged, which was based on a data-driven modelling approach (data driven fuzzy systems). In fact, fuzzy systems can be used as non-linear models, whose complexity increases with the number of rules specified in the rule base. In [438] [72], it could be proven that fuzzy systems are universal approximators, which means that they are able to approximate any non-linear function within the range of a pre-defined desired accuracy. This property could be exploited for developing some constructive approaches for building up fuzzy systems, see e.g. [468]. Together with the aspect that fuzzy sets can be used to model uncertainties in the data (e.g. caused by noise levels in measurements or uncertain facts in nature) [73], it serves as an important reason for the usage of fuzzy systems in today's data-driven modelling, control [28] [1], non-linear system identification [320], visual quality control [382] and fault detection tasks [13] as well as in bioinformatics applications [199] and classification systems [236]. Together with the aspect, that fuzzy systems are providing rules which can be read in an 'if-then' context, they have some superiority over other types of data-driven models such as neural networks or support vector machines, which usually appear as complete black box models without any interpretable meaning. Indeed, when generating the fuzzy systems completely from data, often precise modelling approaches are applied which do not (or only a little) take into account any transparency and interpretability aspects of the final achieved models. Then, fuzzy systems may also appear nearly as black box models (for instance, consider fuzzy systems with a high number of antecedent parts in the rules or with fuzzy partitions containing 20 fuzzy sets or more). Hence, during the last years a lot of concentration has been placed on a trade-off between precise fuzzy modelling and linguistic modelling [69] [469], either by including several mechanisms directly in the data-driven learning phase such as constraints on the number of fuzzy sets or on their knot points [128], or by applying post-processing steps such as fuzzy sets merging and rule pruning [393]. These approaches are guaranteeing that data-driven fuzzy systems are nowadays not black box models, but can be rather counted to (at least) dark-gray box models: depending on the models' expressiveness regarding interpretability the 'dark-gray' may also change to a lighter grey.

It now lies at hand to combine the favorable properties of (data-driven) fuzzy systems (especially universal approximation capabilities in connection with comprehensability and understandability aspects) with the concept of evolving modelling approaches in order to widen the applicability of fuzzy systems to on-line processing and modelling tasks. Under this scope, fuzzy systems inherit all the benefits and merits evolving models have over batch modelling approaches (see previous section). Therefore, during the last decade many approaches have emerged to tackle this issue, most of these appeared under the name 'evolving fuzzy systems' (EFS) and some others under the rubric 'incremental learning of fuzzy systems'. This book is dedicated to this new emerging field of research by demonstrating the most important EFS approaches, basic (incremental and adaptive learning) algorithms and methods which are important and necessary in EFS approaches, advanced concepts regarding process safety, improved performance, user-friendliness and enhanced interpretability as well as applications, where EFS play(ed) an important role within on-line operation modes and phases.

1.2 Definitions and Goals

We first define various types of fuzzy system architectures widely used in literature and also in some EFS approaches (see Chapter 3). Then, we give a problem statement about the goals of evolving fuzzy systems respectively about the learning algorithms achieving evolving structures and converging parameter update schemes in fuzzy systems.

1.2.1 Mathematical Definition of Fuzzy Systems

Let us recall the following notations (from the list of mathematical symbols):

p	# of input variables/features, dimensionality of the input space
$\{x_1,...,x_p\}$	input variables/features
y	output, target variable
$\hat{y}$	estimated output (model output), target
l_i	consequent function of the ith rule
$\{\mu_{i1},...,\mu_{ip}\}$	antecedent fuzzy sets of the ith rule
m	polynomial degree in consequent functions
$\{w_{i0},...,w_{ij_1...j_m}\}$	linear weights of the polynomial regressors in the ith rule

The basic difference between input variables and input features is that variables are usually used in raw form as recorded (plus ev. down-sampled and de-noised) and sent into the learning algorithm, whereas features represent a summarized, compact information/characteristics of the data, mostly of context-based information in the data . For the model building algorithms, features and variables (once pre-processed) are handled in the same way.

1.2.1.1 Takagi-Sugeno-Kang Type Fuzzy Systems

The ith rule of a (single output) Takagi-Sugeno-Kang (TSK) type fuzzy system [417] is defined in the following way:

$$\text{Rule}_i: \quad \text{IF } x_1 \text{ IS } \mu_{i1} \text{ AND...AND } x_p \text{ IS } \mu_{ip} \text{ THEN} \tag{1.1}$$
$$l_i = w_{i0} + \sum_{j=1}^{p} w_{ij}x_j + \sum_{j_1=1}^{p} \ldots \sum_{j_m=j_{m-1}}^{p} w_{ij_1 \ldots j_m} x_{j_1} \ldots x_{j_m}$$

This means that the antecedent part is completely linguistic as combining fuzzy sets, usually representing a kind of linguistic meaning, within IF-THEN constructs. AND denotes a special kind of conjunction, also called t-norm which possesses specific properties, most commonly used are minimum, product and Lucasiewicz t-norm [387]. Please refer to [220] for an extensive discussion on t-norms, their characteristics and influence on fuzzy systems. The consequent part represents a polynomial function with maximal degree of m and linear weights on the regressors. The estimation of these weights from training or on-line data is one key issue in evolving fuzzy systems. How this can be achieved in an adaptive incremental manner, will be shown in Chapter 2. Let us assume that we have C rules as defined above (e.g. identified by a data-driven learning approach): then the rules are combined by a weighted inference scheme (fuzzy inference), where each rule membership degree, obtained by applying the t-norm among the corresponding antecedent part, is normalized with the sum of the membership degrees from all rules. These normalized values serve as additional weights in the consequent function of the corresponding rule. This leads us to the definition of a Takagi-Sugeno-Kang fuzzy system in a closed functional-type form:

$$\hat{f}(\mathbf{x}) = \hat{y} = \sum_{i=1}^{C} l_i \Psi_i(\mathbf{x}) \tag{1.2}$$

where

$$\Psi_i(\mathbf{x}) = \frac{\mu_i(\mathbf{x})}{\sum_{j=1}^{C} \mu_j(\mathbf{x})} \tag{1.3}$$

and

$$\mu_i(\mathbf{x}) = \operatorname*{T}_{j=1}^{p} \mu_{ij}(x_j) \tag{1.4}$$

where x_j is the j-th component in the data vector, hence reflecting the value of the j-th variable. The symbol T denotes a t-norm in general. The l_i's are the consequent functions of the C rules and are defined by:

$$l_i = w_{i0} + \sum_{j=1}^{p} w_{i,j} x_j + \sum_{j_1=1}^{p} \dots \sum_{j_m=j_{m-1}}^{p} w_{i,j_1 \dots j_m} x_{j_1} \dots x_{j_m} \tag{1.5}$$

In this sense, it is remarkable that a TSK fuzzy system can be inspected in two ways: the linguistic form which provides transparency and interpretability also for EFS under certain conditions (see Chapter 6) and the functional form which has some benefits when developing a training algorithm based on some optimization criteria, see Chapter 2.

The higher m is, the more flexibility is already included in the consequent parts for approximating highly non-linear dependencies in the data. On the other hand, the more likely is an over-fitting or even an over-shooting of the data by the fuzzy system. For an example, see Figure 1.6 (a), where a highly fluctuating relationship is badly approximated by a polynomial function with degree 11 (note the over-shooting in the left part of the functional relationship). Furthermore, the more complex the consequent function already is, the more the antecedent part gets superfluous, finally ending up in a pure non-linguistic polynomial model. Due to these reasons, a widely applied form of the Takagi-Sugeno-Kang fuzzy system is a Takagi-Sugeno (TS) fuzzy system [417], which reduces the high complex consequent function to a linear hyper-plane and therefore puts more interpretation about the data characteristics into the antecedent parts. According to the definition of the Takagi-Sugeno-Kang fuzzy system as outlined above the Takagi-Sugeno fuzzy system can be defined in the same manner, by substituting

$$l_i = w_{i0} + w_{i1}x_1 + w_{i2}x_2 + \dots + w_{ip}x_p \tag{1.6}$$

in (1.5). In fact, this fuzzy model type represents a kind of step-wise linear regression model, where the single linear consequent hyper-planes are combined by the non-linear fuzzy sets and t-norm operators in the antecedent parts of the rules to form a smooth non-linear model. Figure 1.6 (b) represents the approximation by a TS fuzzy system, clearly achieving a more accurate model. A further less complex model is achieved when using only singleton (constant) consequent parameters $\mathbf{w}_0$; this type of fuzzy system is then called Sugeno fuzzy system and usually has lower

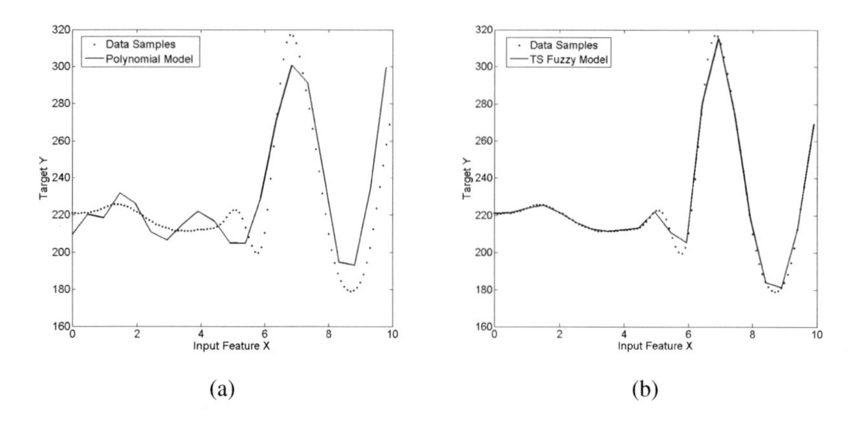

Fig. 1.6 (a): Approximating a highly fluctuating relationship with a polynomial regression model, please note the bad approximation (especially over-shooting in the left part of the functional relationship); (b): approximating the same relationship with a Takagi-Sugeno fuzzy system

predictive power in regression problems than a Takagi-Sugeno fuzzy system [342] (as a local part is represented just by one single value rather than a hyper-plane).

1.2.1.2 Choices for Fuzzy Sets

The most common choices for fuzzy sets are trapezoidal fuzzy sets or Gaussian fuzzy sets. Trapezoidal fuzzy sets are defined by

$$\mu_{ij}(x_j) = \begin{cases} \frac{x_j - a_{ij}}{b_{ij} - a_{ij}} & \text{if } a_{ij} < x_j < b_{ij} \\ 1 & \text{if } b_{ij} \leq x_j < c_{ij} \\ \frac{d_{ij} - x_j}{d_{ij} - c_{ij}} & \text{if } c_{ij} \leq x_j < d_{ij} \\ 0 & \text{otherwise} \end{cases}$$

where for the breakpoints the equational chain $a_{ij} \leq b_{ij} \leq c_{ij} \leq d_{ij}$ holds. They enjoy a great popularity, as they can be easily interpreted as having a compact support and a clearly defined range of maximum membership degrees. On the other hand, they are not steady differentiable (usually causing non-smooth model surfaces resp. decision boundaries) and may not cover the input space sufficiently well, especially when applying a data-driven learning approach for extracting the fuzzy sets directly from the data. This may lead to holes in the fuzzy partition and hence to undefined input states when performing a prediction on new values, especially when the data is sparse or not well distributed over the complete input feature range. An example of such an occurrence is visualized in Figure 1.7 showing two non-overlapping

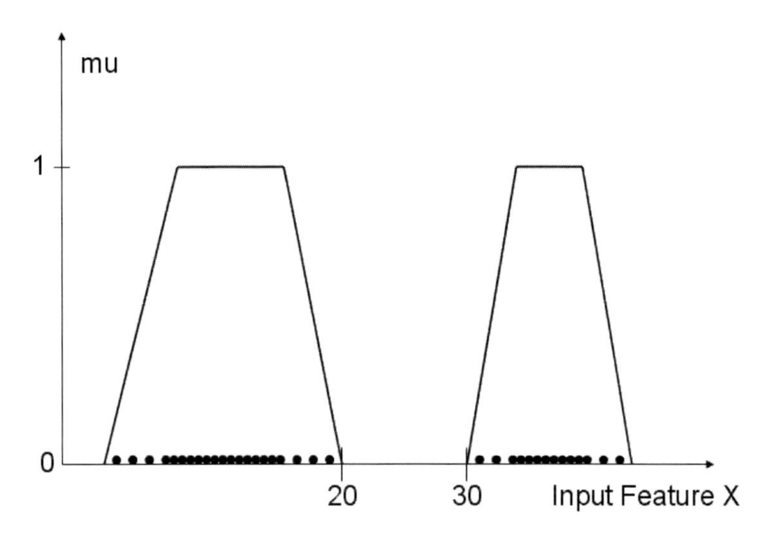

Fig. 1.7 A fuzzy partition not sufficiently covering the input feature due to missing data in the range [20,30]

trapezoidal fuzzy sets, causing a hole between 20 and 30 in the input feature X. Data (from which the fuzzy sets are extracted) is indicated by dots.

Gaussian fuzzy sets are defined by

$$\mu_{ij}(x_j) = e^{-\frac{1}{2}\frac{(x_j-c_{ij})^2}{\sigma_{ij}^2}}$$

where c_{ij} denotes the center and σ_{ij}^2 denotes the variance of the Gaussian function appearing in the j-th premise part of the i-th rule. They possess infinite support (membership degree is nowhere 0), so the problem with undefined input states may not appear, however always all rules fire to a certain degree, which may have computational shortcomings. Furthermore, they are steady differentiable, usually leading to smooth model surfaces. Their interpretability is weaker than for trapezoidal functions, but can be improved when the focus of interest is restricted to the main body of the function around the center $(c_{ij} \pm 2\sigma_{ij})$. Figure 1.8 presents two typical fuzzy partitions, each one containing five fuzzy sets with overlap 0.5 between two adjacent fuzzy sets (also referred as Ruspini partition [376]); Figure 1.8 (a) visualizes trapezoidal where Figure 1.8 (b) shows Gaussian fuzzy sets.

Other types of fuzzy sets are sigmoid functions (achieving an equivalency to feed forward neural networks [175] as the activation function there is typically a logistic function), defined by

$$\mu_{ij}(x_j) = \frac{1}{1+e^{(-a_{ij}(x_j-b_{ij}))}} \tag{1.7}$$

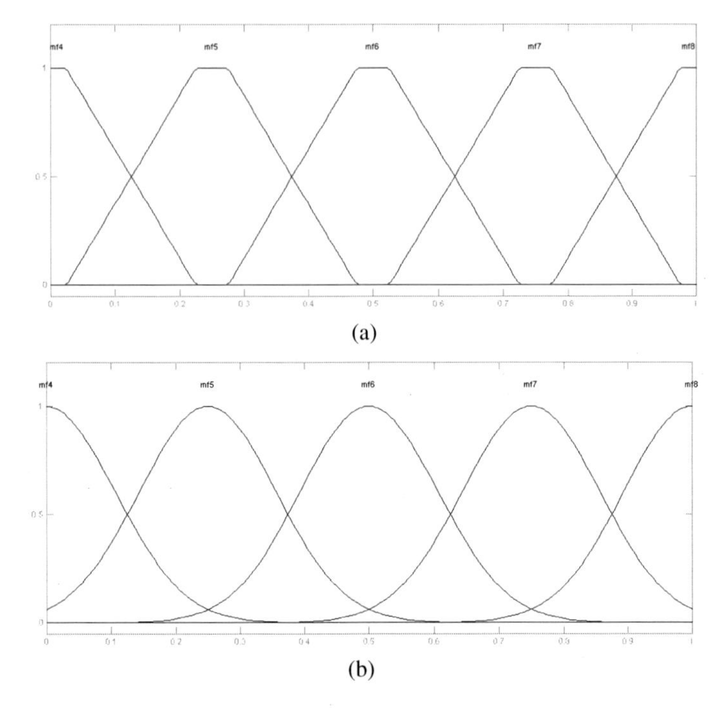

(a)

(b)

Fig. 1.8 (a): fuzzy partition with five trapezoidal fuzzy sets (b): fuzzy partition with five Gaussian fuzzy sets

where b determines the characteristic coordinate, at which the grade of membership in the fuzzy set amounts to 0.5, and a determines the slope of the function at this point. A great attraction and wide usage also enjoy the B-splines of various orders, recursively defined by:

$$B_{j,1}(t) = \begin{cases} 1 & \text{if } t_j \leq t \leq t_{j+1} \\ 0 & \text{otherwise} \end{cases} \tag{1.8}$$

$$B_{j,k}(t) = \omega_{j,k}(t)B_{j,k-1}(t) + (1 - \omega_{j+1,k}(t))B_{j+1,k-1}(t)$$

for $k > 1$ where

$$\omega_{j,k} = \begin{cases} \frac{t-t_j}{t_{j+k-1}-t_j} & \text{if } t_j < t_{j+k-1} \\ 0 & \text{otherwise} \end{cases}$$

$t = (t_j^0, ..., t_j^{k_j+1})$ a knot sequence with

$$t_j^0 = a_j < t_j^1 \leq ... \leq t_j^{k_j} \leq b_j = t_j^{k_j+1}.$$

1.2.1.3 Fuzzy Basis Function Networks and Neuro-Fuzzy Systems

Gaussian fuzzy sets are used in many methods and algorithms for building up Takagi-Sugeno-(Kang) fuzzy systems, also in many EFS approaches (see Chapter 3), there usually connected with product t-norm. In this case, the obtained specific Takagi-Sugeno fuzzy systems are called fuzzy basis function networks, going back to Wang and Mendel [439]:

$$\hat{f}(\mathbf{x}) = \hat{y} = \sum_{i=1}^{C} l_i \Psi_i(\mathbf{x}) \tag{1.9}$$

with the basis functions

$$\Psi_i(\mathbf{x}) = \frac{e^{-\frac{1}{2}\sum_{j=1}^{p}\frac{(x_j-c_{ij})^2}{\sigma_{ij}^2}}}{\sum_{k=1}^{C} e^{-\frac{1}{2}\sum_{j=1}^{p}\frac{(x_j-c_{kj})^2}{\sigma_{kj}^2}}} \tag{1.10}$$

and consequent functions

$$l_i = w_{i0} + w_{i1}x_1 + w_{i2}x_2 + \ldots + w_{ip}x_p \tag{1.11}$$

It is remarkable that these types of Takagi-Sugeno fuzzy systems are more or less equivalent to normalized radial basis function networks (see [195] [225]). This follows since the product of several univariate Gaussian fuzzy sets results in one multi-dimensional Gaussian radial basis function (which represents one neuron). So, for each neuron i the following identity holds:

$$\prod_{j=1}^{p} e^{-\frac{1}{2}\frac{(x_j-c_{ij})^2}{\sigma_{ij}^2}} = e^{-\frac{1}{2}\sum_{j=1}^{p}\frac{(x_j-c_{ij})^2}{\sigma_{ij}^2}} \tag{1.12}$$

The consequent functions can be seen as weights in the RBF function network, mostly there they are reduced to singletons, i.e. to w_{i0}. This result can be further extended and adopted to local model neural networks [187]. Furthermore, most of the neuro-fuzzy systems[134] can be interpreted as Takagi-Sugeno(-Kang) fuzzy systems. Typically, the fuzzy model is transformed into a neural network structure (by introducing layers, connections and weights between the layers) and learning methods already established in the neural network context are applied to the neuro-fuzzy system. A well-known example for this is the ANFIS approach [194], where the back-propagation algorithm [446] (rediscovered by Rumelhart et al. [375] and widely-used in the neural network community) is applied and the components of the fuzzy model (fuzzyfication, calculation of rule fulfillment degrees, normalization, defuzzification), represent different layers in the neural network structure, see Figure 1.9. In Chapter 3 this architecture is also applied to some evolving fuzzy systems approaches.

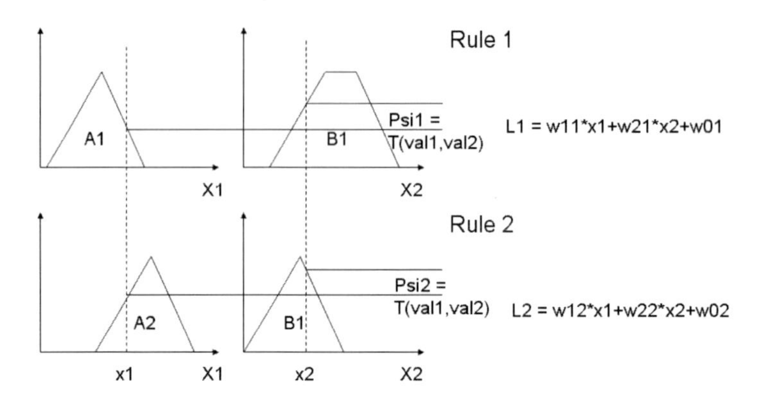

(a)

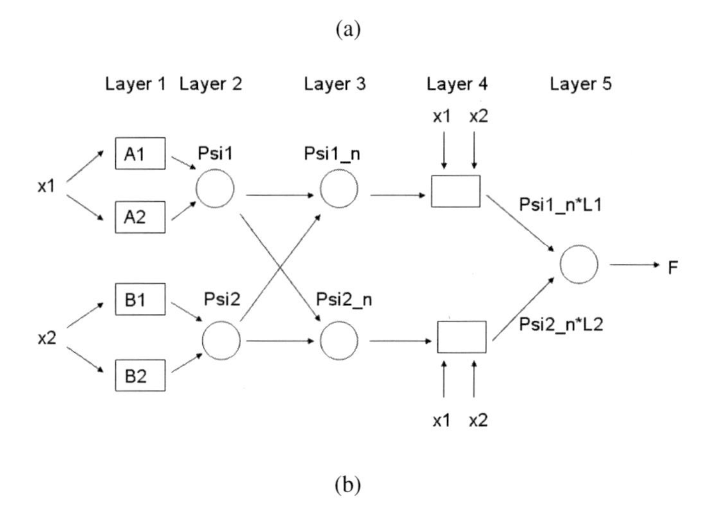

(b)

Fig. 1.9 (a): Takagi-Sugeno type fuzzy system, (b): Equivalent neural network structure

1.2.1.4 Mamdani Fuzzy Systems

Opposed to TSK type fuzzy systems and neuro-fuzzy approaches, which exploit model architectures supporting precise fuzzy modelling, Mamdani fuzzy systems [297] are fuzzy systems that are providing more interpretability and hence are used for linguistic fuzzy modelling tasks. The basic difference to TSK type fuzzy systems is, that they contain fuzzy sets in the consequent parts of the rules, which makes a single rule fully readable in a linguistic sense. The definition of a (single output) Mamdani fuzzy system is thus as follows:

$$\text{Rule}_i : \text{IF } x_1 \text{ IS } \mu_{i1} \text{ AND...AND } x_p \text{ IS } \mu_{ip} \text{ THEN } l_i \text{ IS } \Phi_i$$

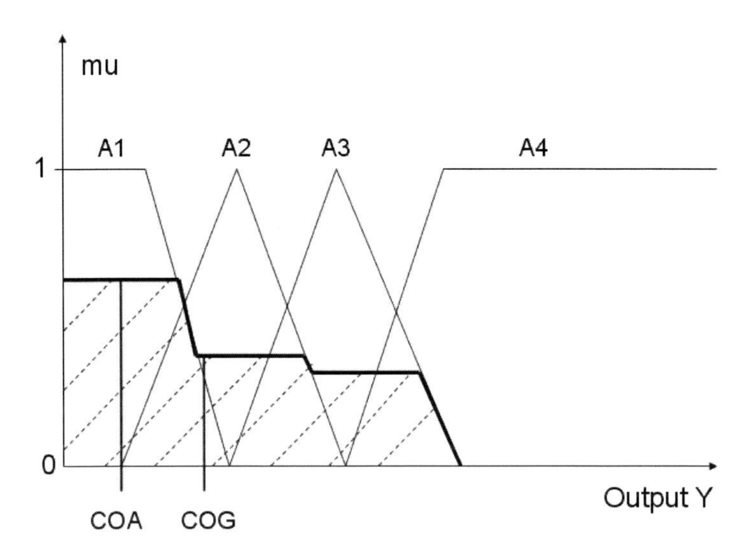

Fig. 1.10 Center of maximum (COM) and center of gravity (COG) defuzzification for a rule consequent partition (fuzzy partition in output variable Y) in a Mamdani fuzzy system, the shaded area indicates the joint consequents (fuzzy sets) in the active rules; the cutoff points (alpha cuts) are according to the membership degrees obtained from the rule antecedent parts of the active rules

with Φ_i the consequent fuzzy set in the fuzzy partition of the output variable used in the ith rule. It may happen, that $\Phi_i = \Phi_j$ for some $i \neq j$. Hence, a t-conorm [220] is applied which combines the activation levels of the antecedent parts of those rules having the same consequents to one output set. Most common choice is the maximum operator. This is done for the whole fuzzy partition where the various output sets are joined to one fuzzy area. In order to obtain a crisp output value, a defuzzification method is applied, most commonly used are the mean of maximum over the whole area, the centroid of the area or the bisector which is the vertical line that will divide the region into two sub-regions of equal area. This may coincide with the centroid, but not necessarily. Figure 1.10 demonstrates a defuzzification example for a typical fuzzy area inducted from the fuzzy rules (shaded area): the height of the shaded area among the fuzzy sets indicate the aggregated membership fulfillment degrees of the antecedent parts (single fuzzy set membership degrees combined with a t-norm) in the corresponding rules (those rules where the fuzzy sets appear in the consequents).

Even though Mamdani fuzzy systems usually have a lower predictive power than Takagi-Sugeno-Kang fuzzy type systems [342], they enjoy a wide attraction and application in linguistic modelling (especially when designing fuzzy systems directly from expert knowledge rather than from data, see e.g. [247] [181]). However, the linguistic interpretability may not always be better than for Takagi-Sugeno(-Kang) type fuzzy models. Here it depends more or less on the kind of application and which type of interpretability is wished by the user. For instance, the consequent functions in Takagi-Sugeno fuzzy systems may show a certain behavior of the model in certain circumstances (e.g. being constant in certain parts) based on the values of the weights $\mathbf{w}$. This may give a more important insight for an operator than a complete linguistic rule. A further discussion of this issue will be done in Chapter 6 which deals with techniques for improving the interpretability in evolving fuzzy systems.

1.2.1.5 Fuzzy Classification Models

Fuzzy classifiers, short for fuzzy classification models, differ from fuzzy regression models in their response to the system: opposed to a predictive numerical value, the response is usually a classification resp. decision statement represented by a class to which a sample belongs (for instance 'woman' versus 'man' in case of gender classification). Here, we demonstrate two types of model architectures which will be used in some of the EFS approaches mentioned in Chapter 3.

Single Model Architecture

The first one is the classical single model architecture with single output [373] [236] [319], most commonly used in applications where fuzzy approaches are applied. There the ith rule is defined in the following way:

$$\text{Rule}_i : \text{IF } x_1 \text{ IS } \mu_{i1} \text{ AND...AND } x_p \text{ IS } \mu_{ip} \text{ THEN } l_i = L_i \tag{1.13}$$

where L_i is the crisp output class label from the set $\{1,...,K\}$ with K the number of classes. Now, for each new incoming sample $\mathbf{x}$, the final crisp class label can be obtained by taking the class label of that rule with the highest activation degree, i.e. by calculating

$$L = L_{i^*} \text{ with } i^* = \text{argmax}_{1\leq i\leq C}\ \mu_i \tag{1.14}$$

with C the number of rules, μ_i the activation degree of rule i defined by $T_{j=1}^{p}\ \mu_{ij}(x_j)$ Furthermore, K rule weights $conf_{i1},...,conf_{iK}$ are introduced, which denote the confidence level of the ith rule in all the K classes, where

$$conf_i = max_{k=1,...,K} conf_{ik} \tag{1.15}$$

corresponds to the output class L_i. These rule weights are important in case of modelling the uncertainty of local regions, i.e. one rule for itself is not completely confident in one class label (as for instance samples from different classes appearing in the same local region). In an ideal case, a weight vector for one rule contains one entry with value 1 and the remaining entries with value 0. This is also referred to as a clean classification rule. Based on the rule weights an overall confidence $conf_L$ of the overall output class label $L \in \{1,...,K\}$ is calculated through the following weighted average:

$$conf_L = \frac{conf i^* L}{\sum_{k=1}^{K} conf_{i^*k}} \tag{1.16}$$

with i^* as defined in (1.14). This definition is quite intuitive as $conf_L$ measures the predominance of the most frequent class over the other classes in the nearest rule. Note that confidence levels are of great help for a better interpretation of the model's final output. They are quite often demanded by operators, especially in case of incorrect or uncertain model feedback (see Section 4.5.2). Moreover, confidence level are necessary when applying classifier fusion techniques for boosting performance, see Sections 5.4 and 9.1. In case of a two-class classification task, the single-model architecture can be made even slimmer, as the confidence levels (instead of the class labels) can be directly encoded in the consequent part of the rules, serving as continuous output values between 0 = class#1 and 1 = class#2. If for example a rule has consequent value of 0.6, this means that it is confident in class#2 with a degree of 0.6 and in class#1 with a degree of $1-0.6=0.4$. Consequently, if the value of the rule with highest activation degree is greater or equal 0.5 the current data sample belongs to class#2, otherwise to class#1.

A softening of the above classical crisp rule structure is achieved by introducing a confidence vector $conf_{i1},...,conf_{iK}$ as multiple consequents. This means the rules as in (1.13) are extended by the following consequent part:

$$\text{Rule}_i : ... \text{ THEN } l_i = 1(conf_{i1}), l_i = 2(conf_{i2}), \ldots, l_i = K(conf_{iK}) \tag{1.17}$$

Then the final crisp class label is obtained by calculating

$$L = L_{i^*} \quad L_{i^*} = \text{argmax}_{k=1,...,K} conf_{i^*k} \quad i^* = \text{argmax}_{1 \leq i \leq C} \mu_i conf_i \tag{1.18}$$

instead of (1.14) and L_{i^*} the output class label. This may have some advantages, especially when one of two nearby lying, overlapping rules has a significantly higher confidence in one class label than the other. For instance, consider a one-dimensional case as shown in Figure 1.11: the left rule associated with fuzzy set μ_1 covers three classes with confidences 0.4, 0.3, 0.3, hence has a majority confidence of 0.4 in its output class L_i (class '□'). The other rule associated with fuzzy set μ_2 covers only samples from class 'X', hence has a confidence of 1.0 in its output class. Now, if a new sample to be classified lies in the overlapping region of the two rules, i.e. correspond to the left rule by 0.6 and to the right one by 0.4 membership degree, the sample would still be classified to class 'X', although lying nearer to the core part of left rule (as $0.6*0.4 < 0.4*1.0$), finally achieving a more plausible class response. The confidence in the output class label L includes the membership function degrees of all rules:

$$conf_L = \frac{\sum_{i=1}^{C} conf_{iL}\mu_i}{\sum_{i=1}^{C} \mu_i} \tag{1.19}$$

For example, if a new sample is described exactly by the i^*th rule and no other rule fires, then the confidence is $conf_{i^*}$, i.e. the confidence of the i^*th (the winner) rule to the output class L; if more rules are firing to a certain degree, then a weighted

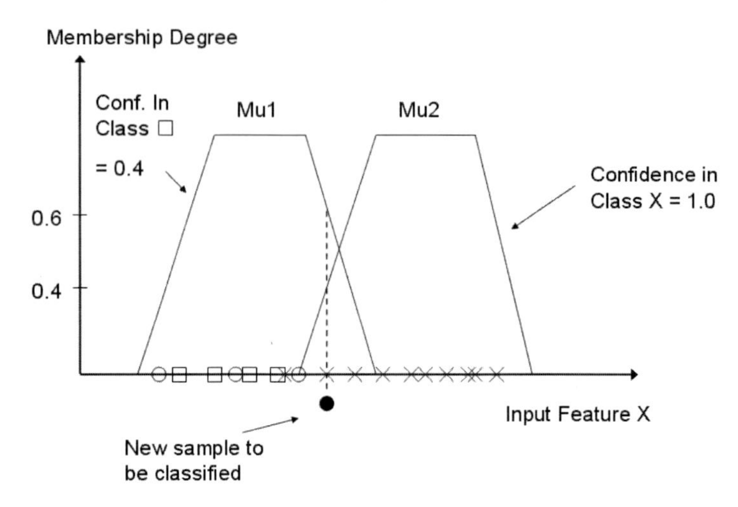

Fig. 1.11 New sample to be classified lying in the overlapping regions of two fuzzy sets (=rules), the confidences of the two rules in the majority classes indicated; when using classical class label output function as in (1.14), the sample will be classified to the class '□', when using extended approach through (1.17) and (1.18) the sample is classified to class 'X'

average of all the rules' confidences in the output class L is produced. This is in consistency to the classification output calculated in (1.18).

Multi Model Architecture

The multi model architecture for classification problems is considered to exploit the model architecture of a Takagi-Sugeno fuzzy regression model as defined in (1.2) and which was first introduced in [276] when designing *FLEXFIS-Class* algorithm for an evolving fuzzy classifier approach (see Section 3.1). A straightforward regression on the class labels may end up in severe problems in case of multi-class classification tasks. This is because a TS fuzzy model tries to fit a polynomial regression function between two classes for a smooth transition. If a new sample falls near the boundary between class#1 and class#3 such a classifier would deliver result 'class#2', which does not necessarily appear between class #1 and class #3. This is outlined in Figure 1.12. In order to circumvent this problem and furthermore to have an even better flexibility with respect to extensions of the number of classes during on-line operation, multiple Takagi-Sugeno fuzzy regression models are proposed to be trained on K indicator matrices for the K classes. These are generated from the original feature matrix, containing all the input features in different columns and a specific label entry in the last column. For the ith matrix each label entry is set to 1, if the corresponding row belongs to the ith class, otherwise it is set to 0, see Table 1.1, where four different classes trigger four label vectors (the columns in the table), which are appended as last columns to the feature matrix. The input feature

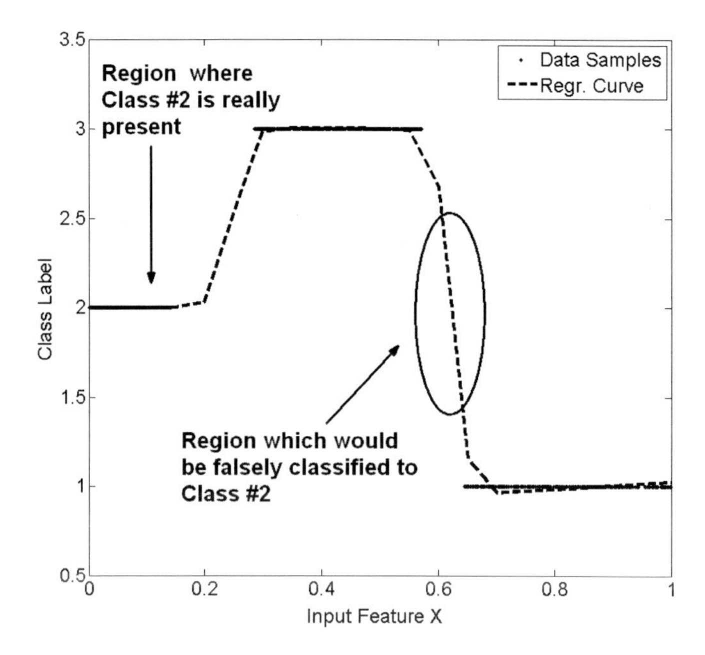

Fig. 1.12 A regression on three class labels (1,2, and 3) performed by a TS fuzzy model (dotted line) based on some synthetic training data (dark dots); samples falling in the transition between class#1 to class#3 would be falsely produced as class #2 (as a value around 2 would be produced by the fuzzy model).

Table 1.1 Original label vector including four classes resulting in the label vectors for four indicator matrices

Orig. Labels	Ind. entries for class#1	Ind. entries for class#2	Ind. entries for class#3	Ind. entries for class#4
1	1	0	0	0
3	0	0	1	0
1	1	0	0	0
2	0	1	0	0
4	0	0	0	1
1	1	0	0	0
3	0	0	1	0
3	0	0	1	1
4	0	0	0	1
...	...	...	...	...

columns remain the same for all indicator matrices. In this sense, it is guaranteed that the regression surface is forced towards 1 in the region of those samples belonging to its corresponding class and otherwise it approaches 0.

In fact, in [174] (Chapter 4, Section 4.2) it is maintained that (linear) regression by an indicator matrix [108] only works well on two-class problems (0/1) and can have problems with multi-class problems, as then a complete masking of one class by two or more others may happen. However, opposed to linear regression models, the TS fuzzy models are non-linear, i.e. not only the (locally linear) rule consequents but also the non-linear antecedent parts are trained based on the indicator matrix information, i.e. by taking the label column as the (regressing) target variable. In this sense, the approximation behavior is non-linear which forces the surface of a TS fuzzy model f_k going to 0 more rapidly in regions apart from class k samples as in the case of inflexible pure linear hyper-planes. Therefore, the masking effect is much weaker than in the pure linear case. To underline this statement, Figure 1.13 (a) shows the masking effect on a multi-class classification problem obtained when using linear regression by indicator matrix $\rightarrow$ the middle class is completely masked out as nowhere maximal, whereas in Figure 1.13 (b) the masking effect is solved due to the stronger flexibility of the models (going down to zero more rapidly where the corresponding class is not present). This type of classification is called *fuzzy regression by indicator matrix* (or vector for on-line case) [276].

The overall output from the fuzzy classifier is calculated by inferencing a new multi-dimensional sample $\mathbf{x}$ through the K Takagi-Sugeno models

$$\hat{f}_m(\mathbf{x}) = \sum_{i=1}^{C} l_i \Psi_i(\mathbf{x}) \qquad m = 1,...,K \tag{1.20}$$

and then eliciting that model producing the maximal output and taking the corresponding class as label response, hence

$$L = class(\mathbf{x}) = \text{argmax}_{m=1,...,K} \hat{f}_m(\mathbf{x}) \tag{1.21}$$

This is in accordance to a one-versus-rest (or one-against-all) classification approach[344]. The confidence $conf_L$ of the overall output value $L = m \in \{1,...,K\}$ is elicited by normalizing the maximal output value with the sum of the output values from all K models:

$$conf_L = \frac{\max_{m=1,...,K} \hat{g}_m(\mathbf{x})}{\sum_{m=1}^{K} \hat{g}_m(\mathbf{x})} \tag{1.22}$$

where $\hat{g}_m(\mathbf{x}) = \hat{f}_m(\mathbf{x}) + |\min(0, \min_{m=1,...,K} \hat{f}_m(\mathbf{x}))|$. This assures that all output values from all TS fuzzy systems are forced to be positive, and hence the confidence level well defined. In case when all output values are positive $\hat{g}$ equals $\hat{f}$.

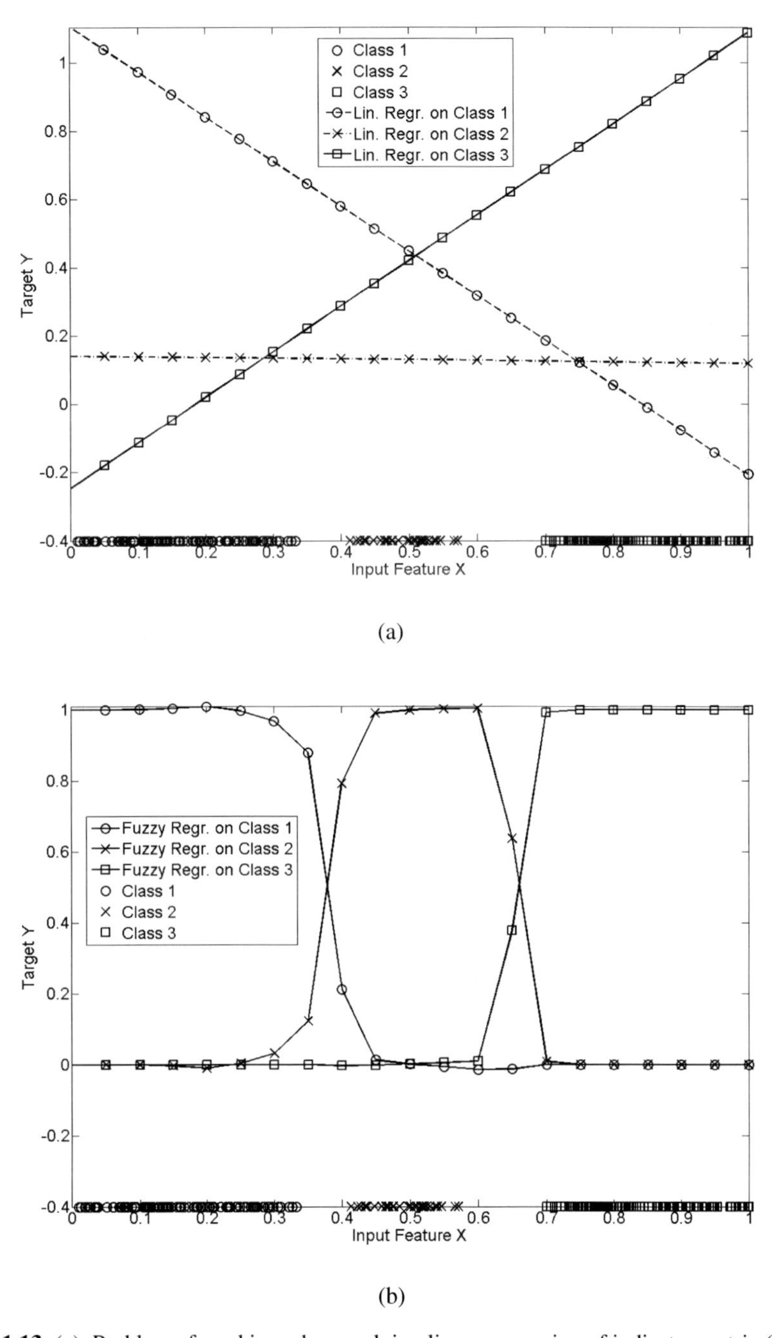

Fig. 1.13 (a): Problem of masking when applying linear regression of indicator matrix (light line with class 'X' samples is nowhere maximal); (b): masking problem solved by fuzzy regression of indicator matrix

1.2.1.6 Relational and Implication-Based Fuzzy Models

A short note is made here on relational and implication-based fuzzy models. The latter are widely used in theory of fuzzy sets and systems [31] [421]. They extend the classical Boolean implications to many-valued implications by exploiting the fuzzy definition of an implication through a t-norm [220]:

$$\mathbf{T}(x,y) = \sup\{u \in [0,1] | T(x,u) \leq y\} \tag{1.23}$$

They are of great interest for a detailed mathematical analysis to get a deeper insight of dependencies/equivalences/synergies between the Boolean logic and the fuzzy logic resp. between crisp and fuzzy rules [109]. Furthermore, implication-based fuzzy systems [140] support a mechanism for checking and preventing contradictions in the fuzzy implication sense [223].

Fuzzy relational models [337, 322] encode associations between linguistic terms defined in the system's input and output domains by using fuzzy relations. They possess the property that a rule can consist of more consequent parts, each part fulfilled to a certain degree, so linguistic rules are not considered to be fully true, but more to be partly true, so for instance a typical fuzzy relational rule has the following form:

IF x is *HIGH* AND y is *MEDIUM* THEN z is *LOW* with 0.6
z is *MEDIUM* with 0.4

Due to the lack of uniqueness in the conclusion parts of the rules, they are basically applied for knowledge-based modelling tasks, where this splitting up of the consequent parts can lead to a better interpretation of the models.

Neither types of model architectures are hardly ever used for data-driven design purposes of fuzzy systems and hence will be not considered further in this book.

1.2.1.7 Type-2 Fuzzy Systems

Type-2 fuzzy systems were invented by Lotfi Zadeh in 1975 [466] for the purpose of modelling the uncertainty in the membership functions of usual (type-1) fuzzy sets as were defined in Section 1.2.1.2. In particular, the crisp values are indeed 'fuzzified' through usual fuzzy sets, however the membership degree of a fuzzy set to which one single sample belongs to is still a crisp value. This reaped some criticism among fuzzy logic scientists before 1975 as uncertainty is only expressed by single membership values. The distinguishing feature of a type-2 fuzzy set $\tilde{\mu}_{ij}$ versus its type-1 counterpart μ_{ij} is that the membership function values of $\tilde{\mu}_{ij}$ are blurred, i.e. they are no longer a single number from 0 to 1, but are instead a continuous range of values between 0 and 1, say $[a,b]$. One can either assign the same weighting or a variable weighting to the interval of membership function values $[a,b]$. When the former is done, the resulting type-2 fuzzy set is called either an interval type-2 fuzzy set or an interval valued fuzzy set (although different nomenclatures may be used,

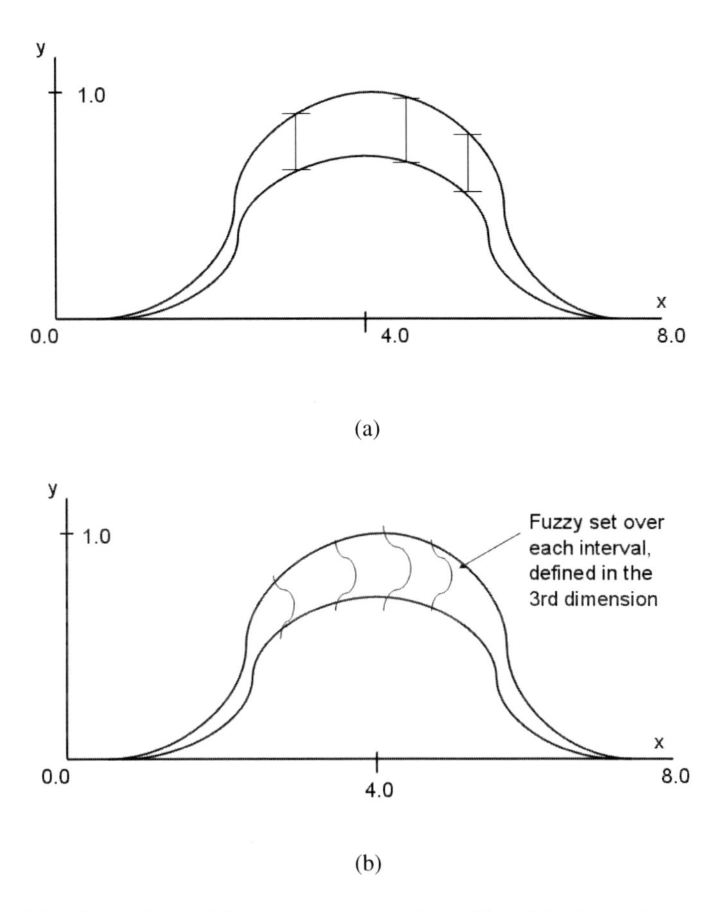

Fig. 1.14 (a): Interval type-2 fuzzy set, note that the widths of the intervals are usually not constant over the whole region; (b): general type-2 fuzzy set, additional fuzzy sets are defined over the intervals in the third dimension

they are the same) [304]. When the latter is done, the resulting type-2 fuzzy set is called a general type-2 fuzzy set [304]. Figure 1.14 (a) visualizes an example of a interval type-2 fuzzy set (mind the interval between the type-1 fuzzy sets), whereas Figure 1.14 (b) represents its extension to a general type-2 fuzzy set by defining an additional set over each interval between the two type-1 fuzzy sets. Note that the space between the two type-1 fuzzy sets is also called the footprint of uncertainty (FOU).

Similarly to a classical Takagi-Sugeno-Kang type fuzzy system, the ith rule of an interval-based type-2 fuzzy system is defined in the following way [255] [302]:

$$\begin{aligned} \text{Rule}_i: \text{IF } x_1 \text{ IS } \tilde{\mu}_{i1} \text{ AND...AND } x_p \text{ IS } \tilde{\mu}_{ip} \text{ THEN} \\ l_i = \tilde{w}_{i0} + \tilde{w}_{i1}x_1 + \tilde{w}_{i2}x_2 + \ldots + \tilde{w}_{ip}x_p \end{aligned} \quad (1.24)$$

with $\tilde{w}_{ij}$ an interval set (instead of a crisp continuous value), i.e.

$$\tilde{w}_{ij} = [c_{ij} - s_{ij}, c_{ij} + s_{ij}] \quad (1.25)$$

There are various ways of performing the inference of an interval-based type-2 fuzzy system [302]. One of the most common choices is to calculate two fuzzified values for a crisp input value **x** by using the upper and lower bound fuzzy set, achieving two firing degrees for each rule, i.e. for the ith rule $\bar{\Psi}_i(\mathbf{x})$ and $\underline{\Psi}_i(\mathbf{x})$, denoting the upper and the lower bound of membership degree. The same is done for the consequent part, leading to upper bound $\bar{l}_i$ and lower bound $\underline{l}_i$. A final output value y can be achieved by multiplying upper as well as lower bounds of rule membership degrees and consequent values, achieving $\bar{y}$ and $\underline{y}$, and taking the average of these two. However, in this case the interval-based type-2 fuzzy system would simply reduce to two parallel type-1 fuzzy systems with different definitions of the (type-1) fuzzy sets. Hence, usually an enhanced approach for eliciting the final output is applied, the so-called Karnik-Mendel iterative procedure [207] where a type reduction is performed before the defuzzification process. In this procedure, the consequent values $\bar{l}_i, i = 1,...,C$ and $\underline{l}_i, i = 1,...,C$ are sorted in ascending order denoted as $\bar{y}_i, i = 1,...,C$ and $\underline{y}_i, i = 1,...,C$. Accordingly, the membership values $\bar{\Psi}_i(\mathbf{x})$ and $\underline{\Psi}_i(\mathbf{x})$ are sorted in ascending order denoted as $\bar{\psi}_i(\mathbf{x})$ and $\underline{\psi}_i(\mathbf{x})$. Then, the outputs $\bar{y}$ and $\underline{y}$ are computed by:

$$\bar{y} = \frac{\sum_{i=1}^{L} \bar{\psi}_i(\mathbf{x})\bar{y}_i + \sum_{i=L+1}^{C} \underline{\psi}_i(\mathbf{x})\bar{y}_i}{\sum_{i=1}^{L} \bar{\psi}_i(\mathbf{x}) + \sum_{i=L+1}^{C} \underline{\psi}_i(\mathbf{x})} \qquad \underline{y} = \frac{\sum_{i=1}^{R} \underline{\psi}_i(\mathbf{x})\underline{y}_i + \sum_{i=R+1}^{C} \bar{\psi}_i(\mathbf{x})\underline{y}_i}{\sum_{i=1}^{R} \underline{\psi}_i(\mathbf{x}) + \sum_{i=R+1}^{C} \bar{\psi}_i(\mathbf{x})} \quad (1.26)$$

with L and R positive numbers, often $L = \frac{C}{2}$ and $R = \frac{C}{2}$. Taking the average of these two yields the final output value y.

For type-2 fuzzy systems with Mamdani inference the consequent parts are simply substituted by type-2 fuzzy sets. The extension to a general type-2 fuzzy system is achieved by substituting the $\tilde{\mu_{ij}}$s and the intervals in the consequent functions with general type-2 fuzzy sets, see [303].

1.2.2 *Goals of Evolving Fuzzy Systems*

First we will provide some general definitions about incremental learning and the spirit of model evolution during incremental learning. Then, we will demonstrate the demands to the incremental learning/evolving concept from the machine learning and mathematical sense. This should yield a guideline how to design algorithms for evolving models in general. Finally, we will adopt these concepts to fuzzy systems, i.e. explain which parameters, components in a fuzzy system should be incrementally trained and evolved in order to guarantee flexibility and stability.

1.2.2.1 Definition of Incremental Learning and the Evolving Concept

Incremental learning is the engine for evolving modelling of processes par excellence. This is, because newly appearing operating conditions, i.e. system states causing new dependencies in the system are automatically integrated into the already built up models in form of incremental learning steps, as the usage of older data would slow down the training process significantly. In this sense, evolving models automatically learn new coherences and interrelations among certain components in the system by the usage of incremental learning methods. From a technical point of view, the definition of incremental learning is as follows:

> *Incremental learning* can be defined as a sample-wise or block-wise training of a model, either from scratch or starting with an already available one.

This means, that a model is built up step-wise with parts of data rather than a whole data set being sent into the training algorithm at once and (usually) iterating over this data set a multiple times in order to achieve convergence of parameters (as done in batch learning) and rather than data being collected and models being re-built from time to time (requiring a huge computational effort). Unsupervised incremental learning deals mainly with incremental clustering approaches, where clusters are updated and evolved on demand based on new incoming data. Supervised incremental learning deals with the incremental update of data-driven models (regression models and classifiers) by either by-measured targets (regression) or exact class responses (classification) on the newly loaded samples. Hence, the learning algorithm knows exactly, in which direction the models need to be updated. This is opposed to reinforcement learning [413], where a detailed response is not provided, but a reinforcement signal, providing information about the quality of a model action/response. So, it is usually applied for reaching long-term goals, i.e. maximizing a long-term reward (as e.g. used in game strategies). Furthermore, incremental learning usually focuses on updating a global model, and not updating the data base as is done in instance-based learning schemes (e.g. see [44]). A specific case of incremental learning is the so-called single-pass incremental learning, which is defined as follows:

> *Single-pass incremental learning* can be defined as a specific form of incremental learning, where new incoming samples are sent into the learning algorithm and immediately discarded, afterwards.

This means that no prior samples (e.g. within a sliding window) are used for updating the models and only the model's structure and parameters together with some help information (like for instance the variance or the mean of input features, variables etc.), usually updated synchronously to the models, is used for further processing. Most of the approaches discussed in this book are single-pass incremental learning approaches. The big advantage of these approaches is that they only

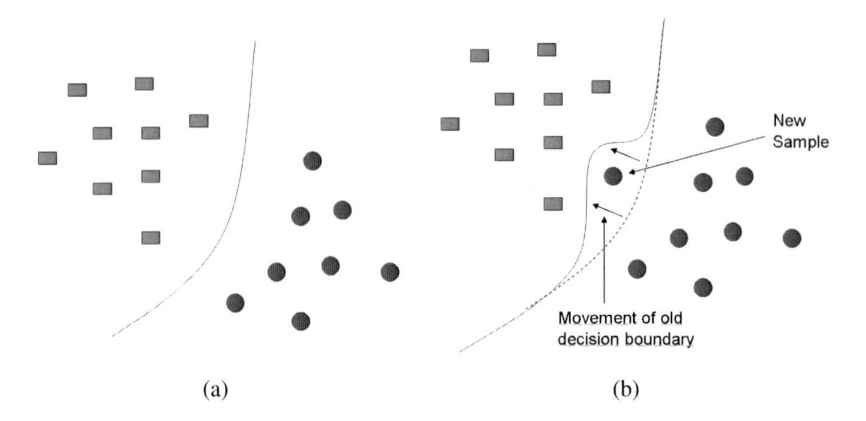

Fig. 1.15 (a): initial decision boundary between two classes (rectangular and circle); (b): updated decision boundary according to a new incoming sample

demand a low virtual memory and computational effort. In literature incremental learning sometimes also refers to the incremental nature of a batch algorithm; however, there the whole batch of data is used at once for training, and by more and more iterations the model incrementally expands (iterative incremental learning).

In literature, incremental learning is also often referred to as life-long learning, as incremental learning in its nature tries to preserve previously learned knowledge. In most cases, this is an important aspect for guaranteeing stability in the models (see also Section 1.2.2.2 below). In a few cases, however, this is an undesired characteristic, especially when drifts or shifts in the data stream occur. This aspect will be handled in Section 4.2, where incremental learning schemes for fuzzy systems will be extended in order to account for these situations.

An example for underlining the incremental learning process is shown in Figure 1.15. There, the impact of incremental learning on the decision boundary of a classifier is demonstrated. The left image visualizes the original class distribution (two classes, one represented by rectangular samples, the other by circle samples) together with the decision boundary of the corresponding trained classifier. Now, a new sample comes in falling into the circle class (marked by an arrow), appearing on the wrong side of the decision boundary: the decision boundary is moved by an incremental learning step based on just using the old decision boundary and this new sample. In the ideal case, the shift of the decision boundary is large enough as shown in the image, in order to classify other samples similar to this new sample as correct (as 'circle' in this case). In some other cases, i.e. when using specific types of training procedures, the decision boundary may be moved a bit towards the new sample. However, if more data samples in the local neighborhood are presented from the circle class, this usually ends up again with a decision boundary as shown in Figure 1.15.

When speaking in terms of mathematical models which should be incrementally trained, an incremental model update is defined in the following way: Let therefore be $\mathbf{x}_{N+1,\ldots,N+m}$ be m new observations of variables/features $x_1, x_2, \ldots, x_n$. An incremental model update I of the former model f_N (estimated from the N initial samples) is defined by

$$f_{N+m} = I(f_N, \mathbf{x}_{N+1,\ldots,N+m}) \quad (1.27)$$

So, the incremental model update is done by just taking the new m samples and the old model, but not using any prior data. Hereby, the whole model may also include some additional statistical help measures, which needs to be updated synchronously to the 'real' model (see Chapter 2).

If $m = 1$, we speak about *incremental learning in sample mode* or *sample-wise incremental learning*, otherwise about *incremental learning in block mode* or *block-wise incremental learning*.

Now, two update modes in the incremental learning process are distinguished:

1. Update of the model parameters: in this case a fixed number of parameters $\Phi_N = \{\phi_1, \ldots, \phi_l\}_N$ of the original model f_N is updated by the incremental learning process and the outcome is a new parameter setting Φ_{N+m} with the same number of parameters, i.e. $|\Phi_{N+m}| = |\Phi_N|$. Here, we also speak about a model adaptation resp. a model refinement with new data.
2. Update of the whole model structure: this case leads to the evolving learning concept, as the number of the parameters may change and also the number of structural components may change (e.g. rules are added or pruned in case of fuzzy systems) automatically according to the characteristics of the new data samples $\mathbf{x}_{N+1,\ldots,N+m}$. This means that usually (but not necessarily) $|\Phi_{N+m}| \neq |\Phi_N|$ and $C_{N+m} \neq C_N$ with C the number of structural components. The smaller m, the more likely equality holds. The update of the whole model structure also includes an update of the input structure, i.e. input variables/features may be exchanged during the incremental learning process. This issue is a hard topic, as changing dimensionality and structural dependencies in the model may cause discontinuities in the whole update process. An appropriate concept will be treated in a separate section in Chapter 5.

From the second itemization point, it is easy to see that the evolving learning concept directly follows from a specific action which takes place in the incremental learning mechanism.

Another definition in this regard is dedicated to on-line learning.

On-line learning is defined as a model training process which takes place during on-line operation modes (typically in on-line applications), where the models are permanently trained, expanded with new data collected from the on-line processes.

This is opposed to off-line training, where models are trained on specific data sets (pre-collected from on-line operation and stored onto media devices) and then implemented to operate in an on-line environment (for performing predictions, classifications, control statements etc.) without changing their parameters, structures. Often, model training during on-line learning should terminate in real-time in order to guarantee models which are up-to-date as early as possible (referred to as *real-time learning*). Intuitively, it is clear that a re-building of models from time to time slows down the model update step too significantly, especially when using all the data seen so far as performed in 'full instance memory' systems [366]. If using only a small portion of the data (the latest data samples) as applied in 'partial instance memory' systems [7], the model may forget older learned situations which may be repeated at the system in a later stage. This finally has the consequence that incremental learning is the only feasible option in today's on-line learning scenarios.

1.2.2.2 Demands to the Evolving Concept

In order to guarantee correct and reliable models, there are two important demands for an incremental learning and evolving concept of models [273]:

- The performance (predictive power) of the evolved models should be as close as possible to that one achieved by their (hypothetical) batch versions (machine learning point of view)
- Convergence to optimality (with respect to an optimization criterion) (mathematical point of view)

The performance close to the batch solution is defined in the following way:

Let therefore f_{N+m}_batch be a model trained in batch mode based on $N+m$ samples and let f_{N+m}_inc be the incrementally updated model of f_N_batch (trained from initial N samples) with m new incoming samples. Then, the major goal of the incremental learning algorithm should be to come as close as possible to the performance (accuracy) of the (corresponding) batch trained model using all $N+m$ samples, i.e.

$$acc(f_{N+m}\_inc) \approx acc(f_{N+m}\_batch) \tag{1.28}$$

If this is fulfilled, we can speak about a *stable incremental learning algorithm*.

The reason why this is an important issue is that the batch learning algorithm 'sees' the whole data set at once and may also perform more iterations of the data samples in order to converge its parameters to an optimality criterion and hence usually comes to a stable solution with a high predictive power.

In terms of the mathematical demand, the convergence of at least some parameters to the optimal batch mode solution should be achieved. This is only possible, when an optimization problem is defined (e.g. least squares optimization problem), whose solution leads directly to the parameter estimation of the model. Then, optimality can be defined 'in the optimization function sense' (e.g. 'in the least squares sense'). In Chapter 2, we will see that this optimality is guaranteed for fuzzy systems when performing an appropriate update of consequent parameters without changing the structure of the models. In Section 3.1, we will present approaches which are able to update and evolve the structure without loosing this favorable property. A generalization of this concept is often referred to as plasticity-stability dilemma [155]: stability is achieved when components in the models (rules in case of fuzzy systems) are not changed or are changed very slightly. Therefore stability is related to parameter convergence, as small changes in the structure usually do not disturb this convergence. Plasticity is related to significant movements of structural components. Hence, a reasonable incremental learning approach tries to find a compromise between stability (in order to guarantee convergence) and plasticity (in order to guarantee that new information in the data is included in the models) at the same time.

In terms of robustness and accuracy, an initial batch mode training step with the first amount of training samples is, whenever possible, often preferable to incremental learning from scratch, i.e. a building up of the model sample per sample from the beginning. This is because within a batch mode phase, it is possible to carry out validation procedures (such as cross-validation [411] or bootstrapping [116]) in connection with search techniques for finding an optimal set of parameters for the learning algorithm (such as best parameter grid search) in order to achieve a good generalization quality. The obtained parameters are then usually reliable start parameters for the incremental learning algorithm to evolve the model further. When performing incremental learning from scratch, the parameters have to be set to some blind default values, quite often not appropriate for the given data stream mining problem. An initial batch mode step can be performed within an off-line pre-initialization phase, where the model is built up based on some prior collected recordings or based on some simulated data, or also during on-line phase, after the first few dozens or hundreds of samples have been recorded. The applicability of the latter depends on the frequency of the data sampled on-line, the dimensionality of the variable resp. feature space and on the speed of the batch learning procedure.

1.2.2.3 What to Adapt and Evolve in Fuzzy Systems?

Now, we are dealing with the question: which components of a fuzzy system should be updated/evolved during the incremental learning and evolution phase? Even though this issue is dependent

- from the chosen type of model (regression versus classification models)
- from the chosen model architecture (TSK, Mamdani, neuro-fuzzy (NF) approaches, type-2 fuzzy systems)

some answers can be given, which are applicable to all types of fuzzy systems defined in Section 1.2.1:

- In order to enhance a model trained originally with only a handful of data samples, it is surely necessary to adapt and refine linear and non-linear parameters appearing in the fuzzy system. This refinement guarantees more confident and more exact parameters and finally more accurate models. Linear parameters occur in the consequent functions of TSK type as well as weights in neuro-fuzzy type systems, whereas non-linear parameters appear within the fuzzy sets in the rules antecedent parts, see Section 1.13. From an algorithmic point of view, this task is split into two parts, the adaptation of linear parameters and the update of nonlinear parameters, as they usually require different algorithms.
- In order to extend a fuzzy system to new system states (e.g. by expanding it to its former extrapolation region in the feature space), it is necessary to evolve new core components within the fuzzy system. This means that new rules and new antecedent fuzzy sets needs to be evolved in order to cope with these changing system states.
- In order to incorporate new upcoming (fault) classes during on-line operation mode in case of classification problems (e.g. an operator overrules a classifier's decision and defines a new class label), it is necessary to evolve new core components (rules and fuzzy sets) within the fuzzy classifier in the single model architecture case resp. to open up a new partial TS fuzzy model in the case of multi model architecture, please compare Section 1.2.1.5.
- In order to gain more accuracy in case of high-dimensional feature spaces, a connection of incremental training of a model with adaptive feature selection/calculation component is preferable; this may require adaptive changes in the input structure of the fuzzy system.

In Chapter 2 we will demonstrate some methods for dealing with adaptation of linear consequent resp. weight parameters in TSK and neuro-fuzzy type systems. In Chapter 3, we will describe several evolving fuzzy systems approaches which present solutions for the demands in the second and third itemization points above. In Section 5.2 (Chapter 5) we will demonstrate how to tackle dimensionality problems in a dynamic incremental and smooth interaction concept with evolving fuzzy systems.

1.2.2.4 Enhanced Goals for Evolving Fuzzy Systems

Enhanced goals in evolving fuzzy systems include the important aspects 1.) 'process-safety', 2.) 'interpretability' and 3.) 'increased performance and usability'. The first issue tackles specific occurrences in the data streams, which may lead to undesired behavior of the training mechanisms, such as faults and outliers, extrapolation cases, ill-posed learning tasks or drifts/shits in the data streams. The first aspect can be seen as a strict necessity for a proper run-through of the on-line learning algorithms (preventing system crashes) and assuring evolving fuzzy systems with a reasonable predictive accuracy and smooth feedback. The second issue is one of the key reasons for experts at industrial systems to use fuzzy system instead of other model architectures for machine learning and data-driven model identification purposes (see also Section 1.1.5). Therefore, it is very demanding to have mechanisms available which guide the evolved fuzzy systems to more transparency and interpretable power (ideally linguistically and visually), and this within the incremental on-line learning context (i.e. with reasonable computation time and not using any past data). The third issue tackles approaches for enhancing the predictive power of the evolving fuzzy systems and increasing their usability for researchers, experts and operators, finally boosting their overall performance: not strictly necessary, but certainly nice-to-have for many users.

1.3 Outline of the Book

The book gives an overview about the most important evolving fuzzy systems approaches and the basic incremental and on-line learning concepts, techniques applied in these. The book emphasizes intuitive explanations and basic principles illustrated by figures and examples, and tries to keep things as simple as possible and also offers perspectives from the engineering side (especially in the application section dealing with on-line scenarios where EFS have been successfully applied over the last several years). The book also provides all detailed machine learning and mathematical aspects (formulas, equations, algorithms) which are necessary for motivating and understanding the approaches. Detailed (analytical) deductions as well as proofs of these formulas are in large parts omitted. The book is dedicated to researchers from the field of fuzzy systems, machine learning as well as system identification, but also to practical engineers and people employed in and at industrial processes where data-driven modelling techniques serve as important components, especially in online operation modes.

The book is divided into three parts:

- The first part of the book comprises two chapters, where the core methodologies for evolving fuzzy systems from data are explained. Chapter 2 describes basic incremental learning algorithms which are useful or even necessary in various evolving fuzzy systems approaches for updating statistical help measures and for different variants of recursive learning of linear and non-linear parameters. Specific emphasis is made to enhanced methods for increasing accuracy, stability and computation times; the latter is essential in the case of on-line learning scenarios. The last section in Chapter 2 contributes to new challenges in the field of parameter optimization in incremental manner, especially using alternative optimization functions, other than conventional least squares error measure (which tends to be prone to over-fitting). Chapter 3 provides a comprehensive overview and in some parts comparison of the most important EFS approaches which have emerged during the last 10 years section-wise, also pointing out their weaknesses and strengths.

- The second part of the book contains three chapters, which all deal with enhanced concepts in evolving fuzzy systems, leading the evolved models to increased robustness and reliability during the incremental (on-line) training phase for guaranteeing process-safety (Chapter 4), achieving improved performance and enlarging their field of useability and user-friendliness (Chapter 5) as well as improving the interpretability and understandability of the evolved models (Chapter 6). Higher robustness can be achieved with techniques from regularization theory, specific methodologies to handle drifts and shifts in data streams and aspects regarding an appropriate treatment of outliers and extrapolation cases. Increased reliability can be achieved with the concepts of adaptive local error bars (regression case) and confidence regions (classification case). The enlarged field of useability and user-friendliness is achieved by integrating machine learning and data mining concepts such as active and semi-supervised learning, dynamic data mining, lazy learning or incremental model fusion. Improved performance is achieved by rule splitting-and-merging techniques as well as integration concepts on how to decrease curse of dimensionality in a dynamic manner. Complexity reduction and enhancing the transparency of the evolved models are key steps in achieving improved interpretability.
- The third part of the book contains four chapters, the first three dealing with concrete application scenarios where evolving fuzzy systems were successfully applied, implemented and also evaluated, including on-line system identification and prediction (Chapter 7), on-line fault and anomaly detection in high-dimensional (multi-channel measurement) systems (Chapter 8) and on-line visual inspection (image/texture classification) problems (Chapter 9). The fourth chapter (Chapter 10) describes some application fields which have not been explored so far with incremental learning methods in the fuzzy community and which therefore we see as strong potential candidates for future usage of evolving fuzzy systems. The book is concluded with an epilogue demonstrating important open issues and problems in the field of evolving fuzzy systems, which have not been tackled with full satisfaction so far. This should give the readers some inspiration and food for thoughts about the next generation of evolving fuzzy systems.

Part I
Basic Methodologies

Chapter 2
Basic Algorithms for EFS

Abstract. This chapter deals with fundamental algorithms for updating some parameters and statistical values as used in several evolving fuzzy systems approaches, demonstrated in the next chapter. These algorithms include update mechanisms for several important statistical values such as mean, variance, covariance etc. (Section 2.1) and a recursive incremental adaptation of linear consequent parameters in TS(K) fuzzy systems, also applicable to linear weights in neuro-fuzzy type systems. Hereby, we describe two different learning schemes, global learning (Section 2.2) and local learning (Section 2.3), provide an analytical and empirical comparison of these two approaches (Section 2.4) and also demonstrate why an adaptation of linear consequent (weight) parameters alone may not sufficiently perform if the behavior of the non-linear process changes (Section 2.5). We also outline some enhanced aspects and alternatives for adapting consequent parameters (Section 2.6) and will describe ways on how to incrementally learn non-linear antecedent parameters (Section 2.7). The chapter is concluded by discussing (the necessity of) possible alternatives for optimization functions based on which recursive (incremental) learning of antecedent and consequent parameters are carried out (Section 2.8).

E. Lughofer: Evolving Fuzzy Systems – Method., Adv. Conc. and Appl., STUDFUZZ 266, pp. 45–91.
springerlink.com

2.1 Incremental Calculation of Statistical Measures

This section deals with the incremental calculation of various statistical measures such as mean, variance, covariance and model quality measures, which are quite commonly used in evolving fuzzy systems approaches (see Chapter 3), sometimes as part of the model updates, sometimes synchronously and independently from the model updates serving as (necessary) auxiliary quantities during the on-line learning process.

2.1.1 Incremental Update of the Mean

In batch mode the mean of a variable x is defined by

$$\bar{x}(N) = \frac{1}{N}\sum_{k=1}^{N} x(k) \tag{2.1}$$

with N the number of samples. For m newly loaded samples this can obviously be updated in an exact (recursive) way by:

$$\bar{x}(N+m) = \frac{N\bar{x}(N) + \sum_{k=N}^{N+m} x(k)}{N+m} \tag{2.2}$$

Hence, the old mean is multiplied by the older number of samples, the new samples added in sum and this divided by the whole number of samples $(N+m)$ seen so far. Exact means that, if using all the $N+m$ samples to calculate the mean in batch mode, the same results are achieved as when updating through (2.2). In this case, we speak about an exact incremental learning algorithm. Note that (2.2) includes sample-wise update when setting $m = 1$.

2.1.2 Incremental Update of the (Inverse) Covariance Matrix

The covariance matrix (denoted as Σ) is an important auxiliary quantity, as it measures the scatter of the data samples in-between and among different variables. It is used in Fisher's interclass separability criterion [111] which can be exploited for a filter feature selection approach, see [159] (filtered means that the feature selection is performed before hand and totally decoupled from the model training phase). This separability criterion is also exploited in Section 5.2 when developing a concept for incremental dynamic dimensionality reduction in evolving fuzzy systems. The covariance matrix is also applied when extracting ellipsoidal clusters in general position from the data. Hereby, the inverse covariance is used in the Mahalanobis distance measure, for more information see also Section 3.1.4. Furthermore, the covariance matrix is needed to build up linear discriminant functions [298], which serve as linear decision boundaries in classification problems:

$$\delta_k(\mathbf{x}) = \mathbf{x}^T \Sigma^{-1}\mu_k - \frac{1}{2}\mu_k^T \Sigma^{-1}\mu_k + log(\pi_k) \tag{2.3}$$

with k the class label, μ_k the mean value of class k and π_k the relative frequency of class k. Incremental updating of such decision functions can be achieved by incremental updating the covariance matrix Σ, which is defined in the following way:

$$\Sigma = \frac{1}{N}\sum_{k=1}^{N}(X_{k,.} - \bar{X}(N))^T(X_{k,.} - \bar{X}(N)) \tag{2.4}$$

with X the regression matrix containing N samples as rows and p variables as columns, $\bar{X}(N)$ the mean value of all the variables over the last N samples. Similarly, the covariance matrix Σ_k over the kth class resp. the kth cluster (or kth region in the data space) can be defined in the same manner. In [334] an incremental variant of discriminant analysis both, in sample as well block mode, is demonstrated. There, also an update of the covariance matrix is derived whose step can be summarized in the following way: when including m new samples someone obtains:

$$\begin{aligned}\Sigma(new) =& \frac{1}{N+m}\sum_{k=1}^{N}(X_{k,.} - \bar{X}(N+m))^T(X_{k,.} - \bar{X}(N+m)) + \\ & \frac{1}{N+m}\sum_{k=N+1}^{N+m}(X_{k,.} - \bar{X}(N+m))^T(X_{k,.} - \bar{X}(N+m))\end{aligned} \tag{2.5}$$

Then, the following is obtained:

$$\begin{aligned}\Sigma(new) =& \frac{1}{N+m}\sum_{k=1}^{N}((X_{k,.} - \bar{X}(N)) + c(\bar{X}(N) - \bar{X}(m)))^T \\ & \times ((X_{k,.} - \bar{X}(N)) + c(\bar{X}(N) - \bar{X}(m))) + \\ & \frac{1}{N+m}\sum_{k=N+1}^{N+m}((X_{k,.} - \bar{X}(N)) + c(\bar{X}(N) - \bar{X}(m)))^T \\ & \times ((X_{k,.} - \bar{X}(N)) + c(\bar{X}(N) - \bar{X}(m)))\end{aligned} \tag{2.6}$$

with $c = \frac{m}{N+m}$. By developing the products within the sum terms using $(A+B)^T(A+B) = A^TA + B^TB + A^TB + B^TA$ one can finally derive the update of the covariance matrix with m new samples (proof is left to the reader):

$$\Sigma(new) = \frac{1}{N+m}(N\Sigma(old) + m\Sigma_{pnew} + \frac{Nm}{N+m}(\bar{X}(N) - \bar{X}(m))^T(\bar{X}(N) - \bar{X}(m)) \tag{2.7}$$

with Σ_{pnew} the covariance matrix on the m new samples. Note that in case of $m = 1$, i.e. feeding the $N+1$th sample $\mathbf{x}_{N+1}$ into the update, the covariance matrix update becomes:

$$\Sigma(new) = \frac{1}{N+1}(N\Sigma(old) + \frac{N}{N+1}(\bar{X}(N) - \mathbf{x}(N+1))^T(\bar{X}(N) - \mathbf{x}(N+1))) \tag{2.8}$$

as the covariance matrix on one new sample is equal to 0. In [350] it is reported that an even more stable calculation can be achieved by including the effect of mean changes over the new data samples $\Delta\bar{X}(N+m) = \bar{X}(N+m) - \bar{X}(N)$ as a rank-one modification. This leads us to the following formula for the (recursive) variance update of variable x with $m = 1$, i.e. with a single new sample $\mathbf{x}(N+1)$ (which will be also used in Section 3.1.4):

$$\sigma^2(new) = \frac{1}{N+1}(N\sigma^2(old) + (N+1)\Delta\bar{X}(N+1)^2 + (\bar{X}(N+1) - x(N+1))^2) \tag{2.9}$$

respectively for the covariance update of two variables x_1 and x_2:

$$\begin{aligned}(N+1)cov(x_1,x_2)(new) = \\ Ncov(x_1,x_2)(old) + (N+1)\Delta\bar{X_1}(N+1)\Delta\bar{X_2}(N+1) \\ + (\bar{X_1}(N+1) - x_1(N+1))(\bar{X_2}(N+1) - x_2(N+1))\end{aligned} \tag{2.10}$$

In some constructs, the inverse covariance matrix is required, e.g. when calculating the Mahalanobis distance of new samples to already existing cluster prototypes, achieving clusters in arbitrary position, see Section 3.1.4.4. In these cases, the update of the inverse covariance matrix instead of the covariance matrix in each incremental step would essentially speed up the learning phase, as no matrix inversion (after each update of the covariance matrix) is required. In [30], the update formula for inverse covariance matrix with one new sample is defined by:

$$\begin{aligned}\Sigma^{-1}(new) = & \frac{\Sigma^{-1}(old)}{1-\alpha} \\ & - \frac{\alpha}{1-\alpha} \frac{(\Sigma^{-1}(old)(\mathbf{x}(N+1) - \bar{X}(N)))(\Sigma^{-1}(old)(\mathbf{x}(N+1) - \bar{X}(N)))^T}{1+\alpha((\mathbf{x}(N+1) - \bar{X}(N))^T\Sigma^{-1}(old)(\mathbf{x}(N+1) - \bar{X}(N)))}\end{aligned} \tag{2.11}$$

with $\alpha = \frac{1}{N+1}$.

2.1.3 Incremental Calculation of the Correlation Coefficient

The correlation coefficient represents an important statistical measure for finding redundancies in the features set or on-line measurement data. In case of input variables/features for approximating a target or a class distribution, redundant variables/features can always be deleted without losing any information and accuracy of the further trained models resp. classifiers (see [159] for a detailed analysis of this claim). This helps in reducing the feature space and hence decreasing the curse of dimensionality effect [39] (see also Section 2.8 for a more detailed analysis on the curse of dimensionality). Furthermore, redundancies can be exploited for supervision and fault detection scenarios. For instance, consider two sensors at a measurement system which deliver redundant data streams, e.g in form of a relation $sensor1 = a * sensor2$, i.e. the values of the first sensor can be simply supervised by multiplying the values of the second sensor with a fixed constant a; then, if a fault in one of these sensors happens, this relationship can be exploited to detect this fault automatically as the relationship will be violated. The (empirical) correlation coefficient measures exactly how close two variables meet such a linear relationship. In fact, if the correlation coefficient between two variables/features is high in its absolute value (usually greater than 0.99), one of these two can be either deleted or the dependency of the two variables/features exploited. Note that, although this is restricted to linear relationships, it is a perfect indicator for truly redundant variables. Higher non-linear dependencies between two variables may be contained in the data set, but usually should not be deleted, as someone cannot be sure whether the model (e.g. a classifier) is able to resolve the modelling task with sufficient accuracy (e.g. finding an adequate decision boundary) if one of the two variables (and therefore information) is discarded — on the other hand, in case of real linearities, a deletion of one variable does not decrease the information gain as can be fully substituted by the other variable. Furthermore, moderate non-linear dependencies usually cause a significantly higher correlation coefficient as if there exists no dependency at all. This also means that such relations can be detected with the correlation coefficient. An example for underlining this statement is presented in Figure 2.1:

- in (a) a strictly linear relation is shown for which the correlation coefficient is 1.
- in (b) a moderate non-linear relation in form of a parabola in $\mathbb{R}^2_+$ is presented, for which the correlation still reaches a high value of 0.968.
- in (c) a more complex non-linear relation in form of a sinusoidal function is shown and the correlation coefficient has a medium value of 0.78.
- in (d) variable Y appears randomly over variable X (hence no dependency exists), the correlation coefficient has a value of 0.015.

The correlation coefficient between two variables x and y is defined in the following way:

$$r_{yx} = \frac{\sum_{k=1}^{N}(x(k)-\bar{X})(y(k)-\bar{Y})}{\sqrt{\sum_{k=1}^{N}(x(k)-\bar{X})^2}\sqrt{\sum_{k=1}^{N}(y(k)-\bar{Y})^2}} \tag{2.12}$$

with $\bar{X}$ the mean value of x, $\bar{Y}$ the mean value of y and $x(k)$ the kth sample of x. It is a corollary of the Cauchy-Schwarz inequality [107] that the correlation coefficient cannot exceed 1 in its absolute value. An incremental calculation of the correlation coefficient may be demanded during on-line modelling scenarios — e.g. for dynamic deletion of redundant features or the permanent update of the degree of a linear dependency between features serving as a kind of justification level for either a linear or a non-linear model. The empirical correlation coefficient possesses the favorable property that it can be divided into three parts of variance-like calculations, namely $\sum_{k=1}^{N}(x(k)-\bar{X})(y(k)-\bar{Y})$, $\sqrt{\sum_{k=1}^{N}(x(k)-\bar{X})^2}$ and $\sqrt{\sum_{k=1}^{N}(y(k)-\bar{Y})^2}$, which all are modified variance-like calculations, i.e. two standard deviations and one co-variance with the difference not to divide the sums through N. In this sense, someone can update each part of the denominator with recursive variance formula as in (2.9) resp. the numerator with recursive covariance formula as in (2.10), and multiply their values with the number of samples seen so far, omitting the division through $N+1$ in (2.9) and (2.10).

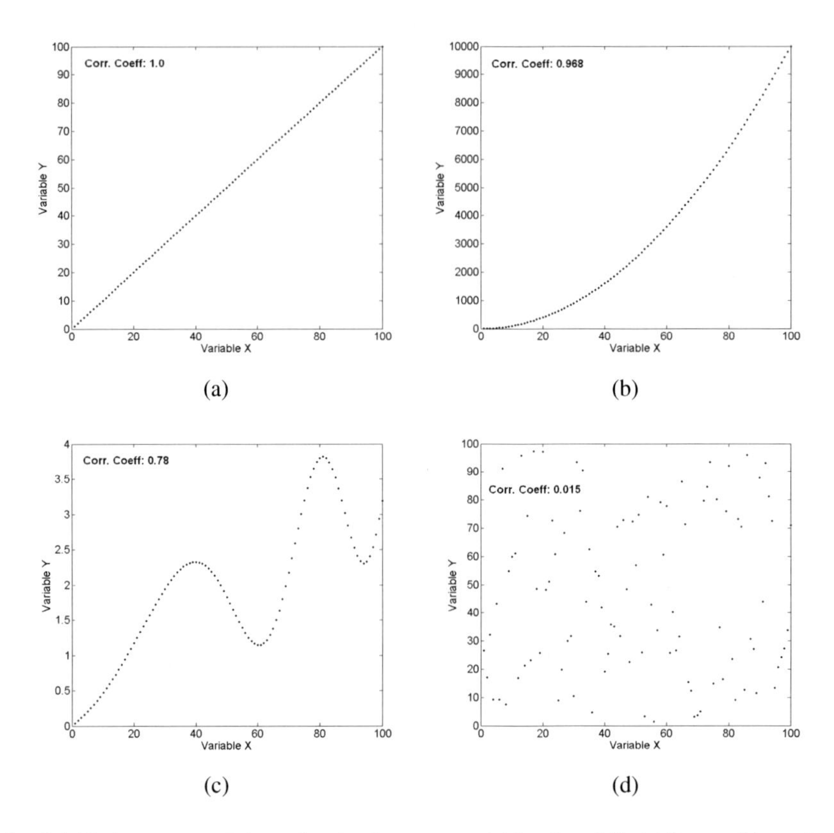

Fig. 2.1 Various types of dependencies between variables X and Y: (a) linear, (b) quadratic (parabolic), (c) sinusoidal, (d) none (random), and the corresponding correlation coefficients indicated in the images.

Given p variables in a system $x_1,...,x_p$, the correlation matrix is a $p \times p$ matrix whose entries (i, j) are the correlation coefficients between x_i and x_j. This means that all variables are cross-correlated among themselves and stored into a big matrix. An example of a correlation matrix for measurement channels at an engine test bench (including 97 channels) is presented in Figure 2.2: lighter rectangles represent higher correlations, the diagonal of this matrix represents the correlation of the variables to themselves and hence are always white (=1).

2.1.4 Updating Model Qualities

Model quality measures give rise whether models are reliable and their predictions can be trusted. This is absolutely necessary when models are either trained incrementally from scratch or further adapted in an on-line system, as unreliable models

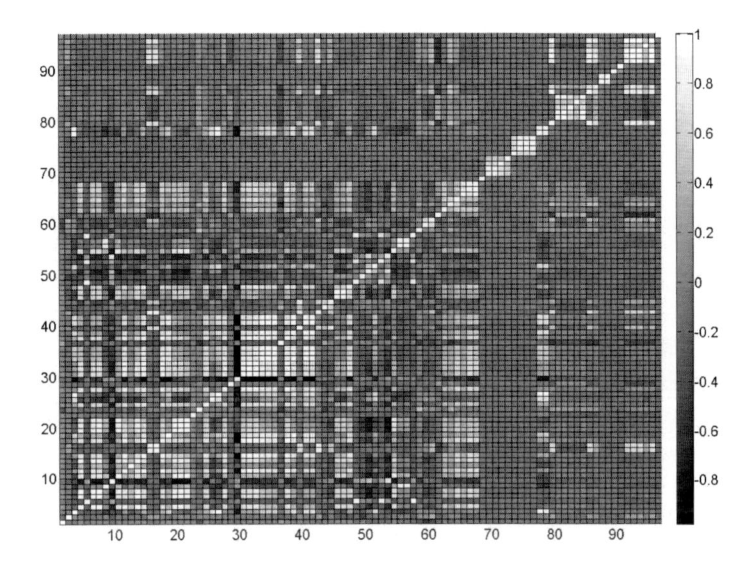

Fig. 2.2 Example of a correlation matrix, lighter rectangles represent higher correlations

may deteriorate the quality of products or even cause severe failures in an industrial system, finally resulting into breakdowns of some system components etc. For instance, consider a quality control system, where unerring classifiers are used for sorting out bad production items. In this case, a drop in the accuracies of classifiers during on-line production should be recognized early and reported. This demands a permanent adaptation of the miss-classification rates. Another example underlining the necessity of model qualities is an on-line fault detection framework, e.g. as reported in [13] (for details see Chapter 8), where model qualities are used first for identifying if there is a useful relationship between a given target variable and a subset of input variables at all (so no a priori knowledge about possible relationships is used). So identifying, which model will yield a substantial contribution for fault detection further is an important aspect $\rightarrow$ those with low qualities will be simply not used in the fault detection logic [287]). Second, the model qualities may be directly applied as internal quality measures in the fault isolation process (=finding responsible variables when a fault alarm is triggered) [115] and supervision purposes, and may serve as confidence values for reliability criteria indicating how much a user can trust the predictions of the models — see also Section 4.5. In an on-line identification process, the model quality measures should be permanently kept up-to-date. First, an assortment of possible quality measure in batch is presented, then their extension to the incremental case is demonstrated.

2.1.4.1 Quality Measures in Batch Mode

First of all, we inspect different quality measures, which are also quite common choices when applying regression modelling. The first one, which is widely used in various application, is the so-called *mean squared error*, abbreviated with *MSE*:

$$mse = \frac{1}{N}\sum_{k=1}^{N}(\hat{y}(k) - y(k))^2 \tag{2.13}$$

where $y(k)$ are the measured values of the target and $\hat{y}$ the estimated ones out of the fuzzy model. Hence, the *MSE* describes an absolute error measure between model outputs and measured target values. In literature often the *root mean squared error* (RMSE) is applied, which is nothing else than the square root of the *MSE* and therefore causes no significant difference when performing relative comparison among model training methods.

The great drawback when using *MSE* is the fact that as an absolute error measure its value can neither be compared among different models having different target variables nor can an average *mean squared error* (over all models) be computed, which stands for an overall performance of a model training method. This is because different target variables span up different ranges, causing completely different *MSE* values, although the quality of the models might be the same. This means that *MSE* is quite un-interpretable when inspecting it as a quality for a model in an isolated way. For this reason, either the columns in the training data matrix including input and output variables should be normalized (e.g. to the unit interval) and fuzzy models trained from this normalized data, or as an alternative a normalized error measure should be used. The latter one is preferable as training with normalized data causes normalized models which may get un-interpretable as usually variables are better understood by the experts within their original ranges. The so-called normalized *MSE* is defined as follows:

$$mse\_norm = \frac{1}{N}\sum_{k=1}^{N}\frac{(\hat{y}(k) - y(k))^2}{(\max(y) - \min(y))^2} \tag{2.14}$$

Due to the normalization with the range, comparable quality values among several models with different target variables are guaranteed.

Another possible normalized error measure is the so-called *average percent error* = *APE*, which is defined as follows:

$$ape = \frac{1}{N}\sum_{k=1}^{N}\frac{|\hat{y}(k) - y(k)|}{|y(k)|}100\% \tag{2.15}$$

As a division through the absolute values of the target variables gives also a normalization and hence a comparable quality measure. However, *APE* has a drawback in the case when values of the target variable approaches 0, because this causes numerical instabilities. Hence, this measure cannot be used for all relationships, so normalized *MSE* (and *MAE* = *mean absolute error*) is often preferred over *APE*.

Although normalized *MSE* provides a better interpretation of the model error and can be used for comparing errors of different models using different target variables, it is in general (without input from experts) not clear which values denote a good model, which ones an average model and which ones a bad model (this often depends on the application resp. regression problem). In this case a quality measure is desirable, where a value near 0 means the model is not reliable and a value near 1 means the model is reliable. R^2, also stated as R-squared, is a widely used quality measure, which fulfills this requirement and is defined as follows:

$$R^2 = \frac{ssreg}{ssreg + ssres} \tag{2.16}$$

where

$$ssreg = \sum_{k=1}^{N} (\hat{y}(k) - \bar{y}(k))^2 \qquad ssres = \sum_{k=1}^{N} (y(k) - \hat{y}(k))^2 \tag{2.17}$$

and $\bar{y}$ the mean value of output target channel y. This measure is obviously normalized to the unit interval and is nearer to 1 the smaller $ssres$ is, i.e. the *mean squared error* of the approximation to the measure target values. If $ssres$ approaches to $ssreg$ it causes an R^2 of almost 0.5 and means that the approximation is not much better than just taking a constant mean value of the target variables as model (which is simulated in the $ssreg$ value), which is trivial, of course. Hence, from this point of view a model with a quality measure of $R^2 \leq 0.5$ can be surely rejected.

A modified version of R^2, the so-called R^2-adjusted[108], takes also into account the number of degrees of freedom, i.e. the number of model parameters which were estimated within the training process. Thus, it is more robust against over-fitting and therefore does not require a separate test data set for model evaluation and circumvents the splitting of an original small set. R^2-adjusted is defined as follows:

$$R^2_{adjusted} = 1 - \frac{(N-1)(1-R^2)}{N - deg} \tag{2.18}$$

where deg denotes the degrees of freedom and N the number of data points used for model training. Obviously, over-fitting of a model is punished by the denominator as the second term gets big, when the degrees of freedom approaches to the amount of training data leading to a small R^2-adjusted. In fuzzy systems the degrees of freedom can be estimated by $2Cp + kC$ with C the number of rules, p the number of inputs, i.e. the number of antecedent parts in the rules, and k the number of parameters in the consequent functions (see Section 2.8 for a justification of this formula).

In the classification case, the number of correct classifications in proportion to the number of samples is used as quality measure of a fuzzy classifier (also called classification rate):

$$Class_{rate} = \frac{|\{x_k | class(x_k) = L_r(k), k = 1, ..., N\}|}{N} \tag{2.19}$$

with N the number of samples, $L_r(k)$ the true class response for the sample x_k and $class(x_k)$ the classification output of a classifier for the kth input sample.

2.1.4.2 Calculating Quality Measures in Incremental Manner

When applying the quality measures in on-line mode, an incremental calculation is required, which updates the model qualities sample per sample or at least block-wise. A conventional procedure along updating the fuzzy models is to first make the prediction, compare with the real measured value (or operator's feedback), then to update the model quality (as prediction is done with the old model) and finally to update the model. In the next cycle, it can be chosen whether a specific model is further processed for predictions based on the updated model quality (if the model quality gets too low, a model may be switched off or at least a warning to the user should be given).

The update for the quality measures *MSE*, *normalized MSE* and *APE* can be carried out by simply calculating the absolute or squared error of the new incoming samples to the estimation values from the models and adding these deviations to the old quality estimate. So, in case of *normalized MSE* this leads us to the *incremental normalized MSE*:

$$mse\_norm(N+m) = \frac{N(\max(y)-\min(y))^2 mse\_norm(N) + \sum_{k=N}^{N+m}(\hat{y}(k)-y(k))^2}{(N+m)(\max_{new}(y)-\min_{new}(y))^2} \tag{2.20}$$

where $\max(y)-\min(y)$ denotes the old range of the target variable and $\max_{new}(y) - \min_{new}(y)$ the new one extended with m new samples. When doing it in this way, it leads to a slightly different outcome of the model quality compared to the batch mode case (i.e. calculating the model quality with all samples loaded so far at once), as the model itself changes over time through the incremental learning process: when calculating the deviations at time instance $N+m$ with m new samples, a different model estimation $\hat{y}$ takes into account in the above formula than when calculating the deviations at the end of the complete incremental learning process. Hence, at the current time instance any calculated deviation in all preliminary time steps is not fully exact.

For the R^2 and R^2-adjusted the same procedure as above can be performed, as R^2 can be divided into the two parts *ssres* and *ssreg*, which both incorporate a deviation from the estimated target variable values to the real ones respectively the mean of the real values; *ssreg* can be updated by using (2.9). An alternative approach for updating R^2 (and hence also R^2-adjusted), which leads finally to an almost exact (recursive) incremental computation, can be deduced from the statistical knowledge, that a multiple correlation coefficient can be calculated by means of a partial correlation coefficient. For instance, for the three-dimensional case the following holds:

$$1 - mult\_corr^2_{y\bullet x_1 x_2 x_3} = (1 - r^2_{yx_1})(1 - r^2_{yx_2\bullet x_1})(1 - r^2_{yx_3\bullet x_1 x_2}) \tag{2.21}$$

where $mult\_corr_{y.x_1x_2x_3}$ denotes the multiple correlation coefficient between target y and inputs x_1, x_2 and x_3 and $r_.$ the partial correlation coefficient defined as follows:

$$r_{yx_3\bullet x_1x_2} = \frac{r_{yx_3\bullet x_1} - r_{yx_2\bullet x_1} r_{x_3x_2\bullet x_1}}{\sqrt{(1-r^2_{yx_2\bullet x_1})(1-r^2_{x_3x_2\bullet x_1})}} \tag{2.22}$$

(2.23)

with

$$\begin{aligned} r_{yx_3\bullet x_1} &= \frac{r_{yx_3} - r_{yx_1} r_{x_3x_1}}{\sqrt{(1-r^2_{yx_1})(1-r^2_{x_3x_1})}} \\ r_{yx_2\bullet x_1} &= \frac{r_{yx_2} - r_{yx_1} r_{x_2x_1}}{\sqrt{(1-r^2_{yx_1})(1-r^2_{x_2x_1})}} \\ r_{x_3x_2\bullet x_1} &= \frac{r_{x_3x_2} - r_{x_3x_1} r_{x_2x_1}}{\sqrt{(1-r^2_{x_3x_1})(1-r^2_{x_2x_1})}} \end{aligned} \tag{2.24}$$

where $r_{yx_1}, r_{yx_2}, \ldots$ are the partial correlation coefficients between y and x_1, y and x_2 and so on. A generalization of this formula to p input dimensions is left to the reader. It is known that R^2 can be seen as a multiple correlation coefficient between the target y and its inputs, hence above formula (2.21) can be directly applied to R^2.

In the case of a linear dependency the partial correlation coefficient can be simply calculated through the empirical correlation coefficient as defined (2.12), which can be exactly updated as mentioned in Section 2.1.3 by exploiting recursive variance and covariance formulas (2.9) and (2.10). In the case of fuzzy systems, however, this adaptation strategy cannot be directly applied as it is predominantly applicable for linear models. However, in the special case of Takagi-Sugeno fuzzy systems the existence of the linear part appearing as local linear models within separate rules can be exploited in order to compute the R^2 with respect to these linear functions. Please note that this is basically only true in connection with local learning of linear consequent functions, as it, opposed to the global estimation/adaptation, triggers linear functions which snuggle along the real curve (see Sections 2.4 and 6.3.1.3). Under the consideration that the local linear consequent functions are only valid in a local region (and give there a good piecewise linear estimate of the real nonlinear function), the overall R^2 has to be divided into partial R^2 for each rule. Hence, the following algorithm can be applied for an incremental calculation of R^2 in the case of Takagi-Sugeno fuzzy models:

Algorithm 1 : Incremental Calculation of R^2 and $R^2_{adjusted}$

1. **Input**: Initial Takagi-Sugeno fuzzy model (as defined in (1.2)) with C rules
2. Calculate C R^2 values for the C rules, where the calculation of R^2_i belonging to the ith rule is carried out in batch mode through (2.12) via (2.21) and (2.24) with those samples nearest to rule i (i.e. nearest to the ith cluster), also achieving partial correlation coefficients $r_{y.}(i)$.

3. Calculate $R^2_{adjusted\_i}$ through (2.18) with $N = k_i$ equal to the number of samples belonging to rule i and deg the degrees of freedom in the initial model for rule i ($3p+1$ with p the dimension)
4. Calculate the overall R^2 respectively $R^2_{adjusted}$ by computing the weighted average over all R_i^2 respectively $R^2_{adjusted\_i}$:

$$R^2 = \frac{k_1 R_1^2 + k_2 R_2^2 + \ldots + k_C R_C^2}{k_1 + k_2 + \ldots + k_C}$$
$$R^2_{adjusted} = \frac{k_1 R^2_{adujusted\_1} + k_2 R^2_{adjusted\_2} + \ldots + k_C R^2_{adjusted\_C}}{k_1 + k_2 + \ldots + k_C} \tag{2.25}$$

where k_i the number of data samples belonging to cluster (rule) i (i.e. for which the ith rule was the nearest one among all rules).
5. **For** any new incoming data sample do
6. Calculate the nearest rule to the current point, say rule i
7. Update the three variance-like parts of all partial empirical correlation coefficients $r_{y.}(i)$ for R_i^2 with (2.10) and (2.9)
8. Update all the partial empirical correlation coefficients for R_i^2 using (2.12)
9. Update R_i^2 by calculating (2.21) using (2.24)
10. Update $R^2\_adjusted_i$ with (2.18), where $k_i + 1$ instead of N
11. Calculate the overall R^2 respectively $R^2_{adjusted}$ by computing the weighted average over all R_i^2 respectively $R^2_{adjusted\_i}$ as in (2.25):
12. **end for**
13. **Output**: incrementally calculated R^2 and R^2-adjusted

Note: taking the weighted average in Step 4 instead of the non-weighted one is justified due to the following example: considering two rules within a 4-dimensional fuzzy model, where to Rule#1 10 data samples and to Rule#2 100 data samples belong and the data points are approximated with the same quality for both rules, so $R_1^2 = R_2^2 = 0.7$. Then $R^2_{adjusted}$ will punish the approximation with the first rule consequent function because of the danger of over-fitting as the denominator becomes $k_1 - deg = k_1 - (p+1) = 5$, whereas for the second rule it becomes $k_2 - deg = k_2 - (p+1) = 95$. So, after formula (2.18) $R^2_{adjusted\_1} = 0.46$, whereas $R^2_{adjusted\_2} = 0.69$. If taking the average of these two $R^2_{adjusted}$ values an average model quality of 0.575 is obtained, which is biased, as in the case of a joint computation the denominator gets $N - deg = 110 - 10 = 100$ and therefore the $R^2_{adjusted}$ with joint rules equal to 0.673. In the case of a weighted average an overall $R^2_{adjusted}$ of 0.669 is obtained, yielding almost the same value as for $R^2_{adjusted}$ with joint rules. Concluding, the weighted average of R^2 respectively $R^2_{adjusted}$ values for separate linear consequent functions delivers a higher weight to the quality of those rules, which have a higher significance and relevance due to the nature of the data samples.

The update of the classification rate for measuring the quality of fuzzy classifiers is done in straightforward manner by adding the rate of the m new samples to the old classification rate based on N samples, hence:

$$Class_{rate}(new) = \frac{N * Class_{correct}(old) + |\{x_k | class(x_k) = L_k, k = N+1, ..., N+m\}|}{N+m} \tag{2.26}$$

with $Class_{correct}(old)$ the number of correct classifications on the first N samples. L_k the true class responses of the inputs x_k and $class(x_k)$ the classification output of a classifier for the kth input sample.

2.2 Global Learning of Consequent Parameters

We are starting with the global learning approach of linear consequent parameters in Takagi-Sugeno type fuzzy systems, as defined in (1.6). Please note that this also applies to the fuzzy classifiers exploiting multi-model architecture as defined in (1.20), Section 1.2.1.5. For fuzzy classifiers using single model architecture the incremental update of the consequents = singleton class labels usually reduces to counting how many samples per class fall in each rule. For Mamdani fuzzy systems the update of the consequents follows from the update and evolution of the rule structure in the joint input/output space (see Chapter 3). First, we define the optimization problem and describe its solution for the batch mode case and then extend it to the incremental case. Hereby, it is assumed that the non-linear parameters Φ_{nonlin} as well as the whole rule structure are already estimated (in previous batch mode or for the current incremental learning step) resp. were pre-defined (e.g. according to expert input in form of linguistic rules and fuzzy sets). In Chapter 3, we will see how non-linear parameters and the whole rule structure are incrementally learned and evolved by the various EFS approaches and how this effects the consequent learning.

2.2.1 Global Learning in Batch Mode

For the global learning approach, the parameters of all rules are collected to a set $\mathbf{w} = \Phi_{lin}$ and perform a simultaneous estimation of these. The motivation of this approach is given by the fact, that the fuzzy system provides an output based on inferencing over all rules simultaneously and this global output should be close to the real (measured) output. Hereby, a conventional choice is the least squares error criterion as optimization function. This means that the goal of global least squares estimation is now, to minimize the error between the measured values y_k and the estimated values $\hat{y}_k$ of the output variable y, where $k = 1, ..., N$ and N the number of available data samples. This leads to the following general minimization problem

$$J = \|y - \hat{y}\| = \min! \tag{2.27}$$

When applying this error measure to the Takagi-Sugeno fuzzy system formulated in (1.2), it follows

$$J = \|\mathbf{y} - \sum_{i=1}^{C} l_i \Psi_i(\mathbf{x})\| = \min_{\mathbf{w}}! \tag{2.28}$$

Applying the quadratic distance as norm over the discrete values of y and $\hat{y}$ yields a quadratic optimization problem

$$J = \sum_{k=1}^{N} (y(k) - \sum_{i=1}^{C} l_i \Psi_i(\mathbf{x}(k)))^2 = \min_{\mathbf{w}}! \tag{2.29}$$

which can be uniquely solved by derivation the objective function after all linear parameters and solving the resulting linear equation system. Taking into account, that $\hat{y}$ can be rewritten as $R\mathbf{w}$ (this follows directly from the formulation in (1.2)), where $\mathbf{w}$ contains all the linear parameters for all rules to be estimated:

$$\mathbf{w} = \begin{pmatrix} w_{10} \\ w_{11} \\ \vdots \\ w_{1p} \\ w_{20} \\ \vdots \\ w_{Cp} \end{pmatrix}$$

and the matrix R denotes the global regression matrix

$$R = [R_1 R_2 \ldots R_C] \tag{2.30}$$

containing the global regressors

$$\mathbf{r}_i(k) = [\Psi_i(\mathbf{x}(k)) \;\; x_1(k)\Psi_i(\mathbf{x}(k)) \;\; \ldots x_p(k) \;\; \Psi_i(\mathbf{x}(k))] \tag{2.31}$$

for all C rules and $k = 1, \ldots, N$ data samples (note that $\mathbf{x}(k)$ denotes the multi-dimensional kth data sample), hence

$$R_i = \begin{bmatrix} \Psi_i(\mathbf{x}(1)) & x_1(1)\Psi_i(\mathbf{x}(1)) & x_2(1)\Psi_i(\mathbf{x}(1)) & \ldots & x_p(1)\Psi_i(\mathbf{x}(1)) \\ \Psi_i(\mathbf{x}(2)) & x_1(2)\Psi_i(\mathbf{x}(2)) & x_2(2)\Psi_i(\mathbf{x}(2)) & \ldots & x_p(2)\Psi_i(\mathbf{x}(2)) \\ \vdots & \vdots & \vdots & \vdots & \vdots \\ \Psi_i(\mathbf{x}(N)) & x_1(N)\Psi_i(\mathbf{x}(N)) & x_2(N)\Psi_i(\mathbf{x}(N)) & \ldots & x_p(N)\Psi_i(\mathbf{x}(N)) \end{bmatrix}$$

$x_i(k)$ is the ith column of the row vector x in sample k. The error function in the minimization problem above becomes $J = \|y - \hat{y}\| = \|\mathbf{y} - R\mathbf{w}\|$. Using the quadratic distance norm as in (2.29) the minimization problem can be rewritten as

$$(\mathbf{y} - R\mathbf{w})^T (\mathbf{y} - R\mathbf{w}) = \min_{\mathbf{w}}! \tag{2.32}$$

and the derivative with respect to the linear parameter vector $\mathbf{w}$ is reflected by

$$\frac{\partial}{\partial \mathbf{w}} (\mathbf{y} - R\mathbf{w})^T (\mathbf{y} - R\mathbf{w}) = -2(R^T \mathbf{y} - R^T R\mathbf{w}) \tag{2.33}$$

Setting the derivative to zero, the linear parameters can be obtained by

$$\hat{\mathbf{w}} = (R^T R)^{-1} R^T \mathbf{y} \tag{2.34}$$

with $\mathbf{y}$ the vector containing the values of the output variable. Solution (2.34) is also called the least squares solution. This system of linear equations can be solved for example by Gauss-Jordan elimination method including the search of a pivot element and the security check of some singular matrix criteria as described in [348].

Note that various well-known batch learning methods for fuzzy systems exploit this global batch training of linear parameters, for instance *FMCLUST* [28], *genfis2* [454], *ANFIS* [194] or *RENO* [62].

2.2.2 *Global Learning in Incremental Mode*

The major shortcoming of the global batch learning approach is that all data samples $(\mathbf{x}(1), y(1)), (\mathbf{x}(2), y(2)), ..., (\mathbf{x}(N), y(N))$ have to be sent into the least squares estimation formula at once, which is not feasible for fast on-line processes. In principle, data can be collected within a virtual memory over time and the fuzzy models trained again with all the data samples recorded and collected so far: however, this strategy would slow down the identification process more and more over time because of a growing data matrix. The computational effort of the LS method is $O(NM^2 + M^3)$ with N equal to the number of data samples collected and $M = C(p+1)$ the number of regressors (usually $N \gg M$ to achieve stable estimations). In case of N re-estimations for N new samples (so for each new incoming sample the least squares solution is re-calculated) an overall complexity is $O(NM^2 + (N+1)M^2 + ... + (N+N)M^2 + NM^3) = O(N^2M^2 + NM^3)$. In Chapter 7 we will see that this is too much in order to cope realistically with real-time demands. There, re-estimation of the models is compared with incremental sample-wise update in terms of computation time for on-line system identification problems.

2.2.2.1 Basic Formulation

A recursive formulation of the LS method for estimating linear consequent parameters, the so-called *recursive least squares (RLS)* [264, 25] overcomes this drawback as it calculates a new update for the linear parameters $\mathbf{w}$ each time a new data sample comes in and moreover requires a constant computation time for each parameter update. This is in accordance to the definition of an incremental learning procedure as done in Section 1.2.2.1. The basic idea of the *RLS* algorithm is to compute the new parameter estimate $\hat{\mathbf{w}}(k+1)$ at time instant $k+1$ by adding some correction vector to the previous parameter estimate $\hat{\mathbf{w}}(k)$ at time instant k. This correction vector depends on the new incoming measurement of the regressors

$$\mathbf{r}(k+1) = [\mathbf{r}_1(k+1) \;\; \mathbf{r}_2(k+1) \;\; \ldots \;\; \mathbf{r}_C(k+1)]^T \tag{2.35}$$

with

$$\mathbf{r}_i(k+1) = [\Psi_i(\mathbf{x}(k+1)) \;\; x_1(k+1)\Psi_i(\mathbf{x}(k+1)) \;\; \dots \;\; x_p(k+1)\Psi_i(\mathbf{x}(k+1))] \quad (2.36)$$

and the process output $\mathbf{y}(k+1)$. In the following the derivation of the *recursive least squares* algorithm is given. Let therefore be

$$\hat{\mathbf{w}}(k) = P(k)R^T(k)\mathbf{y}(k) \quad (2.37)$$

the nonrecursive estimator at time k after (2.34) where

$$P(k) = (R^T(k)R(k))^{-1} \quad (2.38)$$

denoting the inverse Hessian matrix and $R(k)$ the regression matrix at time instant k, hence

$$R(k) = \begin{pmatrix} \mathbf{r}^T(1) \\ \mathbf{r}^T(2) \\ \vdots \\ \mathbf{r}^T(k) \end{pmatrix}$$

Accordingly the estimator equation for time instant $k+1$ is:

$$\hat{\mathbf{w}}(k+1) = P(k+1)R^T(k+1)\mathbf{y}(k+1) \quad (2.39)$$

Obviously, equation (2.39) can be rewritten as:

$$\begin{aligned} \hat{\mathbf{w}}(k+1) &= P(k+1)\begin{pmatrix} R(k) \\ \mathbf{r}^T(k+1) \end{pmatrix}^T \begin{pmatrix} \mathbf{y}(k) \\ y(k+1) \end{pmatrix} \\ &= P(k+1)[R^T(k)\mathbf{y}(k) + \mathbf{r}(k+1)y(k+1)] \end{aligned} \quad (2.40)$$

Substituting $R^T(k)\mathbf{y}(k) = P^{-1}(k)\hat{\mathbf{w}}(k)$ due to (2.37) in (2.40), and adding and subtracting $\hat{\mathbf{w}}(k)$ on the right hand side, we achieve:

$$\hat{\mathbf{w}}(k+1) = \hat{\mathbf{w}}(k) + [P(k+1)P^{-1}(k) - I]\hat{\mathbf{w}}(k) + P(k+1)\mathbf{r}(k+1)y(k+1) \quad (2.41)$$

where according to (2.38):

$$P(k+1) = \left(\begin{pmatrix} R(k) \\ \mathbf{r}^T(k+1) \end{pmatrix}^T \begin{pmatrix} R(k) \\ \mathbf{r}^T(k+1) \end{pmatrix}\right)^{-1} = (P^{-1}(k) + \mathbf{r}(k+1)\mathbf{r}^T(k+1))^{-1} \quad (2.42)$$

Taking the inverse on both sides in (2.42), we obtain:

$$P^{-1}(k) = P^{-1}(k+1) - \mathbf{r}(k+1)\mathbf{r}^T(k+1) \quad (2.43)$$

Substituting (2.43) in (2.41), the recursive estimator equation is obtained by:

$$\hat{\mathbf{w}}(k+1) = \hat{\mathbf{w}}(k) + P(k+1)\mathbf{r}(k+1)(y(k+1) - \mathbf{r}^T(k+1)\hat{\mathbf{w}}(k)) \tag{2.44}$$

which can be interpreted as the new estimate is achieved through adding a correction vector, multiplied by the difference of the new measurement and the one-step-ahead prediction of the new measurement, i.e. $\mathbf{r}^T(k+1)\hat{\mathbf{w}}(k) = \hat{y}(k+1/k)$, to the old estimate. Thus, the amount of correction is proportional to the prediction error.

Recursive least squares estimation is an exact method, which means, that for each time instant the algorithm converges within each iteration step, hence no convergence criteria for stopping the iteration are needed. This favorable property stems from the fact that the loss function surface is a hyper-parabola as in (2.29) and the RLS algorithm has the same form as any gradient-based nonlinear optimization technique with $-\mathbf{r}(k+1)(y(k+1) - \mathbf{r}^T(k+1)\hat{\mathbf{w}}(k))$ as the gradient for the new measurement (compare (2.33)); therefore the RLS correction vector is the negative of the inverse Hessian matrix $P(k+1) = H^{-1}(k+1)$ times the gradient — the second derivative after $\mathbf{w}$ in (2.33) gives exactly $H = R^T R$. This is equivalent to the Newton optimization method, which converges in one single step if the optimization function is a hyper-parabola. In this sense, it can be expected that the parameters converge to the same solution as in the batch mode case, which gives us a stable incremental learning procedure (as outlined in Section 1.2.2.2).

Note: the matrix $H = R^T R$ is proportional to the covariance matrix of the parameter estimates and denoted as Hessian matrix; hence, P is further denoted as inverse Hessian matrix — sometimes this matrix is denoted as inverse covariance matrix in literature, however we use the term inverse Hessian matrix in order to not get confused with the inverse covariance matrix on the original samples as was introduced in Section 2.1.

2.2.2.2 Speeding Up the Algorithm

The RLS algorithm requires the inversion of the Hessian Matrix H for each incremental learning step (i.e. each new incoming sample), hence the complexity is still high ($O(M^3)$ for each sample). Utilizing the matrix-inversion theorem (also known as Sherman-Morrison formula [395]), the inversion of the Hessian Matrix can be avoided and the computational complexity reduced to $O(M^2)$ resp. $O(NM^2)$ for N update steps on N new samples (compared to $O(N^2M^2 + NM^3)$ for N re-estimations with batch learning):

Lemma 2.1. *Let A, C and $(A+BCD)$ nonsingular matrices, the following holds:*

$$E = (A+BCD)^{-1} = A^{-1} - A^{-1}B(C^{-1} + DA^{-1}B)^{-1}DA^{-1} \tag{2.45}$$

Proof. Is straightforward by applying some matrix multiplications on the left and right hand side.

With this matrix inversion theorem, taking into account that vectors can be seen as one-dimensional matrices and setting $C = 1$, equation (2.42) becomes:

$$P(k+1) = P(k) - P(k)\mathbf{r}(k+1)(1+\mathbf{r}^T(k+1)P(k)\mathbf{r}(k+1))^{-1}\mathbf{r}^T(k+1)P(k) \quad (2.46)$$

so since the term $\mathbf{r}^T(k+1)P(k)\mathbf{r}(k+1)$ is a scalar expression, no matrix inversion is necessary. After multiplying equation (2.46) with $\mathbf{r}(k+1)$ we obtain

$$P(k+1)\mathbf{r}(k+1) =$$
$$\frac{P(k)\mathbf{r}(k+1)}{(1+\mathbf{r}^T(k+1)P(k)\mathbf{r}(k+1))}((1+\mathbf{r}^T(k+1)P(k)\mathbf{r}(k+1)) - \mathbf{r}^T(k+1)P(k)\mathbf{r}(k+1)) \quad (2.47)$$

and therefore

$$P(k+1)\mathbf{r}(k+1) = \frac{P(k)\mathbf{r}(k+1)}{1+\mathbf{r}^T(k+1)P(k)\mathbf{r}(k+1)} \quad (2.48)$$

After the substitution of equation (2.48) in equation (2.44), the following result is obtained:

$$\hat{\mathbf{w}}(k+1) = \hat{\mathbf{w}}(k) + \gamma(k)(y(k+1) - \mathbf{r}^T(k+1)\hat{\mathbf{w}}(k)) \quad (2.49)$$

with the correction vector

$$\gamma(k) = P(k+1)\mathbf{r}(k+1) = \frac{P(k)\mathbf{r}(k+1)}{1+\mathbf{r}^T(k+1)P(k)\mathbf{r}(k+1)} \quad (2.50)$$

$P(k+1)$ can be computed recursively by:

$$P(k+1) = (I - \gamma(k)\mathbf{r}^T(k+1))P(k) \quad (2.51)$$

which follows directly from (2.46) applying $\gamma(k)$. Updating the parameters $\hat{\mathbf{w}}$ through the three equations above, is called *recursive least squares (RLS).*

2.2.2.3 Appropriate Start Values

Choosing an appropriate starting point is a crucial thing for the speed of the convergence of RLS. More or less, three possibilities for the determination of the starting values exist:

- Usage of a-priori known estimates due to expert knowledge: an expert initializes start parameters for the linear consequents, which are refined due to on-line data. This would result in so-called gray-box models (see also Section 1.1.4). Especially, in the field of control theory, the behavior of the output variable within some local parts is known quite often — e.g. in a SISO (i.e. single input single output) system it is often known, in which circumstance the reaction of a state signal y onto an input control signal u is strong ($\rightarrow$ high gradient with respect to y for the linear consequent function which captures this circumstance) or weak ($\rightarrow$ low gradient with respect to y for the linear consequent function which captures this circumstance). If no sufficient knowledge about the underlying functional relationship between some variables is available, one of the two subsequent variants have to be chosen.

- Computation of conventional LS algorithm for the first N data samples and starting RLS from the $N+1$th data sample on with initial parameters obtained from LS. This approach requires an initial model generated from the first few dozen of data samples, where the exact number or required samples depends on the complexity = number of parameters of the system, of course.
- Setting the initial parameter vector $\hat{\mathbf{w}}(0)$ to any value and setting $P(0) = \alpha I$, where α should be a large positive value, because this leads to high correction vectors and therefore to fast convergence of the consequent parameters to the optimal solution in the least squares sense. This approach includes the possibility to perform incremental training from scratch without generating an initial model. The choice of $P(0) = \alpha I$ can be justified as follows:

$$\begin{aligned} P^{-1}(1) &= P^{-1}(0) + \mathbf{r}(1)\mathbf{r}^T(1) \\ P^{-1}(2) &= P^{-1}(1) + \mathbf{r}(2)\mathbf{r}^T(2) \\ &= P^{-1}(0) + \mathbf{r}(0)\mathbf{r}^T(0) + \mathbf{r}(2)\mathbf{r}^T(2) \\ \vdots \quad &= \quad \vdots \\ P^{-1}(k) &= P^{-1}(0) + R^T(k)R(k) \end{aligned} \tag{2.52}$$

Hence, for very large values of α

$$\lim_{\alpha \to \infty} P^{-1}(0) = \frac{1}{\alpha} I \approx 0 \tag{2.53}$$

holds and equation (2.53) is in agreement with the definition of $P(k)$ in the non-recursive estimator (2.38). Furthermore, from (2.44) follows

$$\begin{aligned} \hat{\mathbf{w}}(1) &= \hat{\mathbf{w}}(0)P(1)\mathbf{r}(1)(y(1) - \mathbf{r}^T(1)\hat{\mathbf{w}}(0)) \\ &= P(1)(\mathbf{r}(1)y(1) + (-\mathbf{r}(1)\mathbf{r}^T(1) + P^{-1}(1))\hat{\mathbf{w}}(0) \end{aligned}$$

and with (2.52):

$$\hat{\mathbf{w}}(1) = P(1)(\mathbf{r}(1)y(1) + P^{-1}(0)\hat{\mathbf{w}}(0)) \tag{2.54}$$

furthermore with (2.44)

$$\begin{aligned} \hat{\mathbf{w}}(2) &= \hat{\mathbf{w}}(1)P(2)\mathbf{r}(2)(y(2) - \mathbf{r}^T(2)\hat{\mathbf{w}}(1)) \\ &= P(2)(\mathbf{r}(2)y(2) + (-\mathbf{r}(2)\mathbf{r}^T(2) + P^{-1}(2))\hat{\mathbf{w}}(1) \end{aligned}$$

and with (2.52)

$$\hat{\mathbf{w}}(2) = P(2)(\mathbf{r}(2)y(2) + P^{-1}(1)\hat{\mathbf{w}}(1)) \tag{2.55}$$

and with multiplying (2.54) with $P^{-1}(1)$ from the left hand side we can obtain

$$\hat{\mathbf{w}}(2) = P(2)(\mathbf{r}(2)y(2) + \mathbf{r}(1)y(1) + P^{-1}(0)\hat{\mathbf{w}}(0))$$

so, by generalizing this result to k we obtain:

$$\hat{\mathbf{w}}(k) = P(k)(R^T(k)\mathbf{y}(k) + P^{-1}(0)\hat{\mathbf{w}}(0)) \tag{2.56}$$

From equation (2.53), for large α and arbitrary starting estimations of the linear parameters $\hat{\mathbf{w}}(0)$, (2.56) agrees approximately with the nonrecursive estimator in (2.37)

2.2.2.4 Numerical Steps in Incremental Global Learning

Numerically, the following steps have to be carried out for new samples when applying RLS for linear parameter adaptation of a TS fuzzy system:

1. Split the new incoming data buffer into separate samples: the reason for this is, that RLS can be only carried out sample-wise as in the formula of the correction vector γ a division through a vector would be impossible, but would occur when blowing up the regression vector to a regression matrix containing a complete new data buffer $\rightarrow$ RLS in block mode does not exist. As a splitting of the buffer can always be performed, this is not a real restriction.
2. If the Takagi-Sugeno fuzzy system contains multiple output variables, split the system into more Takagi-Sugeno fuzzy systems each capturing exactly one output variable. This can always be carried out, whenever two output variables do not depend on each other when calculating the overall output of the system. If this situation occurs, a cascading of the fuzzy system can be done as shown in Figure 2.3: fuzzy systems number 1 and 3 only depend on the original input variables $x_1, \ldots, x_n$ and are completely independent of the outputs of any other systems (this situation appears quite often in reality) while the fuzzy systems number 2 and 4 demand outputs of preliminary systems as inputs $\rightarrow$ cascading.

   Then, for each Takagi-Sugeno fuzzy system and for the $k+1$th data sample in the buffer perform the following steps:
3. Fetch out the linear consequent parameters of the FIS structure obtained from the first k data samples. These parameters can be achieved by an initial model generated by usual least squares method or by an expert due to coding of linguistic relationships. Also a coding of the fuzzy sets alone and then learning the linear consequents from scratch in incremental manner is possible. Suitable starting values for the inverse Hessian matrix as well as the linear parameters have to be set (see previous subsection).
4. Calculate $\mathbf{r}(k+1)$ from (2.35) and (2.36) by using the inference mechanism in (1.10) for calculating the normalized membership values $\Psi_i(\mathbf{x}(k+1))$, where $\mathbf{x}(k+1)$ captures the $k+1$th data sample for all input variables.
5. Calculate $\gamma(k)$ from (2.50) and using the previous estimate for the inverse Hesse matrix $P(k)$.
6. Calculate the update for the inverse Hessian matrix from (2.51).
7. Calculate the update for the linear consequent parameters from (2.49).
8. Store back the updated linear consequent parameters into the FIS structure, overwrite previously estimated consequent parameters.

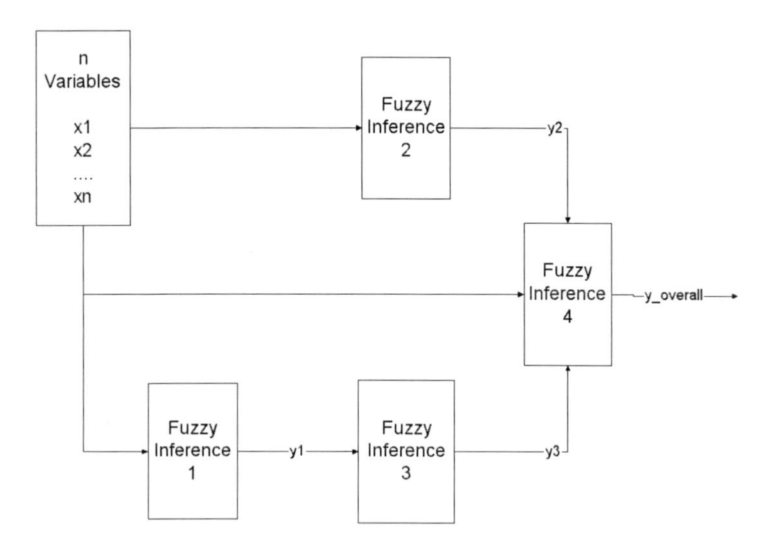

Fig. 2.3 Cascaded fuzzy system including various fuzzy inference systems possessing various input and output variables

2.3 Local Learning of Consequent Parameters

Similarly as we have done for the global learning approach, we are starting with the local learning approach of linear consequent parameters in Takagi-Sugeno type fuzzy systems, as defined in (1.6). Again, as in the case of global learning, this also applies to the fuzzy classifiers exploiting multi-model architecture as defined in (1.20), Section 1.2.1.5. First, we define the optimization problem and describe its solution for the batch mode case and then extend it to the incremental case. Hereby, we again assume that the non-linear parameters Φ_{nonlin} as well as the whole rule structure are already estimated (in batch mode or for the current incremental learning step). In Chapter 3 we will see how non-linear parameters and the whole rule structure are incrementally learned and evolved by the various EFS approaches and how this effects the consequent learning.

2.3.1 Local Learning in Batch Mode

Opposed to global learning, for the local learning approach C separate local estimations (and consequently adaptations) are carried out for the $p+1$ parameters of each local linear model, hence for each rule. This means that an independent local linear fit is performed for each local region (=rule) in the feature space. This has the remarkable effect of maximal flexibility when adding or deleting rules on demand during incremental learning (see Section 2.4). The basic difference to the famous lazy local learning approach [6] is that here the local regions are already given (e.g. according to a pre-defined rule grid or by a cluster partition obtained from a clustering=grouping algorithm) and therefore a kind of fixed model structure obtained. In lazy learning, reference samples are selected on demand (usually the nearest ones according to some distance measure and with different weights) to be used as input into the local or polynomial fits for each new incoming sample separately and independently, inducing a new model for each new incoming sample to be predicted (so maximal flexibility). However, this procedure makes lazy learning usually quite slow [50]. On the other hand, local learning in fuzzy systems enables a more reasonable trade-off between flexibility and computation time.

The model parameter vector for each of the $i = 1,...,C$ estimations is

$$\mathbf{w}_i = [w_{i0} w_{i1} ... w_{ip}]^T$$

The corresponding regression matrices are given by

$$R_i = \begin{bmatrix} 1 & x_1(1) & x_2(1) & ... & x_p(1) \\ 1 & x_1(2) & x_2(2) & ... & x_p(2) \\ \vdots & \vdots & \vdots & \vdots & \\ 1 & x_1(N) & x_2(N) & ... & x_p(N) \end{bmatrix} \tag{2.57}$$

Obviously, the regression matrices of all local linear models $i = 1,...,C$ are identical, since the entries of R_i do not depend on i, but only on the past N data samples (like in linear regression models [170]). A local linear model with the output $\hat{\mathbf{y}}_\mathbf{i} = [\hat{y}_i(1) \;\; \hat{y}_i(2) \;\; ... \;\; \hat{y}_i(N)]^T$

$$\hat{\mathbf{y}}_\mathbf{i} = R_i \mathbf{w}_\mathbf{i} \tag{2.58}$$

is valid only in the region where the associated basis function $\Psi_i(.)$ is close to 1, which will be the case close to the center of $\Psi_i(.)$. Data in this region is highly relevant for the estimation of $\mathbf{w}_\mathbf{i}$. As the basis functions decreases, the data becomes less relevant for the estimation of $\mathbf{w}_\mathbf{i}$ and more relevant for the estimation of the neighboring models. Consequently, it is straightforward to apply a weighted least squares optimization (WLS) where the weighting factors are denoted by the basis function values, i.e. the optimization function in (2.29) is expanded to C optimization functions (for the C rules):

$$J_i = \sum_{k=1}^{N} \Psi_i(\mathbf{x}(k)) e_i^2(k) \longrightarrow \min_{\mathbf{w}_\mathbf{i}} \qquad i = 1,...,C \tag{2.59}$$

where $e_i(k) = y(k) - \hat{y}_i(k)$ represents the error of the local linear model in the kth sample. With the weighting matrix

$$Q_i = \begin{bmatrix} \Psi_i(\mathbf{x}(1)) & 0 & \ldots & 0 \\ 0 & \Psi_i(\mathbf{x}(2)) & \ldots & 0 \\ \vdots & \vdots & \vdots & \vdots \\ 0 & 0 & \ldots & \Psi_i(\mathbf{x}(N)) \end{bmatrix}$$

a weighted LS method (WLS) can be performed in order to estimate the linear consequent parameters $\hat{w}_i$ for the ith rule: setting the derivation of $(\mathbf{y} - R_i\mathbf{w_i})^T Q_i(\mathbf{y} - R_i\mathbf{w_i})$ after $\mathbf{w_i}$ to 0, compare to (2.33), it leads to the so-called *weighted least squares* approach

$$\hat{\mathbf{w}}_\mathbf{i} = (R_i^T Q_i R_i)^{-1} R_i^T Q_i \mathbf{y} \tag{2.60}$$

which is also called *fuzzily weighted least squares (FWLS)* as the weights are the normalized membership degrees of the fuzzy rules in each data sample. For a computer implementation, it is more efficient to multiply each row of R_i with the square root of the weighting factor for each sample leading to the matrix

$$\tilde{R}_i = \begin{pmatrix} \mathbf{r}^T(1)\sqrt{\Psi_i(\mathbf{x}(1))} \\ \mathbf{r}^T(2)\sqrt{\Psi_i(\mathbf{x}(2))} \\ \vdots \\ \mathbf{r}^T(N)\sqrt{\Psi_i(\mathbf{x}(N))} \end{pmatrix}$$

and the weighted output vector

$$\tilde{\mathbf{y}} = \begin{pmatrix} y_1\sqrt{\Psi_i(\mathbf{x}(1))} \\ y_2\sqrt{\Psi_i(\mathbf{x}(2))} \\ \vdots \\ y_N\sqrt{\Psi_i(\mathbf{x}(N))} \end{pmatrix}$$

and furthermore to the *weighted least squares* estimation of

$$\hat{\mathbf{w}}_\mathbf{i} = (\tilde{R}_i{}^T \tilde{R}_i)^{-1} \tilde{R}_i{}^T \mathbf{y} \tag{2.61}$$

The efficiency arises because of storing the weighting matrix Q_i and also two matrix multiplications can be avoided [28].

2.3.2 Local Learning in Incremental Mode

Local learning in incremental mode can be solved by extending the weighted least squares approach to a recursive variant, similarly as done for the global learning approach with least squares. In fact, the deduction of the incremental learning version of the *weighted least squares*, the so-called *recursive weighted least squares = RWLS* is the same as in the previous section for the least squares approach, when doing the following:

1. Writing $Q_i(K)\mathbf{y}(k)$ instead of $\mathbf{y}(k)$ and $q_i(k+1)y(k+1)$ instead of $y(k+1)$ in (2.37) and (2.40)
2. Taking $(R_i(k)^T Q_i(k) R_i(k))^{-1}$ as $P_i(k)$ leading to $\mathbf{r}(k+1)q_i(k+1)\mathbf{r}^T(k+1)$ instead of $\mathbf{r}(k+1)\mathbf{r}^T(k+1)$ in the right hand side of (2.42)
3. Taking a scalar for the matrix C equal to $q_i(k+1) = \Psi_i(\mathbf{x}(k+1))$ instead of $C = 1$ which is applied in the matrix lemma as $C^{-1} = \frac{1}{\Psi_i(\mathbf{x}(k+1))}$ and effects the update formula for $P_i(k)$.

With these substitutions it is straightforward to see that this leads to the following update formulas for the linear consequent parameters of the ith rule, which is called *RFWLS = Recursive fuzzily weighted least squares*:

$$\hat{\mathbf{w}}_{\mathbf{i}}(k+1) = \hat{\mathbf{w}}_{\mathbf{i}}(k) + \gamma(k)(y(k+1) - \mathbf{r}^T(k+1)\hat{\mathbf{w}}_{\mathbf{i}}(k)) \tag{2.62}$$

$$\gamma(k) = P_i(k+1)\mathbf{r}(k+1) = \frac{P_i(k)\mathbf{r}(k+1)}{\frac{1}{\Psi_i(\mathbf{x}(k+1))} + \mathbf{r}^T(k+1)P_i(k)\mathbf{r}(k+1)} \tag{2.63}$$

$$P_i(k+1) = (I - \gamma(k)\mathbf{r}^T(k+1))P_i(k) \tag{2.64}$$

with $P_i(k) = (R_i(k)^T Q_i(k) R_i(k))^{-1}$ the inverse weighted Hessian matrix and $\mathbf{r}(k+1) = [1\ x_1(k+1)\ x_2(k+1)\ \dots\ x_p(k+1)]^T$ the regressor values of the $k+1$th data sample, which is the same for all i rules. The concepts for appropriate start values and numerical steps can be easily adopted from the global learning approach.

Note that the solutions for global and local learning (in batch and incremental mode) can be applied to linear parameters in TSK type fuzzy systems (1.5) in a straightforward manner by extending the regression vectors and matrices (according to the various polynomial terms appearing in the TSK type fuzzy systems); for instance, if using a TSK-type fuzzy system with polynomial degree up to 2, where only parabolic terms in form of $x_1^2, x_2^2, \dots, x_p^2$ are allowed, the regression matrix in the global learning case is set to

$$R_i = \begin{bmatrix} \Psi_i(\mathbf{x}(1)) & x_1(1)\Psi_i(\mathbf{x}(1)) & x_1(1)^2\Psi_i(\mathbf{x}(1)) & \dots & x_p(1)\Psi_i(\mathbf{x}(1)) & x_p(1)^2\Psi_i(\mathbf{x}(1)) \\ \Psi_i(\mathbf{x}(2)) & x_1(2)\Psi_i(\mathbf{x}(2)) & x_1(2)^2\Psi_i(\mathbf{x}(2)) & \dots & x_p(2)\Psi_i(\mathbf{x}(2)) & x_p(2)^2\Psi_i(\mathbf{x}(2)) \\ \vdots & \vdots & \vdots & \vdots & \vdots & \\ \Psi_i(\mathbf{x}(N)) & x_1(N)\Psi_i(\mathbf{x}(N)) & x_1(N)^2\Psi_i(\mathbf{x}(N)) & \dots & x_p(N)\Psi_i(\mathbf{x}(N)) & x_p(N)^2\Psi_i(\mathbf{x}(N)) \end{bmatrix}$$

and the kth regression vector (used in the recursive update for the kth sample) to

$$\mathbf{r}_i(k) = [\Psi_i(\mathbf{x}(k))\ x_1(k)\Psi_i(\mathbf{x}(k))\ x_1(k)^2\Psi_i(\mathbf{x}(k)) \dots x_p(k)\Psi_i(\mathbf{x}(k))\ x_p(k)^2\Psi_i(\mathbf{x}(k))]$$

For the local learning approach the regression matrix/vectors (the same for all C rules) get

$$R_i = \begin{bmatrix} 1 & x_1(1) & x_1(1)^2 & x_2(1) & x_2(1)^2 & \dots & x_p(1) & x_p(1)^2 \\ 1 & x_1(2) & x_1(2)^2 & x_2(2) & x_2(2)^2 & \dots & x_p(2) & x_p(2)^2 \\ \vdots & \vdots & \vdots & \vdots & & & & \\ 1 & x_1(N) & x_1(N)^2 & x_2(N) & x_2(N)^2 & \dots & x_p(N) & x_p(N)^2 \end{bmatrix}$$

resp. $\mathbf{r}_i(k) = [1\ x_1(k+1)\ x_1(k+1)^2\ x_2(k+1)\ x_2(k+1)^2\ \dots\ x_p(k+1)\ x_p(k+1)^2]^T$, extending the number of parameters to $2p+1$.

This straightforward extension for the regression vectors is also the case when adopting the approach to linear weights in neuro-fuzzy systems. Please also note that the solutions are directly applicable to fuzzy classification models exploiting the multi-model architecture as defined in Section 1.2.1.5 by using K Takagi-Sugeno fuzzy systems for K classes (see also Section 3.1.2).

2.4 Comparison of Global and Local Learning

In this section we demonstrate an analytical and empirical comparison between global and local learning, where we point out their advantages and disadvantages with respect to different types of important aspects during batch and incremental learning.

- Improvement of numerical stability of LS and RLS method with local learning, as dealing with smaller Hessian matrices (size $(p+1) \times (p+1)$ instead of $C(p+1) \times C(p+1)$). This means that local learning already implements an inherent regularization effect, as it divides the original optimization problem into smaller sub-problems (for each rule) and hence reduces the size of the matrices to be inverted. This effect is called regularization by discretization [120], and in general, problems of smaller size do not necessarily need to be regularized, because a reasonable lower bound on the singular values of the matrices can be found. However, in case of high-dimensional data sets, the matrices will be still quite large, so the likelihood of a badly conditioned matrix is still high for local learning. If such a matrix occurs, numerical stability of the parameter estimation step can be guaranteed by applying enhanced regularization methods (see Chapter 4, Section 4.1) or adding some model complexity terms (see Section 2.8 below).
- The inherent regularization effect of local learning may lead to a more accurate model when dealing with medium to high noise levels in the data, as such a case often triggers an ill-posed learning problem which is more likely in case of larger matrices. This property is examined, when extracting a two-dimensional fuzzy model for the relationship $f(x) = \sin(x^2) - x$, adding a Gaussian noise in the medium range. For both variants the same number of clusters (using the same setting of the radius of the range of influence in subtractive clustering) is extracted. Figure 2.4 compares both approximations: obviously, the global learning method has some severe problems in regions with low x-values, as a strongly fluctuating approximation curve can be observed, which is not the case when using local learning.

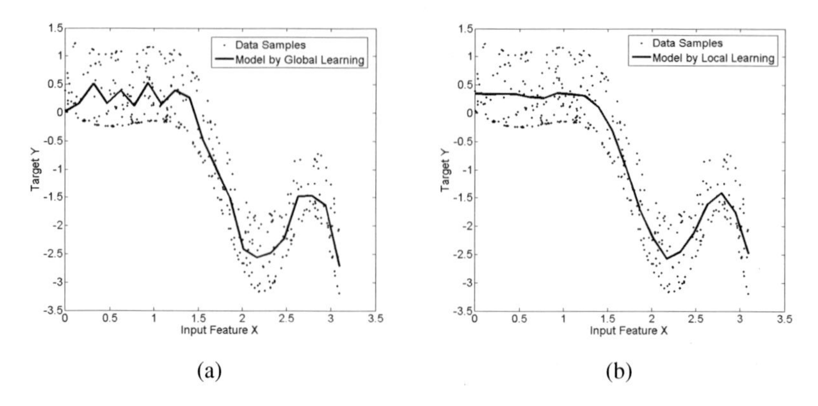

Fig. 2.4 (a): approximation of a sinusoidal type relationship with global learning (note the fluctuation at the left most region); approximation of the same relationship with same number of rules when using local learning

- Acceleration of LS and RLS with local learning, mainly for systems with a high dimensionality and a large number of rules. Global estimation approach takes all the consequent parameters joined together in a matrix as input, hence the number of parameters is $C(p+1)$, where C is the number of rules and p the number of inputs. Therefore, the computational complexity is $O(N(C(p+1))^2+(C(p+1))^3)$ for LS on N samples and $O((C(p+1))^2)$ for RLS on each new incoming sample. Local learning approach performs C LS estimations or RLS adaptations with matrices of dimension size $(p+1)$. Therefore the computational complexity is just $O(NC(p+1)^2+C(p+1)^3)$ for LS on N samples and $O(C(p+1)^2)$ for RLS on each new incoming sample.
- A lower bound on the number of samples for (initial) model generation when using local learning: $C(p+1)$ linear consequent parameters are estimated simultaneously, when applying global learning, whereas for local learning this number reduces to $p+1$ for each rule. It directly follows, that the theoretical lower bound of demanded data samples is reduced to $p+1$, which can not be achieved in the case of global learning as this would lead to an under-determined equational system. However, it should be noted that in case of $p+1$ samples, the danger is high that only a few samples are lying near the center of a rule whose consequent parameters are learned and hence cause a very low diagonal entry in the weighting matrix Q_i, thus affecting the condition of the matrix $(R_i^T Q_i R_i)$. So, with $p+1$ samples or just a bit more the likelihood of an instable estimation of the consequents is still quite high.
- Local learning provides a higher flexibility for adjoining fuzzy sets and rules on demand based on the characteristics of new incoming data samples — note that generating new rules and fuzzy sets on demand is applied in all evolving fuzzy systems approaches discussed in Chapter 3. When global learning approach is used, the global inverse Hessian matrix in RLS formulation has to be set back to

αI (or at least to be modified as done in [11]), because it completely changes in size and entries if a new rule is added (this demands an addition of $p+1$ consequent parameters and hence rows and columns in the inverse Hessian $P = R^T R$). The setting back, which can be avoided by the local learning approach, usually results in a worse approximation (as examined in [285] and [286]), also the dynamic modification has some shortcomings [11]. In local learning, a new rule triggers a new set of $p+1$ parameter which can be synchronously and independently updated by the others. This means that the convergence of the parameter updates for the other rules is not disturbed by a new rule. This also means that new relationships can be learned in one operating regime while the old information is conserved in all other already trained local linear models.
- The interpretability of the rule consequent functions, which represent multidimensional hyper-planes, is significantly improved with local learning approach, as they tend to snuggle along the regression surface rather than being positioned in the space in quite chaotic manner (as in the case of global learning). Global and local learning are compared in terms of interpretability in [462] for batch off-line case and in [281] for incremental on-line case. This comparison will be extended and further analyzed in Chapter 6.
- Online merging of rules can only be applied reasonably in incremental training steps, when using local learning approach. This will be analyzed in more detail in Chapters 5 and 6.

2.5 Success and Failure of Consequent Adaptation

In this section, we demonstrate occasions where the consequent adaptation of the linear parameters succeeds and where it fails. In order to demonstrate this visually to the reader, we exploit simple two-dimensional approximation examples, train an initial fuzzy model on the first portion of the data (samples shown as grey dots, initial models shown as dotted lines) and update the fuzzy model with new incoming samples (the samples shown as crosses, the updated models shown as solid lines). This examination is basically motivated by the fact, that users may be satisfied with incremental learning of consequent parameters and keeping the structure fix (e.g. due to interpretability or computation speed reasons). In fact, originally many approaches in literature (see e.g. [28], [1]) used the consequent adaptation of the linear parameters in their application scenarios such as control or k-step ahead prediction tasks without conducting any change in the structure of the fuzzy models. For instance, in [28] the consequent parameters of Takagi-Sugeno fuzzy models are adapted within a control loop in order to compensate the mismatch between the fuzzy model and the process. Here, we underline that in some cases this will work properly, but in other cases this may have significant shortcomings. The success or failure, of course, depends strongly on the type and degree of the changing characteristics in the underlying process.

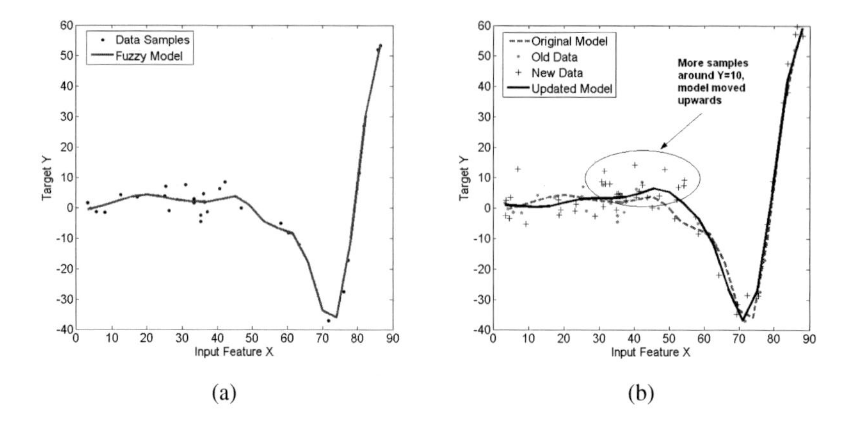

Fig. 2.5 (a): initial fuzzy model (solid line) for approximating the two-dimensional nonlinear relationship between X and Y (dark dots) represented by only 30 samples; (b): updating the initial fuzzy model (dotted line) to the new samples (dark pluses) densifying the already known behavior with 50 new samples: please note that in the middle range of X (between 30 and 55) more values appear for Y around 10 than for Y around 0 causing a move of the model in that part.

Success Cases

A success of a fuzzy model update by just using consequent adaptation is basically achieved when the change in the data distribution is more or less small. Three cases can be distinguished:

1. A densification of local regions in the input/output data space: in this case the first portion of samples already represents the basic tendency of the dependency between the variables to be approximated. An example is given in Figure 2.5, where in (a) an initial fuzzy model is trained (shown as solid line) based on the first 150 data samples (dark dots). The next 150 samples (dark pluses in (b)) densify the first portion by just confirming the same dependency. This leads to a refinement of the consequent parameters in the fuzzy model, which is moved slightly (compare dotted line = original model dark solid line = updated model).
2. A closure of data holes in the original data distribution: in this case the first portion of samples do not significantly cover the range of the input/output space $\rightarrow$ holes may occur which leads to uncertainties in the fuzzy models (see also Chapter 4): an example of this is given in Figure 2.6 (a), where a hole occurs in the middle of the input feature X and the fuzzy model tries to interpolate this based on the tendencies of the nearby lying rules. When new data comes in (crosses in Figure 2.6 (b)), the hole is closed and the fuzzy models properly updated by just using the consequent parameter update (see the significant movement of the updated model (solid line) down to the occurrence of the new samples).

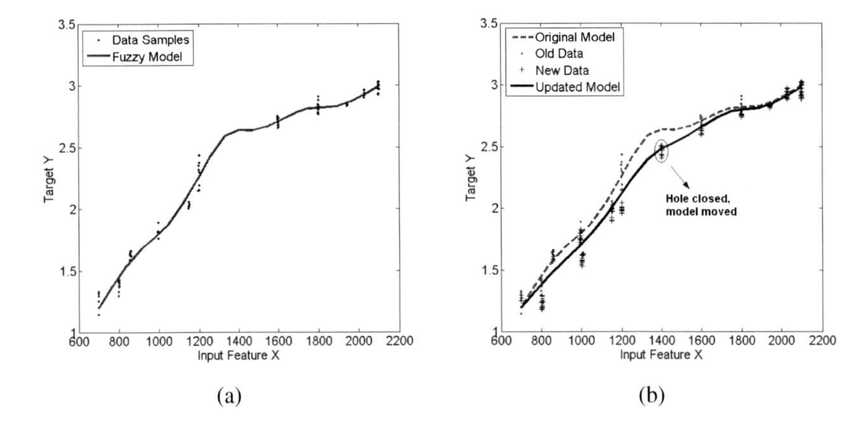

Fig. 2.6 (a): initial fuzzy model (solid line) for approximating the two-dimensional nonlinear relationship between X and Y (dark dots) including a hole in the middle range of X ($\rightarrow$ model uncertain); (b): updating the initial fuzzy model (dotted line) to the new samples (crosses) densifying the already known behavior and closing the hole $\rightarrow$ the model is able to move to the new sample distribution in the the former hole while being properly refined in the other regions.

3. A well-posed extension of the range in the input/output space: in this case the new data samples extend the range of some variables, following the basic trend of the functional dependency at the edge of the old range. An example is provided in Figure 2.7, where in (a) an original model is trained for approximating a Gaussian-type dependency between input feature X and target variable Y. Some new data samples are loaded outside the original range, but following the basic tendency at the right most part in (a). The update of the fuzzy model to this new range performs properly (again shown by a solid line).

Concluding, small changes in the input/output data structure such as a densification of the sample distribution or a closure of an hole in the interpolation domain can be tracked with consequent parameter adaptation alone. Now, we are going to examine what will happen if these holes and range extensions get more extreme.

Failure Cases

In [320] it is maintained that if the nonlinear structure of the process changes strongly, i.e. leading to an original partition in the input space which does not fit to the new data, the fuzzy model may still be adapted appropriately when just using the consequent parameter update by recursive (weighted) least squares (at least better than an update of a pure linear model would do). Here, we give counter examples, where this is clearly not the case and we will demonstrate in Chapter 3 how an evolving structure in the fuzzy models solves these problems. Three cases are distinguished:

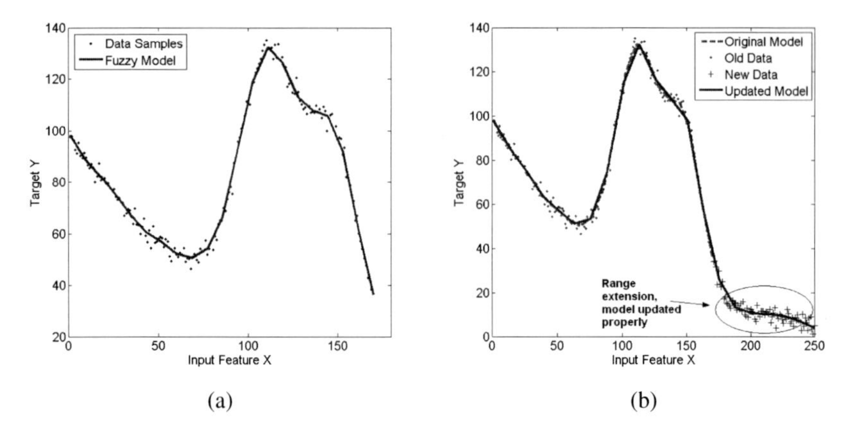

Fig. 2.7 (a): initial fuzzy model (solid line) for approximating the two-dimensional nonlinear relationship between X and Y (dark dots); (b): updating the initial fuzzy model (dotted line) to the new samples (pluses) appearing outside the previous range, the updated model approximates the whole tendency curve (old and new) quite well.

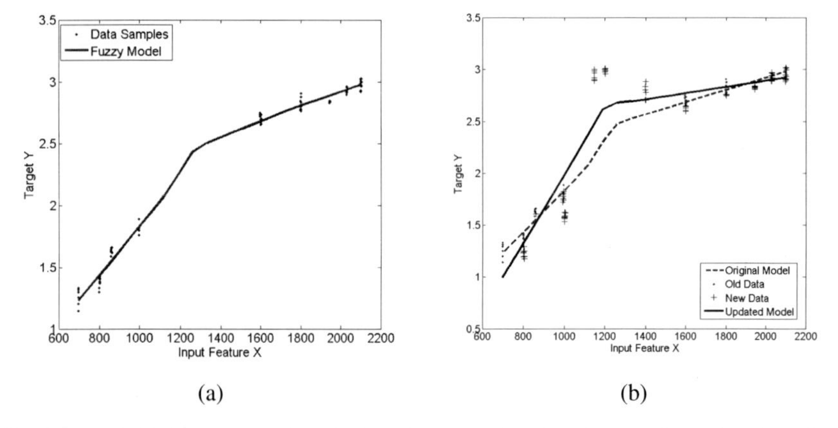

Fig. 2.8 (a): initial fuzzy model (solid line) for approximating the two-dimensional nonlinear relationship between X and Y (dark dots) including a large hole in the middle range of X ($\rightarrow$ model uncertain); (b): updating the initial fuzzy model (dotted line) to the new samples (pluses), closing the hole fails as the model is not able to move to the new sample distribution lying around (1100,3) sufficiently.

1. A strong changing behavior in the interpolation domain of the original data distribution: this case is similar to the case #2 in the previous paragraph about 'success cases', so a hole appears in the interpolation domain; however, the hole is 1.) wider in its range and 2.) closed by samples lying much further away from the principle tendency of the original curve than in the example of Figure 2.6. An example is shown in Figure 2.8, where a similar original data distribution is

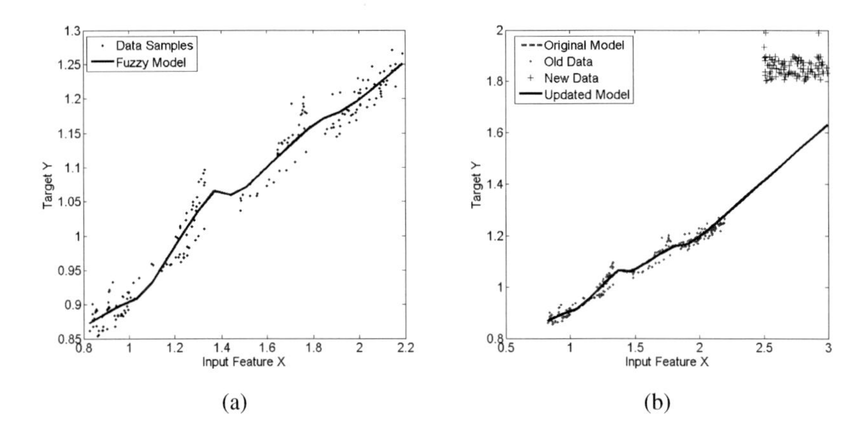

Fig. 2.9 (a): initial fuzzy model (solid line) for approximating the two-dimensional almost linear relationship between X and Y (dark dots); (b): updating the initial fuzzy model (dotted line) to the new samples (pluses) appearing significantly outside the previous range (note the scale of the y-axis), the updated model (solid line) misses the new data cloud completely.

used as in Figure 2.6, but the new samples appear significantly 'above' the basic tendency of the original dependency. When applying the consequent parameter update alone (as in (b)), it is not possible for the updated model to adapt to this change properly (compare solid line with the pluses in (b) — it does not have enough flexibility to go through the new data samples in the input range 1100 to 1400 and lose some of its quality in other regions, especially in the input range from 700 to 1000).

2. A significant extension of the range in the output space: an example is given in Figure 2.9 where the original dependency is basically a linear one with some noise included (a). The new samples appear in a completely different region, extending the range of the input variables, but even more extending the range of the output variable (b). The updated model (in dotted line) misses by far the new data cloud, again by not reaching sufficient flexibility.
3. A significant extension of the range in the input space: is similar to the case #3 in the paragraph about 'success cases', so a range extension of the input variable along the trend of the original dependency, however with a more extreme extension (extending the range more significantly and with a higher nonlinear behavior). Figure 2.10 visualizes such a case, where the tendency curve changes its orientation instead of following an almost linear behavior (a). The updated model is obviously not able to follow the new trend of the dependency, even showing some unstable behavior.

Concluding, larger changes in the input/output data structure such as significant range extension, changing nonlinear behavior or large data holes in the interpolation domain cannot be sufficiently resolved by updating the consequent parameters

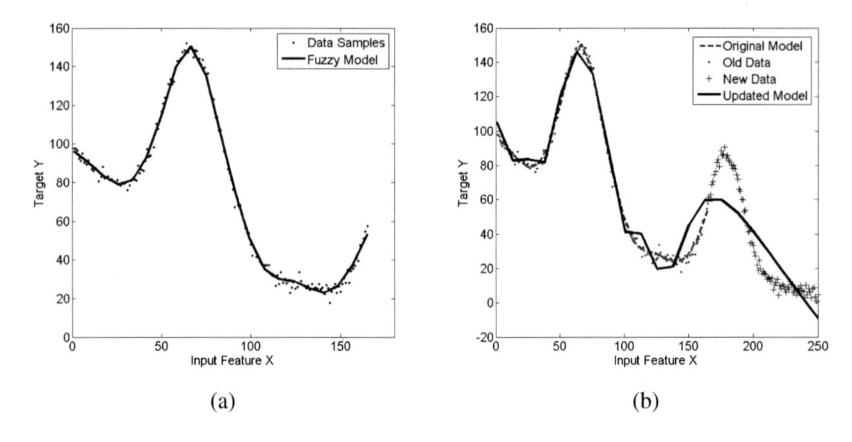

Fig. 2.10 (a): initial fuzzy model (solid line) for approximating the two-dimensional nonlinear dependency between X and Y (represented by dark dots); (b): updating (dotted line) the initial fuzzy model to the new samples (dark dots) appearing outside the previous input range (continuation of the function), the updated model is not able to follow the new trend.

alone. Hence, a change in the structure of the fuzzy models (rule evolution, update of fuzzy sets) is necessary in order to cope with these demands.

In this sense, the *failure cases* are serving as one basic motivation for an enhanced incremental learning scheme for fuzzy systems with the ability to evolve the structure on demand, i.e. finally providing one of the key methodological fundamentals why the field of research for evolving fuzzy systems have emerged over the past decade.

In Chapter 3, we will describe the most important EFS approaches, but first we will point out some enhanced concepts for consequent learning and also demonstrate possibilities for incremental learning of non-linear antecedent parameters.

2.6 Enhanced Issues in Consequent Learning

2.6.1 Weighting of Data Samples

Weighting of data samples during parameter adaptation refers to the problem of coping with dynamic changes in the underlying characteristics of the process and therefore affecting the data stream which is recorded at the process. This is closely related to the drift and shift problematic and will be handled in Chapter 4 (Section 4.2) as it serves as an additional component for more process safety and is coupled with weighting strategies for non-linear antecedent parts in evolving fuzzy systems.

2.6.2 Orthogonal Regressors in Consequent Learning: More Flexibility, Less Computational Complexity

One of the most significant drawbacks in global learning is the low flexibility of adding a new rule: this causes a re-setting or at least a merging of the inverse Hessian matrix, which may lead to solutions which are no longer optimal in the least squares sense (see 2.4). Local learning is able to prevent this shortcoming by using an incremental (recursive) learning approach for each rule separately and independently (hence, older rules are not 'disturbed' when adding a new one). However, the automatic inclusion of new input features is still a challenge. In this section, we demonstrate how orthogonal regressors may solve this problem and whether and how orthogonal regressors are applicable in evolving fuzzy systems using linear consequent parameters.

Orthogonal regressors are regressors which possess the following property:

$$\mathbf{reg}_i^T \mathbf{reg}_j = 0 \quad \text{for} \quad i \neq j \tag{2.65}$$

where $\mathbf{reg}_i$ denotes the ith column of a regression matrix R. This means, that for the global learning approach, where one regressor denotes one column of the regression matrix R (see (2.30)),the Hessian matrix $R^T R$ reduces to

$$R^T R = diag(\mathbf{reg}_1^T \mathbf{reg}_1, \mathbf{reg}_2^T \mathbf{reg}_2, \ldots, \mathbf{reg}_{C(p+1)}^T \mathbf{reg}_{C(p+1)}) \tag{2.66}$$

and therefore the inverse Hessian matrix which is used in the off-line least squares estimation simply to

$$(R^T R)^{-1} = diag(\frac{1}{\mathbf{reg}_1^T \mathbf{reg}_1}, \frac{1}{\mathbf{reg}_2^T \mathbf{reg}_2}, \ldots, \frac{1}{\mathbf{reg}_{C(p+1)}^T \mathbf{reg}_{C(p+1)}}) \tag{2.67}$$

which can be calculated with complexity $O(NM)$ in sum instead of $O(NM^2) + O(M^3)$ complexity for calculating $R^T R$ and its inverse with $M = C(p+1)$ the number of linear consequent parameters to be estimated. So, a significant performance boost is provided which may be significant for allowing a re-calculation of the parameters with all seen samples so far (instead of an incremental update). Furthermore, for the estimation of the parameters $\mathbf{w}$ it follows directly from (2.34):

$$\hat{w}_j = \frac{\mathbf{reg}_j^T \mathbf{y}}{\mathbf{reg}_j^T \mathbf{reg}_j} \tag{2.68}$$

So, each of the parameters $\hat{w}_j, j = 1,...,C(p+1)$ can be estimated separately and regressors can be removed or included to an already selected regressor set, without effecting the other parameter estimates. This finally means, that new rules and also new inputs can be added on demand during the incremental/evolving learning phase of the fuzzy systems without affecting the training of the other rule consequent parameters. In this sense, we can speak here about a maximal flexibility in terms of rule inclusion and input space extension. Although re-training steps during the evolving phase could be considered here, the recursive least squares is still more attractive to apply as it reduces to the *Recursive Orthogonal Least Squares = ROLS*, which is carried out for each linear parameter separately (compare with equations (2.49) to (2.51)):

$$\hat{\mathbf{w}}_j(k+1) = \hat{\mathbf{w}}_j(k) + \gamma(k)(y(k+1) - \mathbf{reg}_j(k+1)\hat{\mathbf{w}}_j(k)) \tag{2.69}$$

with the correction vector

$$\gamma_j(k) = \frac{P_j(k)\mathbf{reg}_j(k+1)}{1 + \mathbf{reg}_j^T(k+1)P_j(k)\mathbf{reg}_j(k+1)} \tag{2.70}$$

and $P_j(k+1)$:

$$P_j(k+1) = (1 - \gamma_j(k)\mathbf{reg}_j^T(k+1))P_j(k) \tag{2.71}$$

with $P_j(k) = \frac{1}{\mathbf{reg}_j^T \mathbf{reg}_j}$ the inverse diagonal entry at the kth sample (starting with 1). Note that all these are scalar operations (so scalars are updated instead of vectors and matrices), which makes the recursive learning approach very fast.

The same procedure can be applied to the local learning approach, where one regressor is simple one input feature in its original values, compare with matrix in (2.57) (note that the Ψ_is are just weights and hence do not violate (2.65), when two original input features are orthogonal. Hence, for the ith rule the following Hessian matrix is obtained:

$$R_i^T Q_i R_i = diag(\mathbf{reg}_{i,1}^T < \mathbf{reg}_{i,1}, \Psi_i >, \mathbf{reg}_{i,2}^T < \mathbf{reg}_{i,2}, \Psi_i >, \ldots, \mathbf{reg}_{i,p+1}^T < \mathbf{reg}_{i,p+1}, \Psi_i >) \tag{2.72}$$

Note that $\mathbf{reg}_{i,k} = \mathbf{reg}_{j,k}$ for all i, j. Again, this can be calculated with complexity $O(NM)$ in sum instead of $O(NM^2)$ complexity for calculating $R_i^T Q_i R_i$ plus $O(M^3)$ complexity for computing the inverse of $R_i^T Q_i R_i$ with $M = p+1$ the number of linear consequent parameters to be estimated. Consequently, for the estimation of the jth parameter $\mathbf{w}_i$ of the ith rule it follows directly from (2.60):

$$\hat{w}_{i,j} = \frac{\mathbf{reg}_{i,j}^T < \mathbf{y}, \Psi_i >}{\mathbf{reg}_{i,j}^T < \mathbf{reg}_{i,j}, \Psi_i >} \tag{2.73}$$

So, each of the parameters $\hat{w}_{i,j}$ in each rule can be estimated separately. Hence new input features can be added and removed without affecting the others. The *Recursive Orthogonal Fuzzily Weighted Least Squares = ROFWLS* can be deduced in the same manner as for the global learning approach, leading to scalar operations in (2.62) to (2.64) for each parameter in each rule.

Now, the final question is whether orthogonal regressors are realistic to occur in the fuzzy system regression problems and if not, which mechanisms could be applied and how (along the evolving fuzzy system concept to make them orthogonal).

For the global learning approach and assuming N data samples loaded so far, orthogonal regressors would mean that the following condition holds:

$$\sum_{k=1}^{N}(x_l(k)\Psi_i(\mathbf{x}(k)))(x_m(k)\Psi_j(\mathbf{x}(k))) \approx 0 \quad i \neq j, l \neq m$$

for arbitrary l and $m \in \{0,...,p\}$ and arbitrary i and $j \in \{1,...,C\}$, where x_0 is set to 1. Obviously, this cannot be guaranteed for an arbitrary data stream resp. arbitrary rule setting. For the local learning approach the same problem is caused as there it is required that

$$\sum_{k=1}^{N} x_l(k)x_m(k) \approx 0 \quad l \neq m$$

Hence, in order to be able to apply orthogonal least squares onto fuzzy systems with linear consequent parameters, the regressors $\mathbf{reg}_i = \mathbf{x}_i\Psi_i$ (for global learning) resp. the original input features (for local learning) have to be made orthogonal. A possibility to accomplish this is by applying *PCA = Principal Component Analysis* [402] on the original regression matrix and applying principal component fuzzy regression (PCFR).

For the global learning approach this means transforming the regression matrix R as in (2.30) into its eigenvector space by applying $\mathbf{z}_m = R\mathbf{v}_m$, where $\mathbf{v}_m$ denotes the mth eigenvector (= principal component) of R, according to the descending ordered list of eigenvalues. This means that the largest eigenvalue provides the most important principal component (its corresponding eigenvector). Hereby, $\mathbf{z}_m$ can be seen as weighted linear combination of all the regressors in R (weighted by the components of the eigenvector and including the weighted membership degrees of all rules). Please note that the eigenvalues and eigenvectors can be estimated by applying singular value decomposition (SVD) [146] . Now, the principal component fuzzy regression on the target y is done by (according to the principal component regression (PCR) [201]):

$$\hat{f}_{fuz} = \hat{y} = \sum_{m=1}^{P} \Phi_m \mathbf{z}_m \tag{2.74}$$

where Φ_m are the linear weights (for the principal components) which can be estimated independently from each other according to (2.68) (using $\mathbf{z}_1,...,\mathbf{z}_m$ as regressors), and P is the number of applied principal components. If $P = C(p+1)$, so the number of principal components equals the number of the regressors, the original least squares estimate is obtained, since the columns of $Z_P = [z_1 z_2 ... z_P]$ span the column space of R. Furthermore, the following important relationship holds (see [174]):

$$\hat{w} = \sum_{m=1}^{P} \hat{\Phi}_m \mathbf{z}_m \tag{2.75}$$

This means that all the original linear parameters (in all rule consequent functions) can be back-estimated once having estimated the linear parameters Φ for the corresponding principal component fuzzy regression problem. This means that when updating a new principal component during incremental phase (and $\hat{\Phi}$ by using (2.69) to (2.71)), it is always possible to update the original linear parameters by using (2.75).

Principal components themselves may be updated with new samples and even increased in numbers by the incremental approach demonstrated in [210] (iPCA), allowing us maximal flexibility in terms of the principal component fuzzy regression. A key point for a well-posed update there is to check whether a new incoming sample has almost all energy in the current eigenspace $Z_m = [z_1 z_2 ... z_m]$ or not. This can be done by computing the residual vector h with

$$\mathbf{h} = (\mathbf{y} - \bar{\mathbf{x}}) - Z_m \mathbf{g} \tag{2.76}$$

where $\bar{x}$ denotes the mean vector and $\mathbf{g}$ is calculated by $\mathbf{g} = Z_m^T (\mathbf{y} - \bar{\mathbf{x}})$, and then checking whether the norm of the residual vector is smaller than a threshold value η. If this is the case, the eigen-space is simply updated, otherwise a column is added to Z_m and $m = m + 1$. Updating of the eigen-space is done by solving a specific intermediate eigenproblem after S (see [165]):

$$\left(\frac{N}{N+1} \begin{bmatrix} \Lambda & 0 \\ 0 & 0 \end{bmatrix} + \frac{N}{(N+1)^2} \begin{bmatrix} \mathbf{g}\mathbf{g}^T & \gamma\mathbf{g} \\ \gamma\mathbf{g}^T & \gamma^2 \end{bmatrix} \right) S = S\Lambda^* \tag{2.77}$$

with $\gamma = \mathbf{h}^T (\mathbf{y} - \mathbf{x})$, S the solution of the eigenproblem containing all the eigenvectors as columns and Λ^* the diagonal matrix containing the eigenvalues (Λ contains the eigenvalues after the previous update step), and updating Z_m by

$$Z_m = [Z_m, \hat{\mathbf{h}}] S \tag{2.78}$$

with $\hat{\mathbf{h}} = \frac{\mathbf{h}}{\|\mathbf{h}\|}$ if the norm is greater η, otherwise 0.

Another incremental PCA approach is demonstrated in [254], which also may expand the eigenspace by adding new eigenvectors, as soon as new incoming samples are presented to the algorithm. A nice feature of this approach is that it allows the definition of weights for new samples, i.e. the possibility to outdate older samples over time and hence to be applicable in drift scenarios (see Section 4.2).

However, the problem of adding new rules and new inputs still remains as this would change the size of the eigenvectors z_m, i.e. adding *rows* in Z_m, which cannot be handled by any of the aforementioned approaches. Therefore, we see here a fruitful challenge to develop an incremental Principal Component Analysis, which allows this flexibility (*ePCA = evolving Principal Component Analysis*). Similar considerations also apply to the local learning approach, where the original input features are transformed to the principal component space and for each rule (2.74) and (2.75) are applied separately by including the Ψ_i's as weights. In the case of local learning, adding new rules is handled by opening up a separate recursive weighted least squares estimation, however the problem of coping with a changing number of inputs (in a smooth manner) still remains. Section 5.2 (in Chapter 5) demonstrates an alternative concept by using incremental feature weights denoting the (dynamically changing) impacts of features.

2.6.3 Alternatives for Speeding Up Rule Consequent Learning

Alternative to the demand of orthogonal regressors, fast *RLS* algorithms with computational complexity demands of $O(M)$ respectively $O(NM)$ for N update steps (on N new samples) instead of $O(M^2)$ resp. $O(NM^2)$ can be achieved. Examples of such algorithms are not explained here in detail and can be inspected in [139], [106] or [305]. The basic problem of all these approaches is the increase in numerical instability by removing redundant information in the *R(W)LS* update. An enhanced approach is presented in [361], which is based on generalized eigendecomposition and converges to the exact solution.

2.7 Incremental Learning of Non-linear Antecedent Parameters

This section is dedicated to the learning of non-linear antecedent parameters in Takagi-Sugeno(-Kang) type fuzzy systems. We are again starting with the batch case, assuming N samples are available and then discuss the extension to the incremental case.

As an analytical solution in a closed formula form (as is the case of linear parameters) is not possible when using least squares optimization function, an iterative numeric optimization methods is required. A possibility to achieve this is the so-called steepest descent optimization procedure . Therefore, we define the least squares optimization problem in dependency of the non-linear parameters Φ_{nonlin} as:

$$J = J(\Phi_{nonlin}) = \sum_{k=1}^{N} (y_k - \sum_{i=1}^{C} l_i(\mathbf{x}_k)\Psi_i(\Phi_{nonlin})(\mathbf{x}_k))^2 \tag{2.79}$$

and the residual in the kth sample as

$$e_k = e_k(\Phi_{nonlin}) = y_k - \sum_{i=1}^{C} l_i(\mathbf{x}_k)\Psi_i(\Phi_{nonlin})(\mathbf{x}_k)^2 \tag{2.80}$$

In the following, we present the gradient descent optimization when using product t-norm as conjunction operator and the Gaussian functions as fuzzy sets (having fuzzy set centers $\mathbf{c}$ and fuzzy set widths σ as non-linear parameters). Let therefore $c_{i,j}$ resp. $\sigma_{i,j}$ be the center resp. width of the jth fuzzy set (for the jth dimension) in the ith rule. Then, we get

$$\frac{\partial J}{\partial c_{i,j}} = \sum_{k=1}^{N} e_k \sum_{n=1}^{C} l_n(\mathbf{x}_k) \frac{\partial \Psi_n(\mathbf{x}_k)}{\partial c_{i,j}} \tag{2.81}$$

respectively

$$\frac{\partial J}{\partial \sigma_{i,j}} = \sum_{k=1}^{N} e_k \sum_{n=1}^{C} l_n(\mathbf{x}_k) \frac{\partial \Psi_n(\mathbf{x}_k)}{\partial \sigma_{i,j}} \tag{2.82}$$

where the derivatives of the normalized membership functions are obtained by (proof is left to the reader):

If $i = l$ then we have

$$\frac{\partial \Psi_i(\mathbf{x})}{\partial c_{i,j}} = \frac{(1-\Psi_i)(-2\mu_i)}{\sum_{j=1}^{C}\mu_j(\mathbf{x})}\left(\frac{1}{2\sigma_{i,j}^2}(c_{i,j}-x_j)\right)$$

If $i \neq l$ then we get

$$\frac{\partial \Psi_i(\mathbf{x})}{\partial c_{l,j}} = \frac{(-\Psi_i)(-2\mu_l)}{\sum_{j=1}^{C}\mu_j(\mathbf{x})}\left(\frac{1}{2\sigma_{l,j}^2}(c_{l,j}-x_j)\right)$$

For σ we get similar equations, i.e. if $i = l$, then we have

$$\frac{\partial \Psi_i(\mathbf{x})}{\partial \sigma_{i,j}} = \frac{(1-\Psi_i)(\mu_i)}{\sum_{j=1}^{C}\mu_j(\mathbf{x})}\left(\frac{1}{\sigma_{i,j}^3}(c_{i,j}-x_j)^2\right)$$

If $i \neq l$ then we get

$$\frac{\partial \Psi_i(\mathbf{x})}{\partial \sigma_{l,j}} = \frac{(-\Psi_i)(\mu_i)}{\sum_{j=1}^{C}\mu_j(\mathbf{x})}\left(\frac{1}{\sigma_{l,j}^3}(c_{l,j}-x_j)^2\right)$$

where $\mu_i(\mathbf{x}) = e^{-\frac{1}{2}\sum_{j=1}^{p}\frac{(x_j-c_{ij})^2}{\sigma_{ij}^2}}$. For the steepest descent method, following the negative gradient, the following update in the $m+1$th iteration is achieved:

$$c_{(m+1)_{i,j}} = c_{(m+1)_{i,j}} - \tau\sum_{k=1}^{N} e_k \sum_{n=1}^{C} l_n(\mathbf{x}_k)\frac{\partial \Psi_n(\mathbf{x}_k)}{\partial c_{i,j}} \tag{2.83}$$

and

$$\sigma_{(m+1)_{i,j}} = \sigma_{(m+1)_{i,j}} - \tau\sum_{k=1}^{N} e_k \sum_{n=1}^{C} l_n(\mathbf{x}_k)\frac{\partial \Psi_n(\mathbf{x}_k)}{\partial \sigma_{i,j}} \tag{2.84}$$

where τ (learning step) is a fixed number which is rather small (usual values are around 0.1) in order to avoid fluctuations during the optimization process. The number of iterations depends on the decrease of the residual which is usually intended to get smaller than a pre-defined value.

The problem with steepest descent (SD) is that it is a first order method and usually does not converge with both, a high precision and a low computational complexity at the same time[27]. In this sense, it is recommended to use a second order method such as the Levenberg-Marquardt optimization algorithm [299], which explicitly exploits the underlying structure (sum-of-squares) of the optimization problem on hand (also see [54] [53], where the Levenberg-Marquardt (LM) optimization algorithm was already applied to parameter learning in fuzzy systems). Denoting by J the Jacobian matrix:

$$Jac = \begin{bmatrix} \frac{\partial J(1)}{\partial \Phi_{nonlin_1}} & \cdots & \frac{\partial J(1)}{\partial \Phi_{nonlin_D}} \\ \vdots & \vdots & \vdots \\ \frac{\partial J(N)}{\partial \Phi_{nonlin_1}} & \cdots & \frac{\partial J(N)}{\partial \Phi_{nonlin_D}} \end{bmatrix} \tag{2.85}$$

with $D = 2*C*p$ the amount of nonlinear parameters and the derivatives calculated as in (2.81) and in (2.82), the following LM update procedure is obtained:

$$(Jac^T(k)Jac(k) + \alpha I)s(k) = -Jac^T(k)e(k) \tag{2.86}$$

where $s(m) = \Phi_{nonlin}(m+1) - \Phi_{nonlin}(m)$ denotes the parameter change in the mth iteration step and α the update parameter, which controls both, the search direction and the magnitude of the update. The search direction varies between the Gauss-Newton direction and the steepest direction, according to the value of α. This is dependent on how well the actual criterion agrees with a quadratic function in a particular neighborhood: if the error goes down following an update, it implies that the quadratic assumption is working and α is reduced (usually by a factor of 10) to reduce the influence of gradient descent; on the other hand, if the error goes up, it is reasonable to follow the gradient more and so α is increased by the same factor. The good results presented by the LM method (compared with other second-order methods such as the Quasi-Newton and conjugate gradient methods) are due to the explicit exploitation of the underlying characteristics of the optimization problem by the training algorithm. Furthermore, Levenberg-Marquardt algorithm is a blend between Gradient Descent and Gauss-Newton iteration. Notice that (2.86) can be recast as:

$$\Phi_{nonlin}(k+1) = \Phi_{nonlin}(k) - (Jac^T(k)Jac(k) + \alpha I)^{-1} Jac^T(k)e(k) \tag{2.87}$$

The complexity of this operation is of $O(D^3)$, where D is the number of columns of the Jacobian matrix, i.e. the number of non-linear antecedent parameters in the fuzzy system.

Now, the remaining important question is how either steepest descent or Levenberg-Marquardt optimization algorithms can be adopted to the incremental (single-pass) case. Both, are iterating over the whole batch of samples a multiple times, hence not directly applicable in single-pass mode. On the other hand, both are incremental in the sense that they are performing incremental update cycles of the parameters (also called iterative procedures). These characteristics can be exploited also for the on-line single pass case. In fact, for the Levenberg-Marquardt optimization algorithm a recursive approach exists, which converges to the exact batch solution (see [321] [148]). The basic idea is to incorporate a regularization term in the Gauss-Newton optimization, which allows the update of only one parameter for one single newly loaded sample. This ensures that not the whole parameter space trained on older samples is updated for just one new single sample, as this usually causes a too strong update and holds the risk that older relations are forgotten. The update of only one parameter at each time instance can be assured by, instead of using αI, using $\alpha\Gamma$ as learning gain:

$$\Gamma_{ii} = \begin{cases} 1 & i = k(mod(D)) + 1 \\ 0 & otherwise \end{cases} \tag{2.88}$$

i.e. Γ has only one non-zero element located at the diagonal position $k(mod(D)) + 1$ with D the number of parameter to be optimized. By expressing the inverse Hessian matrix $P(k) = (Jac^T(k)Jac(k) + \alpha I)^{-1}$ and using again matrix inversion lemma (see also (2.45)), the *recursive Levenberg-Marquardt (RLM) algorithm* is given by the following equations (for detailed derivation see [441]):

$$\Phi_{nonlin}(k+1) = \Phi_{nonlin}(k) + P(k)Jac^T(k)e(k) \tag{2.89}$$

with

$$P(k) = \frac{1}{\lambda_k}(P(k-1) - P(k-1)US^{-1}U^T P(k-1) \tag{2.90}$$

λ_m a forgetting factor (set to 1 in case of no forgetting) and

$$S = \lambda_k V + U^T P(k-1)U \tag{2.91}$$

The matrix S is a 2×2 matrix, therefore easily and quickly calculated in each update step; this is because

$$U^T = \begin{bmatrix} Jac^T(k) \\ 0\ldots010\ldots0 \end{bmatrix}$$

and

$$V^{-1} = \begin{bmatrix} 1 & 0 \\ 0 & D\alpha_k \end{bmatrix}$$

with α_k the learning gain in the kth step, and $P(k-1)$ the covariance matrix which is either initialized to $(Jac^T Jac + \alpha I)^{-1}$ after an initial training step of the fuzzy models or to αI with α a big integer when doing incremental learning of non-linear parameter from scratch. In the latter case the start values for the non-linear parameter are all set to 0. Note that this is in accordance to the recursive (weighted) least squares estimator for updating linear consequent parameters. In [441] this recursive optimization algorithm is successfully applied to train the non-linear parameter of an evolving fuzzy systems predictor (EFP) — see also Section 3.7.5; which is combined with the recursive least squares estimator for updating the consequent parameters in a global manner (all parameters in one sweep, see above). For a local estimation of non-linear parameters, the RLM estimator could be carried out for each rule separately by using the part of the Jacobian matrix related to the non-linear parameters of the corresponding rules. Then the gradients of the weighted least squares estimator (2.59) (instead of the least squares estimator J) with respect to each non-linear parameter is needed.

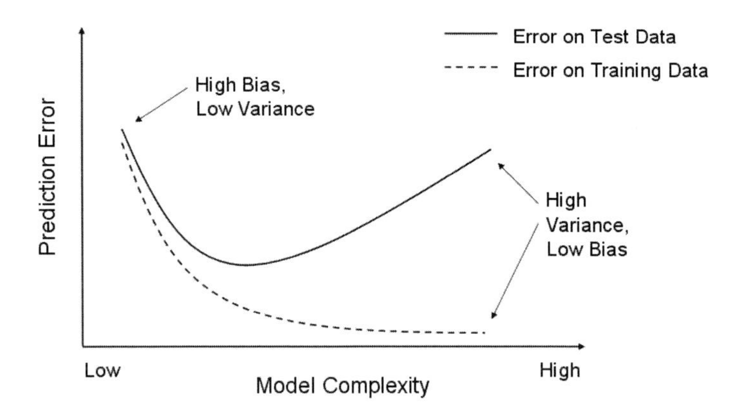

Fig. 2.11 Bias-variance tradeoff: solid line represents the error on the test samples, dotted line the error on training samples with increasing model complexity; note the increasing prediction error when the variance error (error on the test samples) increases too much

2.8 On the Demand of Alternative Optimization Functions

This section deals with alternatives to the least squares optimization problem, commonly applied in data-driven learning approaches for fuzzy systems, neuro-fuzzy approaches and neural networks and as used in this book 1.) throughout the previous sections on consequent adaptation and incremental learning of non-linear antecedent parameters and 2.) for most of the evolving fuzzy systems approaches as will be discussed in Chapter 3.

2.8.1 Problem Statement

The problem with the least squares optimization function is that it represents an overly optimistic estimate of the true generalization error on new unseen data as it only takes into account the error on the training data (also called bias error). Hence, it is prone to over-fit on the training data[385], increasing variance error and hence the whole model error on new test samples drastically. The variance error is due to the sampling variance of the underlying model (usually caused by the noise in the data). Bias and variance error depend on the chosen model complexity, whereas bias error usually decrease and variance error increase with increasing model complexity. This is called bias-variance tradeoff [174] and demonstrated in Figure 2.11. A medium model complexity is able to optimize the prediction error. In the extreme case, someone can achieve zero least squares error on the training data, for instance by defining one rule per data sample, hence exploiting maximal model complexity in fuzzy systems. However, in case of noisy data, this would just approximate the noise rather than the principal tendency of the functional relationship between target and input variables (in regression case) resp. of the decision boundaries between

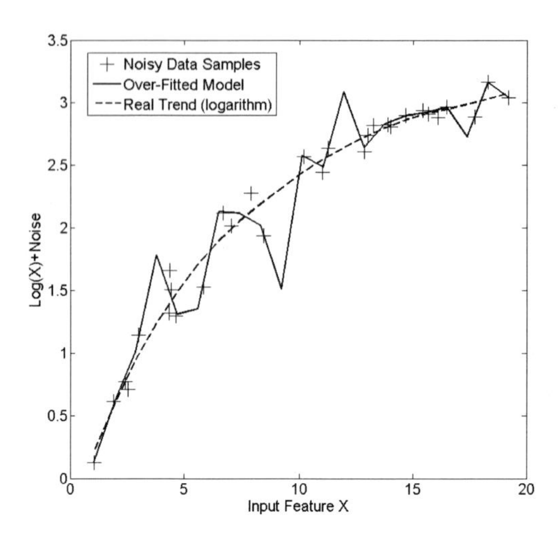

Fig. 2.12 Over-fitting of a logarithmic trend caused by noise in the data (marked as pluses) and high model complexity (solid line) — see the high deviation to the real trend (shown as dashed line)

classes (in classification case). A demonstration of the effect of almost zero bias (an almost perfect approximation of the noise in the data in regression case is given in Figure 2.12. In order to circumvent the sensitivity with respect to over-fitting in case of least squares optimization, N-fold cross-validation (CV) procedure [411] is commonly applied when doing data-driven learning of fuzzy systems. This procedure shuffles the data first and then splits the whole data set into N folds. For each combination of $N-1$ folds training is performed and the remaining fold is used as separate test set for eliciting the mean squared error (which is the error rate as in least squares optimization function divided by the number of test samples, see also Section 2.1.4). The obtained N mean squared error rates are averaged to an overall mean squared error. This procedure can be repeated with different parameter settings, i.e. different number of rules, fuzzy sets and the best model in terms of minimal mean squared error used as final model for further processing. This procedure is also called *best parameter grid search scenario* and the parameters guiding the models to lowest error the *optimal parameter setting*. Often, also the least complex model whose error is no more than one standard error above the error of the best model is chosen as final model (this procedure is for instance applied in *CART (=Classification and Regression Trees)* [58] which is part of the Statistic Toolbox (see "treetest")). Hereby, the standard error can be estimated by the standard deviation of the errors over all N folds. Furthermore, it is recommended to use a separate validation set for the assessment of the generalization error of the final chosen model. The problem of this whole procedure is that

1. it is very time intensive, especially in case of a dense parameter grid (e.g. 1 to 300 rules with step size 1) and a high amount of data — such validation runs using N-fold CV can last over days on regular PCs.
2. it is not applicable for incremental on-line training as applied in evolving fuzzy systems, as it always needs the full data set available at hand.

2.8.2 *Possible Alternatives*

In this sense, we are investigating alternative optimization functions which include sorts of penalization (terms) for over-fitting avoidance in their definitions. These force the optimization algorithm to solutions which are not only minimal in the error sense but also optimal in a combined model error/complexity tradeoff and hence near the global minimum of the generalization error curve as shown in Figure 2.11.

An obvious possibility is to include the model complexity as penalization term directly into the optimization function. The complexity can be measured in terms of the degrees of freedom in the fuzzy system which is

$$deg_{fuz} = 2Cp + kC \tag{2.92}$$

with C the number of fuzzy rules and p the number of inputs and k the number of parameters in the rule consequents. This is quite intuitive, as C defines the degree of granularity in partitioning the input space, whereas p defines the dimensionality of the problem. Each additional granule (in form of a rule) needs to be punished as being a step towards more over-fitting. Also, each additional input needs to be punished as increasing the curse of dimensionality effect [39]. The curse of dimensionality can be understood as high-dimensional spaces are inherently sparse [67], i.e. the higher the dimension gets the less samples are included in a hyper-dimensional sphere with a specific a priori defined radius. This means that we cannot speak about (dense) local regions any longer, which means that clusters of samples (representing rules) are thinned out and therefore quite uncertain. A visualization of this effect when increasing the dimensionality from 1 through 2 to 3. is visualized in Figure 2.13: obviously, the 10 data samples are sufficient for a dense description of one input feature x_1 (left image), whereas in case of three dimensions (right image) those 10 samples are widely spread within the three-dimensional hyper-cube, hence nowhere forming a dense (significant) cluster. Each rule has two essential parameters with respect to each dimension: the core value (center) and the range of influence (width) (therefore the term $2Cp$ in (2.92)). The second term covers the parameters estimated in the output space, therefore $k = 1$ in case of Sugeno fuzzy model (singleton consequent), $k = 2$ in the case of Mamdani fuzzy systems (fuzzy set in the consequents with two parameters: core value + range of influence) and $k = p + 1$ in the case of Takagi-Sugeno fuzzy systems (p weights for the p inputs and one additional parameter for the intercept).

In order to assure a well-defined optimization function both, least squares error and the penalty term have to be normalized to a unique range, let's say $[0,1]$. For least squares error this can be achieved by dividing it with the squared range of the

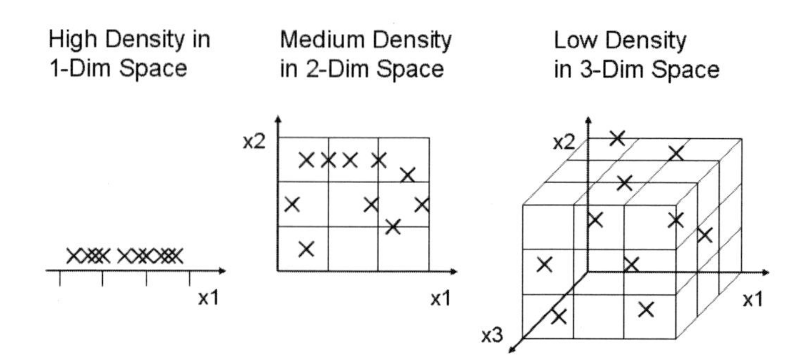

Fig. 2.13 The curse of dimensionality visualized in case of 10 data samples and increasing dimensionality from 1 (left) through 2 (middle) to 3 (right): note the dense distribution of the samples in the one-dimensional case and the quite sparse one in the three-dimensional case.

target variables, for the model complexity by normalizing (2.92) by the maximal allowed model complexity max_{comp}, expressed by the maximal number of allowed rules and the (maximal, eventually reduced) dimensionality of the problem. Hence, the optimization function (punished least squares becomes:

$$J_2 = \frac{\sum_{k=1}^{N}(y(k) - \sum_{i=1}^{C} l_i \Psi_i(\mathbf{x}(k)))^2}{(\max(\mathbf{y}) - \min(\mathbf{y}))^2} + \frac{deg_{fuz}}{max_{comp}} = \min_{\phi}! \qquad (2.93)$$

which should be minimized according to a set of parameters ϕ (included as linear ones in l_i and non-linear ones in Ψ_i).

The problem with this optimization function is that it is not differentiable (due to the second term) which is necessary for applying a feasible optimization algorithm for linear and non-linear parameters (otherwise, only an heuristic search algorithms can be applied, which are quite time-intensive and not possible during on-line learning mode). Therefore, it would be a better choice to include the model complexity in another way into the least squares optimization problem. A possibility would be to use the r-squared-adjusted measure as defined in Section 2.1.4 which includes the degrees of freedom in the denominator of (2.18). This would yield a maximization problem towards an r-squared-adjusted of 1. Another possibility is to apply ridge regression [174] [108], which includes a penalty term on the size of the parameters, hence tries to shrink some parameters towards zero and hence reducing their weights in the final model output. This is achieved by adding $\alpha\|\phi\|$ to the least squares error. This approach is restricted to linear problems, i.e. only applicable for optimizing linear consequent parameters. Section 4.1 will further deal with this approach within the context of regularization.

Another statistically funded alternative (for linear and non-linear parameters) is to add an optimism term [174] to the least squares loss function, which is the expected difference of the squared errors on the training data and the extra-sample error on new test data. In [174] it is shown that for least squares loss functions the optimism can be approximately expressed as

$$op = 2\sum_{k=1}^{N} Cov(\hat{y}(k), y(k)) \tag{2.94}$$

where Cov denotes the covariance and is defined by

$$Cov(x,y) = (x-\bar{x})(y-\bar{y}) \tag{2.95}$$

with $\bar{x}$ the mean of x and $\bar{y}$ the mean of y. Then, the optimization problem called *extended least squares (ELS)* becomes:

$$J_3 = \sum_{k=1}^{N}(y(k) - \sum_{i=1}^{C} l_i \Psi_i(\mathbf{x}(k)))^2 + 2\sum_{k=1}^{N} Cov(\sum_{i=1}^{C} l_i \Psi_i(\mathbf{x}(k)), y(k)) = \min_{\phi}! \tag{2.96}$$

with ϕ a set of parameters. Note that the harder the data is fitted, the greater $Cov(\hat{y}(k), y(k))$ will get hence increasing J_3. Respectively, when considering optimizing parameters in fuzzy systems, the term $Cov(\hat{y}(k), y(k))$ will guide these parameters to values so that a kind of smoothing effect in the approximating curves is achieved, preventing the models to over-fit the data. In [174], it is also maintained that the cross-validation procedure is a direct estimate of the extra-sample error, hence the optimization of (2.96) approximates CV procedure as (2.94) approximates the optimism. Furthermore, J_3 is obviously steadily differentiable whenever the fuzzy sets are steadily differentiable and represent a free optimization problem.

All these aforementioned approaches have the disadvantage that the number of rules are not foreseen to be explicitly optimized within an incremental optimization procedure. This means, the number of rules (C) are defined and then the parameters are optimized according to this fixed set of rules. An alternative to this restriction is presented in [283], where rule weights are included in the definition of the normalized rule membership degrees:

$$\Psi_i(\mathbf{x}(k), \rho) = \frac{\rho_i \mu_i(x)}{\sum_{j=1}^{C} \rho_j \mu_j(x)} \tag{2.97}$$

with ρ_j the rule weight for the jth rule, and the optimization procedure is defined by constraining the number of rule weights greater than 0. In this sense, the optimization function becomes (called *rule-based constrained least squares (RBCLS)*):

$$J_4 = \sum_{k=1}^{N}(y(k) - \sum_{i=1}^{C} l_i \Psi_i(\mathbf{x}(k), \rho))^2 = \min_{\rho}! \quad \text{such that } \#\{\rho_i \neq 0\} \leq K, \tag{2.98}$$

with # the cardinality of a set. Solving this optimization problem in batch mode was achieved in [284] by exploiting a sparsity threshold operator which forces as many as possible rule weights towards 0 during a projected gradient descent optimization algorithm (therefore called *SparseFIS* method). Solving this optimization problem in incremental manner would be a possibility for switching off rules which were previously evolved but are not really necessary later on (by changing rule weights incrementally with new incoming samples). Exchanging the least squares error term in above problem with the extended least squares error term in (2.96) would address a handling of both, parameter-based and rule-based (structural) over-fitting within one optimization problem.

The optimization problems described so far (especially J_3 and J_4 plus a combination of these two) include terms for handling the noise in the output variable y appropriately (omitting over-fitting by punishment terms). However, the noise in the input variables, affecting the exactness of the values of the parameters to be optimized, causing a bias in the parameter estimates themselves, i.e. $E\{\|\hat{\Phi}-\Phi\|\} \neq 0$ with $\hat{\Phi}$ the estimated and Φ the real parameters, is neglected in all of the aforementioned optimization problems. Especially in case of dynamic problems (including regressors in form of time delayed past input and output variables), the noise affects both the target and the regressors. In this sense, it is beneficial to re-construct both, outputs (y) and inputs (regressors r), which is achieved by the *total least squares (TLS)* optimization problem, defined by:

$$J_5 = \frac{1}{2N}\left(\sum_{k=1}^{N}(\mathbf{r}_k - \hat{\mathbf{r}}_k)^2 + \sum_{k=1}^{N}(y(k) - \sum_{i=1}^{C} l_i \Psi_i(\mathbf{x}(k)))^2\right) \quad (2.99)$$

In the linear case, by solving this optimization problem, someone achieves a hyperplane to which the whole bunch of samples has a shortest normalized distance, instead of a shortest quadratic distance with respect to the output y. In [193], this optimization setup was successfully applied in its weighted version for identifying Takagi-Sugeno fuzzy models from data recorded at gas engines, out-performing the conventional least squares error optimization significantly. The weighted version (in order to locally estimate the linear parameter for each rule separately) is called *weighted total least squares (wTLS)* optimization problem and for the ith rule (local linear model) defined by:

$$J_5(i) = \frac{1}{2N}\left(\sum_{k=1}^{N}\Psi_i(\mathbf{x}(k))(\mathbf{r}_k - \hat{\mathbf{r}}_k)^2 + \sum_{k=1}^{N}\Psi_i(\mathbf{x}(k))(y(k) - \sum_{i=1}^{C} l_i \Psi_i(\mathbf{x}(k)))^2\right) \quad (2.100)$$

with $\mathbf{r}$ the regressors which are equal to the original input variables (as defined in (2.57)). An extended version of above optimization problem is demonstrated in [192], which is called *generalized total least squares (GTLS)*. This method is based on a combination of QR factorization, generalized singular value decomposition (GSVD) and Householder transformation and only reconstructs noisy components (inputs/outputs) present in the data.

Concluding, we see it as a challenge to develop resp. apply optimization algorithms for linear and non-linear parameters in fuzzy systems using these extended optimization functions as defined above instead of pure least squares error. This is even more essential in evolving fuzzy systems, as there a time-intensive rebuilding process for model evaluation and assessment (as done in CV procedure) is not possible, as data has to be permanently included into the models as it is incoming. Concretely speaking, a *recursive (combined) extended least squares* or a *recursive total least squares* approach with similar convergence properties as conventional *recursive least squares* would be beneficial and of great importance to achieve enhanced noise treatments (also with respect to inputs/regressors) and to prevent evolving fuzzy systems from severe over-fitting. All these considerations can be also made when using local learning by weighting data samples with membership fulfillment degrees to the various rules (and including these weights in the optimization problems).

Finally, a further (more heuristic) strategy for preventing over-fitting of evolving fuzzy systems within rule merging and pruning concepts will be handled in Section 5.1, Chapter 5.

Chapter 3
EFS Approaches for Regression and Classification

Abstract. This chapter deals with an analytical description and comparison of several well-known evolving fuzzy systems approaches. Even though most of these methods were originally designed for regression problems, some of these were also extended to classification tasks. The list of methods we are dealing with includes

- the *FLEXFIS* family: *FLEXFIS = FLEXible Fuzzy Inference Systems* for regression and *FLEXFIS-Class = FLEXible Fuzzy Inference Systems for Classification* applied on two different classifier architectures plus spin-off *eVQ-Class* as rule-based classifier in the cluster space (Section 3.1)
- the *eTS* family: *eTS = evolving Takagi-Sugeno fuzzy systems* and *eClass = evolving fuzzy Classifiers* coming in four different variants) (Section 3.2
- *ePL = evolving Participatory Learning* (Section 3.3)
- *DENFIS = Dynamic Evolving Neuro-Fuzzy Inference Systems* (Section 3.4)
- *SOFNN = Self-Organizing Fuzzy Neural Network* (Section 3.5)
- *SAFIS = Sequential Adaptive Fuzzy Inference System* (Section 3.6)
- other approaches such as *SONFIN*, *GD-FNN*, *ENFRN*, *SEIT2FNN* and *EFP* (Section 3.7)

Most of these approaches rely on some methods demonstrated in Chapter 2 for calculating statistical measures and (non-)linear parameters in incremental manner. However, each one of these exploits apply different mechanisms for rule evolution and incremental learning of antecedent parameters and also combine these differently with the learning scheme of the consequent parts. In this sense, we will give a detailed description of the specific characteristics of the various methods.

E. Lughofer: Evolving Fuzzy Systems – Method., Adv. Conc. and Appl., STUDFUZZ 266, pp. 93–164.
springerlink.com

3.1 The *FLEXFIS* Family

The *FLEXFIS* family comes with two variants, one for regression which is simply called *FLEXFIS* and short for *FLEXible Fuzzy Inference Systems* [268] and one for classification which is called *FLEXFIS-Class* and short for *FLEXible Fuzzy Inference Systems for Classification* [276]. We're going to start with the description of the regression variant which will then lead to the classification case.

3.1.1 FLEXFIS *for Regression*

FLEXFIS exploits the Takagi-Sugeno model architecture as defined in (1.2) by using a polynomial degree of order 1, hence linear consequent parameters, triggering hyper-planes locally defined for each rule. For an incremental learning of the linear consequent parameters, the local learning approach exploiting recursive fuzzily weighted least squares approach as demonstrated in (2.62) to (2.64) is used. This is because local learning has some significant advantages over global learning, especially the flexibility and the interpretability aspects, see also Section 2.4. Flexibility will be a major aspect in this section, as new rules may evolve on demand based on the characteristics of new data samples. The question is now how to evolve new rules and how to update the antecedent parameters therein in order to find a reasonable tradeoff between stability (convergence to a solution) and plasticity (including new information into the model), also called plasticity-stability dilemma . This will be treated in the subsequent section. Afterwards, it will be explained how to connect rule evolution and antecedent learning with the recursive consequent learning without losing incrementality and stability of the solution obtained from recursive weighted least squares approach. This will lead directly to the *FLEXFIS* algorithm for evolving TS fuzzy systems in incremental on-line manner. In Figure 3.1 an overview of the process chain for *FLEXFIS* is presented. Each newly loaded buffer or single sample is processed first through the pre-processing component (e.g. normalization of the data), then through the antecedent learning and rule evolution part and finally through learning and adaptation of consequent parameters by including correction terms.

3.1.1.1 Incremental Learning of Antecedent Parameters and Rule Evolution

The Basic Concept

The premise parameter and rule learning take place with the usage of clustering the input/output space (also called product space clustering [29]) into local parts and projecting these clusters in order to form the premise parts of the rules (i.e. fuzzy sets) and the rules themselves. A three-dimensional example (including 2 inputs x_1 and x_2 and 1 output y) of this approach is visualized in Figure 3.2: three clusters are projected to two axis (x_1 and x_2) forming the input space and the premise part of the rules, the remaining variable y is that one which should be approximated.

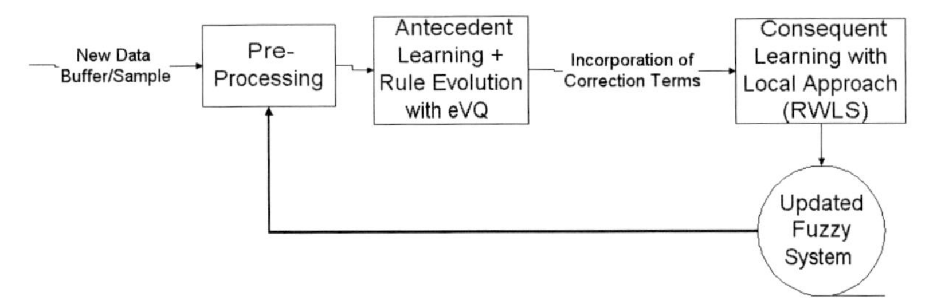

Fig. 3.1 Process chain for incremental learning and evolving Takagi-Sugeno Fuzzy Models with *FLEXFIS*, the updated fuzzy system together with the new data block/sample is sent into the next incremental learning cycle (indicated by the feedback arrow)

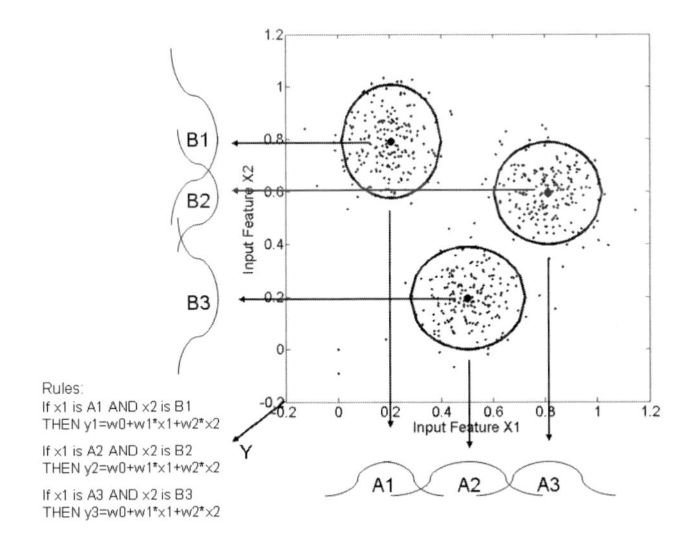

Fig. 3.2 Projection of three clusters onto the input axes x_1 and x_2 to form six fuzzy sets (three per input dimension) and three rules

Here, the focus is placed on regression problems, i.e. the output variable y is continuous, and the clusters are for the purpose to generate local partitions (rules) of the input/output space in order to obtain flexibility in the approximation models, i.e. one cluster is associated with one rule. This projection of clusters to the one-dimensional inputs is accomplished in various batch learning approaches as well, e.g. *genfis2* [454] (using *subtractive clustering* [83] [332]), *FMCLUST* [29] (using *Gustafson-Kessel clustering* [158]) or *genfis3* (based on the *fuzzy c-means* approach [46]) and hence a well-studied possibility for extracting local regions from the complete data space (note that all these approaches are also available in MATLAB's fuzzy logic toolbox). In fact, in many regression problems, in principal there do not occur any distinct clusters, as the data samples follow a basic linear or non-linear

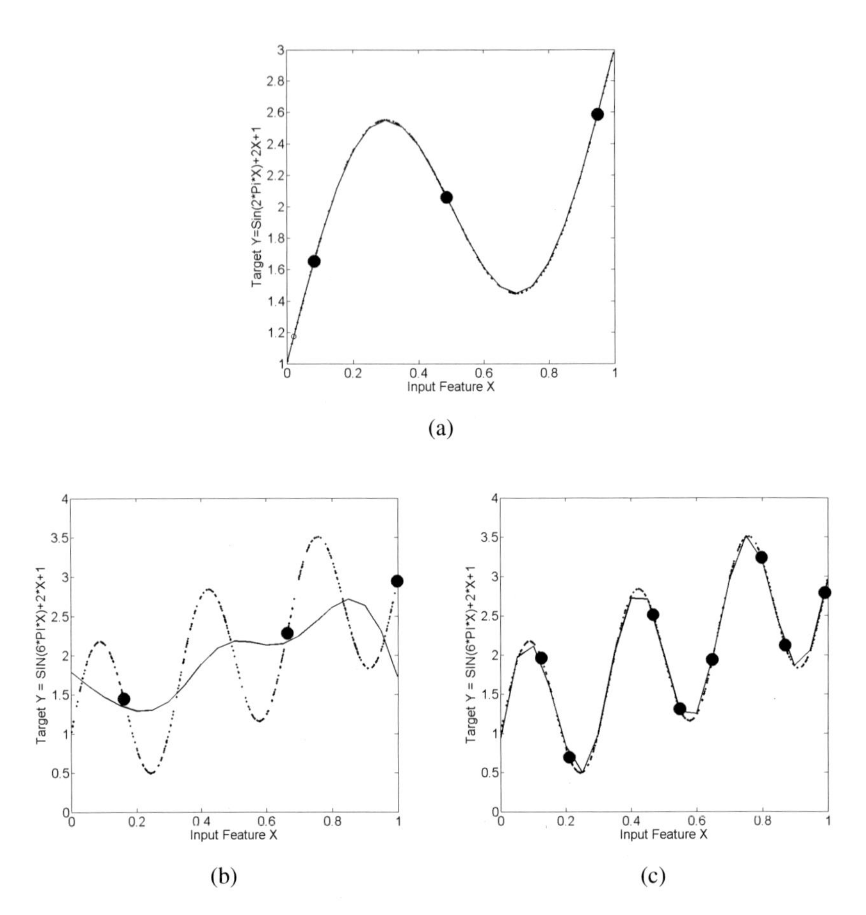

Fig. 3.3 (a): accurate fuzzy model (solid line) obtained with a low number of clusters (3 in sum marked by big black dots) in case of a low non-linear relationship (note that for each up/down-trend of the curvature one rule is sufficient; (b): bad fuzzy model (solid line) obtained with a low number of clusters (3 in sum marked by big black dots) in case of a high non-linear relationship (a lot of up/down-trends are not covered by a rule); (c): increasing the number of clusters = rules to 8 helps improving the quality of the fuzzy model

trend of the relationship(s) to be approximated. In this sense, clustering here is used just for the purpose to get any reasonable partition of the input/output space in form of partial local functions and the higher the non-linearity of the regression problem is, the finer the partition should get, hence the more clusters should be generated. For instance, consider up-and-down fluctuations one time with a low, the other time with a high frequency as shown in Figure 3.3: in the case of a low frequency a low number of clusters/rules are sufficient for an accurate approximation (a) (the three cluster centers marked as big black dots, the approximation curve exactly lying on the samples), whereas in case of a high frequency a low number of clusters/rules leads to a bad approximation (solid line in (b)); hence, increasing the number of clusters/rules is necessary to obtain a good accuracy (solid line in (c)) — achieved by using 8 rules instead of 3.

Vector Quantization and its Characteristics

In *FLEXFIS*, vector quantization [150] is exploited for clustering tasks and extended to an incremental and evolving variant (next paragraph). The purpose of vector quantization originally stems from encoding discrete data vectors to compress data which has to be transferred quickly e.g. for online communication channels. The prototypes, here called code-book vectors, are the representatives for similar/nearby lying data vectors and are shifted through the data space on a sample-per-sample basis. Let the dimensionality of the data to be clustered be $p+1$ (p input, one output dimension). The number of code-book vectors (clusters) C has to be parameterized a-priori, where each cluster has $p+1$ parameters corresponding to the $p+1$ components of each cluster center. With these notations and assuming that the input data is normalized due to its range the algorithm for vector quantization can be formulated as in Algorithm 2.

Algorithm 2: Vector Quantization

1. Choose initial values for the C cluster centers $\mathbf{c}_i, i = 1, ..., C$, e.g. simply by taking the first C data samples as cluster centers (code-book vectors).
2. Fetch out the next data sample of the data set.
3. Calculate the distance of the current data sample to all cluster centers by using a predefined distance measure.
4. Elicit the cluster center which is closest to the data sample by taking the minimum over all calculated distances $\rightarrow$ winning cluster represented by its center c_{win}
5. Update the $p+1$ components of the winning cluster by moving it towards the selected point $\mathbf{x}$:
$$\mathbf{c}_{win}^{(new)} = \mathbf{c}_{win}^{(old)} + \eta(\mathbf{x} - \mathbf{c}_{win}^{(old)}) \tag{3.1}$$
where the step size $\eta \in [0,1]$ is called the learning gain. A value for η equal to 1 would move the winning cluster exactly to the selected data sample in each iteration, a value near 0 would not give a significant change for the cluster center over all iterations. A decreasing learning gain with the number of iterations is a favorable choice for achieving a good convergence, see [320].
6. If the data set contains data samples which were not processed through steps 2 to 5, goto step 2.
7. If any cluster center was moved significantly in the last iteration, say more than ε, reset the pointer to the data buffer at the beginning and goto step 2, otherwise stop

A key issue of the vector quantization algorithm is that for each data sample only the winning cluster, i.e. the cluster nearest to this data sample with respect to a pre-defined distance measure, is moved.

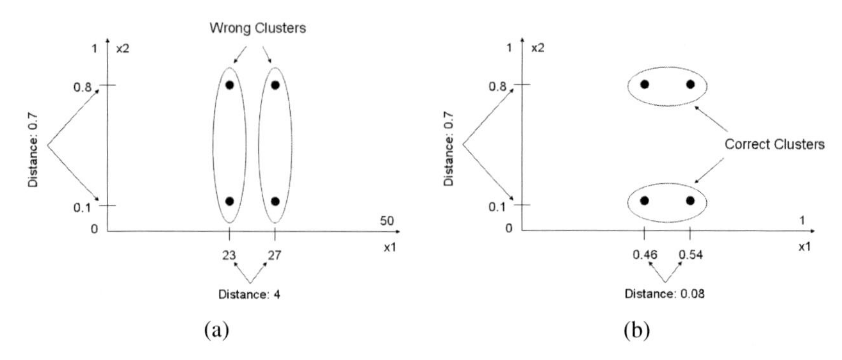

Fig. 3.4 (a): clusters obtained when using original data → two clusters are incorrectly extracted because absolute values are used in the distance calculations, hence feature x1 dominating over feature x2; (b): clusters obtained when using normalized data, dominance of feature x1 lost → extraction of correct clusters as both features are contributing equally in the distance calculations

A priori normalization of the data samples is necessary, as otherwise variables with higher ranges would dominate variables with lower ones when calculating the distance between samples and cluster centers, no matter which distance measure is used. This aspect is called deformation of the data space and is underlined in Figure 3.4 (a): for the calculation of distances in the original data space the values of the x-axis (lying in the range of [0,50]) dominate the values of the y-axis (lying in the range of [0,1]) in such a way that wrong clusters are obtained (the y-coordinate becomes almost irrelevant when calculating the distances). This effect can be prevented when normalizing the variables to $[0,1]$ as shown in Figure 3.4 (b).

From a mathematical point of view, vector quantization is basically a simplified version of k-means clustering [9] with sample-wise adaptation of the cluster centers (as done in (3.1)). However, the adaptation is done in several iterations over the complete data set, hence the method is not a single-pass incremental method. On the other hand, if several iterations were carried out for each incremental learning step i.e. for each actual loaded data block or even for each single sample, the cluster centers would only represent a reliable partition of this data block/sample and forget the older data completely. Furthermore, an appropriate number of cluster centers needs to be parameterized a-priori and kept fixed which means that the method is not able to evolve new clusters on demand based on the characteristics of new data samples. Moreover, only cluster centers/rule centers are learned, but no ranges of influence of the clusters. The latter are necessary in order to be able to project fuzzy sets onto the one-dimensional axis.

Evolving Vector Quantization

The conventional vector quantization is extended to a novel single-pass incremental and evolving variant, called *eVQ = evolving Vector Quantization*, which is characterized by three main issues:

- Ability to update clusters on a single-pass sample-wise basis without iterating over a data buffer multiple times.
- Ability to evolve clusters in a single-pass incremental manner without the need of pre-parameterizing the number of clusters.
- Calculating the range of influence of clusters in each direction incrementally and synchronously to the cluster centers; these ranges are used as widths for the fuzzy sets projected onto the axes.

There are basically two concepts responsible for achieving the first two issues.

The first one is given by a specific handling of the learning gain η as it steers the degree of shifting the centers and is responsible for a convergence of the rule centers. In *eVQ* not a global learning gain decreasing with the number of iterations is applied, but different learning gains $\eta_1, ..., \eta_C$ for the different clusters according to their significance. The significance of the ith cluster can be expressed by the number of samples which formed this cluster, i.e. the number of samples for which the ith cluster was the winning cluster. Hence, $\eta_i, i = 1, ..., C$ are updated by:

$$\eta_i = \frac{init\_gain}{k_i} \tag{3.2}$$

with k_i the number of data samples belonging to cluster i and $init\_gain$ a parameter which is not sensitive and usually set to 0.5. In this sense, monotonic decreasing learning gains are achieved with respect to the number of data samples which formed each cluster so far. This guarantees a convergence of the cluster prototypes to the centers of the local regions they were originally defined in.

Evolution of new clusters is achieved by introducing a so-called vigilance parameter ρ (motivated by the ART approach [66]) steering the tradeoff between generating new clusters (stability as this does not change any already learned cluster) and updating already existing ones (plasticity as adapting to new information according to (3.1)) . The condition for evolving a new cluster becomes:

$$\|\mathbf{x} - \mathbf{c}_{win}\|_A \geq \rho \tag{3.3}$$

where A is a specific distance measure. If (3.3) is fulfilled, the prototype c_{C+1} of the new (the $C+1$th) cluster is set to the current data sample. The condition (3.3) checks, whether the distance of a new data sample to the nearest center is higher than a predefined value ρ within the unit hypercube $[0,1]^{p+1}$. In [267] it is reported that the following choice of this parameter should be preferred:

$$\rho = fac * \frac{\sqrt{p+1}}{\sqrt{2}} \tag{3.4}$$

The dependency of ρ on the $p+1$-dimensional space diagonal can be explained with the so-called *curse of dimensionality* effect (see also Section 2.8): the higher the dimension, the greater the distance between two adjacent data samples; therefore, the larger the parameter ρ should get in order to prevent the algorithm from generating too many clusters and causing strong over-fitting effects. Finally, the parameter fac manages the number of clusters evolved, typical settings for achieving good performance are in the range $[0.2, 0.3]$ (see [267]), however this also depends on the actual application (characteristics of the data stream). In *FLEXFIS* and its classification variants *FLEXFIS-Class* and *eVQ-Class*, an option is provided where the vigilance parameter is automatically tuned based on some initial data samples.

Another important fact is that Condition 3.3 assures that already generated clusters are moved in local areas bounded by the vigilance parameter ρ. Furthermore, a cluster center is never initialized far away from the middle of a new upcoming data cloud (opposed to conventional *VQ* where prototypes could be initialized somewhere in the space and need to converge to the density points over more iterations). Thus, one iteration over an incrementally loaded data buffer or a single sample is quite reliable for a proper performance. This concept can also be seen as a kind of regularization approach in the clustering space.

Combining these concepts yields the (basic) evolving Vector Quantization (*eVQ*) algorithm as defined in Algorithm 3 (see also [267]).

Algorithm 3: Evolving Vector Quantization (Basic)

1. Initialize the number of clusters to 0.
2. Either ranges are pre-known or collect a few dozen data samples and estimate the ranges of all $p+1$ variables.
3. Take the next incoming data sample $\mathbf{x}$ and normalize it to $[0,1]$ according to the ranges of the variables; in the same way, normalize centers and widths of the clusters to $[0,1]$.
4. **If** number of clusters = 0

   a. Set $C = 1$, $k_1 = 1$.
   b. Set the first center $\mathbf{c}_1$ to the current data sample, i.e. $\mathbf{c}_1 = \mathbf{x}$.
   c. Goto Step 10.

5. Calculate the distance of the selected data sample to all cluster centers by using a predefined distance measure.
6. Elicit the cluster center which is closest to $\mathbf{x}$ by taking the minimum over all calculated distances $\rightarrow$ winning cluster represented by its center $\mathbf{c}_{win}$.
7. **If** $\|\mathbf{x} - \mathbf{c}_{win}\|_A \geq \rho$

   a. Set $C = C + 1$
   b. Set $\mathbf{c}_C = \mathbf{x}$, $k_C = 1$
   c. Goto Step 10.

8. Update the $p+1$ components of the winning cluster by moving it towards the selected point $\mathbf{x}$ as in (3.1) with adaptive learning gain as in (3.2).
9. $k_{win} = k_{win} + 1$.
10. Update the ranges of all $p+1$ variables.
11. If new data samples are coming in goto step 3, otherwise stop.

An essential issue therein is that the ranges of the variables either have to be known or to be pre-estimated from the first data blocks. Otherwise, the distances and furthermore the winning clusters are not calculated correctly (same problem as in *VQ* and shown in Figure 3.4). The ranges are further extended in Step 10 in order to compensate an inaccurate initial range estimation and preventing over-clustering within the (initial) small fraction of the data space. Another important issue is the handling of outliers in order to distinguish them from new operating conditions and hence not evolving new clusters whenever Condition (3.3) (in Step 7) is fulfilled. This will be analyzed in more detail in Section 4.4.

Figure 3.5 demonstrates the impact of different settings of the vigilance parameter ρ (0.06, 0.1 and 0.2) for a two-dimensional data set containing 17 distinct and clearly visible clusters: whereas in the first case (a) too many clusters are generated (over-clustering: some compact regions are split up in two or more clusters), in the third case (c) too few cluster are generated (under-clustering: some clearly distinct regions are joined together). Figure 3.5 (b) gives the correct cluster partition. Therein, a light sphere around the bottom left cluster indicates the "movement horizon" of the cluster center; a sample appearing outside this sphere would trigger a new cluster center through Condition (3.3). Figure 3.5 (d) is for comparison purposes with the conventional *VQ*: here, someone can see that the original batch approach (with 17 initial centers) performs even worse than the incremental approach.

Rule Evolution

Rule evolution can be defined as a generation of a new fuzzy rule on demand based on the characteristics of new incoming data samples.

In our case, rule evolution is related to the evolution of new clusters as done in Algorithm 3. In order to be able to perform a reliable rule evolution, the multidimensional clusters are projected onto the input axes to obtain 1-dimensional (Gaussian) fuzzy sets in the antecedent parts of the rules. Hereby, not only the rule (=cluster) centers, but also the ranges of influence of the clusters have to be calculated in incremental manner. When using the Euclidean distance measure (for calculating distances in Algorithm 3) triggering ellipsoids in main position, the ranges of influence can be estimated by calculating the variance of the cluster in each direction by using those data samples for which this cluster was the winning cluster. From a theoretical point of view, this is justified, because the variance is a good estimator for parameter σ in a normal distribution $N(\mu,\sigma)$ described by a Gaussian-shaped function. The incremental mechanism for updating the variance is given by formula (2.9) (please see Section 2.1.2 for a detailed derivation); the only modification on this formula required is that instead of using N denoting the number of samples seen so far, the number of samples forming cluster i, k_i, is applied. In other words, for the ith cluster the recursive range of influence update formula is obtained:

$$(k_i+1)\sigma_{ij}^2 = k_i\sigma_{ij}^2 + (k_i+1)\Delta c_{ij}^2 + (c_{ij}-x_{kj})^2 \quad \forall j=1,...,p+1 \tag{3.5}$$

where Δc_{ij} is the distance of the old prototype to the new prototype of the corresponding cluster in the jth dimension. For each new data sample this update formula

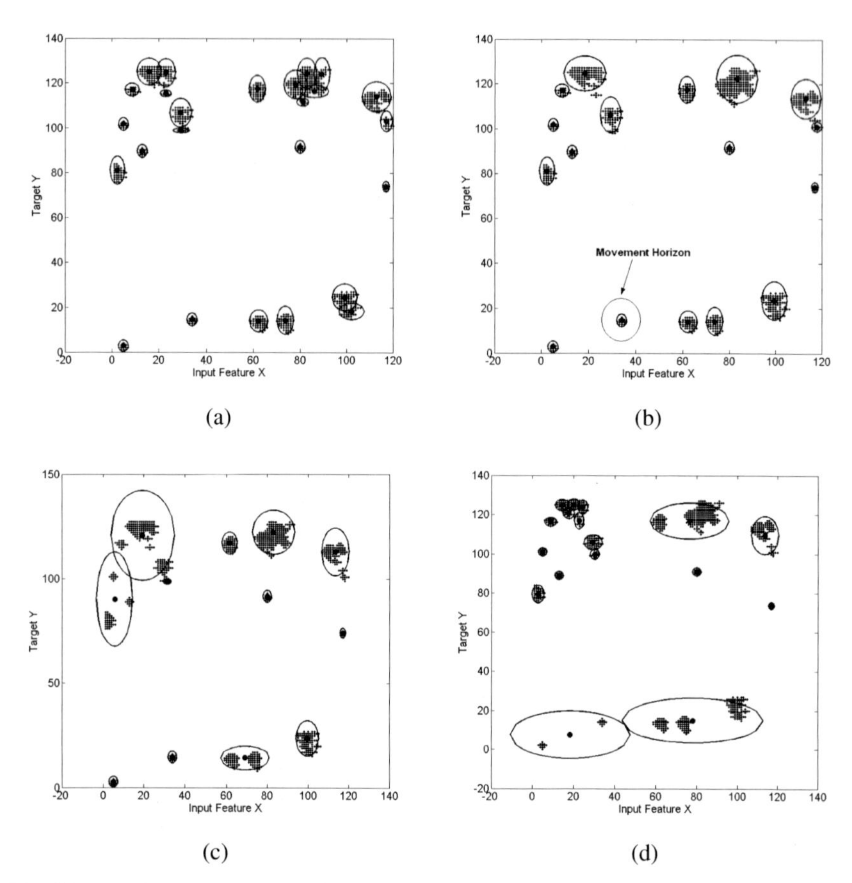

Fig. 3.5 (a): clustering obtained by *eVQ* when setting the vigilance parameter to 0.06 ($\rightarrow$ over-clustering); (b): clustering obtained by *eVQ* when setting the vigilance parameter to 0.1 ($\rightarrow$ appropriate clustering with 17 clusters); (c): clustering obtained by *eVQ* when setting the vigilance parameter to 0.2 ($\rightarrow$ under-clustering); (d): clustering obtained by conventional *VQ* when setting the number of clusters to 17 and the initial centers to the some random data samples; ranges of influence of the clusters in all images are indicated by ellipsoids

is performed only for the winning cluster, hence by substituting index win with i in the above formula. Now, whenever a new cluster is evolved according to Step 7 in Algorithm 3, a new rule is also evolved by initializing the parameters c and σ of the antecedent fuzzy sets $\mu_{(C+1),.}$ of the $C+1$th rule in the following way (fuzzy set innovation):

$$\begin{aligned} c_{(C+1),j}(sets) &= x_j \quad \forall j \in \{1,...,p\} \\ \sigma_{(C+1),j}(sets) &= \varepsilon * range(\mathbf{x_j}) \quad \forall j \in \{1,...,p\} \end{aligned} \tag{3.6}$$

with x_j the jth dimension of the current data sample, and by innovating the new rule as

$$\text{Rule}_{C+1}: \quad \text{IF } x_1 \text{ IS } \mu_{(C+1),1} \text{ AND } x_2 \text{ IS } \mu_{(C+1),2} \text{ AND...}$$
$$\text{AND } x_p \text{ IS } \mu_{(C+1),p} \text{ THEN}$$
$$y = x_{p+1} = w_{(C+1)0} + w_{(C+1)1}x_1 + \ldots + w_{(C+1)p}x_p \tag{3.7}$$

where the linear weights $w_{(C+1)j}$ are set initially to 0 for convergence reasons (see Section 2.2.2). Note that the width of the Gaussian fuzzy sets are not set to 0, as this would lead to instabilities for the first data samples until the width is significantly enlarged through (3.5). However, in order to ensure an exact update of the range of influences, the widths will be initialized with 0 in the cluster space, but during the projection the widths of the fuzzy sets are set by $\sigma_{.,j}(sets) = max(\varepsilon * range(\mathbf{x_j}), \sigma_{.,j}(cluster))$.

Moreover, including the variance update into Algorithm 3 leads us to *eVQ* with variance update as demonstrated in Algorithm 4.

Algorithm 4: Evolving Vector Quantization with Variance Update

1. Initialize the number of clusters to 0.
2. Either ranges are pre-known or collect a few dozen data samples and estimate the ranges of all $p+1$ variables.
3. Take the next incoming data sample (let's call it $\mathbf{x}$) and normalize it to $[0,1]$ according to the ranges of the variables; in the same way, normalize centers and widths of the clusters to $[0,1]$.
4. **If** number of clusters = 0

   a. Set $C = 1$, $k_1 = 1$.
   b. Set the first center $\mathbf{c}_1$ to the current data sample, i.e. $\mathbf{c}_1 = \mathbf{x}$.
   c. Set the range of influence of this cluster to 0, i.e. $\sigma_1 = 0$.
   d. goto step 12.

5. Calculate the distance of the selected data sample to all cluster centers by using a predefined distance measure.
6. Elicit the cluster center which is closest to $\mathbf{x}$ by taking the minimum over all calculated distances $\rightarrow$ winning cluster represented by its center $\mathbf{c}_{win}$.
7. **If** $\|\mathbf{x} - \mathbf{c}_{win}\|_A \geq \rho$

   a. Set $C = C + 1$.
   b. Set $\mathbf{c}_C = \mathbf{x}$, $k_C = 1$.
   c. Set $\sigma_C = 0$.
   d. Goto step 11.

8. Update the $p+1$ components of the center of the winning cluster by moving it towards the selected point $\mathbf{x}$ as in (3.1).
9. Update the $p+1$ components of the range of influence of the winning cluster by using (3.5).
10. $k_{win} = k_{win} + 1$.
11. Update the ranges of all $p+1$ variables.
12. If new data samples are coming in goto step 3, otherwise stop.

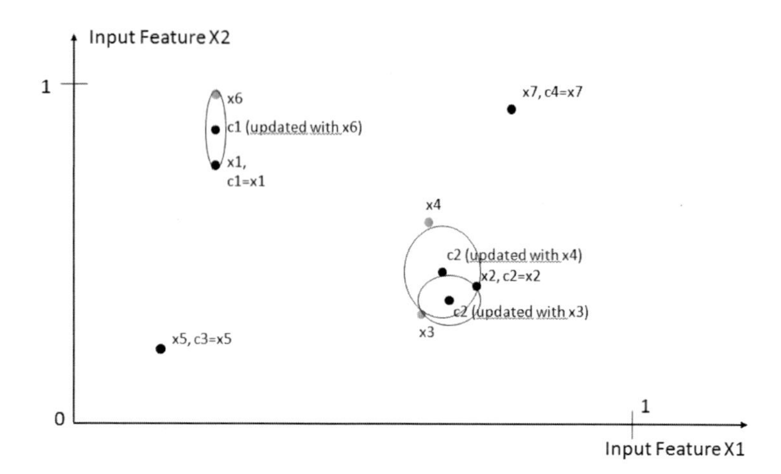

Fig. 3.6 Cluster update progress (centers c1 to c4 and ranges of influence) in *eVQ* approach by loading the first seven samples (x1 to x7) in incremental mode into the algorithm

Figure 3.6 visualizes the update progress of *eVQ* (cluster centers and ranges of influence) for the first seven samples of an artificial data set: four clusters are evolved by samples x1, x2, x5 and x7 as lying significantly away from previous generated ones, the other samples are enforcing movements of already generated clusters and expanding their ranges of influences as updated through variances along the single dimensions; also note that the ranges of influence of clusters containing just one single sample are 0, but will innovate fuzzy sets with ε according to (3.6).

3.1.1.2 Combining Antecedent Learning and Rule Evolution with Consequent Learning

This section deals mainly with an appropriate connection of premise parameter/rule learning and incremental adaptation of rule consequent parameters, which guides the latter towards an optimal solution in the least squares sense. Furthermore, the principal possibility as well as pros and cons for an incremental learning resp. evolution from scratch will be discussed. A formulation of the *FLEXFIS* algorithm will be given in Section 3.1.1.3, followed by a discussion on the correction term problematic in Section 3.1.1.4 and an evaluation of *FLEXFIS* on the two-dimensional 'failure cases' demonstrated in Section 2.5.

Combination Strategy

An obvious and straightforward way for doing a reasonable connection would be first to update the cluster partition with Algorithm 4, to project the updated cluster partition onto the axes updating fuzzy sets and rules, and then to adapt the consequent parameters of the rules. However, structural changes in the antecedents may lead to non-optimal solutions in the consequent parameters (as updated by *RFWLS*). When doing an update step of the premise part, two possibilities may arise (assuming that the currently loaded sample is denoted by $\mathbf{x}$):

1. If a new cluster needs to be set (as condition (3.3) is fulfilled), automatically a new rule is set using (3.7) and (3.6) without disturbing the consequent parameter in any of the other rules. This is because local learning approach is used, which applies a *recursive fuzzily weighted least squares* approach converging to the optimal parameters in the least squares sense for each rule separately (see also Sections 2.4 and 2.3.2): a new rule simply evolves a new recursive weighted least squares estimate for its consequent parameters.
2. If no new cluster needs to be set, the cluster center $\mathbf{c}_{win}$ and range of influence σ_{win} of the winning cluster are updated according to (3.1) and (3.5). Thus, a change of the fuzzy sets corresponding to the winning cluster makes the prior estimated linear consequent parameters always non-optimal for this changed fuzzy partition, as these were optimized for the old fuzzy set positions. For clarification of this point the following explanation is given: Optimality is expressed in terms of the weighted least squares error measure between measured and estimated target values which yields a quadratic optimization problem with respect to linear consequent parameters for each cluster (=rule). Then, the quadratic optimization function as demonstrated in (2.59) is used. In the batch mode case an analytical (exact) solution exists, see (2.61), when applying Q as weighting matrix as in (2.60). In the incremental case, *recursive weighted least squares* possess the favorable property to converge to the optimal parameter solution within each update step, i.e. for each newly loaded data sample. However, this is only true when the weights of older points $\Psi_{.}(\mathbf{x}(1)),...,\Psi_{.}(\mathbf{x}(k))$ remain unchanged. This is not the case for that rule belonging to the winning cluster of a certain new data sample as fuzzy sets in the antecedent part are moved by Algorithm 4 (called *structural change*). Hence, a so-called correction vector δ for the linear parameter update (2.62) and a correction matrix Δ for the inverse Hessian matrix update (2.64) are introduced [268] in order to balance out this non-optimal situation towards the optimal one. In this sense, before updating rule consequent parameters of the winning rule, they are added to linear consequent parameters as well as to the inverse Hessian matrix. The terminus correction term will be used when addressing both (correction vector and matrix) at the same time.

Discussion on Incremental Learning from Scratch

Another consideration deals with the question whether a rule evolution and incremental antecedent and consequent learning from scratch is 1.) possible and 2.) beneficial. Incremental learning/rule evolution from scratch means that no initial data is collected from which an initial fuzzy model is built, but the structure of the fuzzy system is evolved by starting with an empty rule base and adding successively more and more rules.

The possibility in *FLEXFIS* approach is guaranteed as soon as the ranges of the variables are either a-priori known or estimated from the first dozens of data samples as done in *eVQ* approach (Step 2 in Algorithms 3 resp. 4). The necessity of the availability of range estimations is obvious when considering that at the beginning of the incremental learning process too many rules may be evolved in a small fraction of the complete data space. This is because the ranges may be quite small for the first samples, not representing the fully-spanned space; however, the *eVQ* algorithm only sees this tight range and would evolve clusters (rules) in the same way as when having the full range available. Later, when the range is extended, samples representing distinct local regions in the original space may fall together to one cluster. Hence, the initial small fraction of the data space (covered by the first samples) will be over-clustered. An example of such an occurrence is demonstrated in Figure 3.7, where in (a) the original data distribution and cluster partition is shown, whereas in (b) the same features are significantly extended in their range: two new clusters outside the original range are indeed adjoined, but the initial clusters turn out to be too small and too many as dividing the left most cluster into four parts. On the other hand, in the original view (Figure 3.7 (a)) four clusters seemed to be appropriate. In fact, if the real range were available from the beginning, one cluster would be extracted correctly for the left most data cloud (as all samples lie in the sphere around the center spanned up by the vigilance parameter value as a (now correct) fraction of the diagonal), see Figure 3.7 (c), resulting in the correct cluster partition as shown in Figure 3.7 (d). This whole problematic is also called the *dynamic range problematic*.

On the other hand, the benefit of a "from-scratch approach" is controversial. Indeed, this approach yields a nice feature towards more automatization in from of a kind of plug-and-play on-line learning method. However, there are some essential disadvantages to be considered:

- Parameters of the learning method have to be set to some start values before starting on-line process (e.g. from past experiences, test runs). Although there may be some fruitful guidelines, there is no guarantee that they perform properly for new applications resp. new on-line data streams.
- When designing a fuzzy model for a new application scenario, often it is not a priori clear whether there is a feasible solution in form of a high-qualitative model at all. Usually, first off-line data samples are collected and initial models trained in order to see whether the problem is "modelable" at all with sufficient accuracy (often experts require a minimal reliability of the model). This means that a plug-and-play evolving method is often quite hypothetical in practical usage.
- Even though the modelability of a problem is known (due to some pre-studies, -simulations or funded expert knowledge about the process), a fully incrementally built up fuzzy model is quite often too risky for the experts as not allowing sufficient insight into the process models as permanently changing its structure and parameters from the beginning. An initial model with a fixed structure can be first supervised by an expert and upon agreement further evolved, refined etc. to improve its accuracy. Based on our experience with real-world on-line applications, this is a widely-accepted procedure.

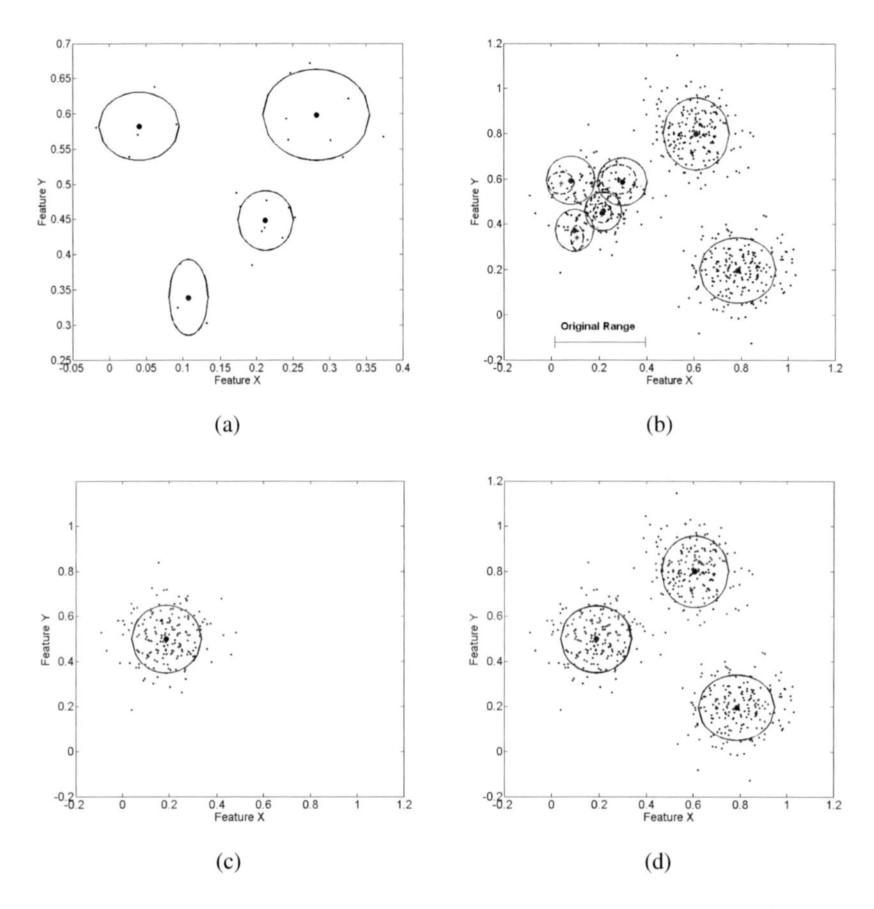

Fig. 3.7 (a): clustering obtained for the first 30 samples based on the initial range $[-0.05, 0.4]$ for x and $[0.25, 0.7]$ for y; (b): clustering obtained when updating old clusters and adjoining new ones in the extended range (note the over-clustering effect for the left most data cloud as the original range was too small, the four clusters in the left most data cloud correspond to those shown in (a), compare dotted ellipsoids (original clusters) versus solid ones (updated clusters); (c): clustering obtained for the left data cloud when using the full correct range $[-0.2, 1.2]$ for both features from the beginning $\rightarrow$ clusters are extracted correctly as shown in (d)

The first disadvantage can be circumvented when training an initial fuzzy model from off-line samples in a cross-validation (CV) procedure [411][1] connected with a best parameter grid search scenario. The essential parameters (here basically the vigilance parameter) can be varied in certain step-sizes and for each parameter setting

[1] The data set is divided into N equal folds, each of this fold is once selected as test data set, whereas the remaining $N-1$ folds are used as training data for learning a model; the accuracies/errors of the N achieved models over each left out fold are averaged yielding the CV accuracy/error.

a fuzzy model trained and evaluated with respect to accuracy in the CV procedure. The model with the smallest error rate can be used for further adaptation and evolution using the same parameter setting (as giving a first valuable guideline about the characteristics of the on-line samples). The other disadvantages are obviously compensated by an initial off-line model generation process.

Despite these disadvantages, in *FLEXFIS* both options are integrated. An initial fuzzy model may be trained either from simulation data in a different off-line process or based on the first dozens of samples collected during the on-line process. The training of an initial fuzzy model can be done with *FLEXFIS batch*, which is by sending all the data through *eVQ* (Algorithm 4), projecting all the obtained clusters onto the axes by (3.7) and then estimating the consequent parameters of all rules by weighted least squares approach (2.61) with all the data in one sweep.

3.1.1.3 *FLEXFIS* Algorithm

Summarizing the approaches and discussion points in the three previous sections, yields the formulation of the (native) *FLEXFIS* approach in Algorithm 5. For the setup phase we mention three options (as discussed in the previous subsection), where the first option is the most recommendable one (if applicable) due to the consideration made above.

Algorithm 5: FLEXFIS = FLEXible Fuzzy Inference Systems

1. **Setup Phase:**

   Option 1 (usage of fully off-line trained fuzzy model): Read in an initial fuzzy model trained fully in off-line mode and evaluated through a CV procedure coupled with a best parameter grid search scenario, read in the range estimates on all variables, set ρ to the optimal value obtained from the parameter grid search.
   Option 2 (initial fuzzy model estimated from first on-line samples): Collect k data samples sufficient enough in order to yield a good approximation without over-fitting; a reasonable value for k is $max(4p^2,50)$ with p the number of input variables, see [268]. From these collected k data samples generate an initial fuzzy model with *FLEXFIS* in batch mode, also estimate the ranges of all input and output variables. If the data is coming-in in a quite slow frequency during online operation, cross-validation with best parameter grid search may be an option as well. Then, use the optimal ρ for further evolution.
   Option 3 (evolution from scratch): Start with an empty rule-base (set $C=0$), either ranges are pre-known or collect a few dozen data samples and estimate the ranges of all input and output $(p+1)$ variables, set ρ to a good guess from past experience (in the range $[0.2,0.3]\frac{\sqrt{p+1}}{\sqrt{2}}$).
2. Take the next incoming data sample: $\mathbf{x}_{k+1}$
3. **If** $C=0$ (Option 3 was used in setup phase), perform the same Steps as in 6(a) to 6(g) without 6(d) and continue with Step 8.
4. Normalize cluster centers and widths as well as the current sample by ($i=1,...,C$, $j=1,...,p$)

$$c_{ij} = \frac{c_{ij} - \min(\mathbf{x}_j)}{\max(\mathbf{x}_j) - \min(\mathbf{x}_j)}, \quad \sigma_{ij} = \frac{\sigma_{ij}}{\max(\mathbf{x}_j) - \min(\mathbf{x}_j)}$$

$$x_{(k+1)j} = \frac{x_{(k+1)j} - \min(\mathbf{x}_j)}{\max(\mathbf{x}_j) - \min(\mathbf{x}_j)} \tag{3.8}$$

where $\min(\mathbf{x}_j)$ denotes the minimum and $\max(\mathbf{x}_j)$ denotes the maximum over all past samples for the jth variable (not including the current sample).

5. Perform Steps 5 and 6 of Algorithm 4.
6. **If** $\|\mathbf{x}_{k+1} - \mathbf{c}_{win}\|_A \geq \rho$

   a. Increase the number of clusters (i.e. rules) C, hence $C = C + 1$, and set $k_C = 1$
   b. Start a new cluster at the $k+1$th sample, hence $\mathbf{c}_C = \mathbf{x}_{k+1}$, where $\mathbf{x}$ includes all dimensions of the current sample.
   c. Set the width of the new cluster to zero, hence $\sigma_C = \mathbf{0}$.
   d. Transfer cluster centers and widths back to original range, hence ($i = 1,...,C$, $j = 1,...,p$)

$$c_{ij} = c_{ij}(\max(\mathbf{x}_j) - \min(\mathbf{x}_j)) + \min(\mathbf{x}_j)$$
$$\sigma_{ij} = \sigma_{ij}(\max(\mathbf{x}_j) - \min(\mathbf{x}_j)) \tag{3.9}$$

   e. Set the centers and widths of the new fuzzy sets (in all input dimensions) as in (3.6).
   f. Set the linear consequent parameter $\hat{\mathbf{w}}_C$ of the new rule C to $\mathbf{0}$.
   g. Set the inverse Hessian matrix $(R_C^T Q_C R_C)^{-1}$ of the new rule to αI.

7. **Else**

   a. Update the p center components of the winner cluster c_{win} by using the update formula in (3.1) with the choice of an adaptive η as described in (3.2).
   b. Update the p width components of winner cluster c_{win} by using variance update formula in (3.5).
   c. Transfer cluster centers and widths back to original range as in (3.9).
   d. Project winning cluster onto the axes to update the antecedent part of the rule belonging to the winning cluster (in the same way as shown in Figure 3.2).
   e. Correct the linear parameter vector of consequent functions and the inverse Hessian matrix of the rule corresponding to the winning cluster.

$$\hat{\mathbf{w}}_{win}(k) = \hat{\mathbf{w}}_{win}(k) + \delta_{win}(k)$$
$$P_{win}(k) = P_{win}(k) + \Delta_{win}(k) \tag{3.10}$$

8. Perform recursive fuzzily weighted least squares as in (2.62) and (2.64) for all C rules.
9. Update the ranges of all input and output variables.
10. If new incoming data samples are still available set $k = k+1$ and goto step 2, otherwise stop.

Please note that the centers and ranges of influence of the clusters are normalized in each incremental learning step in (3.8) as always transferred back to the original range in (3.9) — this is necessary in order to obtain fuzzy sets and rules which are defined in the original range and hence better interpretable as in the normalized range (usually, experts have the linguistic understanding of the process variables in their original range). Another key point is the update of the consequent parameters (Step 8) which is carried out for all C rules and not only for the rule corresponding to the winning cluster, increasing significance of all hyper-planes and finally of the whole model.

3.1.1.4 Discussion on the Correction Term Problematic (Towards Optimality in *FLEXFIS*)

The final open issue is how to estimate the correction vectors and matrices in (3.10). Obviously, these correction terms depend on the change extent of the winning cluster and influences all previously evaluated regressors and therefore can only be calculated by reusing the first k samples again. This would lead to the loss of the benefit of a single pass incremental learning algorithm applicable for fast on-line modelling processes, as the linear parameters have to be re-estimated in each incremental learning step. The following lemma gives a bound to these correction terms and ensures the convergence of the incremental learning algorithm.

Lemma 3.1. *Let δ_i with $i = 1,...,C$ be the correction vector for the independent linear parameter estimations of the C rules, then*

$$\forall i \quad \lim_{k_i \to \infty} \delta_i(k_i) = \mathbf{0} \tag{3.11}$$

i.e. the sequence of correction vectors for all rules tends to 0 when k_i goes to infinity.

The proof of this lemma exploits the fact that movements of cluster centers and widths are bounded by the vigilance parameter, for details see [268]. The extension of above lemma to the correction matrix for the inverse Hessian update is straightforward. The basic aspect of the lemma is that with a movement in a local area (bounded by vigilance parameter ρ) and the increase of significance of a cluster the change in the corresponding rule's premise parameters and therefore the sequence of correction terms converge to 0. This means that the algorithm finally leads to the *recursive weighted least squares* for each rule when k goes to infinity, which solves the least squares optimization problem in each sample. Furthermore, as the convergence is weakly monotonic, i.e. in fact the learning gain is monotonically decreasing and the movement always bounded by a (small) fraction of the feature space, the sum of correction terms is bounded by a constant, i.e.:

$$\sum_{i=1}^{\infty} \delta_i(k_i) < C \tag{3.12}$$

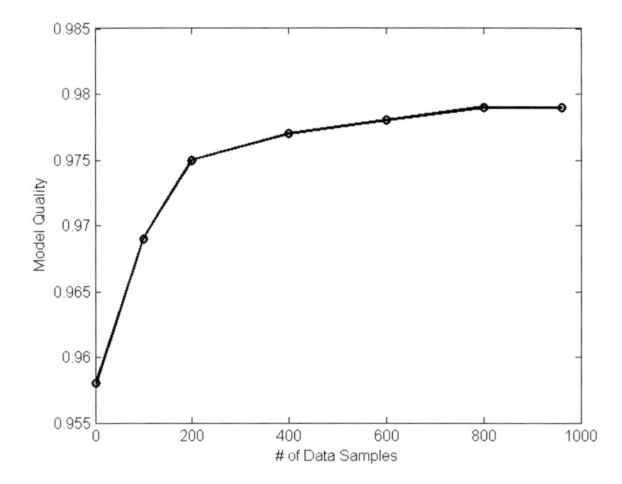

Fig. 3.8 Dependency between block size and average model quality when training with *FLEXFIS*, the circles denote the tested block sizes 1, 100, 200, 400, 600, 800, 960 (all points), the difference of model qualities between sample-wise incremental training (sample size = 1) and batch training (sample size = 960) is in a quite small range

Finally, this means that it is guaranteed that the algorithm does not 'break out' when recursively (sample-wise) applying (3.10), which leads to a near-optimality in the least squares sense, when setting the correction vectors artificially to $\mathbf{0}$.

Furthermore, for data blocks with more than one sample first antecedent learning with all the samples (Steps 3 to 7), then consequent learning with all the samples (Step 8) and then updating the ranges of all variables with all samples (Step 9) in a block are carried out by *FLEXFIS*. This yields a better convergence to the optimum for the linear consequent parameters. In fact, it can be elicited empirically that the larger the data blocks are, the closer optimality can be reached (also called convergence with increasing block size): Figure 3.8 shows an example of the dependency of model qualities (maximum quality = 1) on the number of data samples each block contains loaded into *FLEXFIS*; this example is drawn from a real-world identification problem and the model qualities are measured by r-squared-adjusted measure (2.18). The maximum of 960 data samples denotes the whole batch of data collected and *FLEXFIS batch* delivers the optimal solution in the least squares sense. It can be seen that the model quality of the sample-wise evolved model is not significantly worse than the quality of the model trained in batch mode (0.958 versus 0.978). This means, that the sample-wise evolved solution comes close to the optimal one. In this case, we also speak about a stable incremental learning approach (see also Section 1.2.2.2, equation (1.28) and explanation, or [273]).

From Lemma 3.1 it is easy to see that the larger k (and hence the larger the k_i's) gets for the initial generation of fuzzy models (Step 1 of Algorithm 5), the smaller the correction vector values and their sums get for the adaptation process.

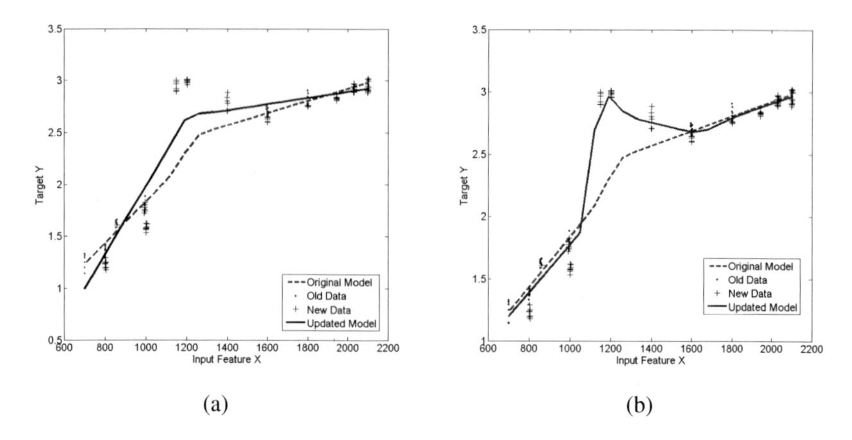

Fig. 3.9 (a): updated fuzzy model (solid line) based on new samples (crosses) by only using adaptation of consequent parameters → inaccurate model around the range of 1200 to 1400 originally not covered by data samples; (b): updated fuzzy model (solid line) by *FLEXFIS* → model has sufficient flexibility to go through the new data samples closing the former data hole.

In this sense, the choice of the number of data samples for initial training is a matter of a tradeoff between closeness to the optimum and an early applicability of the algorithm.

3.1.1.5 Evaluation on Two-Dimensional Data Sets

It is inspected how Algorithm 5 performs when new operating conditions appear in the data set resp. when large data holes are closed with new data samples. In Section 2.5 we demonstrated 'failure cases' when only the linear consequent parameters are updated in Takagi-Sugeno fuzzy systems (see Figures 2.8, 2.9 and 2.10), leading to bad approximations or even incorrect fuzzy models. This was one of the major motivations to design evolving fuzzy systems approaches which also evolve structural components on demand. When applying *FLEXFIS* algorithm as evolving fuzzy system approach (Algorithm 5) this drawback can be omitted, which is underlined in Figures 3.9, 3.10 and 3.11. The left images represent the 'failure cases' when doing conventional adaptation of consequent parameters for the initial generated models (dotted lines) based on the original data (marked by dots) (compare right images in Figures 3.9, 3.10 and 3.11), the right images visualize correct model updates (solid lines) achieved by *FLEXFIS* based on new incoming samples (marked by crosses).

More complex use cases on higher-dimensional data will be examined in the third part of the book.

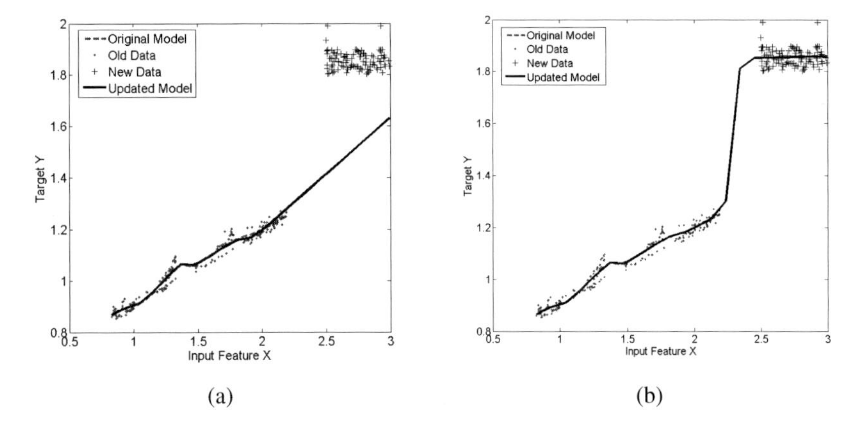

(a) (b)

Fig. 3.10 (a): updated fuzzy model (solid line) based on new samples (crosses) by only using adaptation of consequent parameters → the updated model misses the new data cloud completely. (b): updated fuzzy model (solid line) based on new samples (crosses) by applying *FLEXFIS* → appropriate inclusion of the new operating conditions appearing significantly outside the original ranges of the input and output variable.

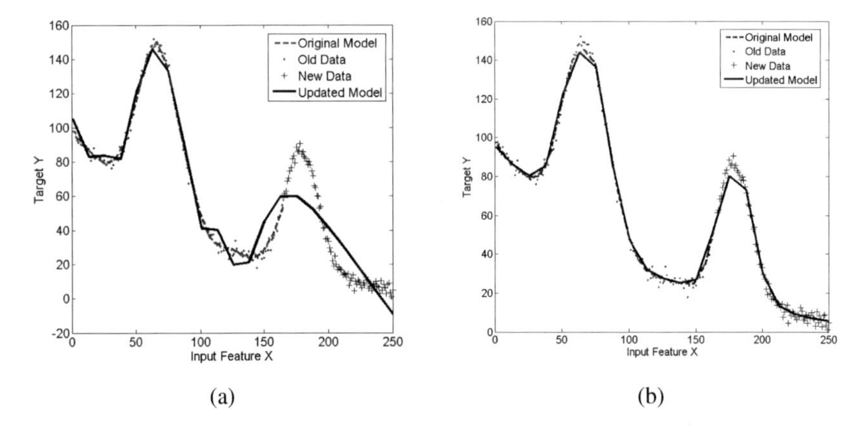

(a) (b)

Fig. 3.11 (a): updated fuzzy model (solid line) based on new samples (crosses) by only using adaptation of consequent parameters → the updated model is not able to follow the new trend. (b): updated fuzzy model (solid line) based on new samples (crosses) by applying *FLEXFIS* → the updated model is able to follow the new trend.

3.1.2 FLEXFIS *for Classification (*FLEXFIS-Class*)*

The classification variant of *FLEXFIS*, *FLEXFIS-Class*, exploits two model architectures for fuzzy classifiers: single model architecture as defined in (1.13) (Section 1.2.1.5) and multi model architecture by using multiple Takagi-Sugeno fuzzy systems, one for each class based on indicator vectors (refer to Section 1.2.1.5). The latter is also called evolving fuzzy regression by indicator matrix. Accordingly, for single model architecture the training method is called *FLEXFIS-Class SM*, whereas for multi model architecture it is called *FLEXFIS-Class MM* (see also [276] [18]).

3.1.2.1 *FLEXFIS-Class SM*

The antecedent part of the rules for this type of fuzzy classifier is the same as for Takagi-Sugeno fuzzy systems (compare (1.13) with (1.1)). Clustering is again a feasible option as usually classes fall into different regions in the feature space. Indeed, classes may significantly overlap, however usually some non-overlapping regions are present which can be extracted by rough grouping (clustering) methods. If this is not the case (i.e. classes fully overlap), no reasonable discrimination among these classes without heavy over-fitting is possible in general (using any classification method). The evolving training procedure for the rules' antecedent parts is done with the help of *eVQ* (Algorithm 4) Again, after each new incoming sample the updated resp. newly generated cluster in the high-dimensional data space is projected onto the axes to form one-dimensional Gaussian fuzzy sets — as shown in Figure 3.2.

The crisp and unique rule class label in the consequent part of the ith rule is elicited by counting the relative proportions of all the samples which formed the corresponding cluster (rule) to all the K classes and taking the maximal value of these proportions (according to the winner-takes-it-all classification scheme as defined in Section 1.2.1.5):

$$L_i = argmax_{k-1}^{K}(conf_{ik}) = argmax_{k-1}^{K}(\frac{n_{i,k}}{n_i}) \tag{3.13}$$

These values can be updated by counting n_i, the number of samples, which formed the antecedent part of the ith rule (for which the ith cluster was the winning cluster in *eVQ*) and $n_{i,k}$ the number of samples forming the antecedent part of the ith rule and falling into class k. Note that the K confidence levels in the classes $conf_{i1},...,conf_{iK}$ are also obtained through (3.13).

Regarding an incremental learning from scratch the same considerations as for *FLEXFIS* (see Section 3.1.1.2) are valid, leading again to three options for the setup phase. Combining these issues together yields the *FLEXFIS-Class SM* algorithm as described in Algorithm 6 (see also [276]).

Algorithm 6: FLEXFIS-CLASS SM (single model architecture)

1. Setup phase with three options as in Algorithm 5 and consequent estimation as in (3.13) instead of applying least squares.
2. Take the next incoming data sample and normalize it as in (3.8): $\mathbf{x}_{k+1}$.
3. Elicit its class label, say L.
4. Normalize already extracted cluster centers and their ranges of influence to $[0,1]$.
5. Proceed through Steps 4 to 10 of *eVQ* with variance update as demonstrated in Algorithm 4.

6. Transfer back the updated cluster resp. newly generated cluster to original range as in (3.9).
7. Project updated or newly generated cluster onto the input feature axes as in (3.7) and (3.6), forming the complete antecedent part of the corresponding rule.
8. **If** $L > K$, $K = K + 1$, $n_{i,L} = 0$ for all $i = 1, ..., C$, **endif**.
9. **If** a new rule is born (as new cluster evolved in Step 7 of Algorithm 4), set $n_C = 1$ and $n_{C,L} = 1$.
10. **Else** Update the consequent label and confidence values in all K classes corresponding to the winning cluster by setting $n_{win} = n_{win} + 1$ and $n_{win,L} = n_{win,L} + 1$ and using (3.13).
11. Update ranges of all p features
12. If new incoming data samples are coming then goto Step 2; otherwise *stop*.

A central aspect in Algorithm 6 is the elicitation of the (target/consequent) class label for new incoming on-line samples in Step 3 as usually these are not pre-labelled. This is a problem which is not specific to evolving fuzzy classifiers but concerns generally all incremental (supervised) classification approaches. A common strategy is that the classifier first outputs a class response for the current data sample (here by using (1.14) and (1.16)) and then a user gives feedback on this response whether it is correct or not. Of course, this causes a quite high supervision effort, especially in systems where the samples are loaded/recorded with a high frequency, making such a scenario impossible in specific cases. Here, active learning strategies [87] [253] [293] help to reduce this workload significantly, where the operator is asked for a response only in special cases. A more detailed analysis of this issue is given in Section 5.3, where balanced learning will also be a hot topic. Another approach handling un-labelled input samples is described in [55], where new clusters/rules evolved by unlabelled samples are not assigned any label until new labelled samples fall within their ranges of influence. The consequent parts of such rules remain 'unknown' for a while.

3.1.2.2 *FLEXFIS-Class MM*

FLEXFIS-Class MM trains K Takagi-Sugeno fuzzy regression models as defined in (1.1) for K different classes separately and independently based on indicator entries (target values) which are 1 for those samples belonging to the corresponding class and 0, otherwise. In this sense, it is guaranteed that the regression surface is forced towards 1 in the region of those samples belonging to its corresponding class and otherwise it approaches 0. This is also called regression by indicator entries and in our case fuzzy regression by indicator entries. A new output response is calculated by inferencing through all K TS fuzzy models and taking the maximal value over these. This is also called as one-versus-rest classification approach [344]. For further details as to why this is a reasonable classification strategy, please see the description of the multi model architecture in Section 1.2.1.5.

The basic characteristics of *FLEXFIS-Class MM* can be summarized in the following itemization points:

- It exploits a multi-model architecture with a one-versus-rest classification approach (1.21).
- Confidence levels can be estimated by a relative comparison of the outputs from all TS fuzzy sub-models (1.22).
- It performs an evolving regression modelling for each class separately with the usage of *FLEXFIS* (Algorithm 5) exploiting first order TSK fuzzy models and based on indicator entries.
- In this sense, the evolution of every sub-model has the same characteristics (with convergence to optimality etc.) as pointed out in Section 3.1.1.4 ($\rightarrow$ stable evolving fuzzy classification approach).
- It does not suffer from the masking problem in case of multiple classes, opposed to linear regression by indicator matrix (see Figure 1.13 in Section 1.2.1.5).
- It has the flexibility to include upcoming new classes during on-line operation mode on demand by opening up a new first order TSK fuzzy sub-model and hence without 'disturbing' already trained fuzzy sub-models for the other classes.

Then, the algorithm for *FLEXFIS-Class MM* becomes as outlined in Algorithm 7 (see also [271] [18] [276]).

Algorithm 7: FLEXFIS-Class MM

1. Setup Phase with three options as in Algorithm 5, and training K TS fuzzy models $TS_i, i = \{1,...,K\}$ for the K classes by estimating K cluster partitions $CL_i, i = \{1,...,K\}$ with *eVQ* (Algorithm 4) and project clusters from each partition on the axes separately and independently.
2. Take the next incoming data samples and normalize it to $[0,1]$ according to the ranges as in (3.8): $\mathbf{x}_{k+1}$
3. Elicit its class label, say L.
4. **If** $L \leq K$ (no new class is introduced)
5. Update the Lth TS fuzzy model by

   a. taking $y = 1$ as response (target) value,
   b. Performing Steps 3 to 8 of Algorithm 5

6. **Else** (a new class label is introduced)

   a. $K = K + 1$
   b. a new TS fuzzy system is initiated as in Step 1 synchronously and independently to the others (after collecting sufficient number of samples); the same setting for vigilance parameter ρ as in the other TS fuzzy models is used.

7. Update all other TS fuzzy models by taking $y = 0$ as response (target) value and performing Steps 3 to 8 of Algorithm 5.
8. Update ranges of all p features.
9. If new incoming samples are still available then goto Step 2; otherwise *stop*.

Regarding elicitation of the class labels during on-line operation mode, the same considerations as in case of *FLEXFIS-Class SM* can be made (see Section 3.1.2.1) by using (1.21) and (1.22) for obtaining an output response plus a confidence level.

3.1.3 eVQ-Class *as Spin-off of* FLEXFIS-Class

The idea of *eVQ-Class*, short for *evolving Vector Quantization for Classification*, is to define a classifier directly in the cluster space without projecting cluster to the axes to form fuzzy sets and rules [266]. Hence, *eVQ-Class* can be seen as a spin-off algorithm from *FLEXFIS-Class*, originally designed for the purpose to classify images during on-line surface inspection processes according to grouping criteria [271], see also Section 9.1. Another specific aspect of *eVQ-Class* is that it comes with two different variants for classifying new incoming samples: variant A performs a winner-takes-it-all classification (majority class of nearest cluster defines the classifier response), whereas variant B takes into account the distance of a new sample to the decision boundaries of the two nearest clusters and the confidence levels in the classes (weighted distance-based classification). First, we describe the training phase and then the two classification variants. The section is concluded with a short description of and references to related approaches.

3.1.3.1 Training Phase

To achieve a classifier directly in the cluster space, a hit matrix H is defined, whose rows represent the clusters and the columns represent the classes. The entry h_{ij} (representing the ith row and jth column) of this matrix simply contains the number of samples falling into cluster i and class j. In this sense, each row of the matrix contains the single relative frequencies of each class falling into the corresponding cluster. Please note that the incremental update of this matrix is straightforward by simply incrementing the entries h_{ij}, whenever a sample is attached to cluster i and class j. If a new cluster is evolved for a sample falling into class j, a row is appended to H, where the single entries are all set to 0, except the jth which is set to 1. If a new class (the $K+1$th) is introduced by a new sample falling into cluster i, a column is appended whose entries are all set to 0, except the ith which is set to 1.

Furthermore, in order to increase the likelihood of *clean clusters*, i.e. clusters containing samples from one class, the class labels are directly included in the feature vectors containing p input features: for the two class classification case this is done by simply appending the label entries (0 or 1) at the end of the input feature vectors and feeding this extended feature vectors into the evolving clustering process, obtaining a supervised clustering approach ; for the multi-class case a K-dimensional vector (with K = the number of classes) is appended at the end of the feature vector, where the K values are all set to 0, except for the class to which the current feature vector belongs (e.g. [0 1 0 0] for a sample belonging to class #2 in a 4-class problem). These extended features are also sent into the evolving clustering process. The main advantage of including the class labels directly in the clustering

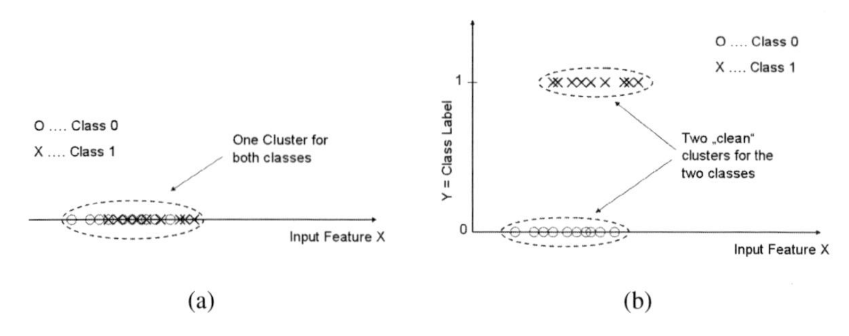

Fig. 3.12 (a): The classes mixed up in one cluster as the class labels are not included in the clustering process; (b): two distinct clusters for two classes are generated as the class labels are included as y-coordinate

process is underlined by the consideration that different classes should trigger different clusters to increase the accuracy of the classifier. This is more likely, when the labels themselves are part of the feature vectors, as then the feature vectors of different classes are stretched away from each other. Figure 3.12 visualizes a one-dimensional (input space) example, including two classes, marked with circles and Xs: in the left image the class label plays no role, hence only one input feature is sent into the clustering process; in this case, one cluster is generated including all the samples from both classes. In the right image the class labels are integrated into the clustering process (as 0s = circles and 1s = Xs), which forces the samples of the different classes to be torn apart, hence two clusters are generated for the two distinct classes, which is beneficial in this case. Note, that this property is weakened in case of high-dimensional feature spaces, as by the addition of one feature vector representing the class labels (or several feature vectors with 0s and 1s in case of multi-class problems) the influence on any multi-dimensional distance measure (used for winner selection and cluster evolution) is quite low. This is another peculiarity of the curse of dimensionality. Therefore, clusters with mixed classes (unpurified clusters) still may happen and hence the hit matrix is still important. The disadvantage of integrating the class label in clustering is that the number of classes has to be known in advance, otherwise (upon an upcoming new class) input vectors with different dimensions may occur. Another disadvantage is that often more clusters are generated, which may deteriorate the computational speed significantly.

Based on these aspects, the algorithm of *eVQ-Class* can be formulated as in Algorithm 8 (assuming the number of classes to be known in advance and the ranges of features either known or estimated from the first samples).

Algorithm 8: eVQ-Class (Evolving Vector Quantization for Classification)

1. Initialization: set the # of clusters to 0 and the # of classes to K, ranges of variables.
2. Take the next incoming data sample and normalize it to $[0,1]$ according to the ranges: $\mathbf{x}$.
3. Elicit its class label, say L.

4. **If** $K = 2$
   a. Set $\mathbf{x_{all}} = (\mathbf{x}, L)$.
   b. $P = p + 1$.
5. **Else**
   a. $\mathbf{x_{all}} = (\mathbf{x}, \mathbf{0}_{1,...,K})$ and $\mathbf{x_{all}}(p+L) = 1$ with p the dimension of the input feature space.
   b. $P = p + K$.
6. **If** number of clusters = 0
   a. Set $C = 1$, set the first center c_1 to the current data point, hence $\mathbf{c}_1 = \mathbf{x_{all}}$, set $\sigma_1 = \mathbf{0}$, $k_1 = 1$.
   b. Introduce the first row in the hit matrix h with $h_{1L} = 1$ and $h_{1k} = 0$ for all $k \in \{1,...,K\} \neq L$, goto step 12.
7. Calculate the distance of $\mathbf{x_{all}}$ to all cluster centers by using Euclidean distance measure.
8. Elicit the cluster center which is closest to $\mathbf{x_{all}}$ by taking the minimum over all calculated distances $\rightarrow$ winning cluster c_{win}.
9. Set min_{dist} as the minimum over all distances.
10. **If** $min_{dist} \geq \rho$
    a. Set $C = C + 1$, $\mathbf{c}_C = \mathbf{x_{all}}$, $\sigma_C = \mathbf{0}$, $k_C = 1$.
    b. Append a new row in the hit matrix h with $h_{CL} = 1$ and $h_{Ck} = 0$ for all $k \in \{1,...,K\} \neq L$.
11. **Else**
    a. Update the P components of the winning cluster by moving it towards the current sample $\mathbf{x_{all}}$ as in (3.1) with η_{win} as in (3.2).
    b. Increase the number of data samples belonging to the winning cluster by 1, i.e. $k_{win} = k_{win} + 1$.
    c. Increment the hit matrix by $h_{win,L} = h_{win,L} + 1$.
12. Update the ranges of all p input features.
13. If new data samples are still coming in goto step 2, otherwise stop.

If the number of classes is not known a-priori, $\mathbf{x_{all}} = \mathbf{x}$ in Step 4 and set $K = K + 1$ in Step 3 in case when a new class label is introduced. For the initialization phase the same approaches as described in Algorithm 5 may be applied (so starting with an off-line trained classifier which is further updated in on-line mode or performing an initial training with the first on-line samples).

For the classification phase, i.e. classifying a new incoming instance based on the evolved clusters and the hit matrix, there are two different approaches leading to *eVQ-Class variant A* and *eVQ-Class variant B* [266], where in both not only the class label response will be elicited, but also a confidence level, indicating the certainty of the classifier in the outputted class label.

3.1.3.2 Classification Phase Variant A (*eVQ-Class A*)

The first variant is a 'winner-takes-it-all' approach, where the nearest cluster is seen as the best representative to label the current sample. Hence, similarly within the training process, the first variant simply calculates the distance of the new incoming sample to be classified to all evolved cluster centers and the nearest cluster elicited, say the ith. However, even though the class labels might be included in the clustering process (in case when the number of classes is known a-priori), one cluster may not be 'clean', i.e. representing one single class. This is especially true for high-dimensional input feature spaces, as then the addition of one or just a few columns for the classes may not tear apart the samples from different classes sufficiently. In this sense, the confidence level for each class $conf_k, k = 1, ..., K$ is calculated by the relative frequencies of each class falling into the ith (the nearest) cluster:

$$conf_k = \frac{h_{ik}}{\sum_{j=1}^{K} h_{ij}} \tag{3.14}$$

The final class response L is given by that class appearing most frequently in the nearest cluster:

$$L = argmax_{k=1,...,K} conf_k \tag{3.15}$$

The drawback of this approach is that it does not take into account if there is a 'clear' nearest cluster or not, i.e. the relative position of the sample to the boundary of the range of influence of two neighboring clusters. We would not say that it is a strict decision boundary (as one cluster does not represent one single class), but it is a kind of soft decision boundary when expecting that different classes mainly appear in different clusters.

3.1.3.3 Classification Phase Variant B (*eVQ-Class B*)

Variant B introduces weight factors w, which are calculated based on the proportion between the distance to the nearest and the distance to the second nearest cluster (say c_{i1} and c_{i2}). The nearer the sample is to the boundary between the range of influence of both clusters, the more equal this weight factor should be for both clusters (i.e. around 0.5); the further away this sample is from this boundary, the more dissimilar the weight factor should be (i.e. near 1 for the nearest cluster and near 0 for the second nearest one), as the second nearest should have low influence when obtaining the final confidence levels in the classes. Hence, the weight factor of the nearest cluster c_{i1} is defined by

$$w_{i_1} = 0.5 + \frac{I}{2} \tag{3.16}$$

and the weight factor of the second nearest cluster c_{i2} by

$$w_{i_2} = 0.5 - \frac{I}{2} \tag{3.17}$$

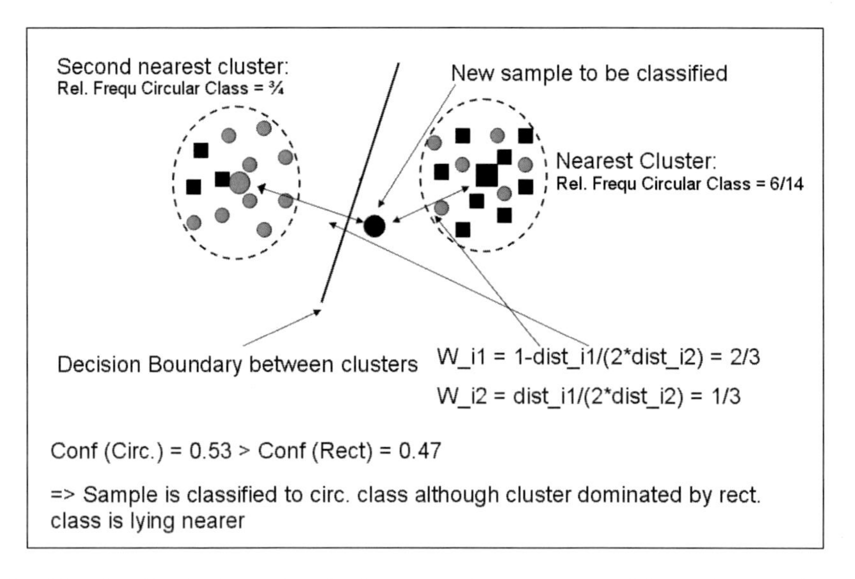

Fig. 3.13 Visualization how classification variant B computes confidences for two classes (rectangular and circled samples) on a new incoming sample (big dot) — note that the new sample is assigned to circular class, although the right cluster dominated by the rectangular class is nearer; this effect is due to the relative frequencies of the classes in both clusters (the circular class is dominating the left cluster more than the rectangular class the right cluster) while at the same time the new sample is close to the decision boundary (moving the sample more to the right cluster would finally result in an assignment to the rectangular class)

with $I = 1 - \frac{dist_{c_{i1}}}{dist_{c_{i2}}}$, where $dist_{c_{i1}}$ the distance to the nearest cluster and $dist_{c_{i2}}$ the distance to the second nearest cluster. These weight factors are multiplied with the relative frequencies as in (3.14) for all classes in both clusters, obtaining the confidences for $k = 1, ..., K$ classes:

$$conf_k = w_{i_1} \frac{h_{i1,k}}{\sum_{j=1}^{K} h_{i1,j}} + w_{i_2} \frac{h_{i2,k}}{\sum_{j=1}^{K} h_{i2,j}} \tag{3.18}$$

The final class response L is given by formula (3.15). A visualization of this classification strategy (weighted classification) is shown in Figure 3.13.

Please note that the confidence levels $conf_k$ as in (3.18) are well-defined in the sense that they are lying in $[0,1]$. This is because $w_{i_1} \frac{h_{i1,k}}{\sum_{j=1}^{K} h_{i1,j}} + w_{i_2} \frac{h_{i2,k}}{\sum_{j=1}^{K} h_{i2,j}} = (1 - \frac{dist_{c_{i1}}}{dist_{c_{i2}}}) \frac{h_{i1,k}}{\sum_{j=1}^{K} h_{i1,j}} + \frac{dist_{c_{i1}}}{dist_{c_{i2}}} \frac{h_{i2,k}}{\sum_{j=1}^{K} h_{i2,j}}$ which is obviously maximal in case of $\frac{h_{i1,k}}{\sum_{j=1}^{K} h_{i1,j}} = 1$ and $\frac{h_{i2,k}}{\sum_{j=1}^{K} h_{i2,j}} = 1$, what leads to $(1 - \frac{dist_{c_{i1}}}{dist_{c_{i2}}}) + \frac{dist_{c_{i1}}}{dist_{c_{i2}}} = 1$. Furthermore, it is obviously minimal in case of $\frac{h_{i1,k}}{\sum_{j=1}^{K} h_{i1,j}} = 0$ and $\frac{h_{i2,k}}{\sum_{j=1}^{K} h_{i2,j}} = 0$ which leads to a value of 0 for the final confidence. Concluding, someone can say that variant B yields a fairer class response in the case when a new incoming sample falls in-between two clusters: the cluster (slightly) further away from the new sample than the other may influence the output response more if it is cleaner and more significant (formed by a larger number of samples) than the other.

3.1.3.4 Related Work and Differences

eVQ-Class focusses on an incremental classification method which is deduced from conventional vector quantization [150] (see [141] for a classical review), moving into a supervised learning context (see 3.1.1.1). This supervised learning context is also exploited by a learning vector quantization (LVQ) approach [226] and its extensions LVQ2, LVQ2.1 and LVQ3 [228] [227] [226]. In [383] the Generalized LVQ (GLVQ) algorithm is proposed, where an explicit cost function (continuous and differentiable) is introduced and the updating rule obtained by minimizing this cost. The main idea of (learning) vector quantization is to build a quantized approximation to the distribution of the input data, using a finite number of prototypes (called code-book vectors which can be seen as cluster centers). These prototypes result from an update procedure based on the training data set; in this sense, a kind of incremental learning scheme is applied, but all these algorithms iterate over the data set multiple times and hence are in principal batch learning approaches. If they had been applied in an on-line learning process for all the data seen so far, they would not terminate in real-time. Extensions to the LVQ approach in order to make it applicable to on-line data samples processed in single-pass manner have been proposed, such as [47] [90] [437] [343]. In [47] they mention an on-line version of the learning vector quantization approach, where they decrease the number of iterations in the training phase and also modify the code-book vectors with a gradual update rule, whose learning gain decreases with the sum of distances between samples and the nearest codebook vectors over the past. However, opposed to *eVQ-Class*, this approach does not evolve new clusters on demand based on the characteristics of new incoming data. In [437] a competitive learning approach for self-creating neural networks with evolvable neurons is mentioned, which is applied in the learning vector quantization context. A basic concept there is the introduction of a winning frequency variable, which gives rise to how often each code book vector was accessed as winning one in the past. In [343] the conventional LVQ approach is extended to an incremental and evolving version (ILVQ) by generating a new cluster, whenever the current sample is lying far away from a cluster having the same class (so a supervised criteria). This is a fundamental difference to *eVQ-class*, where less clusters are generated (as unpurified clusters are allowed). This also means over-fitting is not so likely, especially when having only a few samples per class for training; hence, *eVQ-Class* can be expected to perform better for a low number of samples (e.g. at the beginning of an on-line modelling process). In [49] the dynamics and the generalization ability of various LVQ algorithms (LVQ original, LVQ 2.1, LFM and LFM-W) is studied with the help of on-line learning theory.

All these approaches have in common that classification is carried out by winner-takes-it all approach, whereas in *eVQ-Class* there is also the option to take into account the relative position of a new sample to the nearest clusters (variant B).

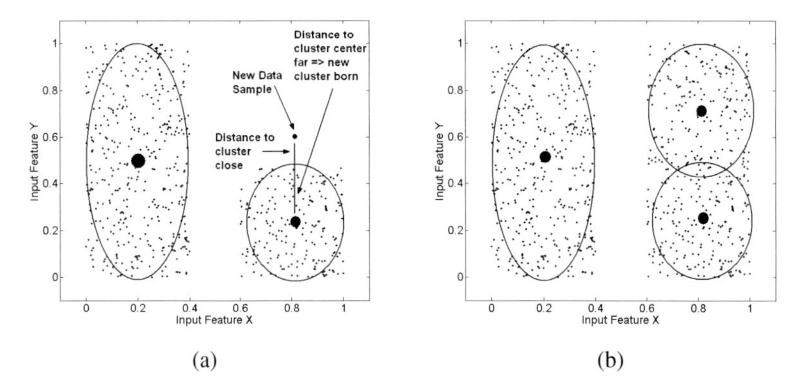

Fig. 3.14 (a): Wrong cluster evolution based on a new incoming sample on the right hand side (as taking distance to the center which is farer away than the vigilance); (b): the final (incorrect) cluster partition (2 clusters for the right data cloud instead of one)

3.1.4 Extensions to eVQ

This section deals with some extensions to *eVQ* for improving its performance in terms of the quality of the generated cluster partitions, especially in case of specific characteristics of (the order of) incoming on-line data samples (see also [267]). This automatically also effects *FLEXFIS(-Class)* and *eVQ-Class*, as *eVQ* is used as learning engine in these.

3.1.4.1 An Alternative Distance Strategy for Avoiding Over-Clustering

The problem of condition (3.3), is that in the case of wider data clouds or samples belonging together widely spread over the input space it is quite likely that more clusters than necessary are generated and hence an 'over-clustering' and incorrect partition of the input space is the outcome. This problem is visualized in Figure 3.14, where two clusters are evolved instead of one for the right data block. On the right hand side in (a) it is underlined, why this happens: a new cluster emerges (represented by its cluster center as bigger dark dot at the bottom), because it is lying too far away from the other cluster center already generated before. This leads to the final cluster partition as shown in Figure 3.14 (b). The clusters' ranges of influence are shown as ellipsoids surrounding the cluster centers. If these ranges of influence had been synchronously updated to the cluster centers and the distances of new samples taken to these ellipsoidal surfaces, then the upper right (superfluous) cluster would not have evolved correctly and two clusters would not have been identified during the incremental update process. In the extreme case, even a new sample lying within the range of influence (ROI) of a (very large) cluster could trigger a new cluster, as the distance to its center becomes greater than ρ.

For omitting such occurrences, the ranges of influence of each cluster in terms of ellipsoidal surfaces are updated and the distance of a new incoming sample to the estimated surfaces along the direction to the center are calculated (→ alternative distance strategy). The ranges of influence in each dimension can be incrementally calculated with (3.5) as done in *FLEXFIS* when eliciting the width of the Gaussian fuzzy sets during cluster projection. To elicit the nearest cluster (winning cluster) the distance to the surface along the direction towards the cluster center has to be calculated. The following lemma gives rise as to how this can be calculated (with P the input dimension):

Lemma 3.2. *Let therefore be* $\sum_{j=1}^{p} \frac{(x_j - c_{ij})^2}{\sigma_{ij}^2} = 1$ *a multidimensional ellipsoid of the ith cluster in main position,* σ_{ij} *the variance of the data belonging to the ith cluster in dimension j, then the Euclidean distance of the new data sample* $(q_1, ..., q_p)$ *to the surface along the direction towards the cluster center* c_{ij} *is given by*

$$dist = (1-t)\sqrt{\sum_{j=1}^{p}(q_j - c_{ij})^2} \tag{3.19}$$

with

$$t = \frac{1}{\sqrt{\sum_{j=1}^{p} \frac{(q_j - c_{ij})^2}{\sigma_{ij}^2}}} \tag{3.20}$$

The proof is straightforward by using vector calculations (computing the crossing point of straight line with ellipsoid and distance of this point to the center). In fact, the distance (3.19) is only computed for the actual data sample $\mathbf{q}$, if it is lying outside the ranges of influence of all clusters, i.e. the condition

$$\exists i \sum_{j=1}^{p} \frac{(q_j - c_{ij})^2}{\sigma_{ij}^2} \leq 1 \tag{3.21}$$

is not fulfilled. Otherwise, the usual distance strategy is applied for all clusters, whose range of influence contains the current data sample (clusters may be also overlapping).

In particular, the modified version of *eVQ* is achieved by simply computing (3.19) instead of the distance to the cluster centers in Step 5 of Algorithm 4, whenever (3.21) is not fulfilled. Otherwise, the distance to all the centers for which (3.21) is fulfilled is calculated. This minimal distance is then compared with the vigilance parameter ρ (condition for opening up a new cluster) as done in Step 7 of Algorithm 4. The modified version of *eVQ* was also applied in connection with *FLEXFIS* leading to the *FLEXFIS-MOD* algorithm [277]. For the *eVQ-Class* algorithm (Algorithm 8) the same strategy can be applied, when also updating the ranges of influence of the clusters synchronously to the centers. Then, for the classification variants A resp. B the nearest distances to the surfaces of the clusters instead of to the centers can be calculated.

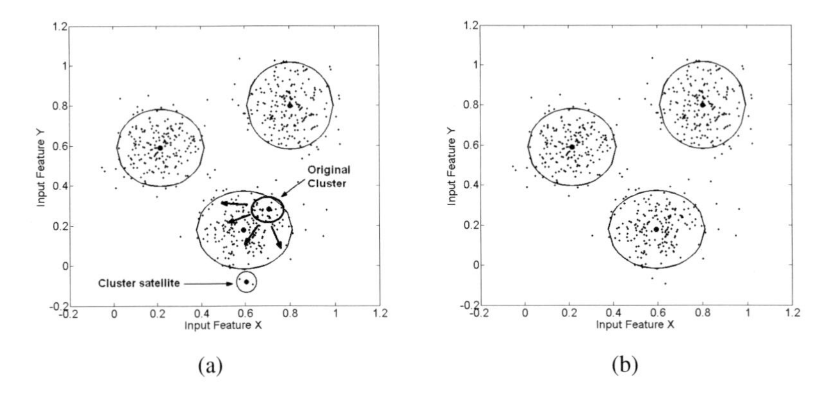

Fig. 3.15 (a): An unpleasant satellite cluster by applying Algorithm 4; (b): The unpleasant satellite cluster removed by applying Algorithm 9

3.1.4.2 Deletion of Cluster Satellites

Algorithm 4 may lead to undesirable tiny clusters, called cluster satellites, which lie very close to significantly bigger ones. An example is demonstrated in Figure 3.15(a), where the tiny cluster is obviously superfluous. The reason for this unpleasant effect is the following: at the beginning of the incremental training process the first samples forming the bottom cluster appear at the upper region of this cluster, the cluster had a very narrow range of influence at this stage (see small ellipsoid inside and surrounding a bigger dark dot). Afterwards, a data sample at the lower region came in and formed a new cluster, as being too far away from the small ellipsoid in the upper region. This cluster forming at this stage of the incremental (online) learning process was correct, as it seemed that a new cluster arises there. Afterwards, newly loaded data filled up the big hole between these two tiny clusters, causing a movement and an expansion of the upper cluster. Finally, it turned out that they in fact belong to one cluster, whereas two clusters are generated by *eVQ*. This should be prevented as it leads to false information and incorrect cluster partition.

Algorithm 9: Deletion of Cluster Satellites

1. **For** all $i = 1,...,C$ clusters generated by an (incremental) clustering algorithm, perform the following steps:
2. **If** $\frac{k_i}{N} < 1\%$ (recall: k_i the number of samples forming the ith cluster and N the number of data samples loaded in sum).
   a. Mark the ith cluster for cutting out, as it is expected to contain only outliers.
   b. Continue with next cluster.
3. **If** $\frac{k_i}{N} < low\_mass$
   a. **If** The ith cluster center lies inside the range of influence of any other cluster
      i. Elicit the closest center c_{win} (like in Step 6 of Algorithm 4).

ii. If $\frac{\sum_{j=1}^{p}\sigma_{ij}}{\sum_{j=1}^{p}\sigma_{win,j}} < \varepsilon$, where σ_{ij} the length of the jth axes of the ellipsoid spanned by the ith cluster, then mark the ith cluster for cutting out.

b. **Else**

i. Calculate the distance of the ith cluster center to the surface of all other clusters. after (3.19) in Lemma 3.2

ii. Elicit the cluster which is closest to the cluster center of the ith cluster, let's say cluster win with $dist_{win}$ the distance.

iii. If $dist_{win} < thr$ and $\frac{\sum_{j=1}^{p}\sigma_{ij}}{\sum_{j=1}^{p}\sigma_{win,j}} < \varepsilon$, where σ_{ij} the length of the jth axes of the ellipsoid spanned by the ith cluster, then mark the ith cluster for cutting out.

4. **End For**
5. Cut out all marked clusters.

The strategy in Algorithm 9 is based on the investigations, what characterizes in fact a cluster satellite , namely 1) a low mass i.e. a small fraction of data samples belonging to it, 2) the cluster center has to be close or inside the range of influence of the (most) adjacent cluster, 3) the cluster has to be significantly smaller in its range of influence than the (most) adjacent cluster. Moreover, clusters with a very low mass will be marked for cutting out immediately (see Step 2 and 3), as they usually denote outliers. The marking and not directly cutting out the other candidates ensures that cluster satellites of cluster satellites are cut out, too.

Note, that in Algorithm 9 cluster satellites are completely cut out. This is opposed to the situation where two or more significant clusters move together and an intrinsic penetration of spheres or ellipsoids can be observed. In this case, an enhanced strategy for cluster merging process during incremental learning is necessary. Section 5.1 deals with this topic. Also note, that the satellite deletion algorithm can be also applied in connection with any other clustering methods and EFS approachs, where cluster prototypes and ranges of influence of the clusters are extracted. In connection with an EFS approach this would lead to a deletion of rules which turn out to be too small and too insignificant. This can be done in straightforward manner without losing the convergence to optimality as in local learning approach each rule is tuned separately (see Section 3.1.1.2).

3.1.4.3 An On-Line Cluster Split-and-Merge Approach

Although Algorithm 9 is able to remove superfluous clusters, it is still possible that over-clustering takes place. This is the case when two (or more) clusters seem to be distinct at the beginning (e.g. during the batch setup phase) and move together with more samples loaded, or when the opposite is the case: at the beginning there seems to be one cluster in a local region of the data space, which turns out later to contain two distinct data clouds. For a detailed problem statement and possible strategies for solving these problems (with automatic rule merging and splitting procedures), please refer to Section 5.1, Chapter 5 (this topic is moved to there as it will be described in a more general context applicable for most of the evolving fuzzy systems approaches demonstrated in this chapter).

3.1.4.4 Alternative Distance Measure

Up till now, Euclidean distance was used when eliciting the winner cluster in *eVQ*, triggering ellipsoidal clusters parallel to the axes. Here, we will examine the application of Mahalanobis distance in *eVQ*, which for a new sample $\mathbf{x}$ to the cluster center $\mathbf{c}$ is defined by [295]:

$$mahal = \sqrt{(\mathbf{x}-\mathbf{c})\Sigma^{-1}(\mathbf{x}-\mathbf{c})} \tag{3.22}$$

with Σ the covariance matrix as defined in (2.4) (Section 2.1.2). Applying this distance measure to the clustering process will trigger ellipsoidal clusters in arbitrary position. Such multi-dimensional ellipsoids in out input/output space can be defined by [239]:

$$E(\mathbf{c},Q) = \{\mathbf{x} \in \mathbb{R}^p | (\mathbf{x}-\mathbf{c})^T Q^{-1} (\mathbf{x}-\mathbf{c}) \leq 1\} \tag{3.23}$$

with $\mathbf{c}$ the center of the ellipsoid and Q a positive definite shape matrix ($Q^T = Q$ and $\mathbf{x}^T Q\mathbf{x} > 0$) for all nonzero vectors $\mathbf{x}$. In our cluster space, the cluster centers can be associated with the centers of ellipsoids and the shape matrix estimated by the covariance matrix Σ in each cluster (see [158]). The latter can be achieved in on-line mode by exploiting recursive covariance matrix update (2.8) (in sample-wise manner) for those samples belonging to the corresponding clusters. This can be even done from scratch.

Then, *eVQ with covariance update* can be simply achieved by modifying Algorithm 4 in the following way:

1. Instead of Step 4c.), the covariance matrix is set to a matrix with zero entries resp. entries with very low positive values.
2. In Step 5 Mahalanobis distance is applied by taking care of covariance matrices which are nearly singular and hence cause a spoiled computation of the inverse in (3.22). This may be the case when starting a new cluster (covariance matrix contains very small or zero entries) and can be solved by adding the identity matrix as regularization term (see also Section 4.1).
3. Instead of (3.5) in Step 9, the update formula for the covariance matrix as defined in (2.8) is applied.

The advantage of this algorithm is that it is able to extract ellipsoids in arbitrary position and hence may yield a cluster partition representing the natural characteristics of the data set better. An example of this effect is presented in Figure 3.16. There, the two left clusters follow a straight linear (correlation) trend, which can be better, represented when using ellipsoids in general position (b) (i.e. by clusters having smaller widths), compared to when using ellipsoids in main position (as triggered by Euclidean measure — see (a)).

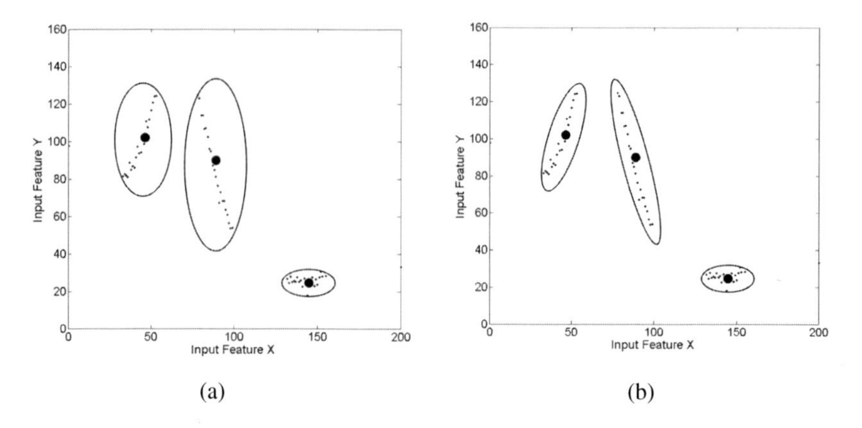

Fig. 3.16 (a): Extracted ellipsoidal clusters in main position by using Euclidean distance, (b) by using Mahalanobis distance (ellipsoidal clusters in arbitrary position are caused); note that in (b) the clusters are able to follow the trend of the clusters better (slimmer clusters yielding a more specific representation and the ability to track local correlations between the two features).

When using alternative distance measures as described in Section 3.1.4.1, the distance to the surface of an ellipsoid in arbitrary position needs to be calculated. This can be achieved in a straightforward manner, yielding the same distance formula as in (3.19) but with (again for the ith cluster):

$$t = \frac{1}{\sqrt{\sum_{j=1}^{p} \left(\sum_{k=1}^{p} (q_k - c_{ik}) cov_{kj} \right) (q_j - c_{ij})}} \tag{3.24}$$

with cov_{kj} the covariance between variable k and j.

Applying Mahalanobis distance in *eVQ-Class* (training and classification) and *FLEXFIS-Class SM* (training), may imply the same advantages as in *eVQ* (more accurate cluster partition). For *FLEXFIS* resp. *FLEXFIS-Class MM* the story is different; there the application of Mahalanobis distance is a 'two-edged sword', as it does not necessarily lead to higher accurate models. This is because of the projection strategy, when forming the antecedent parts of the rules from clusters. Here, most of the additional information the ellipsoids in arbitrary shape provide in the multi-dimensional case is lost, as the projection results are again one-dimensional (widths of the) fuzzy sets. An enhanced fuzzy set projection strategy for ellipsoidal clusters in arbitrary position is used in *FMCLUST* method [28], however leading to multi-modal fuzzy sets which are hardly interpretable.[2]

[2] Multi modal fuzzy sets are fuzzy sets with more than one local maximum.

3.1.4.5 Dealing with Missing Values

In case of missing values present in some input features of a new sample, the (Euclidean resp. Mahalanobis) distance of this sample to all cluster centers resp. ellipsoidal ranges of influence of the clusters is computed by using just the known features. Similarly, the adaptation of cluster centers by (3.1) and the cluster surfaces by (3.5) resp. by (2.8) (in case of Mahalanobis distance) is performed only using the corresponding existing coordinates of the new sample **x**, and not on the missing ones. This concept was proposed and also successfully applied in [243] for conventional (batch mode) *VQ* algorithm. Furthermore, missing values in the current sample may be replaced by the corresponding coordinate of the cluster center nearest to the current sample. This guarantees a usage of samples with missing values in the *FLEXFIS* resp. *FLEXFIS-Class* algorithms.

3.1.4.6 Feature Weighting in *eVQ*

Some approaches such as for instance the weighted version of *VQ* demonstrated in [243] propose a weighting of the features in the original sample space and send the weighted samples into the algorithm. However, when assuming that all features are normalized to the unit interval $[0,1]$, this only causes a compression of the data space to a smaller data space. For instance consider that a feature is weighted by 0.1: then, the range of this feature after weighting is reduced to [0,0.1] and the whole incremental clustering process gets hard to interpret (especially for visualization purposes). Hence, it is suggested using the full spanned data space and include different weights for the features directly in the learning process.

> *Feature weights* can be directly associated with the importance the features currently have for the actual (on-line) data stream in order to guide the training process to accurate cluster partitions, fuzzy classifiers or fuzzy regression models. These weights are also called *feature importance levels.*

The inclusion of feature weights is done two-fold:

1. First, a weight parameter vector $\lambda = \lambda_1, ..., \lambda_p$ is elicited including all the weights for all p input features.
2. Second, this weight vector is included 1.) in the update process of the cluster centers by:
$$\mathbf{c}_{win}^{(new)} = \mathbf{c}_{win}^{(old)} + \eta \lambda I(\mathbf{x} - \mathbf{c}_{win}^{(old)}) \tag{3.25}$$
with I the identity matrix and 2.) when calculating the (Euclidean) distance to the cluster centers resp. surfaces by weighting the quadratic summands (as in (3.19)) with λ_j.

In case of Mahalanobis distance, the diagonal entries of the covariance matrix Σ are multiplied with the weight vector entries. The interpretation for the inclusion of the weights in the center updates is that a shift towards a new data sample with respect

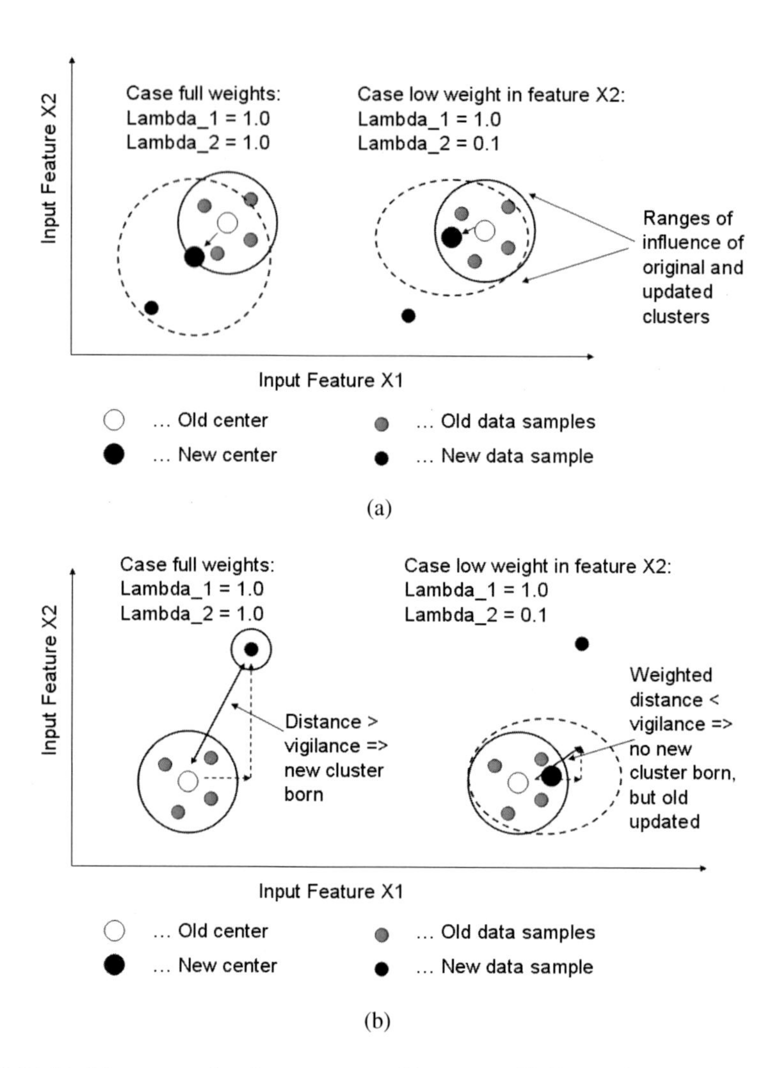

Fig. 3.17 (a): Movement of a cluster center and its range of influence in case of using full feature weights in both directions (left side) and in case of using a small feature weight for feature X2 due to its low importance level in the current learning problem (right side); (b) shows the impact of a low feature weight on cluster evolution, preventing the generation of superfluous clusters (right side) as feature X2 is unimportant

to important features should be higher than a shift with respect to less important ones. Figure 3.17 (a) demonstrates the impact of different weights in the second dimension as indicated: one time using $\lambda_2 = 1$ and one time using $\lambda_2 = 0.1$ (i.e. the second feature almost out-weighted), λ_1 is always set to 1, hence a full update with respect to the first dimension is achieved. Figure 3.17 (b) demonstrates the impact of these two weighting cases on the cluster evolution step (i.e. when calculating the weighted distance of the new sample to already generated clusters and comparing with the vigilance parameter as done in (3.3)). Obviously, when Feature X2 gets less important (right side in (b)), no new cluster is evolved as the distance in this direction is significantly reduced. In this sense, when knowing unimportant variables, also the complexity of the cluster partitions resp. of the evolving fuzzy systems/classifiers generated from these can be reduced.

Now, the central question is how to calculate the weight vector during the incremental learning process, when not a-priori pre-defined, e.g. based on expert knowledge about the importance of system variables. In case of a pure clustering task (without extracting antecedent parts of fuzzy rules), the importance levels of features have to be calculated based on the unsupervised information only (as usually no supervised information is available). One may exploit the approach demonstrated in [336], where the weight vector is updated synchronously to the clusters based on gradient descent steps for a specific cost function, originally motivated by a generalized learning vector quantization approach [383].

For classification and regression problems we will demonstrate a more general approach which is applicable to any evolving fuzzy system technique and hence will be described in detail in Section 5.2.

3.2 The *eTS* Family

Similar to the *FLEXFIS* family, the *eTS* family comes with two variants, one for regression which is simply called *eTS* and short for *evolving Takagi-Sugeno fuzzy systems* [11] and one for classification which is called *eClass*, short for *evolving (fuzzy) Classifiers* [20] [21]. We are going to start with the description of the regression variant which will then carry on to the classification case.

3.2.1 eTS *for Regression*

eTS [11] exploits the Takagi-Sugeno model architecture as defined in (1.2) by using a polynomial degree of order 1, hence linear consequent parameters, triggering hyper-planes locally defined for each rule. The basic difference to the *FLEXFIS* approach lies in the way in which the antecedent parameters in the fuzzy sets are modified respectively new rules are added and evolved during the on-line mode based on new incoming samples. In doing so, a product-space clustering approach is also exploited, but not with the help of a learning gain enforcing local movements of cluster prototypes, but with the help of a recursive calculation of potentials of data samples respectively cluster prototypes.

3.2.1.1 Subtractive Clustering

The method extends the conventional batch version of *Subtractive clustering* [83] to the on-line case which is an improved version of the *mountain clustering* method [455] and is applied in the well-known *genfis2* method in MATLAB(Tm)'s fuzzy logic toolbox. The basic concept of this method is (see [83] [11]) that it uses the data samples themselves as candidates for being cluster prototypes (centers), instead of grid points as done in *mountain clustering*. Thus, it circumvents time-intensive computations of the mountain clustering algorithm which grows exponentially with the number of dimensions. The likelihood of a sample to become a cluster center is evaluated through its potential, a measure of the spatial proximity between a particular sample $\mathbf{x}_k$ and all other data samples:

$$P(x_k) = \frac{1}{N}\sum_{i=1}^{N} e^{-\frac{\|\mathbf{x}_k - \mathbf{x}_i\|}{(r/2)^2}} \tag{3.26}$$

with r a positive constant representing a neighborhood radius and N the number of data samples in the training data set. Hence, a data sample will have a high density value and hence becoming a cluster center if it has many neighboring samples. In this regard, *subtractive clustering* can somehow also be compared to the *DBSCAN* algorithm [123], which also searches for an accumulation of data samples (i.e. dense regions) in the data space. In DBSCAN, no cluster prototypes are extracted, as samples are adjoined to the same clusters according to similarity and density criteria along a density path.

If the first center is found (that one having the maximal potential according to (3.26)), the potential of all other data samples is reduced by an amount proportional to the potential of the chosen point and inversely proportional to the distance to this center. The next center is also found as the data sample with the highest potential (after subtraction). The procedure is repeated until the potential of all data samples is reduced below a certain threshold or a sufficient number of clusters is attained. In this way, *subtractive clustering* can be summarized as in Algorithm 10.

Algorithm 10: Subtractive clustering

1. Set the number of clusters C to 0.
2. Elicit the data sample with the highest potential $P(x_m)$, $\mathbf{x}_m$:

$$m = argmax_{i=1}^{N} P(x_i) \tag{3.27}$$

3. Set the Cth center $\mathbf{c}_C = \mathbf{x}_m$, denote $P(c_1)$ as its potential, increment $C = C + 1$.
4. Reduce the potential of all samples by:

$$P(x_i) = P(x_i) - P(c_C) e^{-\frac{\|\mathbf{c}_C - \mathbf{x}_i\|}{(r_b/2)^2}} \quad i = 1, ..., N \tag{3.28}$$

with r_b a positive constant usually set to $[1, 1.5]r$ [83] and determining the range of influence of one cluster.
5. **If** $max_{i=1}^{N} P(x_i) < thr$, stop, otherwise goto Step 2.

In *genfis2*, this algorithm is used for learning the number of rules and their antecedent parts, as 1.) each cluster center (focal point) can be identified with a rule center where its corresponding coordinates represent centers of fuzzy sets and 2.) the pre-defined neighborhood radius r defines the widths of the fuzzy sets when projecting the clusters onto the axes (according to the approach as shown in Figure 3.2). In this sense, the widths is not directly estimated out of the data but pre-defined and the same for all fuzzy sets. The advantage is that this improves interpretability of the fuzzy system, while the disadvantage is that it does not take into account varying local characteristics of the data clouds and hence may suffer from imprecision with respect to model outputs.

3.2.1.2 Extension to the On-line Case (*eClustering*)

In *eTS* [11], *subtractive clustering* is extended to the on-line case (called *eClustering*) by associating the first data sample with the center of the first cluster and calculating the potential of new incoming data samples recursively. As a measure of potential, a Gaussian kernel is not used (as in (3.26)), but a Cauchy type function of first order (an approximation of a Gaussian kernel):

$$P_k(x_k) = \frac{1}{1+\frac{1}{k-1}\sum_{l=1}^{k-1}\sum_{j=1}^{p+1}(d_{lk}^j)^2} \qquad k=2,3,\ldots \tag{3.29}$$

where $P_k(x_k)$ denotes the potential of the data sample $\mathbf{x}_k$ at time instance k and d_{lk}^j the distance of sample $\mathbf{x}_k$ to the sample $\mathbf{x}_l$ in dimension j. Then the potential of a new data sample recorded at time instance k, $\mathbf{x}_k$ can be recursively calculated by (see [11] for proof):

$$P_k(x_k) = \frac{k-1}{(k-1)(\delta_k+1)+\sigma_k+2u_k} \tag{3.30}$$

where $\delta_k = \sum_{j=1}^{p+1} x_{k,j}^2$, $\sigma_k = \sum_{l=1}^{k-1}\sum_{j=1}^{p+1} x_{l,j}^2$, $u_k = \sum_{j=1}^{p+1} x_{k,j}\beta_{k,j}$ and $\beta_{k,j} = \sum_{l=1}^{k-1} x_{l,j}$. Parameters δ_k and u_k can be directly calculated for each new incoming sample without using older ones; the parameters β_k and σ_k are recursively calculated as: $\sigma_k = \sigma_{k-1}\sum_{j=1}^{p+1}(x_{k-1,j})^2$, $\beta_{k,j} = \beta_{k-1,j}+x_{k-1,j}$. The potential of the cluster centers (also called focal points) are influenced by new samples as well, as they depend on the distance to all data samples (new and old ones) after (3.27). The update of their potentials can be achieved through the following equation (see [11] for proof):

$$P_k(c_i) = \frac{(k-1)P_{k-1}(c_i)}{(k-2)+P_{k-1}(c_i)+P_{k-1}(c_i)\sum_{j=1}^{p+1}(d_{k(k-1)}^j)^2} \tag{3.31}$$

with $P_k(c_i)$ the potential of the ith center (=rule) at time instance k. The rule evolution step during incremental training and recursive calculation of the potentials is done by comparing the potential of a new data sample with the potential of all existing cluster centers (rules). If the former is higher, then the new data sample is accepted as the center of a new rule and the number of clusters (rules) C is incremented. The reason for this step is that in this case the new data sample is more descriptive, has more summarization power than all the other data samples and all the existing cluster centers. This means that a single or a couple of new data samples

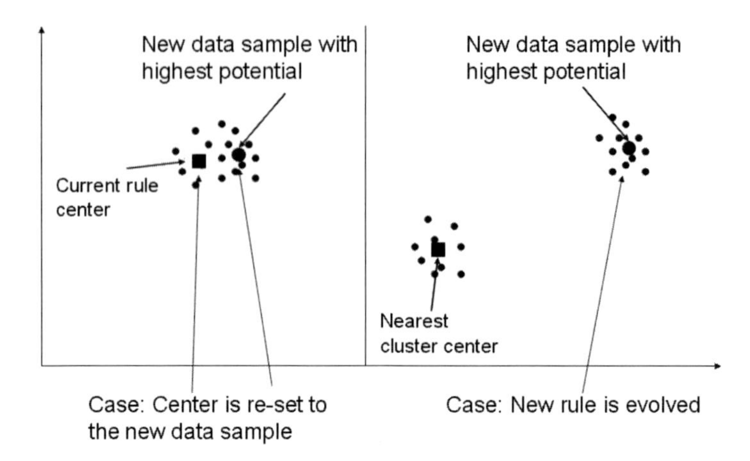

Fig. 3.18 Left part: the case in *eTS* where a cluster (rule) center is replaced by a new data sample which has highest potential among all centers; right part: the case in *eTS* where a new cluster (rule) center is evolved by a new data sample which has highest potential among all centers; note the different distances of the new sample to the nearest centers in both parts.

are usually not sufficient to open up a new rule. Hence, the method is quite robust to outliers in the data. Additionally, it is checked whether the new data sample with highest potential among all centers is close to an already existing center, and if so the existing center is replaced by the coordinates of the new data sample. For closeness, it is checked whether the following condition holds:

$$\frac{P_k(x_k)}{\max_i^C P_k(c_i)} - \frac{dist_{min}}{r} \geq 1 \tag{3.32}$$

with $dist_{min}$ the distance of the current sample to the nearest center. This strategy entails a necessary movement of a cluster center within a data cloud, when the maximal density point in that cloud is drifted/shifted by new incoming data samples (see also Section 4.2). Therefore, *eTS* naturally has the potential to support drift tracking in data streams (with some modifications, in fact, see also Section 4.2). A visualization of these two concepts (evolution of a new cluster center and replacement of an existing one) is shown in Figure 3.18 for a two-dimensional example.

3.2.1.3 Adaptation of Consequent Parameters in *eTS*

For incremental learning of linear consequent parameters, *eTS* exploits both, local learning and global learning [11], one time exploiting recursive fuzzily weighted least squares approach as demonstrated in (2.62) to (2.64) and one time exploiting recursive least squares approach as demonstrated in (2.49) to (2.51). In the former case (allowing more flexibility with respect to replacing or evolving rules, see Section 2.4), when a new rule is evolved, the inverse Hessian is set to αI with α big, according to the investigations in Section 2.2.2.3. The consequent parameters of the new rule are initialized to the weighted average of the parameters in all other rules, where the weights are the normalized firing degrees of the rules in the current (kth) sample, i.e $\Psi_i(\mathbf{x}_k), i = 1,...,C$ as defined in (1.3). This can be achieved separately and independently to the other rules, according to the local learning spirit.

In case of global learning, the evolution of a new rule affects the dimensionality of the parameter space and hence the number of rows and columns in the inverse Hessian matrix, as the consequent parameters of all rules are updated in one sweep. Hence, the linear parameters of the new rule are again set to the weighted average of all the other rules, in the same manner as done for local learning. The inverse Hessian matrix is extended by one row and column containing only zero entries except for the ($C+1$th) diagonal element (note that the new rule is the $C+1$th rule) which is set to α with α big. When a rule center is replaced by the coordinates of the current sample (which is close to this center), the linear parameters as well as the inverse Hessian matrix are simply inherited from the previous time step.

3.2.1.4 The Algorithm

Combining *eClustering* and the concept of consequent parameters and inverse Hessian update together, yields the *eTS* method as demonstrated in Algorithm 11 [11], which starts with an empty rule base and successively adds more and more rules on demand.

Algorithm 11: eTS (evolving Takagi-Sugeno fuzzy systems)

1. Initialization of the rule base structure by forming a rule around the first sample $\mathbf{x}_1$:
$$\mathbf{c}_1 = \mathbf{x}_1 \qquad P_1(c_1) = 1 \qquad \mathbf{w}_1 = 0 \qquad P = \alpha I \tag{3.33}$$
where c_1 the first center, $P_1(c_1)$ its potential at the first time instance, $\mathbf{w}$ all linear consequent parameters joined in one vector and P the inverse Hessian matrix.
2. Collect the next data sample $\mathbf{x}_k$ at the kth time instance.
3. Calculate the potential of the current data sample recursively by using (3.30). Note that δ_k as well as β_k contain accumulated information regarding the spatial proximity of all previous data.
4. Update the potentials of all existing rule centers (focal points) recursively by (3.31).
5. Compare the potential of the new data sample $P_k(x_k)$ to the updated potentials of all existing centers.
6. **If** $P_k(x_k) > P_k(c_i)\ \forall i = 1,...,C$ **And** the new data sample is close to an old center (i.e. (3.32) is fulfilled)
    a. The new data sample replaces the old center, i.e. suppose that the center has index i then $\mathbf{c}_i = \mathbf{x}_k$ and the potential is also inherited: $P_k(c_i) = P_k(x_k)$.
    b. Inherit the consequent parameters and inverse Hessian from the old rule, so these remain unchanged.

7. **If** $P_k(x_k) > P_k(c_i) \;\; \forall i = 1,...,C$ **And** the new data sample is NOT close to an old center (i.e. (3.32) is not fulfilled for any cluster center)

   a. Introduce a new rule by increment the number of rules $C = C+1$, set $\mathbf{c}_C = \mathbf{x}_k$ and $P_k(c_C) = P_k(x_k)$. The width of the fuzzy sets in the antecedent part of the new rule are set to the (pre-defined) radius of influence r.
   b. Set consequent parameters to a weighted average of the consequent parameters from the other rules, where the weights are the normalized firing degrees, i.e. $\Psi_i(\mathbf{x}_k), i = 1,...,C$ as defined in (1.3).
   c. In case of global learning: extend the inverse Hessian matrix by one row and column containing zero entries except for the diagonal element which is set to α with α big.

      In case of local learning: introduce a new Hessian matrix belonging to the new rule by setting it to αI with α big.

8. Update the linear consequent parameter by recursive least squares (2.49) to (2.51) in case of global learning and by recursive fuzzily weighted least squares (2.62) to (2.64) in case of local learning.
9. If new samples come in goto Step 2, otherwise stop.

3.2.1.5 Extensions to *eTS*

In [12] (*Simpl_eTS*) the complexity of the *eTS* method as described in Algorithm 11 is reduced by replacing the notion of potential with the concept of scatter which provides similar by computationally more effective characteristic of the density of data. The scatter is defined by the average distance of a data sample to all data samples:

$$S_k(x_k) = \frac{1}{N(p+1)} \sum_{m=1}^{N} \sum_{j=1}^{p+1} (x_{m,j} - x_{k,j})^2 \tag{3.34}$$

with $p+1$ the dimensionality of the input/output space and $k = 2,3,...,N$. In a similar way as the potential is used in *subtractive clustering* (see (3.26)), the scatter can be used to elicit the cluster (=rule) centers (focal points) when it reaches a minimum. In fact, then a lot of samples coincide and hence are forming a dense cloud (cluster). The scatter of new incoming samples can be calculated recursively by (here for the kth sample, see [12] for proof):

$$S_k(x_k) = \frac{1}{(k-1)(p+1)} \left((k-1) \sum_{j=1}^{p+1} x_{k,j}^2 - 2 \sum_{j=1}^{p+1} x_{k,j} \beta_{k,j} + \gamma_k \right) \tag{3.35}$$

The scatter of a cluster center can be updated recursively by (here for the kth sample, see [12] for proof):

$$S_k(c_i) = \frac{k-2}{k-1} S_{k-1}(c_i) + \sum_{j=1}^{p+1} (x_{k,j} - x_{k-1,j})^2 \tag{3.36}$$

Based on these definitions, a new rule is evolved respectively replaced with a similar strategy as in Algorithm 11: if the scatter of a new sample calculated by (3.34) is either smaller than the scatter of all existing clusters or higher than the scatter of all existing clusters, a new rule is evolved or replaced if also the new sample is close to an already existing cluster center (distance smaller than $0.5r$). Another extension in this approach is that it performs a population-based rule base simplification. This is carried out by monitoring the population of each cluster, i.e. the number of samples which were assigned to each cluster according to the closeness criteria (similar to the concept of k_i = the number of samples forming cluster i in the *eVQ* approach). If $\frac{k_i}{k}$ (with k the number of all samples seen so far) is smaller than a small threshold (usually 0.01=1% is taken) then the corresponding rule will be switched off. The reason for this strategy is that at that moment when a new rule is evolved it seems that a new rule is required, but later on the data characteristic changes and this rule turns out to be somehow a superfluous or representing outlier samples (as representing only a very small fraction of the whole data). The consequent parameters are updated in the same manner as in Algorithm 11.

During the recent years (2006-2010) several additional extensions of the basic *eTS* approach were proposed and formed to an enhanced EFS approach called *eTS+* [10] (short for *evolving Takagi-Sugeno fuzzy models from data streams*). The extensions include adjustments of antecedent parameters and the structure of the rule-bases based on several criteria such as age, utility, local density and zone of influence. The zone of influence is important for adapting the cluster radius to the current data characteristics (which is fixed in case of basic *eTS* and *Simpl_eTS* approaches). This is achieved by [22]:

$$\sigma_{ij}(k) = \alpha\sigma_{ij}(k-1) + (1-\alpha)c_{ij}(t) \tag{3.37}$$

with $c_{ij}(t)$ the center coordinate of the ith cluster in the j input/output feature, k the current data sample (time instance during learning) and α a learning constant (suggested values lying in $[0.3, 0.5]$ [10]). In this sense, σ_{ij} is substituting the fixed radius r when projecting the clusters onto the axes to form the antecedent parts of the fuzzy systems. The utility of a fuzzy rule measures how frequent the fuzzy rule was used after being generated by calculating the accumulated firing level of the ith rule antecedent:

$$U_k^i = \frac{1}{k - I_k^i} \sum_{l=1}^{k} \Psi_i^l \tag{3.38}$$

with I_k^i the index of the data sample that was used as focal point of the ith fuzzy rule. The age of a cluster (rule) indicates how up-to-date the information is which is generalized by the cluster. It is an important property for removing old rules [20] and indicating drifts and shifts in the data streams [275] [274]. The rule age concept will be explained in more detail in Section 4.2.

Furthermore, *eTS+* applies an on-line monitoring and analysis of consequent parameter values: if the values of the consequent parameters belonging to the same input feature i are small over all rules compared to the parameters belonging to the other input features, then someone can conclude that Feature i can be removed.

Comparison takes place by either using the sum over the absolute parameter values belonging to all other input features (for low-dimension problems) or by using the sum over the absolute parameter values of the most influential features (for high-dimensional problems). This is because in case of a high-dimensional input feature space, the sum of contributions becomes large and masks the effect of a particular feature [10]. If the influence of a feature is smaller than the maximal or the summed influence over the other features, this feature can be removed [10]. This causes a change of dimensionality of the whole fuzzy models, as well as corresponding clusters and inverse Hessian matrices (for recursive adaptation of consequent parameters).

3.2.2 eTS *for Classification (*eClass*)*

The classification variant of *eTS*, called *eClass*, comes with four model architectures [21] [18]:

1. *eClass0* is a fully clustering-based classifier, which evolves clusters with the help of potentials as done in Steps 3 to 7 of Algorithm 11 and assigns the most frequent class to a cluster present in those samples forming this cluster (i.e. for which a cluster was the nearest one). If no class labels are provided, then *eClass0* reduces to an unsupervised incremental and evolving clustering algorithm, where different clusters may represent different operation modes respectively system states [470]. Classification is performed by the winner-takes-it all approach , i.e. by taking the most frequent class label in the nearest cluster as classifier response, see also Section 3.1.3.2.
2. *eClassA* is an evolving fuzzy classifier which is based on the classical fuzzy classification model architecture as defined in Section 1.2.1.5, (1.13) (also exploited by *FLEXFIS-Class SM*). The class labels are included in the training process as appended at the end of the on-line (training) samples. Classification is performed by the winner-takes-it all approach as presented in (1.14).
3. *eClassB* uses a first order Takagi-Sugeno-Kang fuzzy system as defined in (1.2) and sends the input features together with the class labels into the update and evolution of clusters according to Steps 3 to 7 of Algorithm 11. The update of the consequent parameters is done by only exploiting the local learning approach. This approach may suffer from the interpolation problem in case of multi-class classification as shown in Figure 1.12 (Section 1.2.1.5).
4. Hence, *eClassB* was extended to the multi-model case in [21] [18], called *eClassM*, which exploits K Takagi-Sugeno fuzzy models for K different classes. These are trained based on indicator entries as done in *FLEXFIS-Class MM* approach (see Section 3.1.2.2), but by applying Algorithm 11 on each Takagi-Sugeno fuzzy system separately.

A comparison of the performance of these four different approaches is done in [21] and [18], in the latter also an empirical comparison with *FLEXFIS-Class SM* and *FLEXFIS-Class MM* is carried out.

Another classifier architecture was introduced in [19], the so-called MIMO (multiple-input multiple-output) architecture, which defines separate consequent functions for each class per rule, i.e. for the ith rule the following consequent matrix for K classes is obtained:

$$l_{i,K} = \mathbf{w}_{i0} + \mathbf{w}_{i1}x_1 + ... + \mathbf{w}_{ip}x_p \quad (3.39)$$

where $\mathbf{w}_{ij}, j = 1, ..., p$ are column vectors with K rows for the K classes. This means that $\mathbf{w}_{i0}$ contains the intercept values for K classes in the consequent part of the ith rule. Classification of a new sample $\mathbf{x}$ is performed by

$$L = argmax_{m=1,...,K}\hat{f}_m(\mathbf{x}) \quad (3.40)$$

where $\hat{f}_m$ is the predicted (regression) value for the mth class (by summing up the contributions to the mth class over all rules). This architecture could finally show the best performance among all *eClass* approaches as pointed out in [18]. Learning is done with Algorithm 11 (hence called *eClassMIMO*) for each row in (3.39), where the multiple consequent functions for each rule are updated by local learning approach (recursive fuzzily weighted least squares) and setting y to either 1 or 0 if belonging or not belonging to the corresponding class. Furthermore, the widths of the fuzzy sets are updated by a linear convex combination of the old widths and the variances of the new samples to the center of the sets, instead of applying a pre-defined radius of influence r. The complete *eClassMIMO* algorithm in modified form can be found in [18].

3.2.3 *Comparison between* FLEXFIS *and* eTS

In this section, we provide a comparison on a methodological basis between *FLEXFIS* and *eTS* (both exploiting Takagi-Sugeno fuzzy model architecture). Hereby, we highlight some important properties when training these systems during on-line operation mode, such as computational complexity, robustness, parametrization aspects and many others. The following paragraphs will treat each of these properties and provide at the same time a summary of the characteristics of both approaches with respect to these properties.

Single-Pass Incremental Learning

Both approaches are exploiting recursive least squares (for global learning) respectively recursive fuzzily weighted least squares (for local learning), which update the consequent parameters in incremental single-pass manner. To learn the antecedent parts, the *FLEXFIS* family exploits an evolving version of vector quantization (*eVQ*), which is able to update and evolve new clusters (=rules) on demand in single-pass incremental manner (so based on single data samples without using prior ones). The same holds for *eTS*, which uses a recursive sample-wise calculation of the potential or scatter and a single-pass rule evolution and replacement strategy based on the updated potentials of samples and already existing clusters (rules). Hence, both methods are potential candidates for real-time on-line learning as requiring a low amount of virtual memory and updating the models with single samples without using computational-intensive re-training phases with older samples.

Incremental Learning/Evolution from Scratch

From the first step in Algorithm 11, it immediately follows that *eTS* is able to evolve the rule base from the first sample on, i.e from scratch. This is a consequence from the fact, that potentials of samples and clusters can be updated from scratch and that no normalization of the data is required in *eTS* (high potentials are representing local dense regions and are scale-invariant). *FLEXFIS*, on the other hand, requires some initial data for eliciting the ranges of the features involved in the training process when not known before hand, see Option 3 in Step 1 of Algorithm 5. In case of knowing these ranges before hand and used as initial parameter setting, *FLEXFIS* allows a full incremental learning and evolution from scratch.

Optimal Parameter Selection

In *eTS* the parameter r needs to be set to a pre-defined value as also responsible for the closeness of a new sample to already existing clusters and hence determining whether a replacement of the center by a new sample is carried out or not. *FLEXFIS* provides a means of eliciting the optimal setting of the vigilance parameter ρ, steering the tradeoff between plasticity versus stability of the learning process. This is done by exploiting a best parameter grid search scenario coupled with a cross-validation procedure in the setup phase of Algorithm 5.

Rule Evolution

Both approaches are able to evolve new rules on demand based on the characteristics of on-line data streams (Steps 6 and 7 in Algorithm 5 and Algorithm 11).

Smooth Rule Movement

In *FLEXFIS* a smooth movement of the rules through the feature space is guaranteed thanks to the decreasing learning gain when updating the centers and the widths, i.e. by using (3.2): this exponentially decreases with the number of samples forming a cluster (=rule) and is also responsible for a convergence to optimality as pointed out in Lemma 3.1. An example of the trajectory of the movement of a cluster center is shown in Figure 3.19. In *eTS* the situation is a bit different as there the movement of a cluster is caused by a replacement of its focal point (center) (Step 7 in Algorithm 11). This is done whenever a new data sample has higher potential as all the existing cluster centers and is close to an already existing center. The closeness aspect guarantees that the replacement shift is bounded in its intensity (distance).

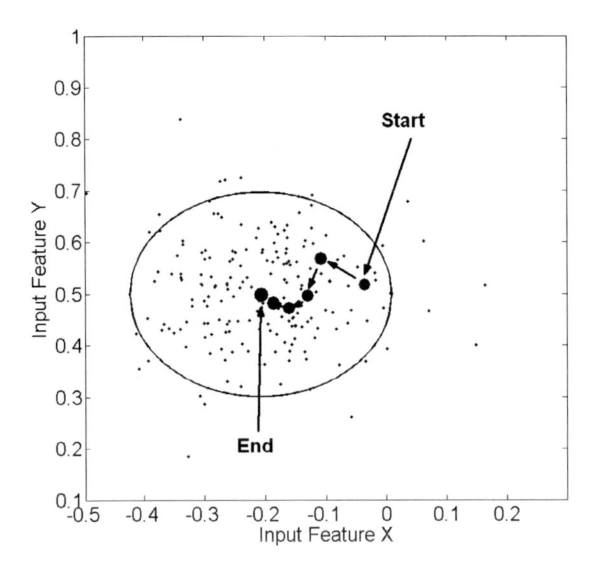

Fig. 3.19 Trajectory of cluster center movement when using *eVQ*, indicated with arrows, cluster centers indicated as big dark dots

However, it is not a smooth movement (centers are reset from one position to the other); in the extreme case centers may jump from one position to another and back, according to dynamically changing densities. To underline this statement, consider a two-dimensional case with features normalized to $[0,1]$, the radius of influence r set to 0.25 and samples appearing at coordinates (0.1,0.1) and (0.15,0.15) in alternative order (so first sample has coordinate (0.1,0.1), second and third samples have coordinates (0.15,0.15), fourth and fifth sample again (0.1,0.1), sixth and seventh sample again (0.15,0.15) and so on — such a case could for instance happen at system operation modes in steady states where the sensor inaccuracy is ± 0.05. Then, the highest data density changes for every two samples between (0.1,0.1) and (0.15,0.15), which also means that the highest potential changes from (0.1,0.1) to (0.15,0.15) for every two samples. As (0.15,0.15) is close to (0.1,0.1) (with respect to r), the center will be replaced instead of a new one evolved.

Rule Growing and Shrinking

In *FLEXFIS*, the widths of the cluster centers in each dimension and hence also the widths of the rules resp. the widths of the fuzzy sets in the rules antecedent parts are permanently updated by (3.5). Hence, the rules may grow and shrink permanently according to the distribution of the data: densifying samples at a small region around the center $\rightarrow$ shrinking, spreading up samples in a larger area around the center (but smaller than the vigilance radius) $\rightarrow$ growing. In *eTS*, the same effect is achieved when applying the idea of updating the widths of the rules by a convex linear combination (as done in its extensions) rather than using a fixed radius of influence r (as done in the basic algorithm).

Rule Pruning

Rule pruning is carried out in extended *eTS* [12] by performing a population-based rule base simplification approach as discussed in Section 3.2.1: rules with low support are deleted. *FLEXFIS* in its native form has no rule pruning technique integrated, however in Sections 5.1 resp. 6.2.2 extensions are presented where clusters resp. rules are merged on-line according to validation criteria for cluster partitions or rule similarity criteria.

Flexibility with respect to Different Local Data Densities

With "different local data densities" it is meant that in different regions of the feature space the groups (clusters) of data may appear with different densities. Hence, it can also be seen as an unbalanced learning problem in a regression setting . An example is demonstrated in Figure 3.20, where at the left hand side the samples forming the sinusoidal trajectory between input feature x and target y are more dense than in the right hand side (compare the densities among dark dots). In such a case, *FLEXFIS* is not affected as the distance to already existing clusters is a criterium for evolving a new cluster rather than a density-based criterium is used. Hence, *FLEXFIS* produces the same number of clusters for the left and the right part (see Figure 3.20 (a)), achieving an accurate approximation of the sinusoidal relationship (solid line). On the other hand, *eTS* evolves new clusters when the potential of new samples (i.e. the density around them) is higher than already existing clusters. This means that samples falling into a less dense region of the feature space may be completely ignored. The effect of this is shown in the right-most part in Figure 3.20 (b), the model extracted based on these clusters shown as solid line. Similar consideration can be carried out for classification problems.

Insensitivity to the Data Order

In *eTS* a new cluster is not evolved as long as there is not a new sample with a higher potential than all the existing centers. In fact, when a specific order of the data is applied to the algorithm, it may happen that sample #i in a dense region becomes a cluster center when this new dense region appears, whereas for another order another sample #j from this dense region becomes a cluster center. However, if these two samples are near to each other (which is usually the case in a compact dense region), then after some time in both cases the center will be replaced to that sample with the highest potential in the dense region (could be sample #i or #j or a different one). In this sense, the replacement property within dense regions assures

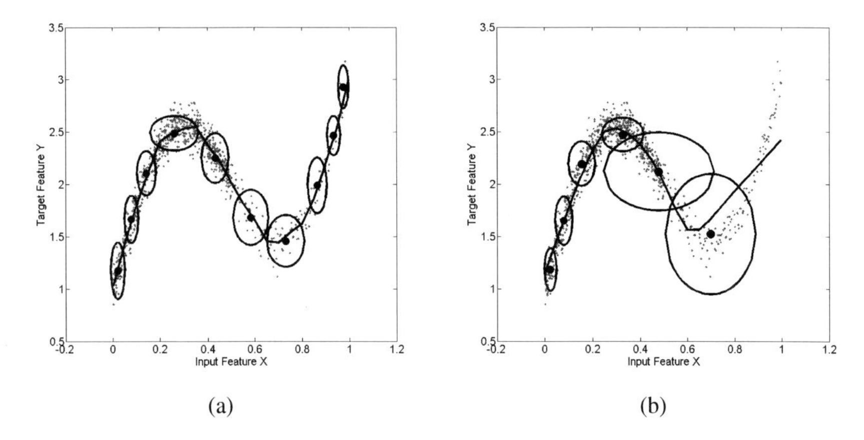

Fig. 3.20 (a): evolved clusters with *eVQ* and obtained approximation (solid line) $\rightarrow$ equally distributed clusters; (b): evolved clusters and obtained approximation (solid line) when using recursive potential (*eClustering*) $\rightarrow$ the right hand side (with lower density) under-clustered; notice the lower data density on the right hand side (from $x = 0.5$ to 1.0) than on the left hand side (from $x = 0.0$ to 0.5)

more or less an insensitivity with respect to the order of the data (although it may happen in a few cases). In *FLEXFIS* the situation is different, as new clusters are evolved based on the distance to already existing centers. This means that in one case (one order of the data) a sample may be far away from an existing center and hence trigger a new cluster, whereas later this gap is filled up with new data joining the two data clouds together to one bigger one. In another case, first, a sample near to the already existing centers comes in (at least not further away than the vigilance parameter), successively growing and expanding the cluster. Figure 3.21 visualizes an example of the impact of the sensitivity with respect to different orders of the data in case of *eVQ* (compare the generated clusters in (a) and (b)). On-line split-and-merge techniques as will be outlined in Section 5.1 will partially solve such effects. In [268] it is demonstrated, even though *eVQ* is quite sensitive to different data orders, *FLEXFIS* remains quite robust (as also underlined in the application part of the book). A reason for this effect is that some differences regarding the appearance of the clusters in the cluster space are compensated by the regression-based learning of consequent parameters (which aiming to form a smooth regression surface). In case of (fuzzy) classifiers using single model architecture (*FLEXFIS-Class SM* and *eVQ-Class*), the impact can be expected as more significant as classifiers are formed directly around the generated clusters.

Convergence to Optimality

Convergence to optimality in the least squares sense is guaranteed by *FLEXFIS* through local learning and the introduction of correction terms, converging to 0 after

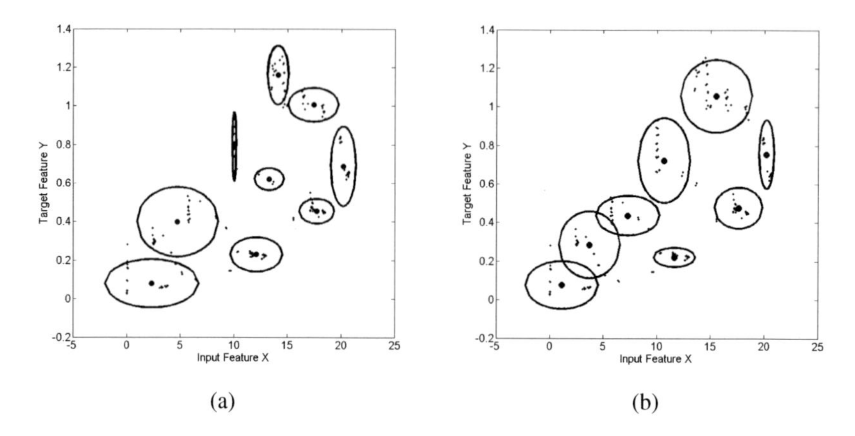

Fig. 3.21 (a): clusters obtained when applying *eVQ* on the original order of the data; (b): clusters obtained when applying *eVQ* on the shuffled order of the data

Lemma 3.1 (see also Sections 3.1.1.2 and 3.1.1.4). *eTS* applies a specific re-setting and merging mechanism for the inverse Hessian matrix and consequent parameters of replaced or newly evolved rules. Also, it is mentioned, that a non-optimal situation may occur in case of a replacement of a rule, but will disappear after a few samples.

Robustness to Outliers

In *FLEXFIS* a new cluster is evolved whenever a new incoming data sample lies further away than the (pre-defined) vigilance parameter ρ from already existing clusters. In this sense, outliers usually cause new rules to be evolved, as it can be not distinguished between a new system state and faults causing outliers. Although these outlier clusters may occur significantly away from the good feature samples, they contribute to the predictions of the evolved fuzzy system, as *FLEXFIS* uses Gaussian fuzzy sets with infinite support. So, *FLEXFIS* in its native form is not robust to outliers. In Section 4.4 (Chapter 4) a (generic) concept will be demonstrated showing how to overcome this problem. *eTS* on the other hand uses the information of potential to decide whether a new cluster should be born or not. The potential of a sample is high when there are a lot of samples near to the samples. One could imagine a dense data cloud (cluster) in the feature space: the center point of this cloud has highest potential (at least, if the density is equally distributed over this cloud, which is usually the case). This means that single outliers or outliers consisting of only a handful of samples would never cause the algorithm to evolve a new cluster, as 1.) too far away from denser regions representing real system states and 2.) not being dense enough on their own. This means that *eTS* is per nature robust with respect to outliers.

Dealing with Missing Values

Missing values can be handled in *FLEXFIS* by using the approach demonstrated in Section 3.1.4.5, which searches the nearest cluster center with respect to known features and assigns the corresponding coordinates of the nearest center to the missing features values. In *eTS* such an approach has not been investigated yet.

Feature Weighting

Giving different features different weights according to their importance levels in the models and hence preventing over-fitting by a kind of continuous feature selection approach is integrated in *FLEXFIS*. This is achieved by applying a weight vector λ in the incremental clustering process when moving centers and updating the widths of the clusters in each dimension, see Section 3.1.4.6. Feature weighting may also serve as an important aspect for gaining more insight into the whole process, as it shows which system variables are more or less important for finding useful interrelations and dependencies. How feature weights will be calculated incrementally, will be demonstrated within the scope of a generic concept in Section 5.2 (also refer to [272] [270]). In *eTS*, such an approach has not been investigated so far.

Invariance with respect to Scaling of the Features

The calculation of potentials of data samples as well as cluster centers is per nature scale-invariant, as it measures the densities of local data clouds rather than distances between samples and centers. When the concept of scatter is applied in the extended version of *eTS* (as in [12]), the samples are normalized to the normal distribution. In *FLEXFIS*, distances between samples play an essential role for deciding whether new rules are evolved or not. Normalization of the samples to $[0,1]$ is done at the beginning of each incremental learning cycle based on the actual ranges of the features. In this sense, scale invariance can be achieved. Both approaches apply the recursive fuzzily weighted least squares for estimating the consequent parameters, which is per nature a scale invariant method.

Computational Complexity

Both methods apply recursive (fuzzily) weighted least squares approach for estimating the consequent parameters, hence we inspect here only the computational complexity of the update of the antecedent parts. In both approaches, whenever the condition for evolving a new rule holds, a new center is formed around the current sample, i.e. the coordinates of the new center become the coordinates of the current data sample, the linear consequent parameters as well as the inverse Hessian matrix are set to an initial value, respectively inherited by the other rules (as done in *eTS*). Whenever the condition for evolving a new rule does not hold, *FLEXFIS* updates the center and widths of the nearest cluster (winner cluster). Eliciting the nearest cluster has complexity $O(C(p+1))+C$, as calculating the Euclidean distance to C clusters requires $p+1$ subtractions, squares and summations (plus one square root) in the case of calculating the distance to the center and $2*(p+1)$ subtractions, squares and summations (plus one square root and p division) when calculating the distance to the surface (as done in (3.19) for modified version); the minimal distance over C clusters has to be elicited (counting for the additional C). Updating the centers in *FLEXFIS* is done by just moving the winner cluster towards the current sample for each of the $p+1$ input/output features, hence having complexity of $O(p+1)$; the same holds for the update of the widths in each direction. For the condition itself no further complexity is required as the distance to the nearest center, already calculated for eliciting the winner cluster, is checked versus the vigilance parameter. In sum, the antecedent learning and rule evolution part in *FLEXFIS* has a complexity of $O(C*(p+1)+2*(p+1))+C$ for updating a fuzzy model with one new incoming sample. In *eTS*, the potential of a new data sample is updated which has a complexity of $O(4*(p+1))$, as δ_k, v_k, σ_k and β_k require $O(p+1)$ summations, see (3.30). Furthermore, for each new incoming sample the potentials of all existing centers are updated, hence requiring a complexity of $O(C(p+1))$. For the condition itself, the updated potential of the current sample is compared with the maximal potential of the existing clusters, hence requiring C comparisons. In sum, the antecedent learning and rule evolution part in *eTS* has a complexity of $O(C*(p+1)+4*(p+1))+C$, which is just a little higher than in *FLEXFIS*. When using the concept of scatter instead of the concept of potential, the same complexity is achieved (as again requiring $O(fac*(p+1))$ summations for updating the scatter with each new incoming sample, with fac a fixed small integer), however then the number and complexity of the divisions are reduced (divisions only require integers instead of real values).

Interpretability

With respect to interpretability of the achieved evolved fuzzy systems, both approaches may suffer from the cluster projection concept: fuzzy sets may strongly overlap when projecting high-dimensional clusters onto the input/output axes, hence causing redundant information in the systems. In this sense, it may even happen that smaller fuzzy sets are (partially) enclosed in larger ones. The whole chapter 6 will be dedicated to this topic and demonstrate approaches on how to guide evolved fuzzy systems to more interpretable power during the incremental on-line learning mode by addressing the accuracy-interpretability tradeoff.

Manifold of Method and Architecture Variants

eTS comes with two variants for evolving new rules, the concept of potential and the concept of scatter, and furthermore includes extended versions *Simpl_eTS* and *eTS+*. The classification version of *eTS*, *eClass*, comes with five different variants (*eClass0*, *eClassA*, *eClassB*, *eClassM*, *eClassMIMO*), exploiting five different model architectures, four fuzzy classification model architectures and one

Table 3.1 A summary of the characteristics of the methods, each property indicated by '+' = fully fulfilled, 'o' = partially fulfilled and '-' = not fulfilled

Property	***FLEXFIS***	***eTS***	**Property**	***FLEXFIS***	***eTS***
Single-pass inc.	+	+	Inc. learning from scratch	o	+
Optimal param. selection	+	o	Rule evolution	+	+
Smooth rule movement	+	o	Rule growing and shrinking	+	+
Rule pruning	-	+	Flex. to data densities	+	-
Insensitivity to data order	o	+	Convergence to opt.	+	o
Robustness to outliers	-	+	Dealing with miss. values	+	+
Feature weighting	+	-	Scale invariance	+	+
Computational speed	+	+	Interpretability	-	-
Manifold	+	+			

clustering-based classifier. *FLEXFIS* applies only one variant for regression problems by using *eVQ* for rule evolution and update of the antecedent parts and some of its extensions (applied in *FLEXFIS-MOD*) and exploits three different model architectures for its classification variant, *FLEXFIS-Class*, two fuzzy classification model architectures (*FLEXFIS-Class SM* and *FLEXFIS-Class MM*) and one clustering-based classifier (*eVQ-Class*).

A summary of the characteristics of the methods with respect to the properties handled in the previous paragraphs is demonstrated in Table 3.1. For each property we assign either a '+' indicating that the method fulfills this property, a 'o' when the property is partially fulfilled, a '-' when the method does not fulfill this property.

For a detailed empirical comparison between *FLEXFIS* and *eTS* on high-dimensional and noisy real-world data sets, please refer to [16] and [17] for regression problems, as well as to [18] for classification problems.

3.3 The *ePL* Approach

Evolving fuzzy participatory learning (*ePL*) , firstly introduced in [258] and significantly extended in [259], exploits the concept of *participatory learning* (*PL*, introduced by Yager in [456]) for learning and evolving the antecedent parts of rules. The *PL* concept can be viewed as an unsupervised clustering algorithm [398] and hence is a natural candidate for finding the structure of the rule base during an evolving fuzzy modelling process. This concept is made incremental and evolvable in the evolving participatory learning variant. *ePL* also uses Takagi-Sugeno type fuzzy systems and hence exploits recursive fuzzily weighted least squares for local learning of the consequent parameters (in order to obtain maximal flexibility — as also done in *FLEXFIS* and *eTS* approaches). For modification of the rule base during the on-line learning process, *ePL* applies the same concept as *eTS*, i.e. in each incremental learning step it is verified whether a new cluster should be created, an old cluster should be replaced or whether redundant clusters should be eliminated.

The basic difference between *ePL* and *eTS* is the procedure on how to update the rule base structure. Differently to *eTS*, *ePL* uses a fuzzy similarity measure to determine the proximity between new data and the existing rule base structure. An essential property of participatory learning is that the impact of new data in causing self-organization or model revision depends on its compatibility with the current rule base structure (=current cluster structure) [259]. More formally, given an initial cluster structure, i.e. a set of cluster centers $\mathbf{c}_1^k,...,\mathbf{c}_C^k$ in the kth update step, the update with a new incoming sample is performed by using a compatibility measure $\rho_i^k \in [0,1]$ and an arousal index $a_i^k \in [0,1]$. Their values depend 1.) on the cluster (associated with the subscript i) and 2.) on the update step (associated with the subscript k). While $\rho_i^k \in [0,1]$ measures how much a data sample is compatible with the current cluster structure, the arousal index $a_i^k \in [0,1]$ is responsible for yielding a warning when the current cluster structure should be revised according to the new information contained in the new sample. If the arousal index is greater than a threshold value $\tau \in [0,1]$, then a new cluster is evolved by setting the center coordinates to the coordinates of the current sample $\mathbf{x}^k$. Otherwise, the ith cluster is adjusted as follows (see [259]):

$$\mathbf{c}_i^{k+1} = \mathbf{c}_i^k + G_i^k(\mathbf{x}^k - \mathbf{c}_i^k) \tag{3.41}$$

with

$$G_i^k = \alpha(\rho_i^k)^{1-a_i^k} \tag{3.42}$$

α the (fixed) learning rate and

$$\rho_i^k = 1 - \frac{\|\mathbf{x}^k - \mathbf{c}_i^k\|}{p} \tag{3.43}$$

where the ith cluster is associated with the winning cluster, i.e.

$$i = argmax_{j=1,...,C}(\rho_j^k) \tag{3.44}$$

The arousal index for the ith cluster is updated by:

$$a_i^{k+1} = a_i^k + \beta(1 - \rho_i^{k+1} - a_i^k) \tag{3.45}$$

The parameter $\beta \in [0,1]$ steers the rate of the change of the arousal index: the closer β is to one, the faster the system is to sense compatibility variations.

The arousal index can be interpreted as the complement of the confidence someone has in the truth of the current belief, i.e. the rule base structure. The arousal mechanism monitors the performance of the system by observing the compatibility of the fuzzy system with the current data samples. Therefore, learning is dynamic in the sense that (3.41) can be viewed as a belief revision strategy whose effective learning rate (3.42) depends on the compatibility between the new samples and the current cluster (=rule) structure as well as on the confidence of the fuzzy system itself. When looking at the update formulas for *ePL* and *FLEXFIS*, i.e. comparing (3.41) and (3.1), there are obviously some synergies as in both methods the centers are moved towards new data samples (in fact, by a fraction of the difference between the new sample and the existing center). This is done for the winning center, i.e. the one which is closest to the current sample in both approaches. In *ePL*, indeed the compatibility index ρ_i is used to elicit the nearest center, whereas in *FLEXFIS* this is achieved by using the Euclidean distance to the center, however both are equivalent in the sense that they select the same winning clusters. The basic difference between these two methods is how the learning gain is set. In *FLEXFIS*, the learning gain for a specific cluster permanently decreases with the number of samples forming this cluster; in this sense, a kind of convergence to a stable center position is achieved, which is favorable for the convergence to optimality aspect (see Section 3.1.1.4). In *ePL* the learning rate is calculated for each sample newly by taking into account the compatibility of this sample to the already existing rule base. Hence, the learning rate may significantly vary from sample to sample, achieving a significantly higher flexibility in the learning process. On the other hand, the stability aspect is diminished and it is currently not fully examined how these more unpredictable movements affect the convergence of the consequent parameters to the optimal least squares solution.

An extension of *ePL* includes the deletion of redundant clusters, which may come up during the incremental learning process (consider that clusters may move together). Therefore, in *ePL*, a cluster is declared as redundant whenever its similarity with another center is greater than or equal to a threshold value λ. If this is the case, the redundant cluster is simply deleted. The similarity between clusters i and j is expressed as the distance between their centers:

$$\rho_{c_{i,j}}^k = 1 - \frac{\|\mathbf{c}_i^k - \mathbf{c}_j^k\|}{p} \tag{3.46}$$

In sum, in *ePL* four parameters need to be tuned (learning gains α and β as well as thresholds λ and τ). Combining the steps described above together yields us Algorithm 12 (see also [259]).

Algorithm 12: ePL (evolving fuzzy Participatory Learning)

1. Initialize the C rule centers and parameters α, β, λ, τ and dispersion (widths of fuzzy sets) r (all fuzzy sets with same widths)
2. Take the next incoming data sample: $\mathbf{x}^k$
3. Compute the compatibility index ρ_i^k using (3.43) with i elicited as in (3.44).
4. Update the arousal index a_i^k using (3.45).
5. **If** $a_i^k \geq \tau, \forall i \in \{1,...,C\}$, then $\mathbf{x}^k$ denotes a new cluster center; increase number of clusters: $C = C + 1$.
6. **Else** Update the center by (3.41) using (3.42)-(3.45).
7. Compute $\rho_{c_{i,j}}^k, \forall j = \{1,...,C\}$ using (3.46).
8. **If** $\exists j,\ \rho_{c_{i,j}}^k \geq \lambda$, then delete $\mathbf{c}_i$ and set $C = C - 1$.
9. Update rule base structure (projection of cluster centers onto input axes).
10. Update rule consequent parameter with local learning approach using (2.62)-(2.64).
11. If new samples are coming in, goto Step 2; otherwise, stop.

3.4 The *DENFIS* Approach

Pioneering work in the field of evolving neural networks and in particular in evolving neuro-fuzzy approaches was done by Kasabov at the beginning of 00s. One of the most important developments in the field of neuro-fuzzy systems was the so called *DENFIS* approach [212], which is short for *Dynamic Evolving Neural Fuzzy-Inference System*. This method was originally deduced from the *EFuNN* approach [211], which is an evolving method for fuzzy neural networks and was originally inspired by the creation of connectionist models in an *ECOS (Evolving Connectionist Systems)* architecture [208]. The *EFuNN* approach was already able to perform an incremental sample-wise life-long learning of its model structure by dynamically creating new structural components, which consisted of five different input layers. Fuzzy quantization was used for fuzzifying input and output neurons. In this sense, fuzzy neural networks are used as connectionist structures that can be interpreted in terms of fuzzy sets and fuzzy rules [262] [135]. On the other hand, fuzzy neural networks are neural networks with all the characteristics therein (such as recall, reinforcement, hidden layers etc.), while neuro-fuzzy inference systems are fuzzy rule-based systems (as defined in Section 1.2.1.3) together with their associated fuzzy inference mechanisms that are implemented as neural networks for the purpose of learning and rule optimization. For instance *ANFIS* [194] is a data-driven learning approach which exploits this model architecture. *DENFIS* is an evolving approach exploiting this model architecture by applying the cross-link to Takagi-Sugeno fuzzy type systems as shown in Figure 1.9, Chapter 1. Hence, in this book dealing with *evolving fuzzy systems*, we neglect the description of the *EFuNN* approach and focus on its successor, the *DENFIS* approach.

A specific property and a principle difference to the other evolving fuzzy systems approaches mentioned in the previous sections is the way how a prediction for a new incoming sample is made. By doing so, *DENFIS* applies the idea of a kind of model-based lazy learning: depending on the position of the input vector in the feature space, a fuzzy inference system for predicting the output is formed dynamically based on the m nearest fuzzy rules that were created during the incremental learning process. Hence, we speak about model-based lazy learning, where the classical lazy learning approach is a sample-based one [400], i.e. taking the local samples nearest to the query point to build a (small) local model on demand. Regarding the learning process itself, there are some synergies to the aforementioned approaches:

1. Usage of Takagi-Sugeno type fuzzy model architecture — indeed, by putting it into a neuro-fuzzy system form, however resulting in the same learning problem, i.e. linear consequent, non-linear antecedent parameters, rules/neurons to be evolved on demand.
2. Rule (neuron) evolution by using a clustering-based approach: evolving a new cluster means evolving a new rule.
3. Local learning of consequent parameters.

The basic differences to the aforementioned methods are that *ECM* is used as *Evolving Clustering Method*, triangular membership functions are used as fuzzy sets instead of Gaussian functions and the local learning of the consequent parameters is done with an alternative weighting scheme (see below).

3.4.1 *Rule Evolution and Update with* ECM

The *ECM*[212] is a fast on-line clustering method, which does not need the number of clusters pre-parameterized as they are automatically evolved from data streams based on their characteristics. It is a distance- and prototype-based clustering method, where each cluster center is associated with a rule center and a cluster radius is synchronously updated defining the widths of the rules and fuzzy sets (when projected). The distance of a new incoming sample to the closest cluster center cannot be larger than a threshold *thr*, otherwise a new cluster is evolved. This is somewhat related to the concept of the vigilance parameter used in *eVQ*. However, here this distance is calculated in a different way by using the distance of a new sample $\mathbf{x}$ to the center minus the radius of influence of a cluster (here for the *i*th):

$$dist_i = \|\mathbf{x} - \mathbf{c}_i\| - r_i \tag{3.47}$$

with $\mathbf{c}_i$ the coordinates of the center of cluster i and r_i the range of influence of a cluster i, which defines spherical clusters (same radius in each direction/dimension). If $\mathbf{x}$ lies inside the range of influence of cluster i, then above distance function is negative. If it is lying close to cluster i, then the distance function is slightly positive. Another difference to *eVQ* is how the radius of influence is updated and how and when cluster centers are moved. In *eVQ*, the center of the closest cluster is always moved towards the current sample, whereas in *ECM* this movement is only carried out when a new sample lies outside the range of influence of the nearest cluster. In this case, the center of the nearest cluster is moved towards $\mathbf{x}$ so that the distance from the new center to $\mathbf{x}$ is equal to the new range of influence of the nearest cluster. The new range of influence $r_i(upd)$ is calculated by $r_i(upd) = \|\mathbf{x} - \mathbf{c}_i\| + r_i$. This finally means that the range of influence of a cluster always depends on only one current data sample and hence may become larger and larger during the learning process (consider samples appearing outside but near the range of influence of a cluster, forcing it to grow larger and larger). In extreme cases, one cluster may fill up the whole feature space. This may be a drawback compared to *eVQ* in its extended version where ranges of influence of clusters are estimated by the recursive variance of the data samples in each dimension, achieving a kind of local convergence of the clusters (ellipsoidal surfaces).

The whole *ECM* is summarized in Algorithm 13 (see also [208]).

Algorithm 13: ECM (Evolving Clustering Method)

1. Create the first cluster by simply assigning the coordinates of its center to the coordinates of the first sample from the data stream: $\mathbf{c}_1 = \mathbf{x}$. Set the number of cluster to 1: $C = 1$.
2. Take the next incoming sample $\mathbf{x}$, if available, otherwise stop.
3. Calculate the distance of $\mathbf{x}$ to the already existing center with a normalized Euclidean distance measure.

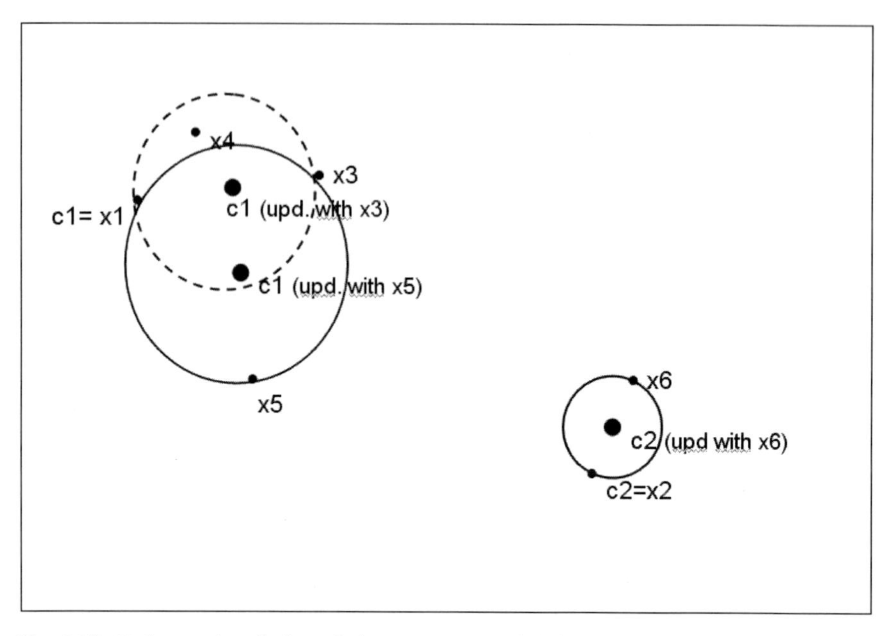

Fig. 3.22 Update and evolution of cluster centers and their ranges of influence; samples x_1 and x_2 create new cluster centers as lying far away from each other, sample x_3 updates c_1 and triggers an initial range of influence (dotted sphere around c_1), which is updated by x_5 (solid sphere around updated c_1); x_4 does not change anything as lying within the range of influence (dotted sphere) of c_1, x_6 updates c_2 and its range of influence

4. Elicit the minimal distance min_{dist} and associate the cluster center which is nearest: $\mathbf{c}_{win}$.
5. **If** $min_dist < r_{win}$ (with r_{win} the range of influence of the nearest cluster), then neither a new cluster is created nor the nearest cluster is updated. Goto Step 2.
6. **Else**

   a. Find a cluster $win2$ which has minimal value of (3.47), associate its center with $\mathbf{c}_{win2}$ and its range of influence as r_{win2}.
   b. **If** $dist_{win2} > 2 * thr$, $\mathbf{x}$ does not belong to any existing cluster and a new one is created in the same way as described in Step 1.
   c. **Else**, the cluster $win2$ is updated by setting its range of influence $r_{win2} = \frac{\|\mathbf{x}-\mathbf{c}_{win2}\|+r_{win2}}{2}$ and by moving the center towards $\mathbf{x}$, so that the distance from the new center to $\mathbf{x}$ is equal to r_{win2}.
   d. Goto Step 2.

An illustration of cluster movement and evolution in *ECM* is presented in Figure 3.22, where data samples x_1 to x_6 are loaded one by one and the cluster centers shown as big dots and their ranges of influence as spherical regions.

An extension of *ECM*, *ECMc*, is presented in [212], where a global optimization procedure is applied to the clusters found a priori by *ECM*. In particular, *ECMc* further minimizes an objective function based on a distance measure (usually Euclidean distance) subject to given constraints and deriving a membership matrix U (similarly as done in fuzzy c-means approach [46]). This can be only done in an off-line iterative method over a batch of data samples, see [212] for details. Another extension applies an adaptive threshold thr steering the evolution versus update of new rules, see [407].

3.4.2 *Adaptation of Consequents in* DENFIS

The consequent parameters in the Takagi-Sugeno type neuro-fuzzy system are again estimated by recursive weighted least squares approach (2.62) to (2.64), exploiting the local learning aspect (for each rule separately). The difference to *FLEXFIS*, *eTS* and *ePL* is that the weights are not the membership values to the corresponding (say the ith) rule ($\Psi_i(\mathbf{x})$), but the (normalized) distance between the current sample to the center of this rule. In this sense, consequent learning in *DENFIS* is independent from the applied fuzzy sets, whereas in the other approaches it depends on the type and widths of the fuzzy sets. On the other hand, in *DENFIS* samples appearing further away from the rule centers are giving higher weights than samples close to these, which may cause linear consequent functions not presenting the core parts of the rules significantly well.

Combining the previously described issues together, gives us the formulation of the *DENFIS* approach in Algorithm 14 (see also [208]).

Algorithm 14: DENFIS (Dynamic Evolving Neural Fuzzy-Inference System)

1. Take the first N samples and apply *ECM* on these to obtain C cluster centers.
2. For every cluster center $\mathbf{c}_i$ find the $k_i < N$ samples which are closer than to any other centers.
3. Associate the C clusters with C rules and create the antecedent parts of these rules by using cluster center coordinates as coordinates for the fuzzy sets in each dimension and using the r_i's as spans of the triangular fuzzy sets.
4. Calculate the linear parameters in the consequents by using the k_i samples for the ith rule in the local learning approach with specific weights (distance to the cluster centers).
5. **Start on-line phase:** take the next incoming sample $\mathbf{x}$.
6. Perform Steps 3 to 6 in Algorithm 13 for updating/evolving the cluster partition.
7. **If** new cluster(s) are created, create new fuzzy sets and rules as in Step 3; create new consequent functions for the new rules by assigning their parameters to the consequent parameters of the nearest (already existing) rules.
8. **Else** Update the antecedent parts of the existing rules (corresponding to the updated clusters)
9. Update the consequent parameters of all rules whose centers are nearer than $2 * thr$ to $\mathbf{x}$.
10. If new samples are incoming, Goto Step 5, otherwise stop.

From this formulation it can be realized that *DENFIS* exploits an initial batch phase (similarly to *FLEXFIS*, but with fixed parameter setting, i.e. no parameter optimization process takes place) to generate a first initial fuzzy model.

DENFIS is compared with *eTS* in analytical and empirical aspects in [14].

3.5 The *SOFNN* Approach

The *SOFNN* approach, short for *Self-Organizing Fuzzy Neural Network*, was first introduced in [245] and later extended by Leng et al. in [246] and particularly in [244] for fuzzy rule extraction. As the name already indicates, the approach exploits a fuzzy neural network structure, which is more or less equivalent to the neuro-fuzzy inference structure as shown in Figure 1.9. In particular, *SOFNN* is one of the decompositional techniques (as e.g. [422]), where the fuzzy rules are directly extracted from hidden layer neurons and are then aggregated to form the global relationship. The on-line learning capability is coming in by adjusting and evolving the neurons and consequent parameters automatically with new incoming samples. Also, the ellipsoidal basis function neuron (i.e. high-dimensional Gaussian functions) do not need to be parameterized with the help of a priori knowledge, but are organized by the method automatically. A specific property of this method is that it guarantees ε-completeness of the fuzzy rules, which helps to alleviate the dilemma between interpretability and accuracy [241]. In fact, ε-completeness of a fuzzy partition means that no individual input can have a membership degree less than a pre-defined low value ε. If applied to the fuzzy partitions for each feature, this ensures a well-defined coverage of the whole input space. Adding and pruning of neurons are based on the assurance of ε-completeness and the generalization quality of the current system when predicting new samples.

3.5.1 Architecture and Consequent Learning

SOFNN exploits a network structure with five layers, similar to the one as shown in Figure 1.9. The only difference is in layer #4 where instead of $\mathbf{x}$ and $\mathbf{y}$ the bias is defined in a more general way by matrix B.

Layer #1 is the input layer, where each neuron represents an input variable $x_1,...,x_p$. Layer #2 is the ellipsoidal basis function (EBF) layer, where each neuron presents the whole antecedent part of a fuzzy rule: each membership function is a Gaussian fuzzy set, defined by (also compare Chapter 1, Section 1.2.1.2)

$$\mu_{ij}(x_j) = e^{-\frac{1}{2}\frac{(x_j-c_{ij})^2}{\sigma_{ij}^2}} \tag{3.48}$$

where c_{ij} denotes the center and σ_{ij}^2 denotes the variance of the Gaussian function appearing in the j-th premise part of the i-th rule. The output of the neurons in layer #2 are computed by applying the product on these functions over all p dimensions; hence, for the jth neuron, someone obtains:

$$\Phi_j = e^{-\sum_{i=1}^{p} \frac{(x_j - c_{ij})^2}{2\sigma_{ij}^2}} \tag{3.49}$$

Layer #3 is the normalization layer, where the membership degree of each neuron is normalized by the sum of all membership degrees — compare (1.10).
Layer #4 is the weighted layer, where the matrix B defines the bias for the linear weights: if it consists of the original data matrix with samples as rows and the inputs $x_1, ..., x_p$ as columns, then it triggers a Takagi-Sugeno fuzzy model with purely linear hyper-planes; if it consists of more complex terms, an arbitrary Takagi-Sugeno-Kang type fuzzy systems can be achieved (compare (1.1)).
Layer #5 defines the output layer, where the normalized membership degrees are weighted by the linear consequent parameters from the 4th layer — compare (1.2).

The adaptation of the consequent parameters during on-line mode is done with a modified version of the recursive least squares algorithm, so global learning of all parameters in one sweep is applied. The modification concerns the introduction of a weight parameter α, which modifies the update to (compare with the original version as demonstrated in (2.49) to (2.51)):

$$\hat{\mathbf{w}}(k+1) = \hat{\mathbf{w}}(k) + \alpha\gamma(k)(y(k+1) - \mathbf{r}^T(k+1)\hat{\mathbf{w}}(k)) \tag{3.50}$$

with the correction vector

$$\gamma(k) = P(k+1)\mathbf{r}(k+1) = \frac{P(k)\mathbf{r}(k+1)}{1 + \mathbf{r}^T(k+1)P(k)\mathbf{r}(k+1)} \tag{3.51}$$

$P(k+1)$ can be computed recursively by:

$$P(k+1) = (I - \alpha\gamma(k)\mathbf{r}^T(k+1))P(k) \tag{3.52}$$

with $\mathbf{w}$ the weights in the consequents, y the target variable, $\mathbf{r}$ the vector containing all regressors and P the inverse Hessian matrix. In *SOFNN* α is set by:

$$\alpha = \begin{cases} 1 & |e(k+1)| \geq |\varepsilon(k+1)| \\ 0 & |e(k+1)| < |\varepsilon(k+1)| \end{cases} \tag{3.53}$$

This means, that at time instance $k+1$, when the estimation error $|e(k+1)|$ is less than the approximation error $|\varepsilon(k+1)|$ as defined in (3.54), the parameters of the system are not adjusted. This ensures an asymptotic convergence of the estimation error and the convergence of the parameters to finite values [244].

3.5.2 Adding Fuzzy Rules

There are two criteria responsible for adding a new fuzzy rule (=neuron) or not. The first one is based on the actual generalization quality and defines an error criterion, where the deviation between the estimated and the given real value of the current sample is measured:

$$|\varepsilon(k)| = |\hat{y}_k - y_k| \tag{3.54}$$

with y_k the measured value and $\hat{y}_k$ the estimated value = output from the network at time instance k. If $|\varepsilon(k)| > \delta$, with δ a pre-defined error a new EBF neuron (fuzzy rule) is considered to be added to the network. The second criterion is called the if-part criterion and checks whether new incoming samples are sufficiently covered by any neurons (fuzzy rules) in the network. When assuming a normal distribution, 95% of the data samples belonging to the jth membership function in the ith rule lie in the range of $[c_{ij} - 2\sigma_{ij}, c_{ij} + 2\sigma_{ij}]$. At the border values of this interval the Gaussian membership degree is 0.1354, which defines the threshold for the minimal coverage of a sample. This automatically gives us the value for ε for assuring ε-completeness of the fuzzy partitions. If $\Phi_L = argmax_{j=1,...,C}\Phi_j < 0.1354$, no fuzzy rule could cover the current input vector sufficiently. Hence, either the width of the nearest neuron should be modified or a new neuron should be considered to be added to the network.

The above two conditions (based on the two criteria for adding or extending neurons) yield four cases when combined together with their opposites:

1. Case 1: $|\varepsilon(k)| \leq \delta$ and $\Phi_L \geq 0.1354$: this means that the network has good generalization quality and some neurons can cover the current input vector sufficiently. In this sense, nothing is done in the antecedent parts of the rules, only the consequent parameters are updated by (2.49) to (2.51).
2. Case 2: $|\varepsilon(k)| \leq \delta$ and $\Phi_L < 0.1354$: this means that the network has good generalization quality, but no neuron can cover the current input vector sufficiently. Hence, the width of the membership function that has the lowest value in the neuron with highest activation degree is enlarged by applying:

$$\sigma_{ij}(new) = k_\sigma \sigma_{ij}(old) \tag{3.55}$$

   where k_σ is a pre-defined constant which is larger than one (usually set to 1.12).
3. Case 3: $|\varepsilon(k)| > \delta$ and $\Phi_L \geq 0.1354$: this means that the network has bad generalization quality, even though some neurons can cover the input vector sufficiently. Hence, a new neuron (fuzzy rule) has to be added as the already existing fuzzy rules are not able to give a prediction with sufficient quality. The distance of the new sample $\mathbf{x}$ to the center of the closest neuron (fuzzy rule) $\mathbf{c}_{win_j}$ is calculated, and this for each dimension $j = 1,...,p$ separately, i.e.

$$dist_j = (|x_j - c_{win_j,j}| \tag{3.56}$$

   where $c_{win_j,j}$ is the jth component of the nearest neuron win_j with respect to dimension j (in general $win_j \neq win_i$ for $i \neq j$). Then, if $dist_j \leq k_d(j)$ with $k_d(j)$ a pre-defined distance threshold of the jth input, then the new (the $C+1$th) neuron is initialized as follows:

$$c_{C+1,j} = c_{win_j,j} \qquad \sigma_{C+1,j} = \sigma_{win_j,j} \tag{3.57}$$

Hence, the center and width of the nearest neuron is assigned to the center and widths of the new neuron. If $dist_j > k_d(j)$, then the new (the $C+1$th) neuron is initialized as follows:

$$c_{C+1,j} = x_j \qquad \sigma_{C+1,j} = dist_j \tag{3.58}$$

4. Case 4: $|\varepsilon(k)| > \delta$ and $\Phi_L < 0.1354$: this means that the network has bad generalization quality and no neuron can cover the input vector sufficiently. In this case, the width of the membership function that has the lowest value in the neuron with highest activation degree is enlarged as in Case 2. If the bad generalization quality of the network remains, a new neuron should be placed as done in Case 3.

3.5.3 *Pruning Fuzzy Rules*

The pruning strategy is based on the optimal brain surgeon approach as demonstrated in [173] and on the research done in [249]. The basic idea is to use second order derivative information to find unimportant neurons (fuzzy rules). This is done with the help of Taylor series expansion for the least squares error in order to be able to calculate the change of this error. The pruning approach is defined in Algorithm 15 (for details see[244]).

Algorithm 15: Pruning Strategy in SOFNN

1. Calculate the training root mean squared error E_{rmse} at time instance k (resp. update this with the new incoming sample).
2. Define the tolerance limit for the root mean squared error as λE_{rmse}, where λ is a pre-defined value in $[0,1]$ (usually set to 0.8).
3. Calculate the change of the squared error ΔE for each fuzzy rule by using Taylor series expansion [3]. The smaller the value of ΔE, the less important is the fuzzy rule.
4. Choose $E = \max(\lambda E_{rmse}, k_{rmse})$ with k_{rmse} the expected root mean squared error which is pre-defined.
5. Select the least important fuzzy rule, delete the corresponding neuron and calculate the root mean squared error E_{rmse} of the new structure. If $E_{rmse} < E$, this fuzzy rule should be definitely deleted. Then, do the same for the second least important neuron and so on. Otherwise, stop and do not delete the neuron.

3.6 The *SAFIS* Approach

The *SAFIS* approach [369], short for *Sequential Adaptive Fuzzy Inference System*, uses the same neuro-fuzzy inference architecture as *SOFNN* and shown in Figure 1.9 (Chapter 1). The only differences are that it exploits a MIMO (=multiple input multiple output) structure and constant (singleton) consequent parameters rather than linear hyper-planes (also called GAP-RBF neural network [184] which is functionally equivalent to a MIMO Sugeno-type fuzzy system with singletons $\mathbf{w}_0$, see Section 1.2.1.1). The concept of sequential learning is defined in the same manner as a single-pass incremental learning method (see Section 1.2.2.1), namely by:

- The training samples are sample-wise = sequentially presented to the system.
- Only one training sample is seen and learned at one time instance.
- After the update of the fuzzy system, the sample is immediately discarded.
- No prior knowledge is available on how many data samples will be sent into the update mechanism.

SAFIS uses a rule evolution and pruning step for dynamically changing the rule base based on the concept of influence of a fuzzy rule, where it starts with an empty rule set (hence incremental learning from scratch is supported). Thereby, the influence of a fuzzy rule is defined as its contribution to the whole fuzzy system in a statistical sense. In particular, the contribution of the ith rule for an input sample $\mathbf{x}$ to the overall output based on all N samples seen so far can be expressed by

$$E(i) = |w_{i0}| \frac{\sum_{k=1}^{N} \Psi_i(\mathbf{x}_k)}{\sum_{j=1}^{C} \sum_{k=1}^{N} \Psi_j(\mathbf{x}_k)} \tag{3.59}$$

When letting the number of samples N go to infinity, the influence of the ith rule becomes [369]:

$$E_{inf}(i) = \lim_{N\to\infty} |w_{i0}| \frac{\sum_{k=1}^{N} \Psi_i(\mathbf{x}_k)/N}{\sum_{j=1}^{C} \sum_{k=1}^{N} \Psi_j(\mathbf{x}_k)/N} \tag{3.60}$$

which is impossible for a sequential learning scheme to calculate as requiring the knowledge of N samples at once. Hence, in [369] a simplification of this formula is derived based on integrals over small parts of the input space, finally leading to

$$E_{inf}(i) = |w_{i0}| \frac{(1.8\sigma_i)^p}{\sum_{i=1}^{C} (1.8\sigma_i)^p} \tag{3.61}$$

with σ_i the width of the ith rule.

Based on this concept, adding a new fuzzy rule is done whenever the following two criteria are both fulfilled (for the kth input sample $\mathbf{x}_k$):

$$\|\mathbf{x}_k - \mathbf{c}_{win,k}\| > \varepsilon_k \qquad e_k \frac{(1.8\kappa \|\mathbf{x}_k - \mathbf{c}_{win,k}\|)^p}{\sum_{i=1}^{C+1} (1.8\sigma_i)^p} > e_g \tag{3.62}$$

hence the distance of the new sample to its nearest center $c_{win,k}$ is bigger than a pre-defined threshold ε_k and the influence of the newly added fuzzy rule is bigger than a pre-defined threshold e_g, usually set to a value in the range $[0.001, 0.05]$. The symbol e_k denotes the error between measured and predicted target value, and κ denotes an overlap factor, determining the overlap of fuzzy rules in the input space; this value is suggested to be chosen in $[1.0, 2.0]$ [369]. The threshold ε_k decays exponentially with the number of samples, allowing more fuzzy rules to be adjoined later.

Pruning of fuzzy rules is achieved whenever the following condition is fulfilled:

$$E_{inf}(i) = |w_{i0}| \frac{(1.8\sigma_i)^p}{\sum_{i=1}^{C}(1.8\sigma_i)^p} < e_p \tag{3.63}$$

with e_p a pruning threshold, which is usually set to a value of around 10% of e_g [369], the threshold responsible for adding a new fuzzy rule, see (3.62).

The update of the constant consequent parameters is done by the global learning approach, i.e. collecting the parameters of all rules to one big vector and exploiting extended Kalman filter approach: linear (singleton consequents) and non-linear (center and widths of neuron) are joined to one vector. When a new rule is added or pruned the inverse Hessian matrix is extended in its dimension in the same way as done in *eTS* approach, see Section 3.2.1.

Combining these aspects together yields the *SAFIS* approach as demonstrated in Algorithm 16 [369] (here for the single output case).

Algorithm 16: SAFIS (Sequential Adaptive Fuzzy Inference System)

1. **Input**: fuzzy system with C rules.
2. Take the next incoming sample $\mathbf{x}_k$.
3. Compute the output of the fuzzy system $\hat{y}_k$ by using the inference in (1.2) with constant consequent parameters $\mathbf{w}_0$.
4. Calculate the parameters required in the criteria for adding rules: $\varepsilon_k = \max(\varepsilon_{max}\gamma^k, \varepsilon_{min})$, $0 < \gamma < 1$, $e_k = y_k - \hat{y}_k$.
5. Check the criteria (3.62) for adding a new rule.
6. **If** these criteria are fulfilled, allocate a new rule by

$$w_{C+1,0} = e_k \quad c_{C+1} = \mathbf{x}_k \quad \sigma_{C+1} = \kappa\|\mathbf{x}_k - \mathbf{c}_{win,k}\| \quad C = C+1 \tag{3.64}$$

7. **Else**
   a. Adjust the system parameters $w_{win,0}$, $\mathbf{c}_{win}$ and $\sigma_{\mathbf{win}}$ with extended Kalman Filter (EKF) method, where *win* denotes the index of the nearest neuron.
   b. Check the criterion (3.63) for pruning the rule.
   c. **If** this criterion is fulfilled, remove the *win* rule, i.e. the rule closest to the current data sample; reduce the dimensionality of the inverse Hessian matrix in recursive least squares formula.
8. If new incoming samples are available, goto Step 2; otherwise stop.

3.7 Other Approaches

This section deals with the description of a collection of further important approaches for on-line learning of fuzzy systems, neuro-fuzzy systems and type-2 fuzzy systems, which have been developed over the last decade and serve as important contributions to the evolving fuzzy systems community.

3.7.1 *The* SONFIN *Approach*

SONFIN [202] which is short for *Self-constructing Neural Fuzzy Inference Network* also exploits a Gaussian-type membership functions as fuzzy sets and a neuro-fuzzy inference architecture as *DENFIS*, *SOFNN* and *SAFIS* do, see Section 3.5.1 for the definition of the layers. One difference is that *SONFIN* splits up the layer #2 in a membership and a fuzzy rule (combined membership) layer, hence achieving six layers in sum. A major difference to the aforementioned methods is that it starts with singleton consequents, associated with the center of a symmetric membership function in the output, and successively adds significant (linear) terms to the network incrementally any time when the parameter learning cannot improve the model output accuracy any more during the on-line learning process. Thus, it expands the dimensionality of the consequent parts by adding more and more input features with weights and can be recognized as a kind of implicit bottom-up on-line feature selection approach. This is denoted as optimal consequent structure identification in [202]. This structure identification process is basically inspired and guided by three strategies: sensitivity calculation method [363], weight decay method [363] and competitive learning [260].

For the adaptation of the parameters occurring in all layers (consequent parameters, antecedent parameters in the Gaussian membership functions and various weights between the layers), an incremental version of the back-propagation algorithm is applied [446] [375]. The basic idea is that, starting at the output layer, a backward pass is used to compute the first derivative of the squared error E between measured and predicted responses with respect to the parameters in the corresponding layers. According to the chain rule, derivatives can be expressed by:

$$\frac{\partial E}{\partial \mathbf{w}} = \frac{\partial E}{\partial a}\frac{\partial a}{\partial \mathbf{w}} \tag{3.65}$$

with a the activation function in the current layer (e.g. in layer #4 in its normalized firing degree). Then, a steepest descent step is applied in the same manner as done in the forward incremental approach for non-linear antecedent parameters (compare Section 2.7), but here a multiple times for each layer separately (starting with the output layer):

$$\mathbf{w}(k+1) = \mathbf{w}(k) + \tau(-\frac{\partial E}{\partial \mathbf{w}}) \tag{3.66}$$

For the specific calculation of (3.65) for each layer, refer to [202].

For an appropriate input-output space partitioning, a clustering approach is exploited, whose basic difference to the aforementioned approaches is that it aligns the clusters formed in the input space, which is a strong aspect of this method. This is done by checking the similarity of newly projected resp. updated fuzzy sets to already existing ones by using an approximate similarity measure as derived in [261]. Thus, it achieves not only a reduction of the number of rules, but also of the number of membership functions used for each partition in the input space and is able to avoid fuzzy sets with high degrees of overlap (as is usually the case when doing clustering in a high-dimensional space and projection, afterwards — also refer to

Chapter 6 which is dedicated to the topic of complexity reduction). A new rule is evolved, whenever the maximal membership degree of a new sample to all rules $\Psi_{max}(\mathbf{x}) = max_{j=1,...,C}\Psi_j(\mathbf{x})$ is smaller than a predefined threshold. Then, the center of a new rule is set to the current data sample and the width initialized in a way so that a certain degree of overlap with already existing fuzzy sets is achieved.

3.7.2 The **GD-FNN** *Approach*

The *GD-FNN* (short for *Generalized Dynamic Fuzzy Neural Network*) approach [449] is a successor of the *D-FNN* approach [450] and hence overcomes some of its deficiencies, for instance:

- Using Gaussian membership functions with arbitrary widths, reflecting the real characteristics of the local data clouds (in *D-FNN* the widths are the same for all fuzzy sets).
- The number of fuzzy sets in the fuzzy partitions are different to the number of fuzzy rules in the whole fuzzy system (in *D-FNN* the number of fuzzy rules is equal to the number of membership functions).
- No random selection of parameters in the learning procedure (as is the case in *D-FNN*).

GD-FNN has some synergies to *SOFNN* and *SAFIS* such as:

- It uses the same layer structure with five layers as in *SOFNN* approach (see Section 3.5.1).
- It applies Gaussian membership functions and linear consequent functions triggering a standard fuzzy basis function network (see Section 1.2.1.3).
- Adding of fuzzy rules is done based on the current error of the system (deviation between measured and predicted value) as is also done in *SOFNN*; the error is compared by a pre-defined threshold which decreases exponentially with the number of samples, allowing more fuzzy rules to be adjoined later (same concept as in *SAFIS*).
- It applies the concept of ε-completeness of fuzzy partitions (as also done in *SOFNN*), which however here is guaranteed by using the concept of semi-closed fuzzy sets $A_i, i = 1,...,m$ [76]. A key issue for assuring semi-closed fuzzy sets is the setting of the widths of the Gaussian functions:

$$\sigma_i = \frac{\max\{|c_i - c_{i-1}|, |c_i - c_{i+1}|\}}{\sqrt{ln(1/\varepsilon)}} \tag{3.67}$$

  whenever (one or two of) the membership function centers c_i, c_{i-1}, c_{i+1} are updated.
- It applies recursive least squares approach for estimating the linear consequent parameters (global learning).

Additional corner stones in *GD-FNN* are:

- A sensitivity analysis of input variables and fuzzy rules: this is done by expressing the significance of the rules resp. of the input variables in the various rules with the help of the error reduction ratio. The ratio can be analytically defined in a single formula when applying a QR-decomposition [145] for transforming the original (global) regression matrix R containing all regressors in all rules (see (2.30), Section 2.2.1) into a set of orthogonal regressors. Finally, this can also be seen as a local feature weighting approach, as the importance of features in each rule can be characterized by its significance. However, the whole concept is explained as a batch off-line learning concept in [449] and not applicable in an incremental learning context. Chapter 5 will present incremental solutions to on-line feature weighting as a kind of soft on-line feature selection in evolving fuzzy systems.
- Dynamic modification of widths in the Gaussian membership functions on demand when a new incoming sample is close to an already existing rule, but triggers a high prediction error. Then, the width of the nearest membership function is decreased by a fraction of its old width.

3.7.3 The ENFRN *Approach*

The approach in [374] (*ENFRN*, which is short for *Evolving Neuro-Fuzzy Recurrent Network*) exploits a series-parallel recurrent neural network [317] for identification of a discrete-time non-linear recurrent system. Therein, one path corresponds to the input state $\mathbf{x}$ and the other to the time-delayed control state $\mathbf{u}$. It evolves a new rule (=neuron) if the smallest distance between the new data sample and an already existing rule is higher than given radius (similar to the vigilance parameter used in *FLEXFIS*, see 3.1.1). Pruning of rules is done based on a density criterium: the rule with smallest density is pruned when its density is below a threshold. A remarkable feature of the *ENFRN* approach is that it applies a modified least squares approach (also called dead-zone recursive least squares), which is able to assure the stability of the whole learning process by providing an upper bound of the identification error. This is done with the help of a Lyapunov function and the matrix inversion lemma as defined in (2.45), Section 2.2.2. For further details see [374].

3.7.4 The SEIT2FNN *Approach*

The *SEIT2FNN* approach [203], short for *Self-Evolving Interval Type-2 Fuzzy Neural Network* , uses an interval-based type-2 fuzzy system as defined in (1.24), Section 1.2.1.7 and extends it with an evolving structure for on-line learning purposes. The components of the fuzzy system are interpreted in terms of layers within a fuzzy neural network type model. The layers are the same as used in *SONFIN* and *SOFNN* approach with the difference of exploiting type-2 fuzzy sets instead of type-1 fuzzy sets. As type-2 fuzzy sets, conventional Gaussian functions are used, whereas the convex hull of two adjacent functions is used as the upper bound and the convex hull of the overlap between two adjacent functions is used as the lower bound, for a detailed definition refer to [203]. Regarding the evolution of new rules, the same concept of rule firing degrees as in SONFIN approach (Section 3.7.1) is applied, where the overall firing degree is calculated by averaging the upper bound and lower bound firing degree, i.e. by

$$\Psi_i(overall) = \frac{\bar{\Psi}_i(\mathbf{x}) + \underline{\Psi}_i(\mathbf{x})}{2} \tag{3.68}$$

If $\max_{i=1,\ldots,C} \Psi_i(overall)$ is smaller than a pre-defined threshold, then a new rule is evolved. Fuzzy sets are projected from the corresponding clusters, whereby highly overlapping sets are omitted. This is carried out by checking whether the membership degree to a type-2 fuzzy set in each fuzzy partition (for each input variable) is higher than a pre-defined threshold. If so, no new fuzzy set is generated as the input sample is sufficiently covered by an already existing fuzzy set. This procedure prohibits time-intensive calculation of similarity measures. Antecedent parameters are tuned and updated by a gradient descent approach, whereas the intervals in the consequent functions (as defined in (1.25)) are updated by recursive least squares approach, joining all the parameters c_{ij} and s_{ij} together to one big parameter vector. This may lose some flexibility and convergence property when adding new rules, see Section 2.4.

3.7.5 *The* EFP *Approach*

The *EFP* approach [441], short for *Evolving Fuzzy Predictor approach* is able to evolve three different fuzzy model architectures, namely Mamdani type fuzzy systems (as described in Section 1.2.1.4) and zero-order as well as first-order Takagi-Sugeno fuzzy systems (see Section 1.2.1.1), whereas in all cases Gaussian membership functions are used in the partitioning the input and output space.

For performing fuzzy rule extraction from on-line data streams, an incremental and evolving clustering procedure is exploited, which processes input and output patterns simultaneously, but in separate feature spaces (so no product clustering in joint input/output spaces is carried out — as done in *DENFIS*, *eTS*, *ePL* or *FLEXFIS*) and relies on two main concepts for deciding whether new clusters are evolved or already existing ones are updated: mapping consistence and cluster compatibility. The mapping consistence holds, if

$$W_I = W_O \tag{3.69}$$

i.e. the winning cluster (nearest cluster to the current data sample) in the input space is the same as the winning cluster in the output space; or, in other words, input and output patterns of the new sample belong to the same cluster. Cluster compatibility is checked by observing the coverage of the new data sample $\mathbf{x} = (x_t, y_t)$ by an already existing cluster, i.e. checking whether there exists an $i = 1, ..., C$ such that

$$\mu_i(\mathbf{x}) > thr \tag{3.70}$$

In [441] $thr = 1/3$ is used.

If both criteria are met, the winning cluster is updated with one new sample by updating its prototype (separately in the input and output space) with the incremental mean formula in (2.2), and updating its spread (separately in the input and output space) by using the recursive variance formula as in (2.9), but omitting the rank-one modification (second term), and increasing the significance of the winning cluster (= # of samples belonging to this cluster) by one. If one of the criteria (3.69) and (3.70) is not fulfilled, a new rule is evolved, where its prototypes (in input and output space) are set to the coordinates of the current data sample (x_t, y_t) and its spreads (in input and output space) are set to 0.1.

For incrementally updating the consequent parameters in case of zero- and first-order Takagi-Sugeno fuzzy systems, the recursive least squares estimator for global learning is exploited (all parameters in a big vector) — as demonstrated in Section 2.2.2. The training of non-linear consequent parameters in Mamdani fuzzy systems and of non-linear antecedent parameters in all three architectures is achieved by using the recursive Levenberg Marquardt algorithm as described in Section 2.7. Hereby, one single parameter is updated with a single new incoming data sample. The choice which one is used, is obtained through a modulo operation acting on the number samples seen so far and in this sense step-wise rotated through a ring-buffer of parameters. In case of Mamdani fuzzy systems, the non-linear parameters are trained by two parallel steps: the consequent parameters are trained in the forward pass, whereas the premise cluster parameters are optimized in the backward pass — for details and concrete formulas see [441].

Part II
Advanced Concepts

Chapter 4
Towards Robust and Process-Save EFS

Abstract. This chapter deals with several issues for guiding evolving fuzzy systems to more robustness during the incremental learning process in order to assure models with high prediction and classification performance. Hereby, we basically see two aspects of defining a robust behavior during on-line modelling:

1. Doing incremental and evolving learning in a way, such that the performance of the evolved models are close to the (hypothetical) batch solution (achieved by collecting all the data and loading them at once into the training algorithms). From mathematical point of view, this means that a kind of convergence of some parameters to an optimality criterion is achieved.
2. Implementing approaches to assure a high performance of the models with respect to accuracy, correctness and stability also in specific cases occurring in the incoming data streams or in the learning procedure itself. The former can be caused by specific situations in the environmental system (e.g. a changing or faulty system behavior) where the data is generated.

The meaning and problematic nature of the former point was already defined and described in more detail in Section 1.2.2.2 (Chapter 1) and dealt with in some parts of Chapter 3 when describing several evolving fuzzy systems approaches. The second point will be handled in detail in this chapter. Therefore, we examine the following problems:

- Instabilities in the learning process (Section 4.1)
- Upcoming drifts and shifts in on-line data streams (Section 4.2)
- Situations causing an unlearning effect (Section 4.3)
- Outliers and faults in on-line data streams (Section 4.4)
- Uncertainties in the models due to high noise levels or lack of data (Section 4.5)
- Extrapolation situations for new samples to be predicted (Section 4.6)

and present solutions to each of these. All these approaches implicitly contain a kind of generic spirit which means they are not specifically dedicated to and linked with one evolving fuzzy system approach, but can be applied to most or at least a group of approaches demonstrated in Chapter 3. These issues finally serve as bases for ensuring process-save implementations and run-throughs of EFS with a high predictive quality in industrial systems.

E. Lughofer: Evolving Fuzzy Systems – Method., Adv. Conc. and Appl., STUDFUZZ 266, pp. 167–212.
springerlink.com

4.1 Overcoming Instabilities in the Learning Process

4.1.1 Problem Statement

Assuming that the complete antecedent part of a fuzzy system is either given or already pre-estimated with any learning method as described in Chapter 3, we concentrate here on a robust estimation of the consequent parameters 1.) with weighted least squares (which is used in some EFS approaches for an initial off-line model building phase) resp. recursive weighted least squares (which is used for the incremental on-line learning phases in most of the EFS approaches described in this book). For the off-line phase the problem of instability in the estimation process becomes clear when inspecting the least squares solution for the whole bunch of consequent parameters (see also Section 2.2.1):

$$\hat{\mathbf{w}} = (R^T R)^{-1} R^T \mathbf{y} \tag{4.1}$$

where the matrix R contains the regression entries for all rules and all dimensions, hence $R^T R$ may get rather big and hence the likelihood to get singular or at least nearly singular (measured in terms of a high condition) is quite high. In this case, the solution in (4.1) becomes unstable, i.e. tend to a wrong solution or is even impossible to calculate. In mathematical terms, this is also called an ill-posed problem, i.e a problem which does not fulfill the following three properties [161]:

- A solution exists
- The solution is unique
- The solution depends continuously on the data, in some reasonable topology.

Clearly, for rank deficient/badly conditioned matrices in (4.1), the second point is violated.

When applying weighted least squares approach for separate local learning of rule consequents (as done in *FLEXFIS*, *FLEXFIS-Class MM*, *eTS*, *eClassM*, *ePL* and *DENFIS*), i.e. by

$$\hat{\mathbf{w}}_{\mathbf{i}} = (R_i^T Q_i R_i)^{-1} R_i^T Q_i \mathbf{y} \tag{4.2}$$

with R_i containing the regression entry just for one rule (see also Section 2.3.1), the dimensionality is reduced and hence the likelihood of singular matrices decreased [120]. So, local learning already implements an implicit regularization effect, as it divides the original optimization problem into smaller sub-problems (for each rule) and hence reduces the size of the matrices to be inverted. This effect is called regularization by discretization [120], and in general, problems of smaller size are more likely to be well-posed, because a reasonable lower bound on the singular value of the matrices can be found.

However, in case of high-dimensional input, which is a usual scenario in real-world on-line identification and modelling applications (refer also to Chapters 7 and 10), the problem is still severe (matrices R_i for each rule are huge). Reducing dimensionality with on-line feature weighting approaches (as will be discussed Section 5.2) may help to increase the likelihood of a stable problem, but is still no

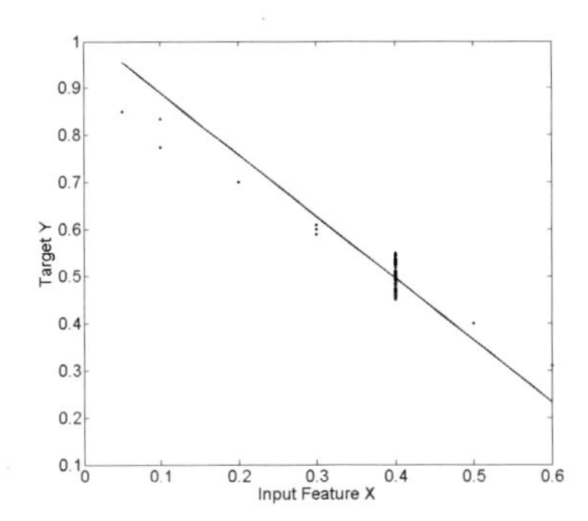

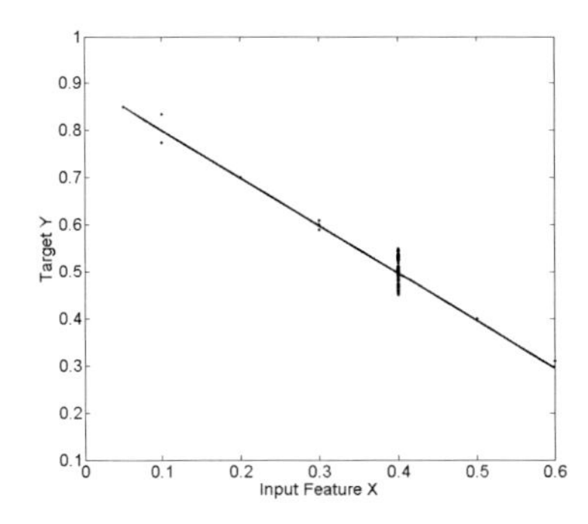

Fig. 4.1 Left: the approximation of a simple two-dimensional relationship with a not-regularized TS fuzzy regression model with one rule (solid line); right: regularization is applied $\rightarrow$ model is correct; note the high noise (spread) in target y at $x = 0.4$ (90% of the samples are lying in this region)

guarantee. In fact, even in case of low-dimensional input, an ill-posed problem may arise, especially when the noise in the output variable is quite high. To underline this claim, in Figure 4.1 a simple first order TS fuzzy model with one rule (so a linear regression model) is shown as a solid line. Even though the noise in the data and the dimensionality are quite low and the tendency of the functional dependency is obvious (linear), the estimated regression model without regularization get very inaccurate (left image), where the same model with regularization fits the data quite well (right image). The reason for this unpleasant effect lies in the fact, that the data samples appear stepwise, whereas at $x = 0.4$ already almost 90% of the data is covered. This means that in the regression matrix R a lot of rows are identical, causing a Hessian matrix $R^T R$ which is nearly singular. For such a small problem and low model complexity, someone cannot blame the curse of dimensionality or the over-fitting effect by choosing too many rules. This problem simply arises only because of the characteristics of the data samples (high noise, and this concentrated in one local region).

During the incremental on-line phase the problem of ill-posedness is strongly diminished as the consequent parameters are updated in a recursive weighted scheme rather than re-estimated by using an inversion of Hessian matrices. However, strong noise may also misguide the solutions, especially forcing the inverse Hessian matrices (updated in local learning case through (2.62) to (2.64)) to numerically instable situations (for instance a whole row in the matrix may become entries which are all near 0). This possible upcoming instability is also underlined by the fact, that recursive fuzzily weighted least squares converges to the solution of batch weighted least squares. An empirical justification of this consideration is demonstrated in [282], where regularization during the on-line training process within the *FLEXFIS* method helped to improve the predictive quality of the evolved fuzzy models by more than 50% in terms of MAE (mean absolute error).

4.1.2 Regularization for Improved Learning Stability

4.1.2.1 Regularization Techniques

Tichonov Regularization

One possibility to overcome this instability problem and to avoid a manipulation of the data, is to apply *Tichonov regularization* [423, 152, 120], which adds a correction term to the Hessian matrix $(R_i^T R_i)$ in form of a Tichonov matrix Γ, hence the matrix inversion in (4.2) becomes $(R_i^T Q_i R_i + \Gamma^T \Gamma)^{-1}$, i.e. the regularized weighted least squares solution becomes:

$$\hat{\mathbf{w}}_\mathbf{i} = (R_i^T Q_i R_i + \Gamma^T \Gamma)^{-1} R_i^T Q_i \mathbf{y} \tag{4.3}$$

By far, the most common choice of the Tichonov matrix is $\Gamma = \alpha_i I$ with I the identity matrix and α_i the regularization parameter (which may be different for different rules). This leads us to the standard form of a Tichonov regularized weighted least squares solution:

$$\hat{\mathbf{w}}_\mathbf{i} = (R_i^T Q_i R_i + \alpha_i I)^{-1} R_i^T Q_i \mathbf{y} \tag{4.4}$$

Sparsity Constraints

Recently, coming from the field of wavelet approximation [104, 80], some regularization methods have been defined, which are similar to Tikchonov regularization but uses a different regularization norm. Tikchonov regularization can also be interpreted as minimizing the Tikhonov functional

$$J_i = \sum_{k=1}^{N} \Psi_i(\mathbf{x}(k)) e_i^2(k) + \alpha \|w_i\|^2 \longrightarrow \min_{w_i},$$

where the least squares functional is additionally penalized by the L^2 (in our case the Euclidean) norm. Instead of this norm, one can also use the L^1 norm (or the sum of the absolute values of $\mathbf{w_i}$) for the penalization term. In contrast to the usual Tichonov approach, this regularization has the tendency to force the entries in the vector $\mathbf{w_i}$ to be zero at best, and hence the resulting regularized vector tends to be sparse. That's why this regularization method is also called *regularization by sparsity constraints* or simply *sparsity constraints*. The resulting solution of the regularization with sparsity constraints can be implemented by a linear programming approach, which has recently been proposed by Daubechies, Defrise and DeMol [96]: The regularized solution $\mathbf{w_i}^k$ is recursively defined as

$$\mathbf{w_i}^k = T_\alpha \left(\mathbf{w_i}^k - \sigma \left((R_i^T Q_i \left(R_i \mathbf{w_i}^k - \mathbf{y} \right) \right) \right) \tag{4.5}$$

where $\sigma < \|R_i^T Q_i R_i\|$ and T_α is the soft shrinkage operator (applied componentwise)

$$T_\alpha(z) = \begin{cases} z-\alpha & \text{if} \quad z>\alpha \\ z+\alpha & \text{if} \quad z<\alpha \\ 0 & \text{else} \end{cases}, \tag{4.6}$$

and where $\alpha > 0$ plays again the role of the regularization parameter. It has been shown that such an iteration converges to a regularized solution enforcing sparsity constraints [96], similar iterations have also been used in nonlinear inverse problems [358] and geometric inverse problems [219]. A more sophisticated iteration than (4.5) by using the concept of active and inactive sets is exploited by the so-called *Semi-Smooth Newton method* [179, 151], which shows excellent convergence [151] and furthermore forces as many consequent parameters as possible towards zero, achieving a kind of local feature selection step (local per rule). In [284, 283] Semi-smooth Newton with sparsity constraints regularization was successfully applied as new (batch, off-line) learning method for rule weights and consequent parameters, called *SparseFIS*, which was among the top performing learning methods for Takagi-Sugeno fuzzy systems.

Figure 4.2 visualizes a high-dimensional regression example, where 19 linear parameters (of the regression plane) are forced towards zero over the number of iteration in Semi-Smooth Newton method starting with a vector of ones as initial guess, (a): after first iteration, (b): after 20 iterations, (c): after 40 iterations and (d): after 60 iterations = final solution. The key point is that Semi-Smooth Newton method tries to find a balance between out-sparsing the parameter space (by forcing as many parameters as possible towards 0), while at the same time still providing a solution which is optimal in the least squares sense.

The combination of regularized and iterative sparsity learning is only applicable in batch mode so far, which reduces the applicability to evolving fuzzy systems using an initial batch learning phase — as e.g. done by *FLEXFIS* (Step 1), *DENFIS* (Steps 1 to 4) or *SAFIS* (Step 1).

4.1.2.2 Regularization Parameter Choice Methods

Now, the final open issue is how to get an appropriate setting of the regularization parameter α in order to obtain a reasonable solution. In fact, this is a matter of a tradeoff between optimality and robustness of the solution: a small value of α leads to a solution which is close to the original and optimal one, however it is not guaranteed whether this improves stability (extended matrix $R_i^T Q_i R_i + \alpha_i I$ can still be near singular); on the other hand, a large value of α causes a significant disturbance of the matrix to be inverted, such that a well-posed problem is enforced; however, this may cause a solution which is quite far away from the real optimal one.

There is a large variety of methods, so-called regularization parameter choice methods, dealing with this topic, and we give a short summary of the most important ones, for a detailed survey on and comparison of these methods, especially with respect to stochastic stability regarding different kinds of noise situations, see [37].

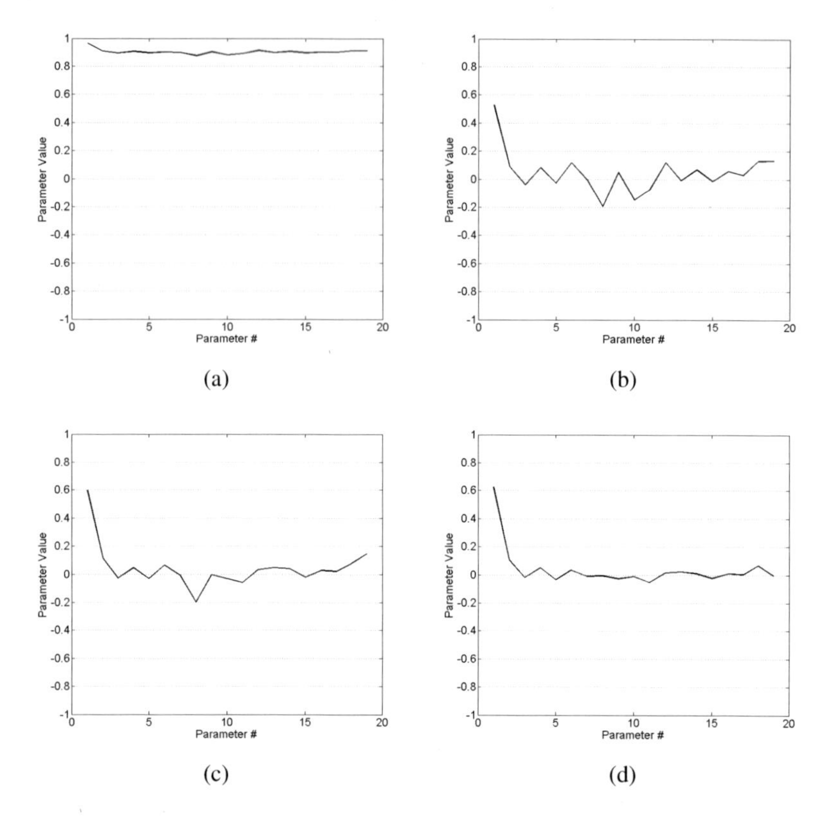

Fig. 4.2 Out-sparsing process of linear parameters in a 19-dimensional regression problem, the parameter values shown after iteration #1 (a), #20 (b), #40 (c) and #60 = final solution (d)

Morozov and Transformed Discrepancy Principle

A method with well-proven convergence to the optimal parameter solution is the Morozov discrepancy principle [314]: this is a parameter choice rule, which requires knowledge of the noise level δ in the data (defined as the norm between unknown true output and observed noise output). If δ is known, then the regularization parameter α is selected as the largest number, where the residual is of order of the noise level:

$$\alpha_{DP} = \sup\{\alpha > 0 \,|\, \|Q_i\mathbf{y} - Q_i R_i \mathbf{w}_{i,\alpha}\| \leq \tau\delta\}. \tag{4.7}$$

with Q_i the diagonal matrix containing the membership values in the N samples to the ith rule. Here $w_{i,\alpha}$ is the Tichonov regularized solution with regularization parameter α, and $\tau > 1$ is a tuning parameter (usually $\tau = 1.5$). In contrast to heuristic parameter choice rules, it can be proven that Tichonov regularization with the discrepancy principle will give a convergent regularization method [120]. It can even be shown that this principle gives rise to optimal order convergence rates under abstract smoothness conditions (so-called source conditions) [120]. The drawback of this method might be that the knowledge of the noise level is needed. Usually, this has to be chosen from experience. On the other hand, for truly ill-posed problems the knowledge of the noise level is necessary as Bakushinskii showed [32].

An extension of the Morozov discrepancy principle, the so-called transformed discrepancy principle (developed by Raus [166] [362]), circumvents this problem, as the noise level in the data only needs to be known approximately as $\hat{\delta}$. For a geometric sequence of parameter values $\alpha(n) = aq^n$ with $q \in [0,1]$, often chosen as $\frac{1}{\sqrt{2}}$, or 0.5 and a the largest eigenvalue of the inverse Hessian matrix, the biggest $\alpha(n)$ is searched which fulfills the following criterion:

$$A_i(n)^{-1}(R_i \mathbf{w}_{\mathbf{i},\alpha(\mathbf{n})} - \mathbf{y}) < \frac{b\hat{\delta}\sqrt{N}}{\sqrt{\alpha_m}} \tag{4.8}$$

with $A_i(n) = R_i^T Q_i R_i + \alpha(n) I$ the regularization operator for the nth iteration in the geometric sequence, $\alpha_n = q^n$, $b > \gamma = 0.3248$ and N the number of data samples. Note that the right hand side in (4.8) is an approximate scaled bound of the propagated noise error. It is shown in [362] that for deterministic noise the method leads to optimal convergence rates when the noise level is known exactly. Hence, it is more stable than the Morozov discrepancy principle.

Hardened Balancing

As usually the noise in the data is not known resp. is hard to estimate in on-line data streams, it is favorable to have a method which does not require the noise level to have available a priori. Such a technique is the hardened balancing principle, recently developed by Bauer [35] as an extension of the famous Lepskij balancing principle [248] and not requiring any tuning parameter. There, the optimal regularization parameter from a sequence of parameter values $\alpha(n) = aq^n, n = 1, ..., N$ with $q \in [0,1]$ is chosen by $\alpha(n*)$, where

$$n* = argmin_{n \leq N}(B(n)\sqrt{\rho(n)}) \tag{4.9}$$

with $\rho(n)$ is an upper bound for the propagated noise and $B(n)$ the balance functional defined by

$$B(n) = \max_{n \leq m \leq N}(b(m)) \quad b(n) = \max_{n < m \leq N}(\frac{\|w_{i,\alpha(n)} - w_{i,\alpha(m)}\|}{4\rho(m)}) \tag{4.10}$$

In [37] it is shown that hardened balancing is among the best performing parameter choice methods due to its high stability. Some care has to be taken, in particular when the regularization parameter gets smaller than the smallest eigenvalue for Tikhonov regularization.

Quasi Optimality and Robust GCV

The quasi optimality criterion is another possible approach, where the noise level in the data does not need to be known in advance. It is one of the oldest and most popular method, originally introduced in [424], later extended in [423] and examined on classical inverse problems in [36]. The optimal regularization parameter from a sequence of parameter values $\alpha(n) = aq^n, n = 1,...,N$ with $q \in [0,1]$ is chosen by $\alpha(n*)$, where

$$n* = argmin_{n \leq N}(\|w_{i,\alpha(n)} - w_{i,\alpha(n+1)}\|) \tag{4.11}$$

so minimizing the distance between two adjacent solutions is sufficient, thus seeking for flat positions in the parameter-error curve (compare Figure 4.3 (c)). Hence, it is assumed that no significant further reduction of the error is possible with higher regularization parameter values (note that the sequence aq^n decreases). Convergence of Tikhonov regularization in combination with the quasi optimality principle could be shown in the discrete case.

Robust GCV, short for robust generalized cross-validation, is an extended version of generalized cross-validation (GCV) developed by [290, 368]. Its strength is the universal applicability to any type of regularization method (hence not only restricted to Tickonov or sparsity constraints as were the others mentioned above). It originates from the older standard cross-validation method where leave-one-out regularized solution is computed and then the total error minimized in using each solution to predict the missing data value. The optimal regularization parameter from a sequence of parameter values $\alpha(n) = aq^n, n = 1,...,N$ with $q \in [0,1]$ is chosen by $\alpha(n*)$, where

$$n* = argmin_{n \leq N}(\frac{\|R_i w_{i,\alpha(n)} - \mathbf{y}\|}{N^{-1} trace(I - R_i A_i(n)^{-1})}(\gamma + (1-\gamma)m^{-1} trace((R_i A_i(n)^{-1})^2))) \tag{4.12}$$

where γ is a robustness parameter, usually set to 0.1. In [37] it is shown that robust GCV is among the best performing parameter choice methods together with hardened balancing.

An example for comparing the performance of Lepskij balancing principle, hardened balancing as well as quasi optimality criterion and robust GCV on satellite data from the EGM96 gravitation field model with uncorrelated noise added is visualized in Figure 4.3. In all plots, the x-axis denotes the regularization parameter, the solid curve the error between the real and regularized least squares solution with respect to the regularization parameter (note that the real solution is known for this example for comparison purposes), the solid vertical line the position of the optimal error (regularization parameter where optimal error is obtained), the dotted line the calculated functional (approximating the error curve between real and regularized solution), the dotted vertical line the chosen regularization parameter from the functional, the solid thick horizontal line the threshold after which the functional is cut off and the regularization parameter chosen. These images provide us the

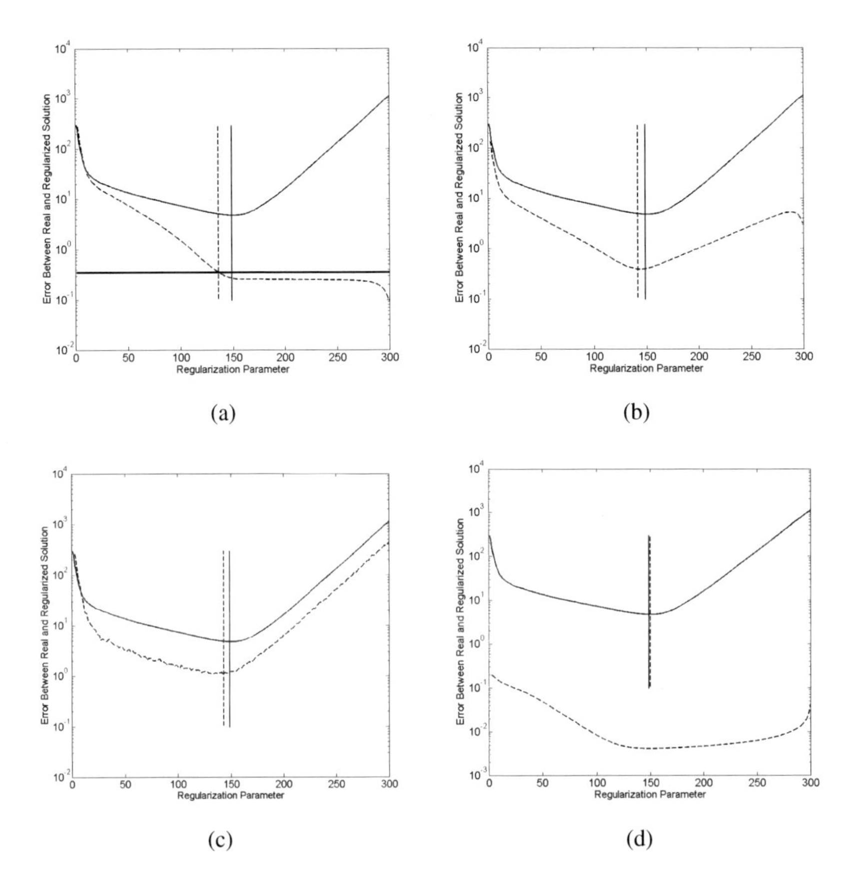

Fig. 4.3 Choice of regularization parameter with (a) Lepskij balancing principle, (b) hardened balancing, (c) quasi optimality and (d) robust GCV, in all plots the solid line represents the real error curve, whereas the dotted line represents the approximation of this curve with the regularization method

following interpretation: Lepskij balancing principle provides the worst regularization parameter in terms that it misses the optimal solution to the largest extent among all methods. Also, its performance depends strongly on a tuning parameter. Quasi optimality does better, but suffers from tiny up-and-down fluctuations in the error curve, which may be misleading. Hardened balancing solves this problem, but is further away from the real solution than robust GCV, which seems to be the best one in choosing the optimal regularization parameter. On the other hand, this method suffers from a quite flat error curve around the optimum. Hence, the robustness of this method with respect to small changes in the (noise of the) data is expected to be quite low.

Heuristic Regularization

All the aforementioned methods have one thing in common: they require the whole data at once in order to elicit an estimation of the optimal regularization parameter. This prevents them from being applicable within a single-pass incremental learning context. Hence, a new method has been designed, simply called heuristic regularization [282], not requiring any past data and whose output is an approximate guess of the optimal parameter approach based on the smallest and largest eigenvalue. This method is applicable for an arbitrary noise level. Hereby, the basic aspect is that the condition of the Hessian matrix is decreased slightly beneath a threshold by exploiting the fact that the addition of $\alpha_i I$ to the weighted Hessian matrix influences the small eigenvalues strongly and the larger ones only slightly (Step 3a).

1. Compute (off-line) $R_i^T Q_i R_i$ resp. update (on-line) $(R_i^T Q_i R_i)^{-1}$
2. Compute the condition of $P_i = R_i^T Q_i R_i$ (off-line) resp. $P_i = (R_i^T Q_i R_i)^{-1}$ (on-line, updated by (2.64)), by applying a singular value decomposition [169] and using the well-known formula $cond(P_i) = \frac{\lambda_{max}}{\lambda_{min}}$, where λ_{min} and λ_{max} denote the minimal and maximal eigenvalue of P_i.
3. If $cond(P_i) > threshold$, the matrix is badly conditioned, hence do the following

    a. Choose α_i in a way, such that the condition of P_i gets smaller than the threshold but not too small due to the considerations above. This can be accomplished by exploiting the fact, that the addition of $\alpha_i I$ to P_i influences the small eigenvalues strongly leading to the approximation formula $cond(P_i) \approx \frac{\lambda_{max}}{\alpha_i}$, hence if desiring a condition of $threshold/2$, α_i can be approximated by

    $$\alpha \approx \frac{2\lambda_{max}}{threshold} \tag{4.13}$$

    b. Compute $(P_i + \alpha_i I)^{-1}$ (off-line) resp. set $P_i = P_i + \alpha_i I$ for further adaptation (on-line).

4. Else, apply weighted least squares approach as in (2.60) (off-line) resp. proceed with P_i (on-line)

Note: Setting the threshold can be carried out from experience with badly conditioned matrices (a value of 10^{15} is suggested [273]) or simply by stepwise trying out from which condition level on the inverse leads to instable results.

An implementation of this heuristic regularization within an evolving fuzzy systems learning approach is straightforward and can be combined with any approach demonstrated in Chapter 3: one step before updating the consequent parameters the following is applied:

For all $i = 1,...,C$ rules:

- calculate the condition of P_i,
- **If** $cond(P_i) > threshold$ elicit α_i by (4.13); **Else** set $\alpha_i = 0$
- $P_i = P_i + \alpha_i I$

If these steps together with regularization in an (optional) initial batch modelling phase are applied, we speak about *regularized evolving fuzzy systems*.

Please note that it is straightforward to apply the whole regularization concept to global learning, when exchanging the P_is with one global matrix $P = (R^T R)^{-1}$. In the application part of this book, a lot of results were achieved by using heuristic regularization during initial training phase in *FLEXFIS* learning engine (as in case of real-world applications significant noise levels are quite often present).

4.1.2.3 Further Considerations

Here, we want to highlight the equivalency of Tikhonov regularization as defined in (4.3) with ridge regression coming from the field of Statistics and motivated by the necessity of gaining a continuous subset selection technique for linear regression problems (see also [174]). In fact, by punishing higher values of the linear parameters $\mathbf{w}_i$ for the ith rule with Lagrange multipliers in the least squares formulation, defined by

$$J_i = \sum_{k=1}^{N} \Psi_i(\mathbf{x}(k)) e_i^2(k) + \alpha_i \sum_{j=1}^{p} w_{ij}^2 \longrightarrow \min_{w_i} \tag{4.14}$$

with $e_i(k) = y(k) - \hat{y}_i(k)$ the error of the local linear model in the kth sample, those directions (inputs) in the columns space of R_i are shrinked most of which have smallest eigenvalues and therefore smallest variance in the data (see also [174]). Hence, the importance of the least important features/variables is squeezed in the least squares solution. Figure 4.4 visualizes the shrinking aspect of the linear parameters when increasing the value of the regularization parameter α_i; the data was normalized before hand to $[0,1]$. Finally, we can also say that an essential aspect of regularization in evolving fuzzy systems is that it provides a concept for decreasing over-fitting in the consequent parts of the rules (as less important variables are forced to 0). Also essential is the potential for a continuous local feature selection per rule, as (4.14) is applied to each rule separately in case of local learning (recursive fuzzily weighted least squares).

The effective degrees of freedom of the regularized regression fit of the ith rule can be estimated by (see [174]):

$$df(\alpha_i) = \sum_{j=1}^{p} \frac{d_{ij}^2}{d_{ij}^2 + \alpha_i} \tag{4.15}$$

with d_{ij} the jth diagonal entry of the diagonal matrix D_i in the singular decomposition of the weighted regression matrix $Q_i R_i$. This has to be taken into account when calculating the effective degrees of freedom for consequent parameter estimation (see Section 2.8.2): in the standard not-regularized case these are obviously

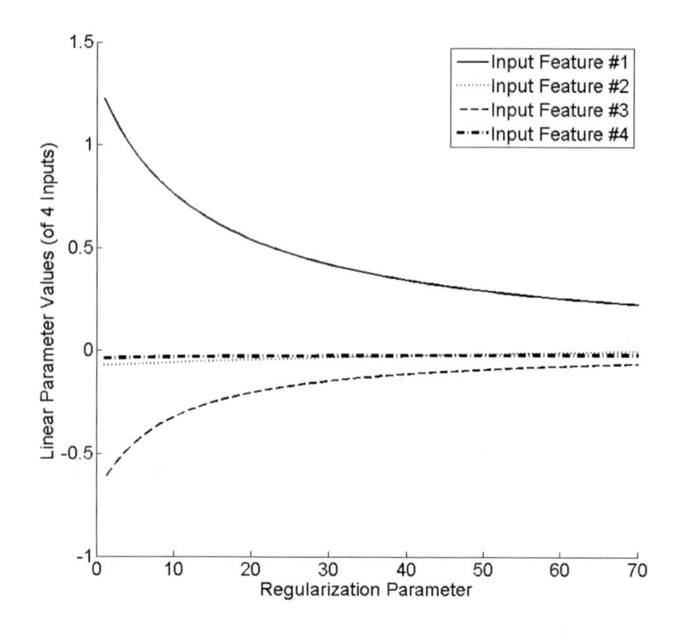

Fig. 4.4 Visualization of the shrinkage effect of linear parameters obtained when learning a regression model for auto-mpg data (from UCI repository) — note the decreasing/increasing values towards zero for the four inputs used in the training process

$C*(p+1)$ with C the number of rules and p the number of inputs, as for each rule $p+1$ parameters are estimated completely independently from each other. In the regularized case ($\alpha > 0$), the effective degrees of freedom for the consequent parts in the evolved fuzzy system are $\sum_{i=1}^{C} df(\alpha_i)$.

4.2 Handling Drifts and Shifts in On-Line Data Streams

4.2.1 Problem Statement

On-line data streams are usually seen as data recorded on-line with a certain frequency and sent permanently into the incremental training process of fuzzy systems. In these streams, it may happen that the underlying data generating process changes (drifts) over time. This can happen because completely new type of data due to different behavior or moods of humans (e.g. consider the behavior of customers in an on-line shop to be predicted), new environmental conditions or various target concepts (e.g. consider weather prediction applications), appear during the on-line learning process, which makes the old data obsolete. Such situations are in contrary to new operation modes or system states which should be included into the models with the same weight as the older ones in order to extend the models to the whole space of possible system states (as discussed in Sections 2.5 and 3.1.1.5, see Figures 2.8 to 2.10, 3.9 to 3.11, respectively).

In machine learning literature, they distinguish between different types of 'concept change' of the underlying distribution of (on-line) data streams: a) drifts, and b) shifts, see [429].

Drift refers to a gradual evolution of the concept over time [447]. The concept drift concerns the way the data distribution slides smoothly through the data/feature space from one region to another.

For instance, one may consider a data cluster moving from one position to another. This concept is closely related to the time-space representation of the data streams.

Shift refers to a sudden, abrupt change of the underlying concept to be learned. In this sense, a shift can be recognized as a more intense drift, where no smooth sliding but discrete jumps of the data samples in the feature space takes place.

While the concept of (data) density is represented in the data space domain, drifts and shifts are concepts in the joint data-time space domain. In machine learning literature they also recognize so called 'virtual concept drift' which represents a change in the distribution of the underlying data generating process instead of the change of the data concept itself, see [447].

In order to demonstrate the impact of a drift onto data samples in the input/output feature space visually, Figure 4.5 presents an (artificial) cluster/rule example. There, the original data distribution (in a 2-D data space) is marked by rectangular samples, which drifts over time into a data distribution marked by circled samples. If a conventional clustering process is applied by weighting all new incoming samples equally, the cluster center would end up exactly in the middle of the whole data bunch (dark gray), averaging old and new data, whereas the correct new center (light gray) would be in the middle of the new data distribution. Such a drift will not only have an impact on the performance of cluster partitions and evolving fuzzy classifiers, but also on the performance of evolving fuzzy regression models, as an approximation curve tends to go through the cluster = rule centers. Hence, in the case of a drift, the approximation curve would be misleaded as going through the center of the whole data appearing in this local region.

Another example is demonstrated in Figure 4.6, where a case of a shift in the output variable is shown: the original trajectory of the 2-dimensional non-linear relationship (consisting of noisy samples) is indicated with dots, whereas the shift is represented by the data samples marked as crosses forming a trajectory below the other one in the middle and right part of the image — please note the abrupt

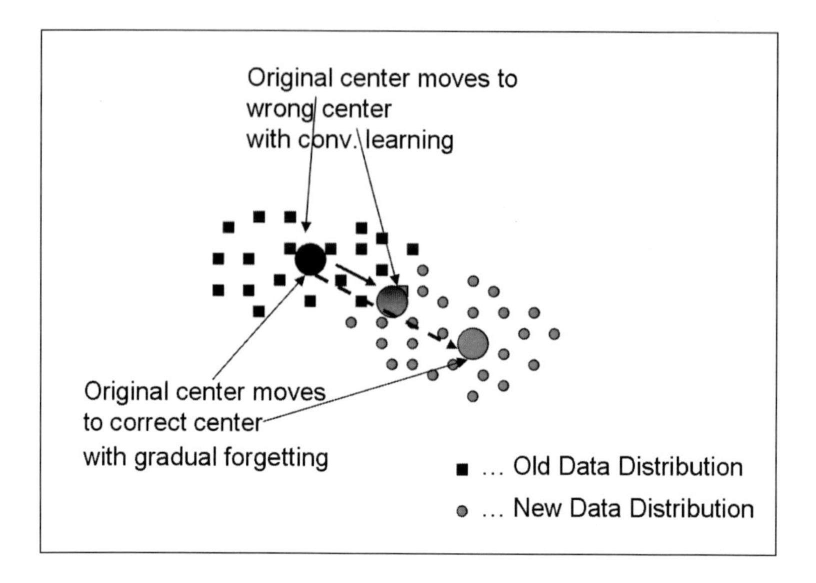

Fig. 4.5 A drift in an evolving cluster (used for learning the antecedent parts of EFS): the samples marked with rectangles represent the underlying data distribution before the drift, the samples marked with circles represent the underlying data distribution after the drift; the big circles represent the cluster center for the original data (left), the wrongly updated center by a conventional learning algorithm (middle) and the center for the new data distribution (right)

change rather than a smooth slide of the samples from one distribution to the other (as in the case of a drift). In the case of just the usual adaptation of the consequent parameters being applied with recursive fuzzily weighted least squares (i.e. local learning by using (2.62) to (2.64)) without forgetting the old data, the approximation curve of the fuzzy model will end up exactly in-between these two (see right image in Figure 4.6, the solid line represents the updated model). This happens because the optimization procedure in consequent adaptation tries to minimize the squared deviation between predicted and real measured values, based on all samples seen so far. Also, if new clusters/rules are evolved in the shifted trajectory, the same problem occurs. Indeed, the consequent parameters and hyper-planes of the newly evolved rules will model the new situation, i.e. snuggle along the surface of the new trend; however, the final output of a fuzzy model will again lie inbetween the two surfaces, as for each input always two (contradicting) rules will fire with high degree.

In order to cope with these problems, drift handling has already been applied in some machine learning approaches, e.g. in connection with SVMs [221][222], ensemble classifiers [357] or instance-based (lazy) learning approaches [44] [97]. In connection with evolving fuzzy systems, this field of research is quite new (a first attempt was made by Lughofer and Angelov in [275] [274]). In fact, all the

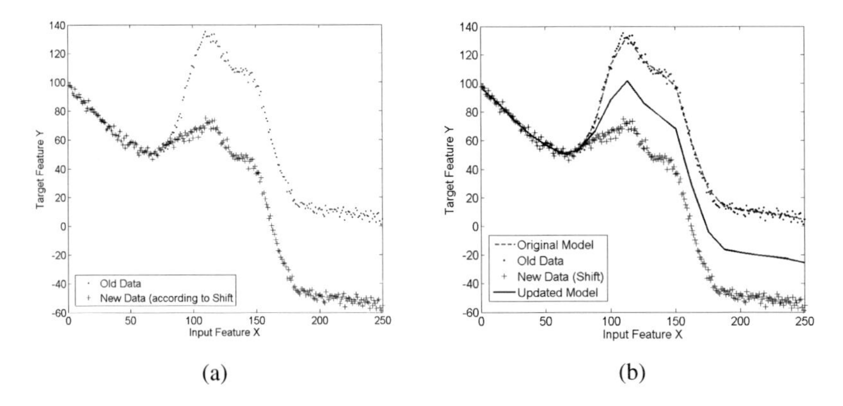

Fig. 4.6 (a): example of a shift in the output variable; compare light dots (original data distribution) with dark dots (data distribution after the shift) on the right-hand side of the graph; (b): the finally adapted fuzzy model (solid line) when including all the samples with equal weights in the incremental update → the model falls exactly in the middle of the two basic sample trends (collected before and after the drift)

approaches described in Chapter 3 have a common denominator in the sense that they are life-long learning approaches.

Life-long learning means that all the data samples are included into the models with equal weights, no older information is discarded or outdated.

Hence, older information is treated equally to newer one and the evolved fuzzy models take into account all the samples seen so far with equal importance. The interesting thing is that mainly this is a beneficial property, especially when a convergence to an optimality criterion or stable state of the model structure is achievable (see for instance the convergence to optimality aspect in Section 3.1.1.4). However, in drift cases this property is counter-productive, as information contained in the older samples is not valid any longer and hence should be discarded from the already evolved models.

In the subsequent section, we will describe approaches for handling of drifts and shifts in evolving fuzzy systems (see also [275] [274]). This embraces two concepts:

1. Automatic detection of drifts and shifts (Section 4.2.2): this will include two approaches, one based on rule ages for tracking changes of the ages of fuzzy rules during the on-line learning process; the other scans for contradicting rules and hence is applicable for detecting *shifts* in the target concept (variable, features).
2. The reaction to drifts and shifts once they are detected: this issue is divided into the reaction in the consequent parts of fuzzy systems and into the reaction of antecedent parts of fuzzy systems. For the former a quite generic concept can be applied for all EFS approaches described in Chapter 3, as these are all using

Takagi-Sugeno (neuro-)fuzzy type systems with hyper-planes and linear consequent parameters in connection with a recursive least squares variant. For the latter, we will present solutions for *FLEXFIS*, *eTS* and *DENFIS* approaches, as applying clustering for antecedent learning.

4.2.2 Automatic Detection of Drifts/Shifts

Automatic detection of drifts and shifts is mainly necessary as in practical use case scenarios there is no input (from users, operators) to the learning system when a drift occurs. The automatic detection of drift is based on the concept of rule ages, originally introduced in [12] and further extended in [275]. Hereby, the age of the ith rule is defined in the following way:

$$age^i = k - \frac{\sum_{l=1}^{k_i} I_l \Phi_{i,l}}{k_i} \tag{4.16}$$

where i is the rule index; k_i denotes the support of rule i (number of samples for which the ith rule was the nearest one) ; I_j denotes the time instance when the data sample was read; k is the current time instance and $\Phi_{i,l}$ the membership degree of the lth sample in the ith rule. Since $\frac{\sum_{l=1}^{k_i} I_l}{k_i}$ can also be accumulated recursively, the age can be easily calculated when necessary. The age of the cluster/rule changes with each new data sample being read. If the newly read data sample does not fall into the ith cluster , resp. supports the ith fuzzy rule with a low or even 0 value of $\Phi_{i,l}$, the age grows by the rate of one sample at a time. That means, if a certain cluster (fuzzy rule) is supported only slightly by any future samples after being initiated (say in the time instant, t_i), then its age at any time instant, k will be approximately $k - t_i$. However, if any new data sample (between t_i and k) supports that fuzzy rule with a high value of $\Phi_{i,l}$, the age does not grow with the same rate, but with a smaller one and one can say that the cluster (fuzzy rule) is being refreshed.

The rule age is important and closely linked to the data streams and to the concept drift, see [275]. There, it is proposed to analyze the age of clusters (respectively, fuzzy rules) on-line by using the gradient of the age curve as well as its second derivative which indicate a change in the slope of the age curve. Indeed, the rate of the aging is equivalent to the gradient of the age in the age curve. Someone can evaluate the average ageing rate by calculating the mean value of the age and can then compare the current ageing rate to the mean ageing to detect a drift in the data stream autonomously on-line. When there is a significant change of the ageing which results in a significant positive change of the slope, then obviously the second derivative of the age curve will be indicative of these inflection points. An example is demonstrated in Figure 4.7, where the drift and activation phases are indicated in (b) for a specific rule. Of course, generally, different rules may cause a different appearance in the age curves. The age curves may have significant local up and down-swing and hence are permanently smoothed on-line with a local linear regression fit. Hence, the second derivative calculated from the change of the

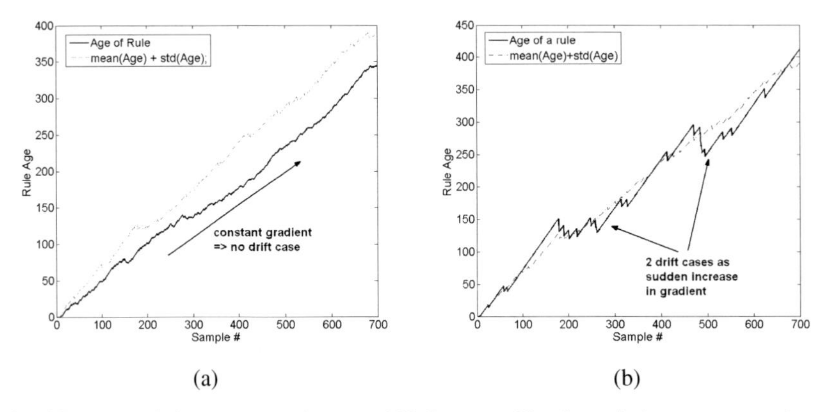

Fig. 4.7 (a): Evolving process where no drift for a specific cluster/rule occurs over time (x-axis) — the gradient in the rule age behaves nearly constantly; (b): evolving process where two drift phases occur for a specific rule — the age suddenly grows around sample # 270 and # 500.

gradients in two consecutive local fits is used as drift indicator. In fact, when this change is high, a drift or shift is quite likely.

Please note that originally the concept of rule age was defined without including the membership degree Φ, as used for deleting old rules from the rule base. In case of small drifts as shown in Figure 4.5, the inclusion is necessary as otherwise the the rule (cluster) is still always attached with data samples from the new distribution as still being the nearest cluster. Hence, no sudden increase in the rule age curves would be visible. Instead, when using the membership degree, the rule age (of the ith rule) will increase as the numerator in (4.16) decreases due to the lower weights $\Phi_{i,l}$.

A shift case represents a stronger drift, which means that an even more abrupt change in the rule age curves is visible. However, the problem with a shift is that it can be hardly distinguished from a new operating mode, which also causes data samples appearing in a different local region in the data space and in older regions not attached for a longer time. This finally means that a medium value of the second derivative of the smoothed rule age curve should trigger the reaction mechanism to drifts (see subsequent section), whereas in the case of a high or very high value forgetting of older learned relationships should be taken with caution. A concept for improving this situation is to detect a shift in the output variable separately (as shown in Figure 4.6) by detecting contradicting (or nearly contradicting) rules. Contradicting rules are rules with similar antecedent parts and dissimilar consequents (see also Section 6.2.2, Chapter 6 for a detailed definition on similarity concepts in EFS). In fact, in the antecedent space they are nearly redundant, whereas in the consequent space they have different fuzzy sets (Mamdani case) hyper-planes or single intercept values. In fact, if such a case as shown in Figure 4.6 appears (surfaces lying over each other), the intercept values in the (parallel) hyper-planes of two rules with nearly the same antecedents will significantly differ from each other. In other cases, the hyper-planes may not even be parallel showing approximation tendencies into different directions: this can be measured in terms of the angle spanning by the normal vectors of the hyper-planes, see Section 6.2.2, (6.15)).

Combining these concepts together, four different cases may arise when doing drift detection in a rule age curve for the ith rule:

- Case 1: the (smoothed) rule age curve shows a constantly growing behavior as shown in Figure 4.7 (b) $\rightarrow$ no drift is indicated, hence the modelling process continues in the same way.
- Case 2: the (smoothed) rule age curve shows a small to medium increasing gradient $\rightarrow$ a small to medium drift is indicated, hence the reaction mechanisms demonstrated in the subsequent section should be triggered.
- Case 3: the (smoothed) rule age curve shows a high increasing gradient and the rule is not contradictory to another rule $\rightarrow$ either a new operation mode or a shift in the input occur $\rightarrow$ reaction mechanism can be triggered, but with caution (slow forgetting resp. interacting with the user(s)).
- Case 4: the (smoothed) rule age curve shows a high increasing gradient and the rule is contradictory (or parallel) to another rule $\rightarrow$ a shift in the output variable occurs (one surface lying over the other), hence the reaction mechanisms demonstrated in the subsequent section should be triggered.

The concept of drift and shift detection with rule ages can be applied to any evolving fuzzy systems approach presented in Chapter 3.

4.2.3 Reaction on Drifts/Shifts in EFS

The reaction mechanism to drifts and shifts is usually triggered in both parts of the evolving fuzzy systems, the antecedent and consequent parts. For the consequent part, a common approach will be described in the subsequent section useable for all EFS approaches.

4.2.3.1 In the Consequents

For the rule consequents, drifts and shifts can be handled in one sweep as it is just a matter of the setting of the forgetting factor as we will see below. Whenever a drift (or shift in the output variable) occurs and is detected by the approach described in Section 4.2.2, it is necessary to apply a specific mechanism in the sample-wise recursive learning steps of the consequent parameters in Takagi-Sugeno type (neuro-)fuzzy systems.

The usual incremental learning scheme in the EFS approaches exploits local learning of the consequent functions (as having some favorable properties among global learning, especially providing much more flexibility for adjoining and deleting rules on demand, see Section 2.4) and defines a recursive fuzzily weighted update scheme, where the membership degrees of the fuzzy rules serve as weights. From Section 2.3.2, we re-call that for the ith rule the recursive update scheme is defined in the following way:

$$\hat{\mathbf{w}}_{\mathbf{i}}(k+1) = \hat{\mathbf{w}}_{\mathbf{i}}(k) + \gamma(k)(y(k+1) - \mathbf{r}^T(k+1)\hat{\mathbf{w}}_{\mathbf{i}}(k)) \tag{4.17}$$

$$\gamma(k) = P_i(k+1)\mathbf{r}(k+1) = \frac{P_i(k)\mathbf{r}(k+1)}{\frac{1}{\Psi_i(\mathbf{x}(k+1))} + \mathbf{r}^T(k+1)P_i(k)\mathbf{r}(k+1)} \tag{4.18}$$

$$P_i(k+1) = (I - \gamma(k)\mathbf{r}^T(k+1))P_i(k) \tag{4.19}$$

with $P_i(k) = (R_i(k)^T Q_i(k) R_i(k))^{-1}$ the inverse weighted inverse Hessian matrix and $\mathbf{r}(k+1) = [1\ x_1(k+1)\ x_2(k+1)\ \ldots\ x_p(k+1)]^T$ the regressor values of the $k+1$th data sample, which is the same for all i rules.

This approach in fact includes different weights for different samples (see the Ψ_i in the denominator of (4.18)), but this depends on the local position of the samples with respect to the rules rather than being motivated by changing characteristics of the sample distribution over time. This means that all newer samples lying in the same local positions as the older samples are included with the same rule weights in the update process. In this sense, for the drift problem as shown in Figure 4.6 (a), the wRLS estimation ends up with fuzzy model approximation curve as shown in the plot of Figure 4.6 (b) (solid line). Obviously, the approximation ends up in the middle of the two trajectories, as trying to minimize the quadratic errors (least squares) of all samples to the curve (also called 'middle way approximation').

Hence, it is necessary to include a parameter in the update process, which forces older samples to be out-dated over time. Gradualism is important here in order to guarantee a smooth forgetting and to prevent abrupt changes in the approximation surface. By doing so, the least squares optimization function for the ith rule is re-defined by

$$J_i = \sum_{k=1}^{N} \lambda^{N-k} \Psi_i(\mathbf{x}(k)) e_i^2(k) \longrightarrow \min_{w} \tag{4.20}$$

with $e_i(k) = y(k) - \hat{y}(k)$ the error of the ith rule in sample k. Assuming N the number of samples loaded so far, this function out-dates the sample processed i steps ago by λ^{N-k}. Usual values of the forgetting factor λ lie between 0.9 and 1, where a value near 1 means a slow forgetting and a value near 0.9 a fast forgetting of former loaded data samples and the exact choice depends strongly on the strength of the drift (see below). From (4.20) it is quite clear that the weighting matrix for rule i becomes

$$Q_i = \begin{bmatrix} \lambda^{N-1}\Psi_i(\mathbf{x}(1)) & 0 & \ldots & 0 \\ 0 & \lambda^{N-2}\Psi_i(\mathbf{x}(2)) & \ldots & 0 \\ \vdots & \vdots & \vdots & \vdots \\ 0 & 0 & \ldots & \lambda^0\Psi_i(\mathbf{x}(N)) \end{bmatrix}$$

Let the weighted error vector for the ith rule at time instant k be defined as

$$e_{k,i} = [e(1)\sqrt{\lambda}^{k-1}\sqrt{\Psi_{1,i}}\ \ \ldots e(k)\sqrt{\Psi_{k,i}}]^T$$

with $\Psi_{k,i} = \Psi_i(\mathbf{x}(k))$ and the objective function J_i for the ith rule at time instant k defined as

$$J_{k,i} = e_{k,i}^T e_{k,i}$$

For the $(k+1)$-th data set it is defined:

$$J_{k+1,i} = \sum_{j=1}^{k+1} \lambda^{k+1-j} \Psi_{j,i} e_i^2(j)$$

respectively:

$$J_{k+1,i} = \lambda \sum_{j=1}^{k} \lambda^{k-j} \Psi_{j,i} e_i^2(j) + \Psi_{k+1,i} e_i^2(k+1) = \\ \lambda e_{k,i}^T e_{k,i} + e_i^2(k+1)$$

or

$$J_{k+1,i} = \begin{pmatrix} \sqrt{\lambda}\Psi_{k,i} e_{k,i} \\ \Psi_{k+1,i} e_i(k+1) \end{pmatrix}^T \begin{pmatrix} \sqrt{\lambda}\Psi_{k,i} e_{k,i} \\ \Psi_{k+1,i} e_i(k+1) \end{pmatrix}$$

Recall the conventional LS estimator at time instant k, which minimizes $J_{k,i} = \sum_{j=1}^{k} \Psi_{j,i} e_i^2(j)$, i.e. $\hat{\mathbf{w}}_{\mathbf{i}}(k) = (R_i(k)^T Q_i(k) R_i(k))^{-1} R_i(k)^T Q_i(k) y(k)$, the modified wLS solution with forgetting at step $k+1$ is:

$$\hat{\mathbf{w}}_{\mathbf{i}}(k+1) = \\ \left(\begin{pmatrix} \sqrt{\lambda}\sqrt{Q_i(k)} R_i(k) \\ \sqrt{Q_i(k+1)} \mathbf{r_i}^T(k+1) \end{pmatrix}^T \begin{pmatrix} \sqrt{\lambda}\sqrt{Q_i(k)} R_i(k) \\ \sqrt{\Psi_i(k+1)} \mathbf{r_i}^T(k+1) \end{pmatrix} \right)^{-1} \\ \begin{pmatrix} \sqrt{\lambda} Q_i(k) R_i(k) \\ \Psi_i(k+1) \mathbf{r_i}^T(k+1) \end{pmatrix}^T \begin{pmatrix} \sqrt{\lambda}\mathbf{y}(k) \\ y(k+1) \end{pmatrix}$$

Following the same recursive deduction scheme as in conventional wLS [264] (defining $P(k+1)$ as the inverse matrix in (4.21), exploiting the fact that $P_i(k+1) = (P_i^{-1}(k) + \mathbf{r_i}(k+1)\Psi_i(k+1)\mathbf{r_i}^T(k+1))^{-1}$, and applying the matrix inversion theorem, also known as Sherman-Morrison formula [395]), the following incremental update formulas with forgetting factor (*RFWLS with forgetting*) λ are obtained:

$$\hat{\mathbf{w}}_{\mathbf{i}}(k+1) = \hat{\mathbf{w}}_{\mathbf{i}}(k) + \gamma(k)(y(k+1) - \mathbf{r}^T(k+1)\hat{\mathbf{w}}_{\mathbf{i}}(k)) \tag{4.21}$$

$$\gamma(k) = \frac{P_i(k)\mathbf{r}(k+1)}{\frac{\lambda}{\Psi_i(\mathbf{x}(k+1))} + \mathbf{r}^T(k+1)P_i(k)\mathbf{r}(k+1)} \tag{4.22}$$

$$P_i(k+1) = (I - \gamma(k)\mathbf{r}^T(k+1))P_i(k)\frac{1}{\lambda} \tag{4.23}$$

Now, the final question is how to set the forgetting factor λ in order to guarantee an appropriate drift tracking. Two possible solutions exist:

1. Define a fixed parameter value of λ in [0,1] according to some prior knowledge about how strong the drifts might become.
2. Introduce a dynamic forgetting factor, which depends on the intensity/speed of the drift.

For the latter, methods exist based on including the gradient LS error with respect to λ for tracking the dynamic properties (see e.g. [404]). In [275], a strategy is deduced directly from the age curves analysis as their outcome is the degree of change in the rule ages which indicates the speed of the drift. In Section 4.2.2, it was mentioned that the age of a rule always lies in $[0,k]$. The age of the ith rule is normalized to $[0,1]$ by

$$age_i\_norm = \frac{age_i}{k} \tag{4.24}$$

in order to achieve gradients of the normalized rule ages Δage_i_norm also lying in $[0,1]$. Whenever the change of the gradient is significant, wRLS with forgetting should be triggered. The following estimation for λ is used [275]:

$$\lambda = 1 - 0.1\Delta^2 age_i\_norm \tag{4.25}$$

This guarantees a λ between 0.9 (strong forgetting) and 1 (no forgetting), according to the degree of the gradient change (1 = maximal change, 0 = no change) (note: to our best knowledge, a value smaller than 0.9 produces instabilities in the models). The forgetting factor is then kept for a while at this level (otherwise, only one single sample would cause a gradual forgetting) and set back to 1, after a stable gradient phase is achieved (usually after around 20 to 30 samples showing a moderate value of $\Delta^2 age_i\_norm$). Setting back to 1 is necessary, as otherwise the forgetting will go on inside the new data distribution. In case of a shift, λ is always set to the minimal value of 0.9. Figure 4.8 (b) shows the impact of the gradual forgetting strategy when learning the consequent parameters recursively $\rightarrow$ the approximation curve is able to follow the new trend denoting the new data distribution (marked with crosses).

Another issue concerning the examination what happens if drifts or shifts are over-detected. In this case, older samples are forgotten and the significance of the parameters and the whole model gets lower. However, the surface of the model will stay at the correct (old) position in the detected drift/shift region as no (significant) move of the samples takes place at all there. In other regions, the update will be quite low, but may become significant with more samples loaded since always all rule fire by a small fraction (as Gaussian fuzzy sets are used). This could cause an unlearning effect (see subsequent section). Hence, keeping the drift phase short (20 to 30 samples) is very important.

In the case of global learning being used for adapting the consequent parameters (as done in *SOFNN*, *SAFIS*, *SONFIN*, *GD-FNN*, *SEIT2FNN* approaches), the

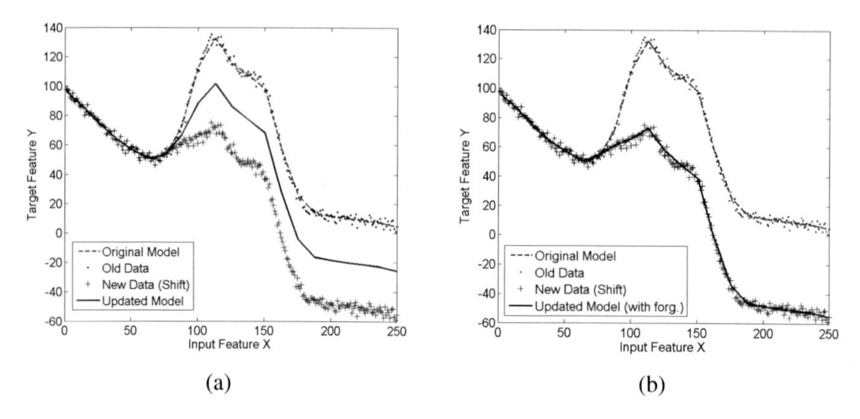

Fig. 4.8 (a): The adapted model (solid line) with the new incoming samples (crosses) when applying conventional recursive weighted least squares approach as defined in (4.17)-(4.19); (b): the adapted model (solid line) by including a forgetting factor of 0.95 and using (4.21)-(4.29), the approximation surface following the trajectory of the data samples denoting the novel data distribution quite well

inclusion of a forgetting factor into the update mechanism can be achieved in the same way as for local learning, leading to recursive least squares with gradual forgetting:

$$\hat{\mathbf{w}}(k+1) = \hat{\mathbf{w}}(k) + \gamma(k)(y(k+1) - \mathbf{r}^T(k+1)\hat{\mathbf{w}}(k)) \tag{4.26}$$

$$\gamma(k) = P(k+1)\mathbf{r}(k+1) = \frac{P(k)\mathbf{r}(k+1)}{\lambda + \mathbf{r}^T(k+1)P(k)\mathbf{r}(k+1)} \tag{4.27}$$

$$P(k+1) = (I - \gamma(k)\mathbf{r}^T(k+1))P(k)\frac{1}{\lambda} \tag{4.28}$$

However, flexibility is lost regarding the reaction on drifts occurring in different local regions of the input/output space: in this case the consequent parameters of the whole model have to be gradually outdated. This could lead to a too strong forgetting, when a drift/shift only takes place in a small area (for instance only in one rule reflected by its age curve). Hence, for the evolving fuzzy systems approaches using global learning it is suggested triggering the forgetting mechanism (4.26) to (4.28) only in the case when most (e.g. more than 75%) of the rule ages show a significant increase in their gradients. Alternatively, a global criterion for drift/shift detection can be applied, e.g. the approach in [136], which uses a significant increase of the actual model error (when doing predictions) as indicator for a drift/shift. The background of this approach is that in non-drift cases the model error decreases over time when more and more samples are included into the model update process (which is a direct consequent of the Stone-Weierstrass theorem [410]).

A generalization concept of the recursive (weighted) least squares with gradual forgetting can be made by using a diagonal matrix Λ whose entries lying in $[0,1]$ can be interpreted as the strengths of the time variances of the single parameters. Thus, if a parameter varies more rapidly than others, the corresponding entry in the diagonal matrix should be chosen larger than others. In this sense, when using local learning, for each single rule the input variables may be assigned different weights according to the intensity with which they influence the drift, rather than using one global forgetting factor value λ. Hence, we also call Λ 'forgetting matrix'. This concept can be included by applying

$$P_i(k+1) = (I - \gamma(k)\mathbf{r}^T(k+1))P_i(k) + \Lambda \tag{4.29}$$

in recursive fuzzily weighted least squares instead of (4.28).

4.2.3.2 In the Antecedents

For *eVQ*

When shifts in data streams occur, usually new clusters are evolved automatically in *eVQ*, as shifts cause new data clouds in explored regions, at least in regions further away than a fraction of the space diagonal in the input space (to which the vigilance parameter ρ is usually set, i.e. 0.1 to 0.3 of the space diagonal, see Section 3.1.1.1). Hence, a new data distribution is covered by new clusters = rules, whereas clusters extracted from the old distribution are usually not really attached any longer, hence having little influence on the overall model output. However, the complexity of the fuzzy systems may increase unnecessarily, which will decrease the transparency and interpretability aspects of the models (and a little the computational complexity when inferencing). An approach on how to tackle this unpleasant effect will be presented in Section 6.2.3 (Chapter 6), as it is related to complexity reduction and interpretability improvement aspects.

On how to react to drifts in the antecedent parts of the rules, re-adjustment of the parameters in the *eVQ* clustering algorithm is proposed which steers the learning gain and furthermore a convergence of the centers and surfaces of the cluster over time (with more samples loaded), namely η. We describe the tracking concepts for the ith rule throughout this section, which can be generalized to any rule in a straightforward manner. Recall that in *eVQ* the learning gain is defined by the following formula:

$$\eta_{win} = \frac{0.5}{k_i} \tag{4.30}$$

which is applied in the update of the prototype of the ith cluster in the following way:

$$\mathbf{c}_i^{(new)} = \mathbf{c}_i^{(old)} + \eta_i(\mathbf{x} - \mathbf{c}_i^{(old)}) \tag{4.31}$$

So, the old center is moved towards the current sample ($\mathbf{x}$) by a fraction of the distance between the current sample and its center coordinates. From formula (4.30) it can be realized that the learning gain decreases with the number of samples forming the ith rule (k_i), which is a favorable characteristic 1.) in order to converge to the center of a local region (data cloud) and 2.) in order to converge to optimality in the least squares sense when updating consequent parameters in the Takagi-Sugeno fuzzy systems (see Section 3.1.1.2). If the learning gain would not be decreased, fluctuating cluster centers and therefore antecedent parts of the rules would be the case, hence the whole evolving fuzzy systems approach would get quite unstable. However, if a drift occurs in a data stream, this favorable characteristics is not favorable any longer as then centers and widths of the cluster(s) should change to the new data distribution (as shown in Figure 4.5). This is also because a drift usually does not trigger a new cluster (like a shift does), just indicates a stronger movement of already existing cluster as expected.

To re-activate the converged clusters, i.e. re-animating them for stronger movements in a drifting case, a sudden increase of the learning gain for the first sample in the drift phase is forced, followed by a gradual decrease for the next samples in order to balance in the new sample distribution in the same manner as is done for original ones. The following mechanisms for the learning gain η are used: first a transformation of the forgetting factor λ, used in the gradual out-weighting when doing consequent learning (see (4.25)) and denoting the intensity of a drift by a value in $[0,1]$. Hereby, 0.9 (minimal value for λ) is mapped to 0.99, whereas 1 is mapped to 0, hence:

$$\lambda\_trans = -9.9\lambda + 9.9 \tag{4.32}$$

Then, when a drift occurs, the number of samples forming the ith cluster (n_i) (used in the denominator of the calculation of η_i) is re-set by

$$k_i = k_i - k_i * \lambda\_trans \tag{4.33}$$

This re-setting mechanism leads to an advanced *eVQ* approach (*eVQ* with drift reaction), where for each new incoming sample first it is checked which cluster is the nearest one ($\mathbf{c}_{win}$) and then whether the start of a drift is indicated in the nearest cluster (based on the analysis of the rule age curve of the winth rule, which has already been updated with the new incoming sample before): if so, k_{win} is re-set by (4.33), if not, k_{win} is further decreased by (4.30) and the remaining update process in *eVQ* continues. This is only done, when the nearest cluster is closer to the current sample than the vigilance parameter ρ (otherwise a new rule is opened up). Summarizing these aspects together, leads us to the advanced *eVQ* approach with drift reaction as shown in Algorithm 17.

Algorithm 17: Evolving Vector Quantization with Drift Reaction

1. Initialize the number of clusters to 0.
2. Either ranges are pre-known or collect a few dozen data samples and estimate the ranges of all $p+1$ variables.

3. Take the next incoming data sample (let's call it $\mathbf{x}$) and normalize it to $[0,1]$ according to the ranges of the variables.
4. **If** number of clusters = 0
   a. Set $C = 1$, $k_1 = 1$.
   b. Set the first center $\mathbf{c}_1$ to the current data sample, hence $\mathbf{c}_1 = \mathbf{x}$.
   c. Set the range of influence of this cluster to 0, hence $\sigma_1 = 0$.
   d. goto step 12.
5. Calculate the distance of the selected data sample to all cluster centers by using a predefined distance measure. Commonly, Euclidean distance is used.
6. Elicit the cluster center which is closest to $\mathbf{x}$ by taking the minimum over all calculated distances $\rightarrow$ winning cluster represented by its center $\mathbf{c}_{win}$.
7. **If** $\|\mathbf{x} - \mathbf{c}_{win}\|_A \geq \rho$
   a. Set $C = C + 1$.
   b. Set $\mathbf{c}_C = \mathbf{x}$, $k_C = 1$.
   c. Set $\sigma_C = 0$.
   d. Goto step 11.
8. Update the rule age of the *win*th cluster by (4.16), normalize it by (4.24)
9. Update the local smoothing regression in the rule age curve (e.g. with RLS) and compute $\Delta^2 age_i\_norm$ by the difference between the gradients in the old smoothing regression fit and the updated one.
10. **If** $high\_thr > \Delta^2 age_i\_norm > low\_thr$, re-set k_{win} by (4.33) using (4.32) and (4.25).
11. Update the $p+1$ components of the center of the winning cluster by moving it towards the selected point $\mathbf{x}$ as in (4.31).
12. Update the $p+1$ components of the range of influence of the winning cluster by moving it towards the selected point $\mathbf{x}$ as in (4.34).
13. $k_{win} = k_{win} + 1$.
14. Update the ranges of all $p+1$ variables.
15. If new data samples are coming in goto step 3, otherwise stop.

Note that this advanced version of *eVQ* together with the gradual forgetting in the consequents automatically triggers advanced versions of *FLEXFIS*, *FLEXFIS-Class* and *eVQ-Class* with reactions on drifts during incremental learning.

The stronger the drift is, the more k_i is decreased and hence the stronger the forgetting effect will be. In Figure 4.9 it is demonstrated how η_i develops (lines) in usual (non-drift) cases (for the first 100 samples), then a drifting scenario is artificially caused with three intensities leading to the three λ values. The development curves of η_i in these three cases are shown as a solid line, dashed line and dashed dotted line. After the drift indicator (at sample 100), it is decreased in the usual way such that the jumped center can converge to the new data distribution.

This resetting of k_i to a significantly lower value also effects the update of the ranges of influence of the clusters, as these are estimated through the recursive variance formula dimension-wise (variance of the data samples forming a cluster is taken for each dimension separately):

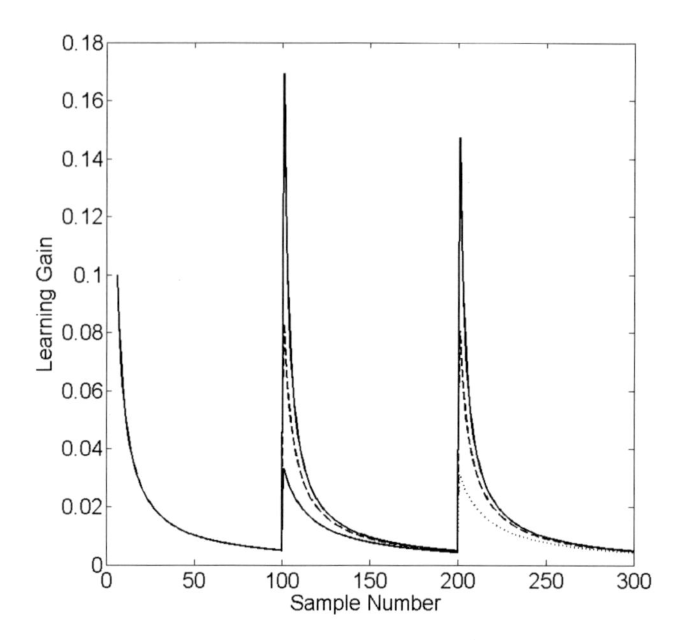

Fig. 4.9 Re-activation of the learning gain η in case of a drift (after 100 samples and again after 200 samples) when applying different values for the forgetting factor λ (indicated by solid = high drift, dashed = medium drift and dotted line = low drift).

$$\sigma^2_{i,j}(new) = \frac{k_i - 1}{k_i}\sigma^2_{i,j}(old) + \Delta c^2_{i,j} + \frac{1}{k_i}(c_{i,j} - x_j)^2 \quad \forall j = 1,...,p+1 \tag{4.34}$$

with p the number of inputs. Obviously, when k_i is reset to a low value after (4.33), the update of $\sigma_{i,j}$ is revived as the impact of $\sigma^2_{i,j}(old)$ gets significantly lower ($\frac{k_i-1}{k_i}$ is lower when k_i is lower). This means that new samples in the new data distribution are represented in $\sigma_{i,j}$ with a much higher impact than the older ones in the old data distribution, achieving a forgetting effect.

Figure 4.10 demonstrates a two-dimensional clustering example with a data drift from one local region (data cloud in (a) and (c), samples indicated as dots) to another one lying nearby (additional cloud in (b) and (c), samples indicated as crosses). In the upper row the trajectories of cluster centers (big dark dots) and the range of influence (ellipsoid in solid line) of the final cluster is shown when doing conventional adaptation of cluster centers without forgetting. The lower row represents the application of the approach mentioned above for reacting on drifts: in this case, the bigger jump from the converged cluster center (as shown in (c)) to the new data distribution, forcing a new compact cluster for the new data distribution instead of an incorrect big joint cluster as achieved without forgetting in (b).

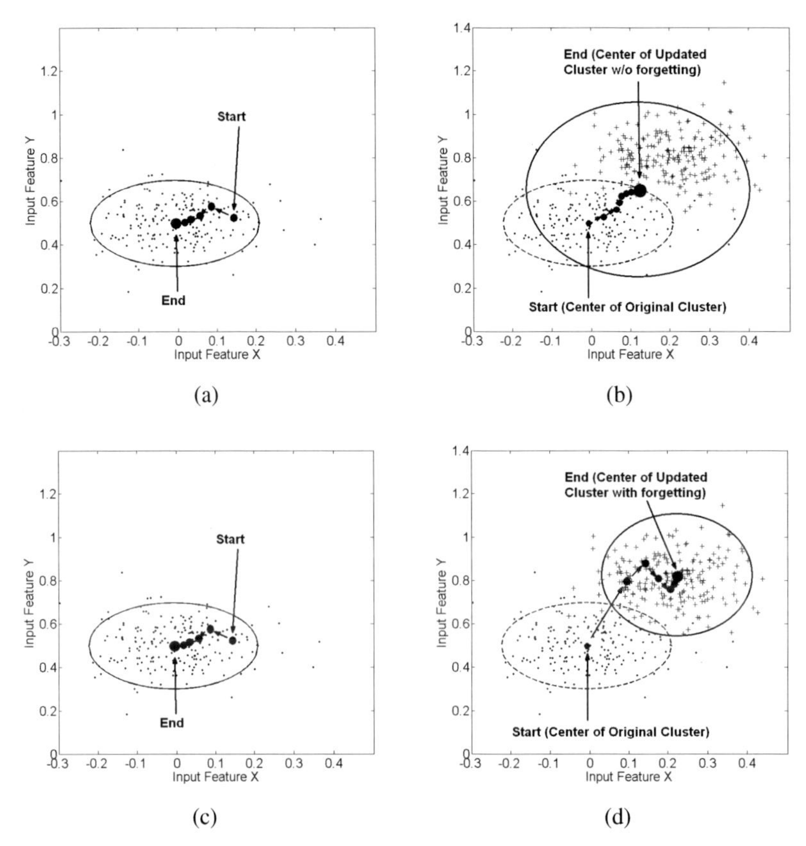

Fig. 4.10 Left: original data cloud, right: drifted data cloud indicated with crosses; (b) represents the trajectory of cluster movements without abrupt decrease of the learning gain (no forgetting), (d) with abrupt decrease (forgetting)

For *eClustering*

In *eClustering*, not a permanent smooth movement of clusters takes place, but an abrupt movement which is triggered based on the potentials of new incoming data samples compared to the potentials of already existing clusters. In fact, when the potential of a new data sample is higher than the potential of all already existing centers, then either a new cluster center is evolved or the nearest one is replaced (when it is close enough to the new sample). In principle, both strategies helps to move clusters from an old data distribution to a new, drifted or shifted one. However, the problem is that a change of cluster centers requires a higher density of samples appearing in a new data cloud than in an already existing one. This means, that in the case when the data distribution slides from an old (heavy) local region to another, it may take quite a long time so that a sample in the new data distribution has a higher potential than the older cluster center. In the extreme case, this situation never happens (for instance when the density and number of samples in the new distribution is lower than in the old one). Hence, whenever a drift or shift is detected by the concept of rule age analysis (Section 4.2.2), the potential of this cluster center where the drift/shift occurs should be re-set to a small value [275]. This triggers a re-setting of centers soon after the detection of drifts.

For *ECM*

In *ECM* (used for rule evolution in *DENFIS* approach), the nearest cluster is not moved when the distance to the current sample is smaller than its range of influence (Step 5 of Algorithm 13). Hence, in the case of a smaller drift, this condition has to be omitted in order to enforce a movement of the nearest center to the new data distribution. In the case of a medium to large drift (or shift), the concept in *ECM* already covers a reaction onto it, as either a new cluster is created (when the distance of the current sample to the nearest center is bigger than $2 * thr$) or the center of the nearest cluster is moved. The movement of the nearest cluster should have a higher intensity when a drift is present than in usual (non-drift) case, outdating the older learned data concept.

4.3 Overcoming the Unlearning Effect

The *unlearning effect* represents the case when an *undesired* forgetting of previously learned relationships reflected in (some structural components of) the models takes place.

This is opposed to a desired (gradual) forgetting of previously learned relationships in the case of a drift/shift as handled in the previous section. In fact, unlearning is caused in non-drift cases by specific characteristics of the data stream (e.g. caused by specific situations at the system where the data is recorded/collected) which should be not included or at least not included with full weight into the models.

A practical example of such a situation is a steady-state behavior of a component in an on-line measurement system. Such a behavior usually causes constant samples, i.e. sample vectors having the same feature/variable values over a longer time. Now, if using fuzzy sets with infinite support (for instance Gaussian functions as commonly used in EFS approaches — see Chapter 3), not only the nearest rules fire, but also all other rules, at least with a small degree. Hence, the parameters of all linear consequent functions are adapted for each incoming data sample (collected during the steady-state mode). In fact, parameters belonging to rules with a very low firing degree (i.e. rules which are far away from the constant region) are adjusted

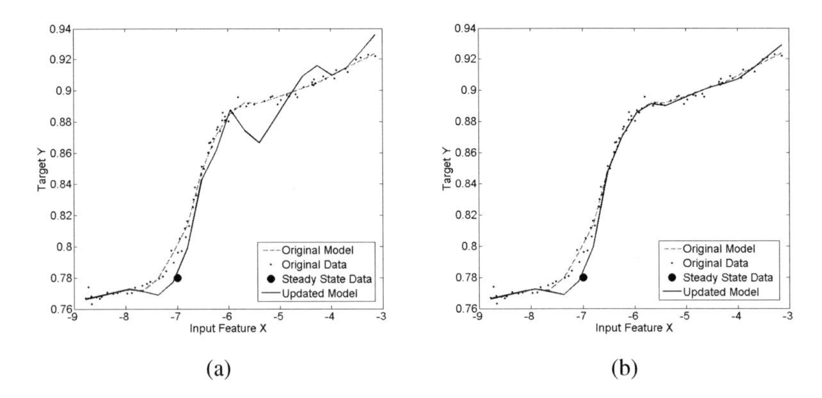

Fig. 4.11 (a): Unlearning effect when adapting all rule consequents for each sample at steady state position $(-7,0.78)$ (dark dot); (b): the unlearning effect prevented when only adapting the significantly firing rules; the solid lines represent the updated models, the dotted lines the original models

very slightly for each sample, but this may sum up when a high number of samples is recorded during steady state. To demonstrate this effect on a simple practical example see Figure 4.11 (a), where the first 200 data samples are indeed well distributed over the whole input space, but the next 1300 samples (represented by the big dark dot) are concentrated around the specific point $(-7,0.78)$. This example represents a situation, whenever a process is in steady state, i.e. staying at one specific operating condition for a long time. When doing an initial learning phase with the first 200 points (no matter if in incremental or batch mode) and performing an adaptation of the fuzzy model with the later 1300 points (in steady state), an undesired 'unlearning' effect of already learned relationships outside this small constant region occurs. This can be recognized in Figure 4.11 (a), where right to the small constant region the shape and behavior of the adapted model (solid line) tend to be completely different to the shape and behavior of the original one (dotted line), even though no new measurements were recorded for that area. The reason for this effect is that the parameters of all linear consequent functions are adapted for each incoming data sample, no matter which firing degree the rules have. Finally, this means that the consequent parameters are forced to minimize the error at $(-7,0.78)$. This even causes the linear consequent hyper-planes to go through the point $(-7,0.78)$.

In principle, a steady-state situation can be seen as a specific drift case, which, however, should be fully ignored in the update process rather than forcing an outweighting of older learned behavior. A possible strategy to circumvent this problem lies in the adaptation of only those parameters which correspond to significantly active rules, i.e. rules which possess normalized membership function values $\Psi(\mathbf{x})$ higher than a certain threshold near 0. This guarantees that rules which represent areas not lying near a constant process state remain unchanged and hence are not disturbed. In the consequent adaptation parts of the evolving fuzzy systems approaches, this can be accomplished by proceeding in the following way for the current sample $\mathbf{x}$:

For $i = 1,...,C$ **If** $\Psi_i(\mathbf{x}) \geq thresh$ perform recursive (fuzzily weighted) least squares for the consequent parameters of the ith rule.

The effect of this procedure in the *FLEXFIS* approach (by replacing Step 8 in Algorithm 5 with the above condition) is demonstrated in Figure 4.11 (b), where a threshold of 0.1 was used. In fact, setting the threshold is a crucial point for a good performing adaptation: by setting it too close to 0, the 'unlearning' effect remains, and setting it too far away from 0, too few new input states are incorporated and the fuzzy system is not able to represent a reliable prediction model. Another possibility is to only update the consequent parameters of that rule which is closest to the current sample (as for instance done in *SAFIS* approach).

Another point of view of the unlearning effect is when considering classification problems, where the data distribution of the classes gets quite unbalanced during the incremental learning process, so one or more classes get significantly more frequent than some others over time. In this case, the danger is high that the less frequent classes are simply overwhelmed and do not contribute to a cluster or a rule with majority any longer, causing a bad classification accuracy for these classes. This problem is also called imbalanced learning problem [349]. A visualization of such an unpleasant situation is presented in Figure 4.12, where (a) represents the data distribution after loading the first block of data. This contains two clusters, where each of these has a different majority class. Then, when new samples are loaded (Figure 4.12 (b)) which are all from the same class (class #1 denoted by samples marked as circles), the other class (class #2 denoted by samples marked as rectangles) is simply overwhelmed in that cluster where it had majority before. This means, when applying the winner-takes-it all classification strategy (as used in *eVQ-Class*, *FLEXFIS-Class SM*, *eClass0* and *eClass A* — see Chapter 3) a new sample will never be classified as class #2. This usually leads to a severe drop in classification performance of the whole system, especially when cases from the under-represented class should be recognized as such with a high accuracy (for instance faulty states in a system).

In order to overcome this unpleasant effect, a so-called 'timing of update' strategy is proposed, where not each single new incoming instance is taken into the evolving process of the fuzzy classifiers, but only in the following situations (also called balanced learning):

- whenever the relative proportion of samples seen so far and falling into all the classes is approximately equally balanced: this is because refining the classifier with additional samples is usually advantageous for guiding it to more accuracy. A usual guess for an approximate equally balanced situation is that the most frequent class includes not more than 3 times the samples which are included in the second frequent class.
- whenever the relative proportion of samples seen so far and belonging to the current class where the sample falls in is lower than the relative proportion of

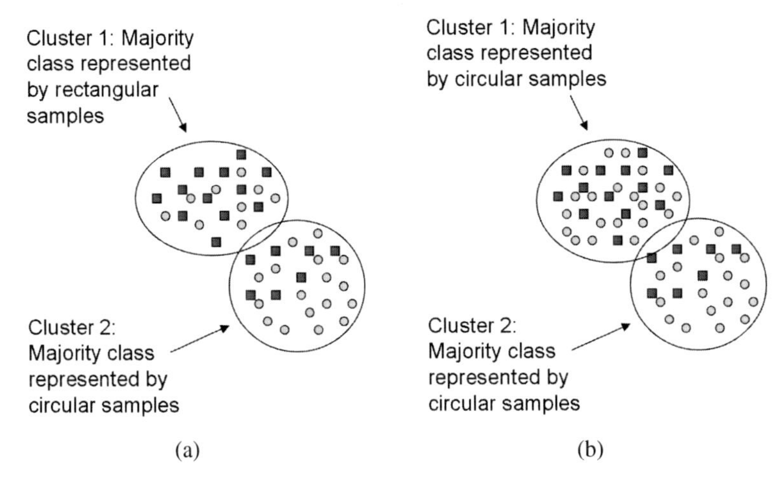

Fig. 4.12 (a): Initial class distribution: 2 cluster each having a majority in one of the two classes, marked as circles and rectangles); (b): new data from circular class fall into the first cluster overwhelming the rectangular class → the rectangular is completely masked out as nowhere the winner

the most significant class (class with highest number of representatives): this is for incorporating more samples from less frequent classes and boost up the classification performance for these.

This automatically means that an update of a (evolving) fuzzy classifier is never performed when a new sample is presented falling into an over-represented class (e.g. a sample belongs to class #1, which covers 80% of the data seen so far). In Section 5.3, we will present an extension of this concept to uncertain class responses (called active balanced learning).

4.4 Outlier and Fault Treatment

Outliers and *faults* are characterized by a significant deviation from the basic trend described by the majority of data samples.

For instance, in a regression modelling task, an outlier or fault may show a significant deviation from the trajectory spanned between system variables, indicating a certain relationship between these. A two-dimensional example of such a situation is visualized in Figure 4.13, where (a) shows a single outlier (marked by a circle around) and (b) a faulty system state, the samples belonging to these state surrounded by an ellipsis.

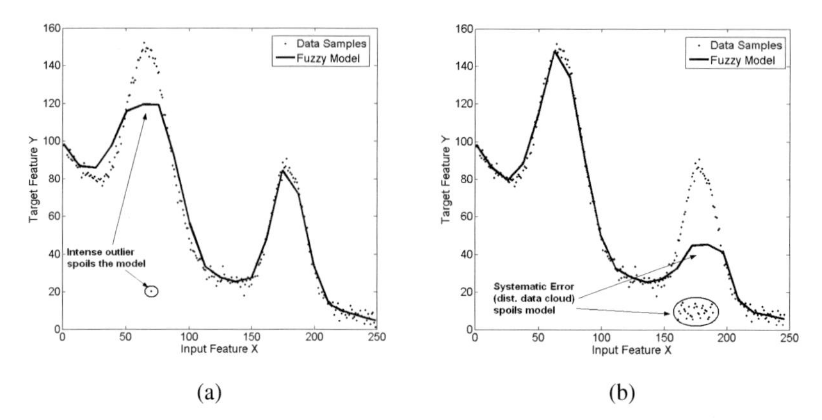

Fig. 4.13 (a): outlier sample (indicated with arrow and circle) lying far away from the real trend of the relationship spoils the model significantly; (b): a systematic error (surrounded by an ellipsis) spoils the model trying to follow the real trend significantly

The distinction between outliers and faults is done based on their systematic behavior: outliers are often unique occurrences in the data stream consisting of only one or at the most a few samples appearing far away from the basic trend. The magnitude of the distance to the basic trend is responsible whether an outlier spoils a model significantly or not. Outliers typically occur from three possible causes [371]: 1.) data entry error, 2.) values from another population or process, 3.) actual unusual values in the data. On the other hand, faults are usually caused by an error in the system where the data was recorded and collected and trigger a much more systematic deviation from the basic trend and often last over a longer time, affecting more samples than outliers. We also want to note that outliers and faults should be not confused with noise in the data. *Noise* is basically referred to tiny deviations along the basic trend, often even following a certain statistical distribution and caused by inaccuracies in measuring devices (e.g. sensors), different experiences of experts when labelling the data (in case of classification problems) or simply background noise (usually caused by white noise) in an industrial system where the data is collected. A lot of the two-dimensional visualized (regression) modelling examples demonstrated so far in this book were affected by noise (see for instance Figures 2.5 to 2.10 or 3.9 to 3.11), as collected from real-world processes. This will also be the case for most of the applications demonstrated in the third part of this book. Note that most of the training algorithms for fuzzy systems and also for evolving fuzzy systems (see Chapter 3) are designed in a way to handle medium to high noise levels appropriately.

In batch off-line cases outliers and faults can be filtered as long as no more than 10 to 20% of samples in the data set are affected by these. Therefore, the basic structure of the whole data set is described by a compact region in the full-dimensional space (using all inputs which are present and available in the data set).

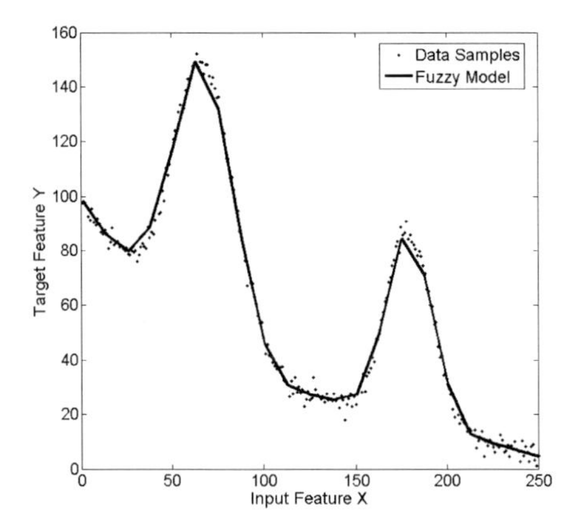

Fig. 4.14 A correct and highly accurate fuzzy model obtained when filtering the outlier and the systematic error shown in Figure 4.13 before hand

The convex hull [347] [60] is one possibility to describe a compact region, but is very computation-intensive, especially when the dimensionality of the problem is very huge. An alternative with less computational effort is the exploitation of Mahalanobis distance [295], which is defined between two single samples **x** and **y** in the following way:

$$mahaldist_{xy} = \sqrt{(\mathbf{x}-\mathbf{y})^T \Sigma^{-1} (\mathbf{x}-\mathbf{y})} \tag{4.35}$$

with Σ the covariance matrix. When calculating the Mahalanobis distance of a data set to itself, outliers can be easily recognized (see MATLAB's Statistics toolbox). In fact, the mean values of all columns (i.e. variables) in one matrix are taken and each sample of the other matrix to the mean value is calculated by (4.35). This vector of distances can be simply checked by exploiting the distance-to-variance strategy, i.e. calculating the mean value μ and variance σ over the distance vector and checking each sample if it lies more than $n\sigma$ away from the mean value: such samples can be omitted for the model training process. Mostly n is chosen as 2 or 3, yielding a 2σ or 3σ confidence interval around the mean value. Another possibility is to use the interquartile range, which is defined through the upper and lower quartiles Q_3 and Q_1, cutting off the 25% highest resp. 25% lowest data samples. Then, someone could detect a sample as an outlier which is not in the range

$$\left[Q_1 - k(Q_3 - Q_1), Q_3 + k(Q_3 - Q_1)\right] \tag{4.36}$$

These concepts of data cleaning can be applied to evolving fuzzy system approaches during initial batch phases. Figure 4.14 shows the correct model which is obtained after filtering the outlier and the systematic error.

For the incremental on-line training process, the situation is a bit more complicated, as single blocks or even single samples of data are loaded into the memory with which the models are adapted. This also means that the training procedure only sees a small snapshot of the whole data set and it is very hard to differ between a new upcoming, developing operating condition and a fault. For instance, in the example visualized in Figure 4.13 (b) the local data cloud marking the systematic error is loaded before the real (Gaussian-like) trend of the curve on the right hand side, causing a new rule in this region and therefore a misguided model (solid line). At least, in case of outliers affecting a single or a couple of samples, a strategy for procrastination of rule evolution can be applied in EFS approaches: whenever a new sample comes in which lies far away from the previously estimated clusters (representing local areas), still a new cluster is born immediately, but not a new rule and hence no new fuzzy sets. Someone has to wait for more data samples to appear in the same region and to confirm the new cluster before it can be projected to form a new rule. This is based on the assumption that the more data samples in a new region occur, the more likely it gets that they denote a new operating condition and no outlier. With this strategy the incorporation of isolated outliers into the fuzzy model can be prevented. Note that some EFS approaches (such as *eTS* or *ePL*) are per nature robust against outliers as using density concepts (e.g. potential) for setting new rules.

In case of systematic faults (an example shown in Figure 4.13 (b)), significantly more samples are affected, such that the rule procrastination strategy is not successful any longer. In particular, it is not sufficient to check just the number of samples forming a new data cloud. A possible extension to systematic faults is to compare the inner structure of a new local data cloud with those of other already existing clusters. This can be accomplished with specific features characterizing the structure, shape and behavior of the clusters (rules). The assumption thereby is, that clusters which are formed in fault-free regions have a different appearance and therefore cause other feature values than clusters generated in faulty regions. This assumption is indeed somewhat vague, but the only chance to discriminate between faulty and fault-free regions: when the clusters are generated in exactly the same manner resp. the generated clusters look exactly the same, then obviously no distinction is possible without using any other external knowledge. A first attempt for doing so includes the extraction of the following features:

- Volume of an ellipsoidal cluster, which can be calculated with the following formula:

$$volume = \frac{2 * \prod_{j=1} p\sigma_j * \pi^{p/2}}{p * \Gamma(p/2)} \quad (4.37)$$

with p the dimension of the input space, σ the width of the cluster in each dimension and Γ the gamma function defined by:

$$gamma(x) = \int_0^\infty t^{x-1} e^{-t} dt \quad (4.38)$$

The volume of a cluster/rule represents a measure for its size and may be an indicator for a systematic fault (e.g. when the size is significantly smaller or larger than that of clusters representing fault-free system states).

- The proportion between the number of data samples k forming the cluster (belonging to the cluster according to the shortest distance to the surface or center) and the volume of the cluster, yielding a kind of density measure for the cluster:

$$density = \frac{k}{volume} \tag{4.39}$$

- The proportion between minimal and maximal widths of the cluster over all the input dimensions indicating the shape tendency of the cluster:

$$shape = \frac{min_{j=1}^{p} \sigma_j}{max_{j=1}^{p} \sigma_j} \tag{4.40}$$

If this value is low, it means that the samples are unequally spread for different dimensions within the cluster (e.g. a drift case may show such a pattern as the samples drift along one dimension whereas the other dimensions have moderate spread). This finally indicates whether a cluster is more spherical (high value of (4.40)) or more a thin ellipsoid (low value of (4.40)).

- The average distance of the data samples forming a cluster to its center, indicating the spread of the samples within a cluster:

$$spread = \frac{\sum_{i=1}^{k} \|\mathbf{x} - \mathbf{c}\|}{k} \tag{4.41}$$

- The relative number of data samples appearing outside the surface of the cluster ellipsoid (but still belonging to the cluster c as it is the nearest one), indicating the degree of cluster anomaly (deviating from a strict ellipsoid) and cluster outliers:

$$anomaly = |\{\mathbf{x}|c = argmin_{i=1}^{C} \|\mathbf{x} - \mathbf{c}_i\| \wedge \sum_{j=1}^{p} \frac{(x_j - c_j)^2}{\sigma_j^2} \geq 1\}| \tag{4.42}$$

In a first experiment these features were used during learning and evolution of rules with the *eVQ* approach (as part of *FLEXFIS*) for measurement data recorded at engine test benches and stored onto disc in the same order as recorded (see Chapter 7 for more detailed information on this application example). Faults in these measurements were marked and it was examined which type of clusters with which characteristics are extracted for the faulty and the fault-free states. Figure 4.15 visualizes two-dimensional plots where the values of feature pairs are plotted either with a circle (= feature values extracted from clusters representing fault-free states/measurements) or with an 'x' (=feature values extracted from clusters representing faulty states/measurements). As someone can realize, in this case by applying only the two features 'spread' and 'volume' (Figure 4.15 (a)) it is possible to discriminate between faulty and non-faulty states: a non-linear decision boundary

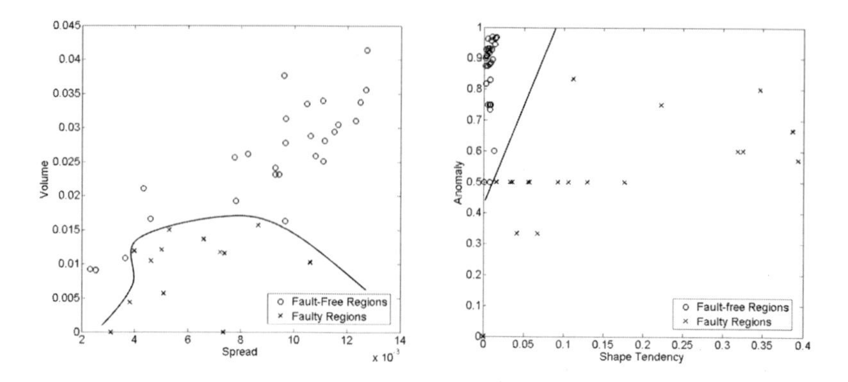

Fig. 4.15 (a): features 'spread' versus 'volume' extracted from faulty and non-faulty regions in measurement data from engine test benches, the regions could be discriminated with a non-linear decision boundary; (b): features 'shape' versus 'anomaly', the regions could be discriminated with a simple linear decision boundary; the decision boundaries shown as solid lines

(solid line) can be placed causing 0 classification error on these samples, in case of 'shape' and 'anomaly' even with a simple, linear decision boundary (Figure 4.15 (b)). An idea is now to train a classifier (for instance an evolving fuzzy or clustering-based classifier) by using these features as input and based on some pre-labelled samples from the process for which models should be generated. An additional valuable option is that the classifier is also able to be adapted and evolved during on-line mode with new incoming samples in order to permanently refine its boundaries. The problem in this case is how to get the information for faulty and non-faulty system states in a fully automatic way (without external operators/users/experts input). Unsupervised criteria such a range effects, various quantiles characterizing the distribution of the features may be a helpful alternative.

Furthermore, the following work-arounds may help to improve the performance of evolving fuzzy systems in case of arising faults in the system:

- Filtering of dynamic measurements with signal analysis methods in intelligent sensors, which also effects a filtering or 'cleaning' of stationary measurements which are elicited through averaging of dynamic ones: this is quite promising, as usually faulty samples cause a more distinct anormal behavior in the one-dimensional signal curves as new operating conditions. A first attempt to this approach was done in [278].
- The exploitation of analytical and knowledge-based models in a complete fault detection framework to deliver a fault detection statement for any loaded or on-line recorded data sample [287]. This statement can be taken as input for the incremental learning method and reacted upon so that samples classified as faulty are not incorporated into the adaptation process (see also Chapter 8 for more detailed information).

Finally, we can summarize that outlier and fault treatment in evolving modelling scenarios is a hot research topic which is still not completely solved, especially the distinction between new operation modes/operating conditions and faults (as both are usually triggering samples in unexplored regions of the input/output data space) and hence requires further examination and investigation in the near future.

4.5 Modelling Uncertainties in EFS, Reliability Aspects

4.5.1 For Regression Problems

Models' decisions may be uncertain due to a high noise level in the training data or in extrapolation regions. In these cases, it is reliable or even necessary to provide confidence regions surrounding the trained models, which should provide an insight, on how trustful a model within a certain region is. This is essential to judge the quality of the models' predictions. These can be either directly weighted with their confidence or a separate confidence value is given as feedback to the user or to the system as additional information. Error bars are a quite useful technique to model such confidence bands [403] and hence can be seen as an additional contribution to more process safety and robustness. This is because error bars indicate one standard deviation of uncertainty and determine whether differences are statistically significant. For instance, error bars serve as an important component in many quality control applications, in order to be able to compare a residual between a new incoming sample and the error bars of an already trained (regression) model, as in the case of noisy (real-world) data new incoming samples do not exactly lie on the approximation surfaces. The deviation to the error bars is then used in order to gain a final decision whether the current point is faulty or not — see Chapter 8, also refer to [279] [280] how error bars are used in connection with fault detection scenarios.

One simple possibility to obtain error bars for a regression model lies in (incrementally) calculating the mean squared error on the past samples and taking a multiple of the mean squared error as error bar. This would have the effect that the error bar denotes a global error band surrounding the model and has the same width throughout the whole model's range. This is shown in Figure 4.16 (a). However, usually the data distribution and density as well as the noise intensity are not equally balanced throughout the whole input/output space: in fact, in some regions less data may occur than in others, see for instance the region between $x = 30$ and $x = 50$ where no data sample appears. In this region as well as in the extrapolation region (below $x = 0$) the model's predictions are obviously less confident than in regions where sufficient training data was available. Hence, the error bars should be wider in these unconfident regions and narrower in the more confident ones, as shown in Figure 4.16 (b). We call these types of error bars *local error bars* as they change their width locally. Please also note that in the extrapolation region on the left hand side the further away the curve is from the data samples, the wider the error bars get. Please note that in literature error bars are often shown as bar charts around crisp values (parameters, data samples, model predictions). Here, as we are dealing with

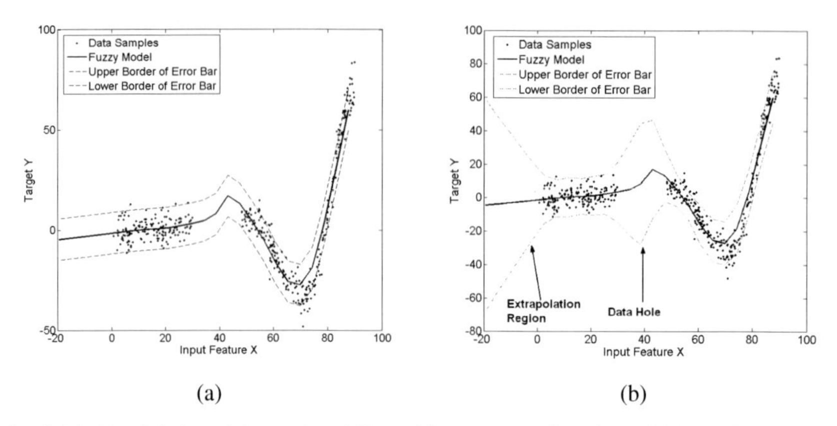

Fig. 4.16 (a): global model error band (dotted lines surrounding the solid line = fuzzy model), (b): (adaptive) local error bars yielding a more precise local information about the confidence of the model, taking into account holes in the data (between $x = 30$ and $x = 48$) and extrapolation regions (from $x = 0$ to $x = -20$)

continuous functions (fuzzy models), the error bars are visualized as continuous bands surrounding the models.

In order to obtain such local error bars for (evolving) TSK-type (neuro-)fuzzy models with linear consequent parameters, first local error bars for linear regression models are described, which are then extended to the non-linear TSK fuzzy regression model case. *Local error bars* for pure linear (regression) models

$$\hat{f}(\mathbf{x}) = \hat{y} = \sum_{i=1}^{deg} w_i * reg_i(\mathbf{x}) \tag{4.43}$$

with reg_i the ith regressor term applied onto input variables $x_1, ..., x_p$, whose linear parameters $\hat{w}$ were estimated by solving the least squares optimization function, can be calculated through the following formula [108]:

$$\hat{y} \pm \sqrt{diag(cov\{\hat{y}\})} \tag{4.44}$$

where $\hat{y}$ the estimated output value from (4.43) and

$$cov\{\hat{y}\} = X_{act} cov\{\sigma^2 (X^T X)^{-1}\} X_{act}^T \tag{4.45}$$

The noise variance σ can be estimated by

$$\hat{\sigma}^2 = \frac{2\sum_{j=1}^{N}(y(j) - \hat{y}(j))^2}{N - deg} \tag{4.46}$$

where N the number of data samples, deg the number of degrees of freedom (=the number of parameters in the linear case), X the regression matrix, X_{act} the matrix containing the actual data samples for which the error bars should be calculated and $y(j)$ the measured values of the target. Note that the smaller the number of data samples and/or the higher the noise in the data, the higher the noise variance gets.

Now, the usage of the error bars as in (4.44) can be exploited for evolving (and also non-evolving) TSK fuzzy systems, as long as the linear parameters of the consequent functions in (1.11) are estimated by the local learning approach (as e.g. done in *FLEXFIS*, *FLEXFIS-Class*, *eTS*, *ePL*, *DENFIS* etc.). Then, a snuggling of the consequent hyper-planes along the model's surface can be observed, which is not the case when using global learning (see also Chapter 6). Hence, this circumstance yields a good approximation of the global model with local linear pieces. In this sense, it is reliable to calculate error bars for each rule consequent function (as one local linear piece) separately and then connect them with weights to form an overall smooth error bar for the whole fuzzy model. Obviously, the rule fulfillment degrees are a feasible choice for the weights, as the degrees get smaller at the borders of each rule. Thus, in case of C rules, the error bar of an evolving (TS) fuzzy model for a specific sample $\mathbf{x}_{act}$ can be calculated by [279]

$$\hat{y}_{fuz} \pm \sqrt{cov\{\hat{y}_{fuz}\}} = \hat{y}_{fuz} \pm \frac{\sum_{i=1}^{C} \mu_i(\mathbf{x}_{act}) \sqrt{cov\{\hat{y}_i\}}}{\sum_{i=1}^{C} \mu_i(\mathbf{x}_{act})} \tag{4.47}$$

where $\hat{y}_i$ the estimated value of the ith rule consequent function, for which cov is calculated as in (4.45) by using inverse weighted matrix $(X_i^T Q_i X_i)^{-1}$ corresponding to the ith rule and the noise variance as in (4.46). The symbol $\mu_i(\mathbf{x}_{act})$ denotes the membership degree of the actual sample to the ith rule and $\hat{y}_{fuz}$ the output value from the TS fuzzy model (see (1.9)) for the actual input sample $\mathbf{x}_{act}$.

Now, the remaining question is how to calculate the local error bars incrementally and synchronously to the evolving TS fuzzy models, in order to obtain the so-called *adaptive local error bars*. The incremental calculation is already guaranteed through (4.47) and (4.45) as the inverse weighted matrix $(X_i^T Q_i X_i)^{-1}$ is updated through the incremental learning process (by RWLS, see (2.64)), and X_{act} usually contains just the current data sample, for which a prediction statement or classification decision is requested. The only open issue is how to update $\hat{\sigma}^2$ in (4.46). This is done with $m \geq 1$ new points by [279]:

$$\hat{\sigma}^2(\text{new}) = \frac{(N - deg(\text{old}))\hat{\sigma}^2(\text{old}) + 2\sum_{j=N}^{N+m}(y(j) - \hat{y}_{fuz}(j))^2}{N + m - deg(\text{new})} \tag{4.48}$$

where $deg(\text{new})$ the new number of parameters, if changed in the model due to an evolution of the structure. So, the squared error on the m new samples is added to the squared error on the N previous samples and normalized with the whole number of samples seen so far minus the new degrees of freedom.

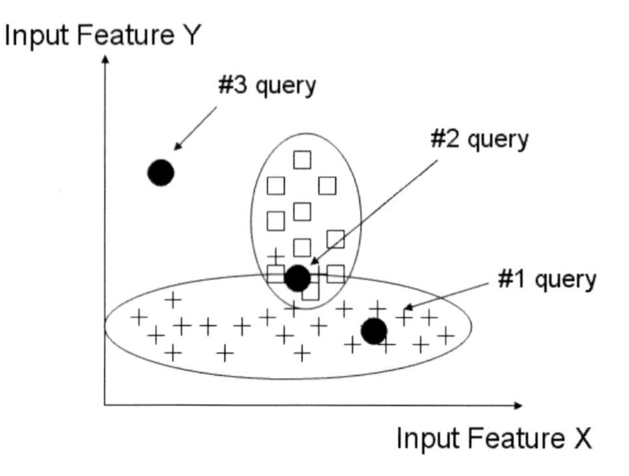

Fig. 4.17 Two clusters (rules) corresponding to two classes; three new query samples triggering three different situations regarding support of the (rule-based) classifier (extracted from the two clusters)

4.5.2 *For Classification Problems*

For classification problems, the uncertainty of models for a given query sample (=sample to be classified) can be expressed by two concepts [185]: conflict and ignorance. Conflict refers to the degree to which two classes are in conflict with each other, as the classifier supports both of them as potential classifications. Conflict is usually caused by a query instance which lies near the decision boundary of the classes. Ignorance represents the degree of extrapolation with respect to the labelled training data (seen so far when used in the incremental learning context). Ignorance is usually caused by a query instance which is either lying (significantly) outside the range of the input features or in regions of the feature space where no training samples occurred before. A visualization of the meaning of conflict and ignorance is demonstrated in Figure 4.17. Two classes ('+' and □) are represented by two bigger clusters (rules), three new query samples are marked as big dark dots: the #1 query sample is lying in a region where only samples from the '+' class are present in the training samples → in this case, the classifier (consisting of two rules extracted from the two clusters) is able to predict the class label with high confidence; the #2 query sample falls in the middle of the area where the two clusters overlap, therefore both classes are supported almost equally by the classifier; the #3 query sample falls in a region where no training samples have been before, therefore it is hard for the classifier to make a decision whether it falls into class '+' or class □. In both of the last two cases, the decision of the classifier will be uncertain. Note that in case of clusters which are not clean, i.e. clusters containing samples from different classes, the degree of conflict is even higher than presented in Figure 4.17.

For the conflict problem, the uncertainty of the classifier can be expressed by its confidence in the decision, compared to the support of the other classes. For evolving fuzzy classifiers, in case of single model architecture (see also Section 1.2.1.5), this confidence can be calculated by

$$conf_L = \frac{\sum_{i=1}^{C} conf_{iL}\mu_i}{\sum_{i=1}^{C} \mu_i} \tag{4.49}$$

with L the output class label, C the number of rules, μ_i the activation degree of the ith rule and $conf_{iL}$ the confidence of the ith rule in the Lth class. In most of the evolving fuzzy classifier approaches (where rules may be extracted from samples belonging to different classes), the latter is calculated by the relative frequency of the Lth class in one local region compared to the other classes. In some approaches, where clean rules or clusters are extracted, this can be set to either 1 (in the case when the consequent of the ith rule contains the final output class label L) or 0. In the case of multi-model architecture, the uncertainty is calculated by (see also Section 1.2.1.5)

$$conf_L = \frac{\max_{m=1,...,K} \hat{g}_m(\mathbf{x})}{\sum_{m=1}^{K} \hat{g}_m(\mathbf{x})} \tag{4.50}$$

where L is the output class label, $\hat{g}_m(\mathbf{x}) = \hat{f}_m(\mathbf{x}) + |\min(0, \min_{m=1,...,K} \hat{f}_m(\mathbf{x}))|$ and $\hat{f}_m(\mathbf{x})$ the output of the mth Takagi-Sugeno fuzzy model for the current sample $\mathbf{x}$.

The uncertainty levels calculated by (4.49) and (4.50) do not account for ignorance, as activation degrees and confidences are relatively compared among the rules, models and classes. To measure the level of ignorance, it is necessary to provide an estimate for the distance to the nearest lying labelled (training) samples. Luckily, fuzzy rule bases are able to resolve this problem, as they deliver (fuzzy) activation degrees of rules. Then, it is quite obvious that if no rule fires significantly, ignorance is present. This means whenever

$$\mu_i < \varepsilon \quad \forall i = 1,...,C \tag{4.51}$$

with $\varepsilon > 0$ a small threshold, no class label is significantly supported, therefore the confidence level of the final output class should be set to a small value, i.e. $conf_L = max_{i=1,...,C}\mu_i$. Finally, this can be seen as a strong argument as to why fuzzy classifiers may be preferred among other crisp rule-based classifiers (here rule activation degrees are either 0 or 1) and also among other machine learning classifiers such as SVMs or statistical discriminant approaches, which are hardly able to resolve the problem of ignorance [185].

The uncertainty levels provided in the classification and regression case can be also seen as one essential step in providing more insight, reliability and interpretability into the fuzzy models (see Section 6.4). This is simply because, additionally to classification or prediction statements, the uncertainty level may tell the operating system or the users how reliable these statements are.

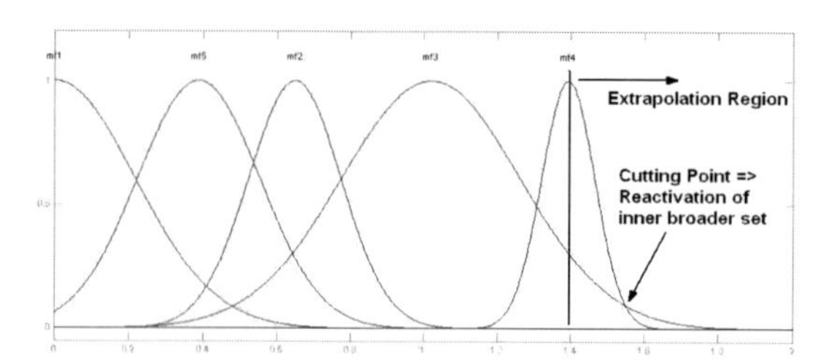

Fig. 4.18 A fuzzy partition where inner sets are broader than outer ones, especially compare the two right most sets → reactivation of the inner set in the extrapolation region

4.6 On Enhancing Extrapolation Behavior of EFS

Extrapolation is a difficult task, which has to be carried out with extreme caution for any model architecture and especially for evolving fuzzy systems as local models (rules) are usually only extracted in regions where (online) data samples occur. If possible, extrapolation should be completely avoided by incorporating samples from all process condition boundaries into the (on-line) training data. However, in practice it is hard to realize such a perfect coverage of all boundaries of the input space. Consequently, the extrapolation behavior of a model during the prediction/classification phase becomes an important issue (note that often prediction/classification statements are requested in off-line setting after having the complete model trained or in on-line setting before extending the model). When doing incremental learning and evolution of premise parts in fuzzy models with the usage of fuzzy sets with infinite support (especially with the most common Gaussian sets) it usually happens that not all fuzzy sets possess the same width. This is due to the nature of online data streams, where local areas (which are projected onto the axes to form the fuzzy sets) reflect different system behaviors with different ranges. This can lead to undesirable extrapolation effects, when the outer most set is not the broadest one, as a kind of reactivation of an inner membership function in the extrapolation region is caused, for an example see Figure 4.18, especially compare the two right most fuzzy sets; the extrapolation region is based on the range of the input variable in the data set (from 0 to 1.4). Inner sets can be reactivated, which is a precarious thing, as an outer most set usually stands for the most confidential information in the case of extrapolation as it is the nearest one to it. In usual cases, the tendency of the (linear) consequent functions belonging to the 'border' rules and fuzzy sets is desirable to be extended to the extrapolation region, as this is the most natural way to trust how the model continues. The impact of an undesired extrapolation behavior onto a two-dimensional approximation function is shown in Figure 4.19, where obviously at the right most part the curve does not follow the latest trend given within the original range of the training samples:

What someone wants to achieve is a nice expected behavior outside the conventional range of the data samples, as is for instance the case of the two-dimensional

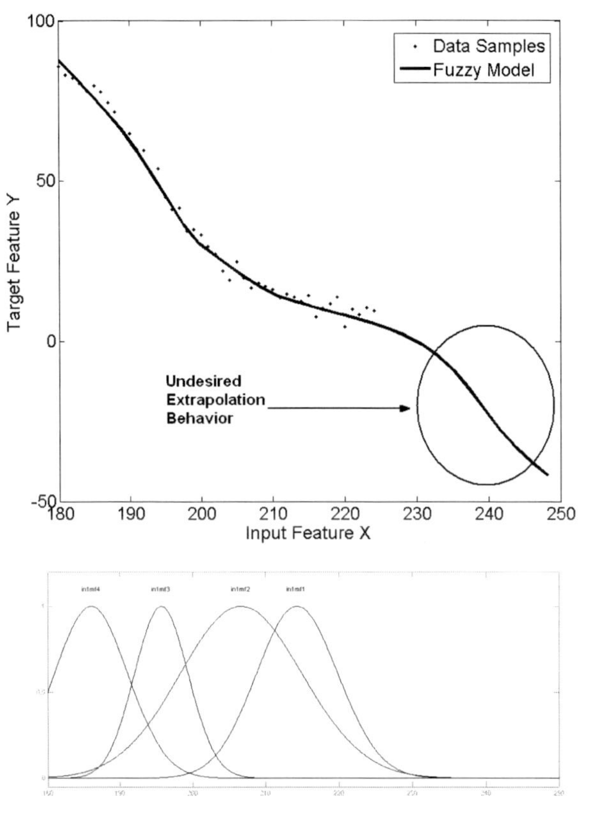

Fig. 4.19 Upper: fuzzy model with bad extrapolation behavior from sample 230 on because of reactivation of inner fuzzy set (rule), lower: the corresponding fuzzy partition (same scale), see the cutting point of outer most set with inner set at around 230

model shown in Figure 4.16 (b). In principle, one may focus on an antecedent learning procedure with constraints (e.g. constrained-based clustering) in order to ensure the same widths for all clusters and consequently fuzzy sets, however this would cause a severe drop of predictive performance in the normal range as the data characteristics would be not modelled as is present there. Hence, another possibility to overcome this drawback is pursued in [265], which is simply keeping all inner fuzzy sets as they are and modifying the (left and right) outer most fuzzy sets by leaving them constant at the membership degree 1 for the second half of their Gaussian membership functions, causing a first order extrapolation behavior over the complete range of the variables [342] (i.e. extending the linear function of the outer most rule to the extrapolation space). This procedure is called full coverage extension and is captured by the following formulas:

$$\mu_{(max)j} = \begin{cases} e^{-\frac{1}{2}\frac{x-c_{(max)j}}{\sigma_{(max)j}}} & x \leq c_{(max)j} \\ 1 & otherwise \end{cases} \quad \text{for the right most fuzzy set}$$

$$\mu_{(max)j} = \begin{cases} e^{-\frac{1}{2}\frac{x-c_{(min)j}}{\sigma_{(min)j}}} & x \geq c_{(min)j} \\ 1 & otherwise \end{cases} \quad \text{for the left most fuzzy set}$$

where $c_{(max)j}$ denotes the center of the right most fuzzy set and $\sigma_{(max)j}$ the width of this set.

In order to overcome a partially defined fuzzy set with two parts (which is not always easy to handle and requires a higher computation time when processing the inference for new (online) predictions), the right and left most set can be transferred to a sigmoid function defined by:

$$\mu_{ij}(x_j) = \frac{1}{1+e^{(-a_{ij}(x_j-b_{ij}))}} \tag{4.52}$$

based on the following two properties (leading to two equations in a and b):

- The position of inflection point of the Gaussian fuzzy set is equal to those of the sigmoid function.
- At the center position of the Gaussian set the value of the sigmoid function is 0.99 (as 1 is not possible).

The solution gives

$$a_{(right)j} = -\frac{\ln 0.01}{\sigma_{(right)j}}$$
$$b_{(right)j} = c_{(right)j} - \sigma_{(right)j} \tag{4.53}$$

for the right most fuzzy set and

$$a_{(left)j} = \frac{\ln 0.01}{\sigma_{(left)j}}$$
$$b_{(left)j} = c_{(left)j} + \sigma_{(left)j} \tag{4.54}$$

for the left most fuzzy set, substituting a_{ij} and b_{ij} in (4.52). The impact of this strategy on the above visualized two-dimensional example is presented in Figure 4.20. Clearly, the approximation curve from sample 225 to 250 follows the same trend as at the border of the original sample range (upper image). The lower image shows the fuzzy partition with the transformed right most fuzzy set (to sigmoid function).

For classification problems (with winner-takes-it-all aspects), it is quite intuitive that such a transformation to sigmoid functions may reduce the number of wrong classifications on left or right most sides of the input/output space.

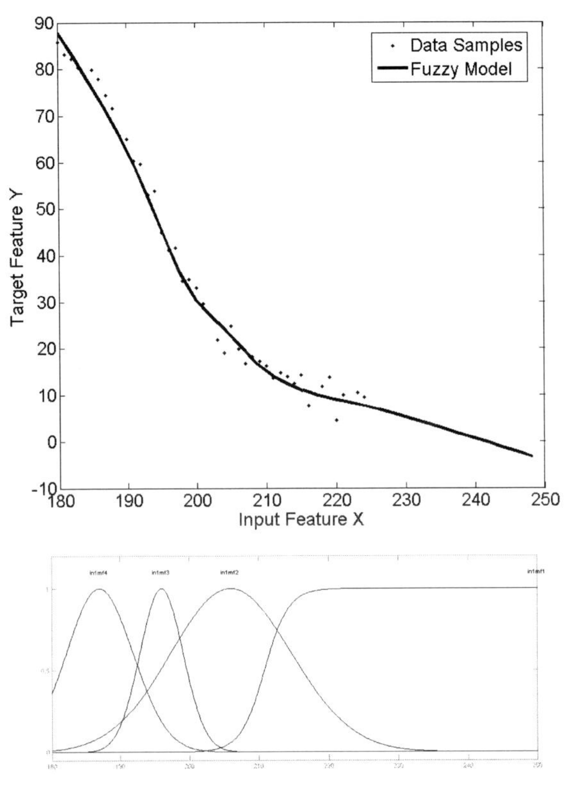

Fig. 4.20 Upper: fuzzy model with good extrapolation behavior (following the same trend as at the border of sample range) because of fuzzy set transformation, lower: the fuzzy partition with the right most Gaussian fuzzy set transformed to a sigmoid fuzzy set by using (4.53), no cutting point with = no reactivation of inner set

An alternative to above mentioned automatic fuzzy set transformation approach is the inclusion of prior knowledge about the process within the extrapolation regions in the form of lower or upper bounds or even slopes. A straightforward way to include such a knowledge is to define additional fuzzy sets at the boarder of the fuzzy partitions in the input space and to define rules including these. The corresponding consequents in the extrapolation regions can be also defined by prior knowledge or estimated from the nearest lying data samples. The fuzzy system output of a Takagi-Sugeno type model will change from (1.2) to

$$\hat{f}(\mathbf{x}) = \hat{y} = \sum_{i=1}^{C} l_i \Psi_i(\mathbf{x}) + \sum_{i=1}^{C_{ex}} l_i \Psi_i(\mathbf{x}) \tag{4.55}$$

with C_{ex} the number of rules defined in the extrapolation region. Rules defined in the extrapolation regions may be adjusted with new data as soon as they appear in these regions in order to refine and confirm the expert knowledge. Therefore, the rules are back-projected to the cluster space with an inverse projection strategy (inverse to the cluster projection as shown in Figure 3.2, simply consider the arrows 180 degrees turned to point into the opposite direction).

Chapter 5
On Improving Performance and Increasing Useability of EFS

Abstract. This chapter deals with further improvements of evolving fuzzy systems with respect to performance and usability. Opposed to the previous section, which demonstrated aspects which are necessary ('must haves') for guiding evolving fuzzy systems to more robustness, process safety and meaningful output responses in case of specific occasions in the data streams (drift, outliers, faults, extrapolation situations etc.), the issues treated in this chapter are 'nice to have' extensions for further increasing prediction power and usability. To achieve this, we demonstrate the following strategies:

- Dynamic rule split-and-merge strategies during incremental learning for an improved representation of local partitions in the feature space (Section 5.1).
- An on-line feature weighting concept as a kind of on-line adaptive soft feature selection in evolving fuzzy systems, which helps to reduce the curse of dimensionality dynamically and in smooth manner in case of high-dimensional problems (Section 5.2).
- Active and semi-supervised learning the reduce labelling and feedback effort of operators at on-line classification and identification systems (Section 5.3).
- The concept of incremental classifier fusion for boosting performance of single evolving fuzzy classifiers by exploiting the diversity in their predictions (Section 5.4).
- An introduction to the concept of dynamic data mining, where the data batches are not temporally but spatially distributed and loaded (Section 5.5.1).
- Lazy learning with fuzzy systems: an alternative concept to evolving fuzzy systems (Section 5.5.2).

Although these aspects can be seen as typical 'add-ons' for the basic evolving fuzzy systems approaches demonstrated in Chapter 3, they may be key stones in some learning context, for instance when dealing with very high-dimensional problems or feedback from experts/operators with different experience or confidences in their decisions.

E. Lughofer: Evolving Fuzzy Systems – Method., Adv. Conc. and Appl., STUDFUZZ 266, pp. 213–259.
springerlink.com

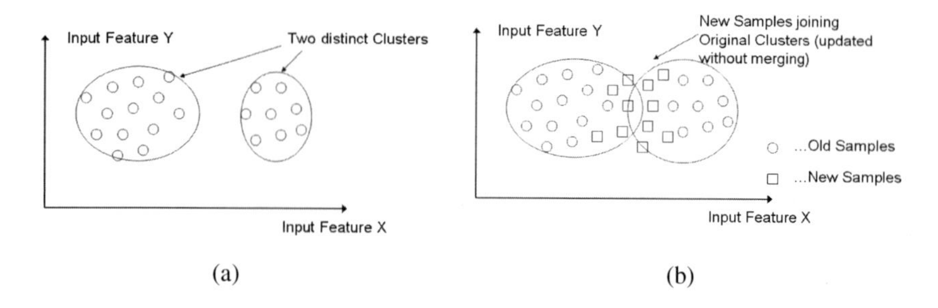

Fig. 5.1 (a): Initial data distribution with two distinct clusters (local regions, rules); (b): New samples (rectangles) filling up the gap between the original clusters and joining them to one big cluster; in both images the surrounding spheres indicate the extracted clusters, which move together and overlap according to the newly loaded samples in (b).

5.1 On-Line Split-and-Merge of Rules

The concept of on-line split and merge of rules is necessary 1.) whenever the essential parameters in the training process of evolving fuzzy systems are inadequately set (based on a priori experience or some off-line batch training phase), tendentially leading to too many or too few local regions (rules); 2.) whenever a varying characteristic, distribution of samples in the data space turns out at a later stage. This section deals with strategies for improving the performance of evolving fuzzy systems, by splitting resp. merging rules on demand in order to prevent under-fitting resp. over-fitting of data and to cope with changing data characteristics on-the-fly.

5.1.1 In the Cluster Space

Problem Statement

When applying (incremental) clustering approaches in evolving fuzzy systems for identifying rules and a projection concept for obtaining fuzzy sets and the premise parts of the rules, it may happen that two (or more) clusters seem to be distinct at the beginning (e.g. during the batch setup phase) and move together with more samples loaded. An (artificial) example of such an occurrence is visualized in Figure 5.1. Figure (a) represents the initial situation where two distinct clusters seem to be present in the data. Then, when new data is loaded (Figure (b)), the gap between two clusters is filled up and visually the whole bunch of data appears as one cluster. The conventional clustering procedures proceed with two clusters (as evolved for the first block of data in (a)) and updates them according to the new samples $\rightarrow$ clusters move together and overlap. In this case, it is recommended to apply a so-called cluster merging process as for instance applied in [233], [367]. However, these are batch approaches and usually require some additional information extracted from the whole data set.

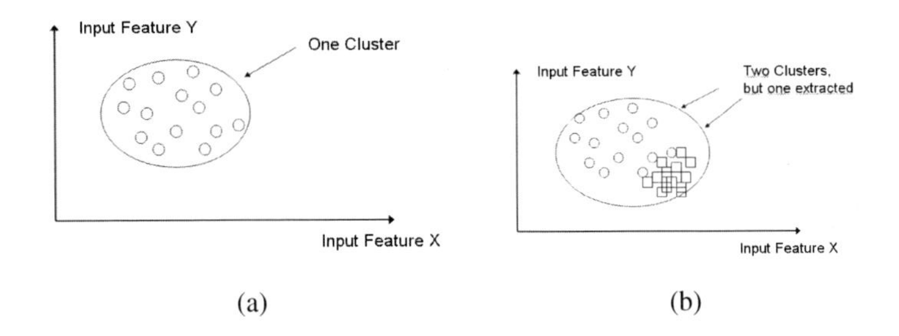

Fig. 5.2 (a): Initial situation (one cluster recognized as such); (b) new samples densifying the lower right region of the initial cluster $\rightarrow$ two distinct clusters may be a better choice.

Another unpleasant situation occurs when the opposite is the case: at the beginning there seems to be one cluster (rule) in a local region of the data space, which turns out later to contain two distinct data clouds. Such an occurrence is shown in Figure 5.2, where (a) visualizes the initial situation.

Split-and-Merge Algorithm

In both situations, it would be beneficial 1.) to detect such occurrences where clusters (rules) move together or internally split up into distinct clouds and 2.) to react on them. To solve this problem for the on-line case, a trial split-and-merge strategy of clusters can be carried out: after each (sample-wise) incremental learning step, the updated cluster is first split into two halves with respect to each axis and the quality of the obtained clustering structures are calculated with a so-called cluster validation index (see [164] for a comprehensive survey of such indices). This index should be either calculated incrementally and synchronously to the cluster update(s) or by its nature not requiring any prior data. Furthermore, the updated cluster is merged with the closest cluster and the new clustering structure obtained in this way is again validated by the same validation index. The same validation is done for the original clustering obtained after the incremental learning step. The clustering structure with the highest quality (in terms of the cluster validation index) is kept and further updated with the next incoming sample. Algorithm 18 gives a concise formulation of this strategy (with C the number of clusters).

Algorithm 18: On-line Split-and-Merge Strategy

1. **Input:** current cluster partition (e.g. after an incremental update step in the clustering procedure) containing C clusters; cluster validation index for calculating the qualities of cluster partitions.
2. **If** $C > 1$, calculate the quality for the current cluster partition (using cluster validation index) $\rightarrow cl\_qual_1$.

3. Split the current winning cluster nearest to the actual sample (represented by c_{win}) into two equal halves with respect to each axis and calculate the quality for the obtained p cluster partitions $\rightarrow cl\_qual_2, ..., cl\_qual_{p+1}$.
4. **If** $C \geq 3$
   a. Merge the current winning cluster (represented by c_{win}) with the closest cluster. The closest cluster can be obtained in the same way as in Steps 5 and 6 in Algorithm 4 resp. by using alternative distance strategy (to surfaces) as pointed out in Section 3.1.4.1 when setting $\mathbf{x} = c_{win}$.
   b. Calculate the quality of the obtained cluster partition $\rightarrow cl\_qual_{p+2}$.
   c. Elicit cluster partition with highest quality: $max\_ind = arg\max_{i=1,...,p+2} (cl\_qual_i)$.
   d. Overwrite the current cluster partition with that cluster partition represented by max_ind, update the number of clusters C if necessary (in case of $max\_ind = 2, ..., p+1$: $C = C+1$, in case of $max\_ind = p+2$: $C = C-1$).
5. **If** $C \leq 2$ [**Optional as maybe slow:**]
   a. Determine if any clustering structure is present at all.
   b. **If** no and $C = 2$, merge the current winning cluster (represented by c_{win}) with the closest cluster, set $C = 1$;
   c. **If** no and $C = 1$, do nothing.
   d. **If** yes and $C = 2$, set $max\_ind = arg\max_{i=1,...,p+1}(cl\_qual_i)$, if yes and $C = 1$, set $max\_ind = arg\max_{i=2,...,p+1}(cl\_qual_i)$; overwrite the current cluster partition with that cluster partition represented by max_ind; update the number of clusters C if necessary, i.e. in case of $max\_ind = 2, ..., p+1$ set $C = C+1$.
6. **Output:** Updated cluster partition for further on-line processing.

Step 5 is an optional step and includes the examination whether a clustering structure is present at all or not (two clusters may move together to one or there was only one cluster extracted so far). Please note that other cluster split-and-merge approaches have been proposed in literature such as [43] for on-line windowed k-means algorithm or [406] for on-line merging of Gaussian mixture models[111] (concurrently seen as clusters) based on multivariate statistical tests instead of cluster validation indices.

Calculation of the Quality of Cluster Partitions

If Algorithm 18 is carried out after each sample which triggered a cluster update (and not a generation of a new cluster) in the incremental clustering process, it is guaranteed that always the clustering structure with the highest quality value is transferred to the next incremental learning step (depending on the cluster quality measure). This means, that if the cluster validation index always prefers the more precise clustering structure, a significant improvement of the performance of the applied incremental clustering approach can be expected. A suitable cluster validation index is the PS-index [458] as in its crisp form it does not require any prior data to calculate the quality of the clustering, just the number of data samples $k_.$ belonging to each cluster (those data points which have formed the different clusters during the incremental learning process so far):

$$PS(C) = \sum_{i=1}^{C} PS_i \tag{5.1}$$

with

$$PS_i = \frac{k_i}{k_{max}} - e^{-\frac{\min_{k \neq i}(\|c_i - c_k\|)}{\beta_T}} \tag{5.2}$$

where

$$k_{max} = \max_{i=1,...,C} k_i \tag{5.3}$$

$$\beta_T = \frac{\sum_{i=1}^{C} \|c_i - \bar{c}\|^2}{C} \tag{5.4}$$

and $\bar{c}$ the mean value of all cluster centers. The higher $PS(C)$ gets, the better the data is partitioned into C clusters. From this definition it is clear, that the PS-index only needs the cluster centers and the number of data samples belonging to each center. Both pieces of information can be updated easily, so that the PS-index can be re-calculated every time after each incremental learning step without requiring any additional update information.

Determination of the Existence of a Cluster Partition

Please note that the merging step in Algorithm 18 is not carried out in the case of two clusters (condition in Step 4). The reason is that the PS-index (and so most of the other well-known cluster validation indices) cannot be calculated for a partition containing one cluster. In this case, a different validation measure has to be applied which is able to give rise whether there is a cluster structure in the data present or not at all (no matter how many clusters there are); this is also called *cluster tendency* of a cluster partition. The cluster tendency can be measured with the Hopkins index [182], whose determination can be carried out in the following way: first, m data samples $R = \{r_1, ..., r_m\}$ are randomly selected from the convex hull of the data space of the data set X, i.e. by generating random points within the ranges of each variables of X. Then, further samples $S = \{s_1, ..., s_m\}$ are randomly selected directly from the data set X, i.e. $S \subset X$. For those sets R and S, the distances $d_{r_1}, ..., d_{r_m}$ respectively $d_{s_1}, ..., d_{s_m}$ of R respectively S to the nearest points in X are determined. Then the Hopkins-index h is defined by:

$$h = \frac{\sum_{i=1}^{m} d_{r_i}^p}{\sum_{i=1}^{m} d_{r_i}^p + \sum_{i=1}^{m} d_{s_i}^p} \in [0, 1] \tag{5.5}$$

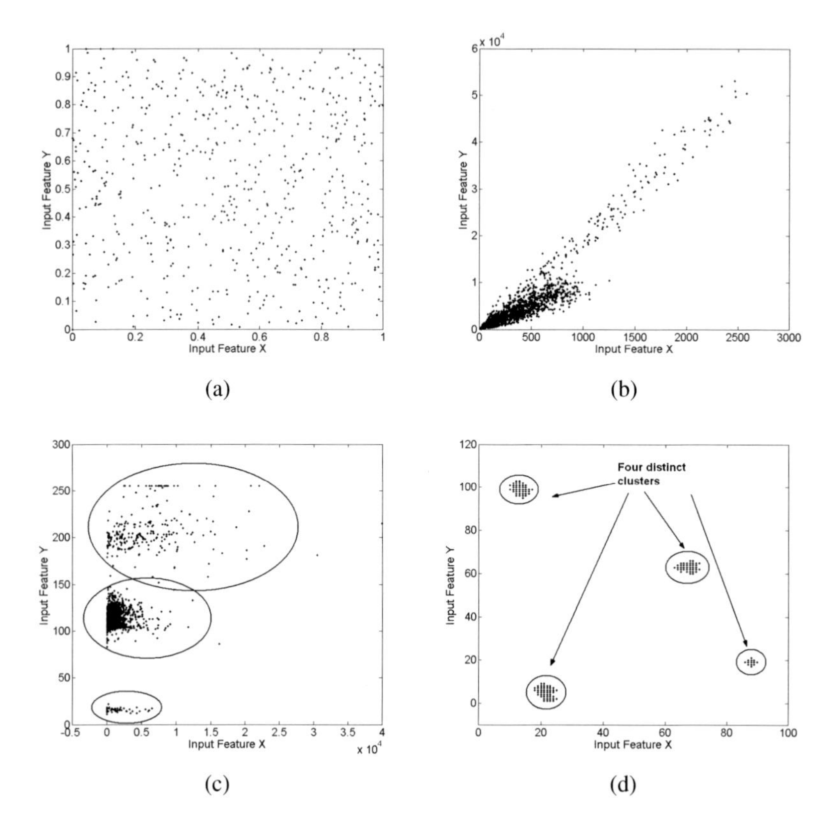

Fig. 5.3 (a): randomly distributed samples, $h = 0.23$; (b): no cluster structure (just a global correlation tendency), $h = 0.31$; (c): a weak cluster structure with 2 larger noisy (broad) and one compact distinct cluster; h=0.66; (d): clear cluster structure with four distinct compact clusters; $h = 0.98$

with p the dimensionality of the data set. The value of h lies between 0 and 1, whereas the following typical cases shown in Figure 5.3 can be distinguished:

1. If the data samples are either randomly distributed (as seen in (a)) or have no clear cluster structure, i.e. containing at least two clusters (as seen in (b)), then usually $h \leq 0.5$.
2. If the data samples contain a weak (noised) clustering structure as shown in (c), then usually $0.5 \geq h \leq 0.75$.
3. If clear distinct clusters are present in the data set (with a number of clusters somewhere between 2 and the number of data samples, as seen in (d)), then usually $0.75 \leq h$.

The incremental calculation of the Hopkins-index is quite straightforward, as for each newly loaded data sample the distances to R and S can be calculated and the new minimal distances obtained. However, although S can be sub-sampled replacing older samples with newer ones, this slows down the process significantly, as the convex hull has to be re-calculated based on the full set of samples seen so far and R re-adjusted Therefore, we pointed out in Algorithm 18 that this is an optional feature (Step 5) and consider only a splitting in case of $C = 2$ (Step 5d) and no merging.

Splitting and Merging Prototype-Based Clusters

Regarding splitting of clusters the currently updated cluster c_{upd} into two equal axis-parallel halves with respect to dimension j, the following formulas are applied:

$$\begin{aligned} c_{upd,j}(new1) &= c_{upd,j}(old) + \sigma_{upd,j}(old) \\ c_{upd,j}(new2) &= c_{upd,j}(old) - \sigma_{upd,j}(old) \\ \sigma_{upd,j}(new1) = \sigma_{upd,j}(new2) &= \frac{\sigma_{upd,j}(old)}{2} \\ k_{upd}(new1) = k_{upd}(new2) &= \frac{k_{upd}(old)}{2} \end{aligned} \tag{5.6}$$

The number of data samples belonging to the original cluster is divided by 2 and assigned to each one of the two new clusters. The symbol σ denotes the variance (widths) of the cluster in each dimension (assuming to process clusters with axis-parallel ellipsoids). In case of ellipsoids with arbitrary shape, the split should be performed along the principal axes. Usually, the range of (statistical) influence of one cluster can be seen as the $c \pm 2\sigma$-area (96% of samples fall in these), therefore the operation $c \pm \sigma$ moves the centers in-between the middle point and the outer point of the ellipses w.r.t to each axis, the operation $\sigma/2$ bisects each semi-principal axis (see Figure 5.4, left part). A reasonable merging of two cluster centers can be carried out by taking their weighted average, whereas the weights are represented by the significance of the clusters, i.e. the number of samples k_{upd} resp. k_{close} forming the clusters. A reasonable merging of the ranges of influence of the two clusters is achieved by updating the range of influence (in each direction) of the more significant cluster with the range of influence of the other by exploiting the width update with recursive variance formula as presented in (3.5), Chapter 3. Here, the second cluster represents a whole collection of points rather than a single sample with which the update is carried out. This means that in the last term in (3.5) the sample x_{kj} is substituted by the center of the less significant cluster (as representing the mean of the 'new' data = the data which formed this cluster). In order to guarantee a good coverage of the original data cloud by the merged cluster, a fraction of the variance of samples belonging to the less significant cluster is added, which is determined by the percentage of samples belonging to the less significant cluster with respect to the samples belonging to both clusters. Hence, the cluster merge is done in the following way ($\forall j = 1, ..., p$):

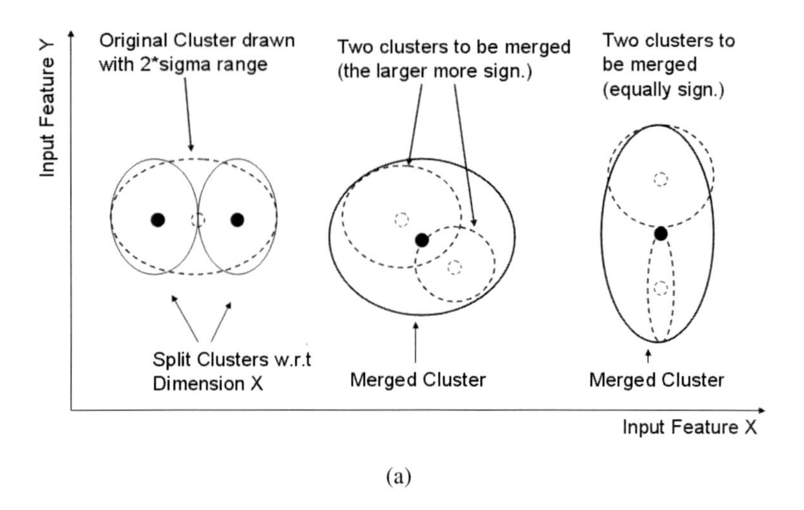

(a)

Fig. 5.4 Left side: splitting case along dimension X, middle part: merging case with two overlapping clusters and where the larger cluster is more significant than the smaller one (center stays closer to the old center of the larger cluster); right side: merging case where both clusters are equally significant and the centers coincide with respect to dimension X; dotted lines represent ranges of influence and centers of the original clusters, solid lines those of the split resp. merged clusters

$$c_{upd,j}(new) = \frac{c_{upd,j}(old)k_{upd} + c_{close,j}(old)k_{close}}{k_{upd} + k_{close}} \tag{5.7}$$

$$\sigma_{upd,j}(new) = \sqrt{\frac{k_{cl_1} * \sigma_{cl_1,j}}{k_{cl_1} + k_{cl_2}} + (c_{cl_1,j} - c_{upd,j}(new))^2 + \frac{(c_{upd,j}(new) - c_{cl_2,j})^2}{k_{cl_1} + k_{cl_2}}}$$
$$+ \frac{k_{cl_2}}{k_{cl_1} + k_{cl_2}} * \sigma_{cl_2,j}$$
$$k_{upd}(new) = k_{upd}(old) + k_{close}(old)$$

where $cl_1 = argmax(k_{upd}, k_{close})$ denoting the (index of the) more significant cluster, and consequently $cl_2 = argmin(k_{upd}, k_{close})$ denoting the (index of the) less significant cluster.

A small (artificial) two-dimensional example on how this splitting and merging of clusters works is presented in Figure 5.4: the original clusters are shown by dotted centers and dotted ranges of influence, the merged resp. split clusters shown by centers with solid filled circles and solid ranges of influence (as 2σ-areas). The left side represents a splitting case with respect to input dimension X, the middle part a merging case where two clusters overlap and the bigger cluster is more significant (contained more samples seen so far than the smaller cluster), but their centers do not coincide in any dimension, the right part a merging case where both clusters are equally significant (hence, the new center is placed in the middle of the older centers) and the centers coincide with respect to dimension X.

Split-and-merge for *eVQ*, *eVQ-Class* and *FLEXFIS-Class*

In the case of *eVQ*, the same procedures as mentioned above can be applied. In the case of *eVQ-Class*, it also has to be considered how to update the hit matrix when splitting resp. merging a cluster. In the case of cluster merging the entries of the corresponding rows in the hit matrix are simply added (recall rows represent clusters, whereas columns represent classes). In the case of cluster splitting, an additional row is evolved with entries of those row belonging to the split cluster divided by two. The entries of the row belonging to the original cluster is also divided by 2. Thus, we assume that there are approximately equal proportions within the two distinct data clouds in the cluster.

In case of *FLEXFIS-Class SM* the split of a cluster automatically evolves a new rule: the antecedent part is simply obtained by projecting the new cluster to form fuzzy sets (as outlined in Figure 3.2); the class label of the original rule is transferred to the consequent labels of the newly formed rules. Merging of two clusters in *FLEXFIS-Class SM* (with the approach mentioned above) causes a merge of two rules, and the consequent label of the merged cluster is obtained by taking that class response having the highest frequency in both (original) clusters together.

Split-and-merge for *eClustering* and *eClass*

In case of *eClass* applying a recursive calculation of potentials of data samples and cluster centers, basically the same strategy as outlined above can be applied. After each update cycle, the cluster partitions are checked with the cluster validation index using Algorithm 18. Merging can be performed by using the two nearest clusters to the current data samples and applying (5.7) (if the widths of all clusters have the same constant value as in original *eTS*, only the centers are updated). Splitting can be performed for the nearest center to the current sample by applying (5.6). The consequent labels can be obtained in the same manner as for *FLEXFIS-Class SM* (see previous section). In case of *eClassB*, a more complicated merging and splitting approach for consequent parts is required, as these are multi-dimensional hyper-planes, see also subsequent section

Split-and-merge for Evolving Fuzzy Regression Models using Clustering (*FLEXFIS*, *eTS*, *ePL* and *DENFIS*)

It is not recommended to apply cluster split-and-merge process to regression problems. This is because there clustering serves for the purpose of obtaining partial local regions for appropriately approximating highly non-linear relationships in order to account for the up-and-down fluctuations of non-linear dependencies (see for instance Figure 3.3 and the discussion in Section 3.1.1.1). This means that Algorithm 18 resp. the cluster tendency analysis with (5.5) would mostly end up in one cluster: for an example, see the relationship in Figure 3.3 (a), where the Hopkins index is 0.3 indicating that no cluster structure is present, however more than one cluster/rule (at least 3) needs to be extracted in order to be able to follow the non-linear trend appropriately. Hence, other rule split-and-merge approaches have to be considered, which are applicable to regression problems. This is also true for other evolving fuzzy classification approaches which do not apply any clustering for eliciting the antecedent parts of the rules (hence disabling the usage of cluster validation indices as quality criterion for local partitions).

5.1.2 Based on Local Error Characteristics

One possibility for a flexible rule split-and-merge approach for regression problems, is to exploit the local error characteristics of a fuzzy system. That is, for new incoming samples the prediction performance is locally elicited with the help of an error measure (for instance mean-squared-error) $\rightarrow$ local error. Local means here that it is not the error of the final global model output that is inspected (including the weighted average over all rules), but the error produced by that rule which is closest to the current sample to be predicted. Hereby, it is assumed that the local linear models provide a solid approximation of the local trend where they are defined. This is assured when using *recursive fuzzily weighted least squares* for local learning, but not true in case of global learning, as the consequent functions can be quite chaotic and therefore provide no useful local piecewise approximation (see Chapter 6), prohibiting a reliable examination of local errors. The rule closest to the current sample $\mathbf{x}$ is that rule with the highest activation degree Ψ, i.e.:

$$R = argmax_{i=1}^{C} \Psi_i(\mathbf{x}) \tag{5.8}$$

The local error to the local output of this rule is calculated as squared deviation:

$$sd(\mathbf{x},R) = (y - y(\mathbf{x},R))^2 = (y - w_{R0} + w_{R1}x_1 + w_{R2}x_2 + \ldots + w_{Rp}x_p)^2 \tag{5.9}$$

with $x_1,\ldots,x_p$ the input components and y the real measured output value of the current sample based on which the fuzzy system is further adapted. Now, if this error exceeds a certain threshold for a certain number of samples for which the Rth rule was the closest one, the fuzzy system is not able to approximate the underlying relationship with sufficient accuracy in this local region (a reasonable threshold would be 10% to 20% error relative to the range of the output variable). In this case, a modification in the local region where the Rth rule is defined is required; whether this should be a splitting of the rule into two rules or a merging with the closest nearby lying rule, depends on whether the current system under-fits or over-fits in this local region. As already outlined in Section 2.8, under-fitting arises when the model flexibility is too low in order to approximate a non-linear functional relation with sufficient accuracy, whereas over-fitting is present when the model flexibility is too high, so that the noise in the data is followed more than the basic trend of the model (see Figure 2.12). The degree of under-fitting can be measured by the so-called bias error , whereas the degree of over-fitting is measured by the variance-error. A reasonable tradeoff between these two is a central goal of each training algorithm, see also Section 2.8. Now, when calculating (5.9), it is essential to know whether a high value of (5.9) is because of (local) under-fitting or over-fitting. In the former case, a split of the local region (rule) is required into two parts in order to increase the flexibility of the whole model. In the latter case, a merge of the local region with the closest one is required in order to decrease the flexibility of the whole model and to decrease over-fitting. A visualization of this point is presented in Figure 5.5, where (a) represents an under-fitting case where one rule has to be split in order to bring in more flexibility of the model and (b) the over-fitting case where two close rules should be merged in order to smooth the regression curve. Decomposing the (local) model error into these two parts is possible in the off-line case, as data can be split into training data and separate and independent test data, whereas the latter can be used to estimate the variance error.

In the on-line case, someone can use new incoming samples as test samples when calculating the local error (5.9) before updating the fuzzy model, and some past samples over a time horizon window as training samples (as the fuzzy model were updated based on these). In case of a high bias error = error on the training samples (for which the Rth rule was the closest one), the local linear model (rule consequent function) of the Rth rule is obviously not flexible enough to follow the basic trend of the functional relationship. Hence, splitting of the Rth rule should be performed which can be done by the same strategy as in (5.6) (assuming Gaussian membership functions), where the dimension in rule R is used for splitting with highest error reduction ratio on past samples. In the case of a high error on (significant) new test samples (for which the Rth rule was the closest one) but low error on past training samples, an over-fitting occurrence takes place, hence the Rth rule should be merged with the closest rule, which can be carried out as in (5.7). The closest rule is that rule Q for which

$$d(\Psi_R, \Psi_Q) \leq d(\Psi_R, \Psi_j) \ \forall j \in \{1, ..., C\} \setminus \{R\} \tag{5.10}$$

holds, where d denotes the distance between two rules, which can be defined by using the distances of the rule prototypes $\mathbf{c}$:

$$d(\Psi_R, \Psi_Q) = \|\mathbf{c}_R - \mathbf{c}_Q\| \tag{5.11}$$

Summarizing these aspects, the split-and-merge strategy for regression problems based on local error characteristics works as demonstrated in Algorithm 19. Application of this algorithm in an evolving fuzzy systems approach can be seen as an optional add-on before each update step.

Algorithm 19: Split-and-Merge Strategy for Regression Problems

1. **Input:** current evolved fuzzy system from previous cycle, current on-line sample
2. Elicit the closest rule to the current sample, say the Rth.
3. Calculate sd by (5.9).
4. **If** $sd > thresh\_error$, $n_R = n_R + 1$ with n_R the number of samples for which the Rth rule was the nearest one and where the local error exceed threshold $thresh\_error$.

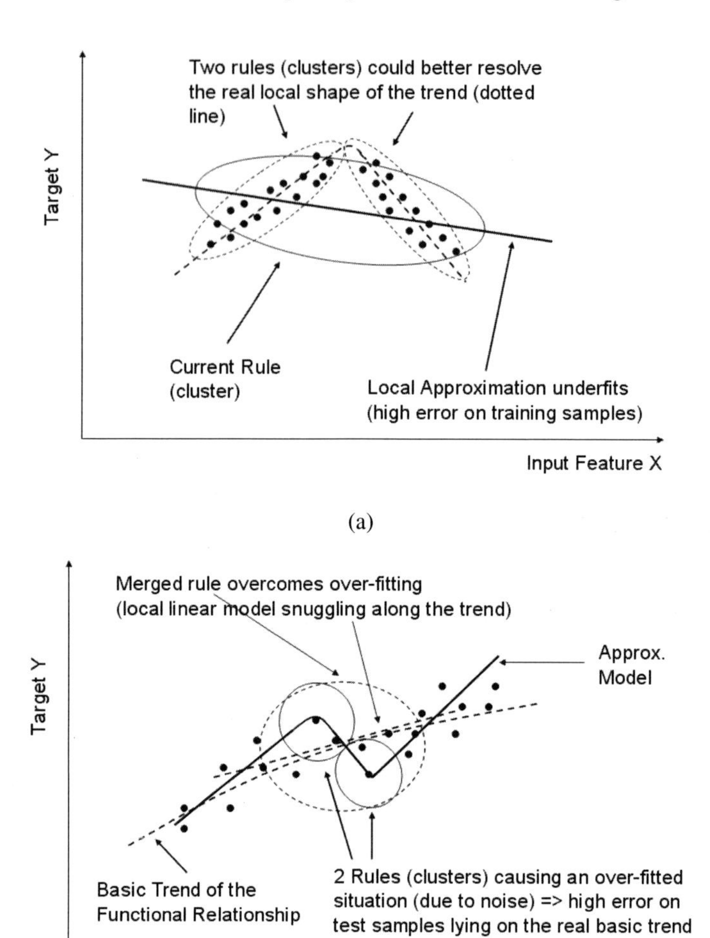

Fig. 5.5 (a): Under-fitting as only one rule is extracted (solid cluster) in a local region where the tendency of the functional relationship is non-linear (shown as dotted line), therefore causing a high error on the training samples (dots); two (split) rules (shown as dotted clusters) can resolve this under-fitted situation (one following the up-, the other following the down-trend); (b): over-fitting as two rules are extracted due to noise in the data, over-shooting the basic (logarithmic) trend of the functional relationship (solid line is the approximation model), therefore causing a high error on new test samples lying on this trend in this local region; a merged rule (dotted cluster) can overcome the over-fitted situation (its local linear model shown dotted straight line)

5. **If** $n_R > thresh\_samples$, decide whether a splitting of the Rth rule or a merging of the Rth rule with the closest rule should be performed:
   a. **If** $\frac{1}{N_R}\sum_{i=1}^{N_R} sd(\mathbf{x}_i, R) > thresh\_error$ on past N_R training samples (sliding time window) for which R was the nearest one then perform splitting (see step below),
   b. **Else**, perform merging (see step below).
6. In case of splitting, elicit that dimension with highest local error reduction ration on past N_R samples. Then, apply (5.6) and assign the consequent function of the Rth rule to the new rule, increase the number of rules $C = C + 1$.
7. In case of merging, elicit the closest rule to rule R by (5.10) and apply (5.7); perform an average of all the consequent parameters of the Rth rule with those of the closest rule, decrease the number of rules $C = C - 1$.
8. **Output:** Updated rule base.

The threshold $thresh\_samples$ controls the sensitivity with respect to outliers and faults in the data. This should be set in accordance to the consideration made in Section 4.4, Chapter 4. The threshold $thresh\_error$ can be set based on expert input (e.g. model should never exceed this maximal error in any local region) or based on a guess of the noise level in the data. The latter can be accomplished with the help of adaptive local error bars which are described in detail in Section 4.5, Chapter 4 (2 or 3 times the local error bar would be an appropriate value of the threshold). It is recommended to apply Algorithm 19 to new incoming data blocks rather than single samples as containing a significant number of new test samples.

5.1.3 Based on Rule Similarity

When clusters are moving together during the incremental learning process (as demonstrated in Figure 5.1), in the extreme case it may even happen that the clusters (rules) get strongly over-lapping (simply consider about an ongoing movement together in Figure 5.1). This will finally lead to redundancy in the cluster space. Such redundant information can always be reduced, no matter whether incremental clustering is exploited for classification or regression problems. Calculating the degree of redundancy can be achieved by analytically computing the overlap degree of two ellipsoids in the high-dimensional feature space [370]. However, this option requires significant computation time and therefore being not applicable in a fast single-pass incremental learning context. A simplification of this is achieved by using the overlap degrees of each pair of one-dimensional Gaussians when both rules are projected onto the single axes and producing an aggregated value over the p overlap degrees (obtained for the p dimensions). A possibility is to calculate the intersection points of the two Gaussians (used as fuzzy sets in (1.10)) which are belonging to the same dimension in the antecedent parts of a rule pair, i.e. for rules i (the modified rule) and k (any other) these are $e^{-\frac{1}{2}\frac{(x_j-c_{ij})^2}{\sigma_{ij}^2}}$ and $e^{-\frac{1}{2}\frac{(x_j-c_{kj})^2}{\sigma_{kj}^2}}$. In general, there are two intersection points whose x-coordinates are obtained by (here for the jth dimension, proof is referred to the reader):

$$inter_x(1) = -\frac{c_{kj}\sigma_{ij}^2 - c_{ij}\sigma_{kj}^2}{\sigma_{kj}^2 - \sigma_{ij}^2} + \sqrt{(\frac{c_{kj}\sigma_{ij}^2 - c_{ij}\sigma_{kj}^2}{\sigma_{kj}^2 - \sigma_{ij}^2})^2 - \frac{c_{ij}^2\sigma_{kj}^2 - c_{kj}^2\sigma_{ij}^2}{\sigma_{kj}^2 - \sigma_{ij}^2}}$$

$$inter_x(2) = -\frac{c_{kj}\sigma_{ij}^2 - c_{ij}\sigma_{kj}^2}{\sigma_{kj}^2 - \sigma_{ij}^2} - \sqrt{(\frac{c_{kj}\sigma_{ij}^2 - c_{ij}\sigma_{kj}^2}{\sigma_{kj}^2 - \sigma_{ij}^2})^2 - \frac{c_{ij}^2\sigma_{kj}^2 - c_{kj}^2\sigma_{ij}^2}{\sigma_{kj}^2 - \sigma_{ij}^2}} \quad (5.12)$$

The maximal y-coordinate = membership degrees of the Gaussian membership functions in these coordinates is then used as similarity degree of the corresponding rules' antecedent parts (in the jth dimension):

$$overlap_{ik}(j) = max(inter_x(1), inter_x(2)) \quad (5.13)$$

Note that in the case when there is one Gaussian fuzzy set fully embedded by another one, there is only one intersection point whose membership degree is 1 ($\rightarrow$ similarity is also 1). The amalgamation over all rule antecedent parts leads to the final overlap/similarity degree between rule i (the modified rule) and k:

$$overlap_{ik} = Agg_{j=1}^{p} overlap_{ik}(j) \quad (5.14)$$

where Agg denotes an aggregation operator. A feasible choice is the minimum operation, as a strong dissimilarity along one single dimension is sufficient that the clusters do not overlap at all (tearing apart local regions).

In case when no cluster is updated but a new one evolved, there is obviously no necessity to perform a rule merging process. This leads us to the steps as described in Algorithm 20 for integrating rule similarity detection and elimination. This can be used in any EFS approach after modifying (the position of) existing rules with new incoming samples.

Algorithm 20: Rule Merging based on Overlap/Similarity Degree

1. **If** a new cluster was evolved, do nothing.
2. **Else** perform the following steps:
3. Check if overlap/similarity of moved/updated cluster (winning cluster) i with any other cluster $m = 1,...,C \setminus i$ calculated by (5.14) is higher than a pre-defined threshold sim_thr
4. **If** yes
   a. Perform rule merging of cluster i with cluster $k = argmax_{m=1}^{C} overlap_{im}$ after (5.7).
   b. Perform merging of corresponding rule consequent functions (hyper-planes) by performing a weighted average of the all consequent parameters, where the weights denote the significance of the rules (number of samples forming the rules).

c. Overwrite parameters $(\mathbf{c}_i, \sigma_i)$ of rule i with the parameters of the merged cluster $(\mathbf{c}_{new}, \sigma_{new})$.
d. Delete rule k belonging to cluster k.
e. Decrease number of clusters/rules: $C = C - 1$.

5.2 On-Line Feature Weighting in EFS (Achieving Dynamic Soft Dimensionality Reduction)

5.2.1 Problem Statement

The concept of on-line feature weighting is another attempt to decrease the overfitting by reducing the curse of dimensionality. Opposed to (crisp) feature selection, where most important features or system variables are selected for training classification and regression models with high predictive qualities[159][160], feature weighting is a concept where the features resp. variables are giving weights according to their importance levels in a modelling scenario. Features with low importance levels are giving low weights (near 0), features with high importance are giving high weights (near 1). In this sense, a kind of regularization effect is achieved, which also tries to shrink the weights of the features (resp. the corresponding linear parameters in regression parts) towards zero, see Figure 4.4 (Chapter 4).

So far, for all training algorithms we have described in this book (except the extension to *eVQ* in Section 3.1.4.6), the features resp. variables were always giving full weights, hence seen as equally important for model building. In some applications, especially with a low dimensional setup, this is sufficient in order to obtain models with good quality. However, in some other applications, it is necessary to reduce the curse of dimensionality, which is often carried out with the help of feature selection. The point is now that in an on-line feature context, on-line feature selection is hardly applicable. Indeed, there exists feature selection methods which incrementally (and this in single-pass mode) can change the list of most important features, see for instance [254] [459] [210], and which therefore could be applied synchronously to the training engines of evolving fuzzy systems (serving as by-information about the most important features); but, the problem is how to change the input structure of the models appropriately without causing discontinuities in the learning process. In fact, features can be dynamically and on-line exchanged in the importance list (e.g. feature #1 replaces feature #2 in the list of five most important features after a new data block), but this requires an exchange in the input structure of the fuzzy models as well (otherwise nothing will happen in the models themselves). There, however, parameters and rules are learned based on the inputs used in the past. As usually two inputs show different distributions and different relations with other features in the joint feature space (especially in drift cases), one input cannot be exchanged with another by simply exchanging it in the model definition and continuing with the old parameter setting learned for the other input: the model will immediately produce many wrong classification or prediction statements. The problem becomes more evident, when considering that the list of important features changes with a high frequency (in extreme case for each single sample). Then, there is not a sufficient number of samples to adapt the parameters and rules to the new input structure.

A concept for overcoming such unpleasant effects is the concept of on-line feature weighting, where features weights (according to the importance of the features) are permanently updated on-line. A new incoming sample will only change the already available weights slightly (especially when these were already learnt on many past data samples), such that, when including the feature weights in the learning process, a smooth learning can be expected without disturbance of a convergence to optimality (resp. to the batch solution) of certain parameters.

As feature weights may approach 0, the feature weighting concept can be also seen as a kind of *soft dimension reduction approach*. If this is carried out during on-line phase for evolving fuzzy classifiers by synchronously and permanently feature weight updates, we call the whole approach *dynamic soft dimension reduction*.

In the following, we will present the feature weighting concept for classification (Section 5.2.2) and regression problems (Section 5.2.3), the way of including the feature weights in the training depends on the chosen evolving fuzzy systems approach: we give guidelines for some of these.

5.2.2 *For Classification Problems*

5.2.2.1 Separability Criterion

Whenever the incremental training of an evolving (fuzzy) classifier takes place, a supervised label information can be used to calculate the importance levels of the features. A well-known technique for feature selection in classification problems can be used, the so-called Fisher's interclass separability criterion which is defined as [111]:

$$J = \frac{det(S_b)}{det(S_w)} \tag{5.15}$$

where S_b denotes the between-class scatter matrix measuring how scattered the cluster means are from the total mean and S_w the within-class scatter matrix measuring how scattered the samples are from their class means. The goal is to maximize this criterion, i.e. achieving a high between-class scatter (variance), while keeping the within-class scatter (variance) as low as possible. S_w can be expressed by the sum of the covariance matrices over all classes, i.e. by:

$$S_w = \sum_{j=1}^{K} \Sigma_j \tag{5.16}$$

with Σ_j as defined in (2.4) by using samples from the jth class. The matrix S_b is defined as:

$$S_b = \sum_{j=1}^{K} N_j(\bar{X}_j - \bar{X})^T(\bar{X}_j - \bar{X}) \tag{5.17}$$

with N_j the number of samples belonging to class j, $\bar{X}_j$ the center of class j (i.e. the mean value of all samples belonging to class j) and $\bar{X}$ the mean over all data samples (for all features). However, this criterion has the following shortcomings [112]:

- The determinant of S_b tends to increase with the number of inputs, hence so does the separability criterion (5.15), preferring higher dimensionality of the input space.
- It is not invariant under any nonsingular linear transformation [133]. This means that once m features are chosen, any nonsingular linear transformation on these features changes the criterion value. This implies that it is not possible to apply weights to the features or to apply any nonsingular linear transformation or projection and still obtain the same criterion value.

Hence, in [272], the following criterion is applied (inspired by Dy and Brodley [112]):

$$J = trace(S_w^{-1} S_b) \tag{5.18}$$

with $trace(A)$ the sum of the diagonal elements in A. A larger value of (5.18) indicates a better discrimination between classes. Regarding incremental capability of this criterion, it is sufficient to update the matrices S_w and S_b and then to compute (5.18). Both matrices can be updated during incremental mode, S_w by applying (2.8) (Section 2.1.2, Chapter 2) for each class separately, and S_b by simply updating N_j through counting and $\bar{X}_j$ and $\bar{X}$ by incrementally calculating the mean as done in (2.2) (Section 2.1.1, Chapter 2).

The within-class scatter matrix may become singular or nearly singular in the case when there are redundant features in an on-line data stream. Then, calculating the inverse in (5.18) leads to an unstable and wrong criterion J. One possibility to circumvent this is to apply Tichonov regularization by adding αI to S_w with α chosen by a heuristic approach not requiring any past samples, see also Section 4.1.2. However, the condition of a matrix and its eigenvalues requires significant computational complexities (and these are computed for each on-line sample). An alternative is to delete redundant features at the beginning of the whole training process (ideally in an off-line pre-training phase) and to continue with the remaining features. Redundant features are identified by finding linear dependencies between two or more features: to do so, one possibility is to build up linear regression models by using one feature as target and a subset of the remaining ones as input (e.g. selected with a variable selection method such as forward selection [309] or its orthogonal version [153]) and inspect their qualities in terms of a quality measure such as r-squared. In case of p features, p models as follows are obtained:

$$\begin{aligned} Feat_1 &= f_1(Feat_{1,1}, Feat_{2,1}, ..., Feat_{L_1,1}) \\ Feat_2 &= f_2(Feat_{1,2}, Feat_{2,2}, ..., Feat_{L_2,2}) \\ &\dots \\ Feat_p &= f_p(Feat_{p,1}, Feat_{p,2}, ..., Feat_{L_p,p}) \end{aligned} \tag{5.19}$$

with f_1 to f_p linear models, where $Feat_{i,j}$ denotes the jth input feature and L_i the number of input features for approximating the ith feature; in general $Feat_{1,i} \neq Feat_{1,j}$, $Feat_{2,i} \neq Feat_{2,j}$,... for $i \neq j$, and $L_i \neq L_j$, so the selection of the L_i features included in the various models is always different (as depends on the target feature) and also using different (number of) inputs. The target features of models with a high quality can be seen as redundant to a linear combination of the input features in these models and hence can be deleted. Cross-deletions have to be checked appropriately such that no target feature is deleted which is needed as input for another regression model in order to achieve a high quality there. This is achieved by the steps as summarized in Algorithm 21 [272].

Algorithm 21: Redundancy Deletion in the Feature Space

1. Input: list of features L.
2. Build up feature relation models as in (5.19) for the p features.
3. Calculate the quality of the p models with quality measure lying in $[0,1]$ (r-squared, r-squared-adjusted).
4. Delete all models which have a lower quality than a pre-defined threshold (0.97 is a feasible option [272]).
5. Sort the p_2 remaining models according to their quality in descending order.
6. Delete the target feature in the model with highest quality from the list of features: $L = L - \{feat_1\}$ (as it can be approximated by a linear combination of others with high quality).
7. For $i = 2, ..., p_2$, take the target feature from ith model and look whether it is an input feature in any of the former $j = 1, ..., i-1$ models (with higher qualities):
   - If no, delete this target feature from the list of features: $L = L - \{feat_i\}$ (as it can be approximated by a linear combination of others with high quality).
   - If yes, keep it in the feature list.
8. Output: new feature list L where the redundant features are omitted.

5.2.2.2 On-Line Feature Weighting Strategies

Now, the question remains on how to use this criterion to assign feature weights. The idea is now to calculate (5.18) p times, each time one of the p features is discarded. In this sense, p different between-class and within-class variances as well as p different values for (5.18), $J_1, ..., J_p$, are obtained which are updated synchronously in incremental mode. This strategy is called *LOFO = Leave-One-Feature-Out* weighting approach (and should be distinguished with the sample-based *leave-one-out* approach used in cross-validation scenarios). A statement on the relative importance of feature can be made, when sorting these values in decreasing order: the maximal value of these indicates that the corresponding discarded feature is the least important one, as the feature was discarded and still a high value is achieved. In the same manner, the minimal value of these indicates that the corresponding discarded feature is the most important one, as dropping the value of the separability criterion more significantly than any of the others. Hence, a relative comparison among the values of $J_1,...,J_p$ obtained in the $p-1$ dimensional space is applied. Then, the weight of feature j, λ_j, is achieved by:

$$\lambda_j = 1 - \frac{J_j - min_{1,...,p}(J_j)}{max_{j=1,...,p}(J_j)} \tag{5.20}$$

This guarantees that the features causing the minimal value of J_j (hence most important one) gets a weight of 1, whereas all others are relatively compared to the minimum: in the case when they are close the weight will also be almost 1, in the case when they are significantly higher, the weight will be forced towards 0. A problem of this feature weight assignment strategy is given in case of a high number of inputs in real-world application scenarios. Then, it is quite likely that the deletion of one feature does not decrease the value of (5.18) significantly compared to when using all features (also not for those with high discriminatory power). This is because the likelihood that a similar or even redundant feature is present in the data set is quite high in case of a high number of features/variables. This finally means that the normalized feature weights according to (5.20) are all close to 1. Hence, in case of a high number of input, it is necessary to normalize the weights to the full range of [0,1] by

$$\lambda_j = 1 - \frac{J_j - min_{1,...,p}(J_j)}{max_{j=1,...,p}(J_j) - min_{1,...,p}(J_j)} \tag{5.21}$$

hence the feature with the weakest discriminatory power (and therefore maximal J_j) is assigned a weight value of 0.

The problem with the LOFO approach is that 1.) S_w may become singular or nearly singular, achieving an unstable solution of (5.18) (requiring regularization steps) and 2.) it requires a quite high computation time, as for each single sample falling into class L the covariance matrix Σ_L and between-class scatter matrix for class L need to be updated, which demands an update of the whole S_w resp. S_b by summing up over all classes. Updating the covariance matrix Σ_L requires $p+2p^2$ operations (mean vector update for each dimension + component-wise multiplication of column vector with row vector to form the new covariance information and summation of two matrices (old covariance matrix with new covariance information)). Updating the between-class scatter for class L requires $p+p^2$ operations due to similar considerations (no matrix addition is required, hence p^2 instead of $2p^2$). To calculate the new S_b and S_w matrices, K p times p matrices are summed up, achieving a complexity of $2Kp^2+3p^2+2p$. Calculating criterion (5.18) from the updated S_w and S_b matrices, requires another $2p^3+p$ operations (matrix inversion as well as matrix multiplication requires p^3, the trace p operations as summing up the diagonal entries of a p times p matrix). The complexity for calculating one J_p is therefore $O((2K+3)p^2+2p^3+3p) \approx O((2K+3)p^2+2p^3)$, hence the overall complexity is $O((2K+3)p^3+2p^4)$.

To reduce this complexity, a greedy-based approach to approximate Dy-Brodley's separability criterion is applied. This is done by calculating the between-class and within-class scatter for each feature separately, so by neglecting possible good interrelations between some features for a better discriminatory power among classes. As reported in [117], this approach gives a quite good approximation of the original separability criterion. At least, features with a high between-class scatter value are also certainly important features when joined with others to build a classifier. Indeed, features with a low proportion of between-class scatter to within-class scatter may improve separability of classes when joined with others, however this improvement is usually stronger for features having higher single discriminatory power [159]. By using the feature-wise approach scalar values for S_b and S_w are obtained, i.e. for the ith feature:

$$S_b(i) = \sum_{j=1}^{K} N_j(\bar{X_{j;i}} - \bar{X}_i)^2 \qquad S_w(i) = \sum_{j=1}^{K} Var_{j;i} \tag{5.22}$$

with $\bar{X}_i$ the mean value of the ith feature, $\bar{X_{j;i}}$ the mean value of the ith feature over class j samples and $Var_{j;i}$ the variance of the ith feature over class j samples. The single separability criterion for feature i is simply calculated by

$$I_i = \frac{S_b(i)}{S_w(i)} \tag{5.23}$$

The update is quite fast as it requires only linear complexity in the number of features ($O(p)$) for each single sample. Note, that the strategy of a feature-wise (independent) calculation of the separability criterion was also successfully applied in [118] and [117]. There, the criterion was used to adapt parameters in a feature extraction component to obtain a better class separability by less complex decision boundaries and alleviating classifier training based on these features. The final feature weights are obtained by:

$$\lambda_j = \frac{I_j}{max_{j=1,\ldots,p}} \tag{5.24}$$

In [272] [270], an evaluation of these incremental feature weighting concepts is presented, where it can be shown when used in combination with *FLEXFIS-Class* that 1.) features with higher discriminatory power are in fact giving higher weights and 2.) that it is a robust method in the sense that it nearly outputs the same weights as its batch learning variant (using the whole amount of data and calculating (5.21) respectively (5.24)) There, it could also be verified that in high-dimensional cases ($p > 20$), usually more than half of the features are assigned weights lower than a small value $\varepsilon > 0$ and therefore are neglected in the training and classification phases, achieving significant reduction of dimensionality.

5.2.2.3 Including Feature Weights in Evolving Fuzzy Classifiers

Basic Strategy

As soon as the feature weights are permanently elicited during the incremental on-line mode, the remaining question is how to use this additional information about the importance of the features in the evolving fuzzy classifiers in order to decrease the curse of dimensionality and boost classification performance as pointed out in Section 5.2.1. In principle, there are three possible strategies:

1. Inclusion of the feature weights during the classification process only, i.e. when classifying new instances.
2. Inclusion of the feature weights during the training process when building up the classifiers.
3. Inclusion of the feature weights during training *and* classification process.

In the first option, the classifiers are trained as in the usual setting (without feature weights) and the weights are only included when classifying new samples. Thus, no decrease of the curse of dimensionality is achieved in the sense that the complexity of the classifiers is reduced by generating less rules. The impact of this classification scheme is visualized in Figure 5.6 for a two-dimensional data set containing two classes and three rules (shown as clusters in the two-dim space): the second feature (y-axis) has almost no discriminatory power, whereas the first feature (x-axis) can discriminate the classes almost perfectly. Hence, when calculating the distance to the centers of the three clusters = rules for a new incoming sample (dark dot), the second feature has almost no impact, so the winning cluster (nearest cluster) turns out to be the bottom left cluster, which only contains samples from class '□'. In this sense, this class is correctly assigned to the new sample. If the feature weights were not included in the classification process (inference of the classifiers), the nearest rule to the new sample would be the upper middle one, where class '+' has a weak majority → this class would be falsely assigned to the new sample (from the distribution of the classes it is somewhat clear that samples appearing at the far left side of the x-axis belong to class '□'). Therefore, even for simple two-dimensional data sets, feature weights can be necessary to guide classifiers to higher correct classification rates.

When using the second option, the classifier can be prevented from generating such rules as indicated by the upper middle cluster in Figure 5.6 — note that within the incremental learning context the bottom two clusters were included in the first data block and the upper middle cluster in the second data block. The weight of the second feature is 1.) either so low that the distance to the lower rule centers is always close enough such that no rule is generated in-between the two clusters (only the already existing ones are updated as shown in Figure 5.7 → in this case a prevention of an unnecessary increase of complexity can be achieved), or 2.) the hole between the lower rule centers is wide enough for a new cluster to be generated (in order to cover the data space sufficiently); however, in this case a new rule will be generated inbetween these two (close to the x-axis) rather than way up along the

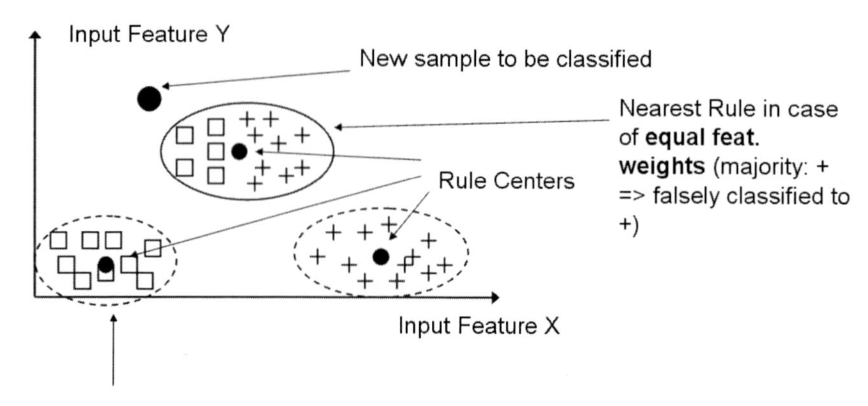

Fig. 5.6 Impact of including feature weights when classifying new samples according to the winner-takes-it-all classification scheme (majority class in the nearest rule is assigned as class label); the original two clusters contained in the older data block shown by dotted ellipsoids, the new cluster contained in the newly loaded data block shown as solid ellipsoid.

second feature. In a winner-takes-it-all concept, a new sample will be closest to this center when falling inbetween the other two rules and not when appearing on either left or right side of the x-axis with a high value in Feature Y.

The third option is a combination of the two above. In usual cases, the second option should decrease the complexity of the classifiers by not generating misleading rules, hence the first option should get superfluous. Nevertheless, in order to be on the safe side, and as inclusion of feature weights usually does not increase computational demands significantly, a combination of both (double-coupled) should be used.

Inclusion in EFC approaches

For *FLEXFIS-Class SM* and its spin-off *eVQ-Class* we already demonstrated in Section 3.1.4.6 how to include the feature weights in their incremental learning engine for antecedent parts, *eVQ*, namely by including them when updating the cluster prototypes:

$$\mathbf{c}_{win}^{(new)} = \mathbf{c}_{win}^{(old)} + \eta \lambda I(\mathbf{x} - \mathbf{c}_{win}^{(old)}) \tag{5.25}$$

and also including them in the distance calculation when measuring the compatibility of a new sample to the current cluster partition, influencing the evolution of new clusters (as compared with the vigilance parameter):

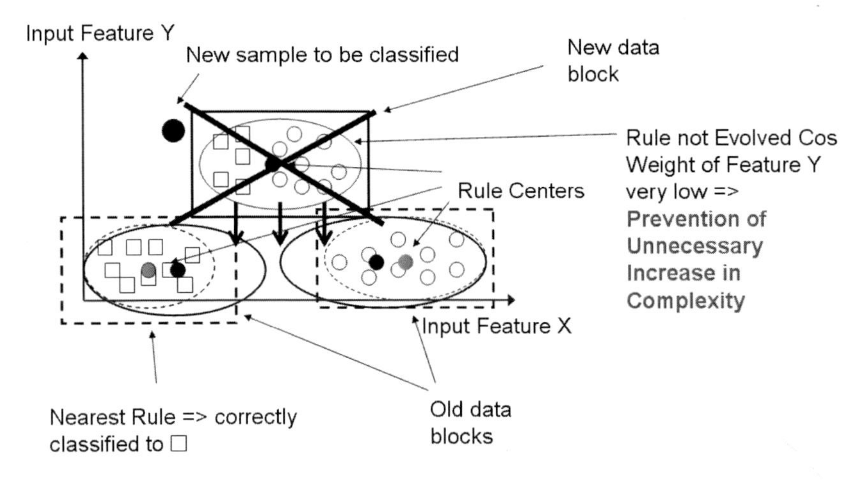

Fig. 5.7 Impact of including feature weights when during incremental learning phase — the second data block does not trigger a new rule when Feature Y is unimportant (out-weighted) → avoidance of unnecessary clusters; the original two clusters are expanded by extending their ranges of influences towards the middle area of Feature X.

$$dist = \sqrt{\sum_{j=1}^{p} \lambda_j (x_j - c_j)^2} \tag{5.26}$$

A similar concept can be applied to *eClass0* and *eClassA*, where the feature weights can be included 1.) into the distance $d_{.,.}$ when updating the potentials for the current sample and all existing centers and 2.) when eliciting the closest cluster center to the current sample (whose potential is higher than all already existing centers = rules) and denoting whether it is close enough. This is simply achieved by multiplying the distance in each dimension with respect to the corresponding feature weight. When using Mahalanobis distance measure, the inclusion of feature weights yields:

$$mahal = \sqrt{(\lambda .* (\mathbf{x}-\mathbf{c}))\Sigma^{-1}(\lambda .* (\mathbf{x}-\mathbf{c}))} \tag{5.27}$$

with Σ^{-1} the inverse of the covariance matrix and $.*$ the component-wise product of two vectors.

For classification purposes, feature weights can be simply included when eliciting the rules' activation degrees, that is by computing the t-norm over all antecedent parts, where the fulfillment degree of each part is multiplied with the corresponding feature weight (belonging to this dimension). In case of Gaussian membership in connection with product t-norm this means:

$$\mu_i(\mathbf{x}) = \prod_{j=1}^{p} e^{-\frac{1}{2}\frac{(x_j-c_{ij})^2}{\sigma_{ij}^2}\lambda_j} \quad i = 1,\dots,C \tag{5.28}$$

with p the dimensionality of the feature space, C the number of rules, c_ij the center of the jth antecedent part in the ith rule and σ_{ij} the width of the jth antecedent parts in the ith rule. This means that the fuzzy set membership degrees of unimportant features (with feature weights near 0) are 1, hence serving a 'don't care parts' as not influencing the whole rule activation when taking the product over these degrees. For important features (feature weights near 1), the (nearly) actual fuzzy set membership degrees are taken, influencing the rule activation degrees. This concept can be generalized to arbitrary membership functions μ_{ij} by using the following expression to calculate the rule activation degrees:

$$\mu_i(\mathbf{x}) = \prod_{j=1}^{p}((\mu_{ij} - 1)\lambda_j + 1) \; i = 1,...,C \tag{5.29}$$

This guarantees 'don't care entries' (values of 1) in case of unimportant features and original μ_{ij}'s, interpolating linearly for features with weights in-between 0 and 1. In this sense, in fact a soft dimension reduction approach is achieved: indeed, the features are still present in the models, but they are switched off and do not contribute to any membership degree calculations (in the classification phase) or any distance calculations (in the training phase). The expression in (5.29) can be used for any other t-norm, for which values of 1 represent 'don't care parts' (for instance minimum would be such an option).

For *FLEXFIS-Class MM* respectively *eClassM* exploiting multiple Takagi-Sugeno fuzzy systems and building up a classifier via indicator vectors, we refer to the subsequent section (the inclusion has to be made for regression problems).

5.2.3 For Regression Problems

In case of regression problems, no discrete class labels are available as targets, but continuous samples (real numbers). This means, the concept demonstrated in the previous section for eliciting the discriminatory power of features is not applicable for regression problems. A possibility for eliciting the feature weights is by exploiting the feature importance information from the consequent parts of the evolved fuzzy systems. In particular, the values of the linear consequent parameters are inspected per rule (=cluster) separately [10], based on which local feature weights are deduced, in case of C rules=clusters: $\lambda_{1,1},...,\lambda_{1,p},\lambda_{2,1},...,\lambda_{2,p},...,\lambda_{C,1},...,\lambda_{C,p}$. In fact, the linear consequent parameters per rule denote the gradient of the features in the local region defined by the corresponding rule. This is especially true when using the local learning approach, as this causes hyper-planes snuggling along the real trend of the functional relationship (see Section 2.4 and [281]) and therefore obtaining a sort of piecewise linear function approximation. An example of this aspect is given in Figure 5.8 (piecewise consequent functions shown in solid line), where in the furthest left part the function remains almost constant and hence the input feature (x-axis) has almost no importance in obtaining a good approximation there (the consequent part could also be well defined by just using the intercept value of around 0.77). The impact of a feature can also be interpreted by its gradient with respect to the target y. For instance, in the middle part a slight change in x causes a

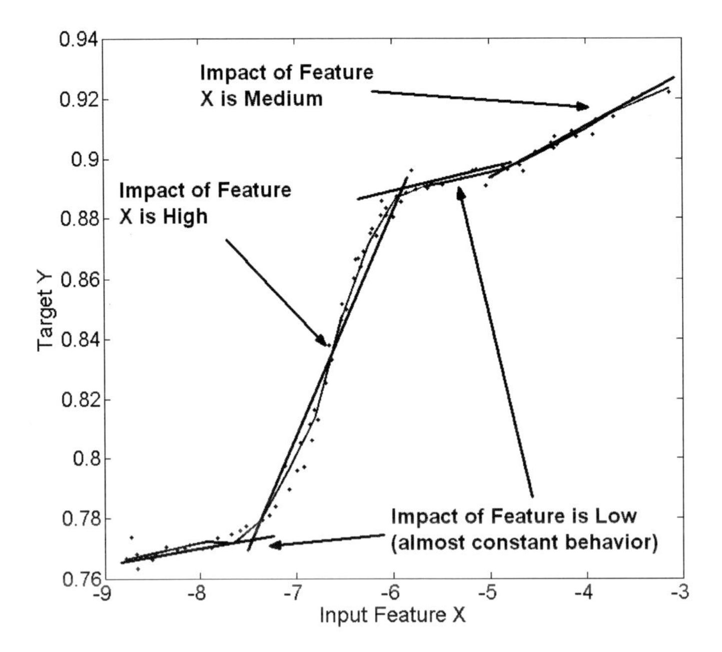

Fig. 5.8 Rule consequent functions (solid straight lines) for a Takagi-Sugeno-type fuzzy system when approximating a two dimensional nearly sigmoid relation; the impact of the input feature X in the different local regions as indicated

quite huge change in the target y; hence, there the impact if x on the target y is quite big.

As mostly features in its original range are applied when estimating the consequent parameters, the linear parameters (gradients) are not comparable in their original form. Hence, each linear parameter is multiplied by the range of the corresponding feature. This gives us the relative impact of the jth feature on the output of the ith local linear model (rule):

$$imp_{i,j} = w_{i,j} * range(x_j) \tag{5.30}$$

Obviously, a feature with the same linear consequent parameter but a significantly higher range than another one has more impact on the output of the ith rule. Whenever a consequent parameter approaches 0, it can be concluded that the behavior of the rule output is almost constant with respect to the corresponding feature, hence the feature gets quite unimportant or even superfluous for modelling the functional behavior in this local region. In this sense, it is obvious to deduce local feature weights for the ith rule by a relative comparison among the most important feature (feature with highest impact):

$$\lambda_{i,j} = \frac{imp_{i,j}}{max_{j=1,...,p}(imp_{i,j})} \quad j = 1,...,p \tag{5.31}$$

This also means that different local feature weights are included when updating resp. measuring the distance to the different rules. This finally provides a more detailed information as in the classification case (see previous section). A global weighting scheme can be achieved as well when summing up the impact of each feature over all rules and normalizing through the maximal impact:

$$\lambda_j = \frac{imp_j}{max_{j=1,...,p}(imp_j)} \quad j = 1,...,p \tag{5.32}$$

with $imp_j = \sum_{i=1}^{C} imp_{i,j}$.

Another possibility for eliciting feature weights is to exploit the so-called t-values which are motivated in statistical regression theory. These include the variance of the linear parameters by multiplying the square-root of the corresponding diagonal elements in the inverse Hessian matrix (the ith element for the ith parameter) with the sum of the squares of regression, measured as the deviation of predicted values from the mean of the measured values, see (2.17). t-values give rise to the degree of influence of the linear parameters when regressing onto the target. For more details please refer to [170].

Incorporation of the feature weights in evolving fuzzy regression models can be again done two-fold: when predicting new occasions only or also directly when training the models from on-line streams, see previous section. In the former case, the approach reduces to inclusion of the weights when processing the inference of the antecedent parts by using (5.29) (resp. (5.28)) in the same manner as in the classification case. In the latter case it is dependent on the used evolving fuzzy systems approach. In case of *FLEXFIS*, it is obvious to use $\lambda_{win,j=1,...,p}$ as feature weights when updating the center resp. range of influence of the winning cluster and $\lambda_{j=1,...,p}$ when calculating the distance of a new sample to all clusters in the antecedent learning part carried out by *eVQ*. The latter enforces lower complexity of models which appear nearly constant w.r.t. certain features.

In *eTS*, similar strategies can be performed as in the case of classification using *eClass* (see previous section). In *ePL*, feature weights can be included 1.) when updating the centers through (3.41) by multiplying the second term with the weight vector, and 2.) when computing the compatibility measure through (3.43) by weighting the features when calculating the norm (distance between current feature vector and cluster/rule center). In *DENFIS*, feature weights can be included by any distance calculation (3.47) performed in Algorithm 13. In *SOFNN*, the maximal membership degree of the current sample to a rule is one of the two criteria whether old rules are updated or new ones are evolved. The inclusion of feature weights in the calculation of the membership degree (as done in (5.29)) contributes to a soft dimensionality reduction and therefore a decrease of over-fitting, as less neurons will be created. This is because it is more likely that the input space is covered by any neuron when some features are (nearly) switched off (product over membership function values to single fuzzy sets increases). In *SAFIS*, the inclusion of feature weights in the distance (norm) calculation in (3.62) helps to prevent creating superfluous rules, as shrinking distances in unimportant dimensions (features).

Semi-smooth Newton method (see Section 4.1.2) would be a promising option for forcing as much feature weights as possible to 0 in each rule consequent part separately, achieving a local feature out-weighting/reduction step. This techniques has already been applied successfully for batch training of fuzzy systems in [284], but has not been investigated for the on-line evolving case (in fact, to our best knowledge, no incremental variant of Semi-smooth Newton method exists so far).

5.3 Active and Semi-supervised Learning

5.3.1 Problem Statement

In classification problems, the main problem is to gather the labelling information for recorded data samples (measurements, images, signals). That is because labelling usually requires a high workload for experts or system operators, which have to look through all the collected samples and annotate them for indication to which classes the samples belong. Finally, this means that for the company where the expert/operators are employed, it may cost significant man-hours and hence significant money. For instance, in a speech/music discrimination system a human has to examine the signals used for training carefully either to segment it into speech, music and silence segments (which usually are 1-2 seconds long). Labelling is even more time-intensive and especially requires deeper expert knowledge in case of physical/chemical/biological measurements without any context. For on-line learning scenarios, such as adapting classifiers during on-line operation modes in order to achieve monotonically predictive performance over time, the labelling problem is even more severe, as often data is recorded with a high frequency which mostly prohibits the operator(s) to give feedback onto the classifiers decision (whether they are correct or not, resp. at least a quality information). This feedback is absolutely necessary as within the incremental learning context it is only possible to further improve the classifier when having the real class label available; otherwise, the classifier would train its own errors again and again into its structure, which after some time would lead to a decreased predictive performance.

In fact, software rating tools equipped with a comfortable user and human-machine interaction interface are more and more replacing traditional simple industrial front-ends. When using these, the user gets a catchy visualization of the data to be classified and can perform the ratings with single mouse-clicks. For instance see [289] where such a rating tool is used for image classification scenarios at surface inspect systems. However, a big effort is still required to label a sufficient number of samples and especially to provide feedback during on-line mode in real-time. This is especially the case for multi-class problems.

For system identification and control problems, often the variables used as targets in the regression models are by-measured, so that the models can be updated automatically without any user intervention — here, models are often used in order to predict future states of the target or for fault detection purposes, see also Chapters 7 and 8. Here, the problem is more a matter of designing the excitation signal appropriately in order to cover the input space sufficiently or at least to have a sufficient number of samples in the relevant regions where the models are defined. This means that decisions about which data to collect are already made within the measurement phase [320].

5.3.2 Direct Active Learning

A possible approach to reduce the problems mentioned in the previous subsection is the so-called active learning strategy [87] [293] (also called direct active learning in order to distinguish it from the hybrid active learning approach demonstrated in the next subsection).

> *Active learning* is understood as any form of learning in which the learning algorithm has some control over the input samples based on which it builds up the model.

In natural systems (such as humans), this phenomenon is exhibited at both high levels (for instance active examination of objects) and low, subconscious levels (for instance the work on infant reactions to "Motherese" speech demonstrated in [126]). In industrial systems this property of controlling the inputs is exactly what someone can exploit to reduce the workload of the operators (labelling effort, designing excitation signals). In particular, in an active learning approach the learning process may iteratively query unlabelled samples to select the most informative samples to annotate and update its learned models. Therefore, the central part of active learning is a data selection strategy in order to avoid unnecessary and redundant annotation and to concentrate on really new and important information to be included into the models.

Regarding an appropriate data selection, there are two basic strategies:

- Certainty-based selection [250] [418]
- Committee-based selection [94] [315]

In the former approach, after training an initial model with some labelled samples, the certainties in the predictions on new (unlabelled) samples obtained from the initial models is measured. In case of (evolving) fuzzy classifiers, this can be achieved by using (1.16) resp. (1.19) (see Section 1.2.1.5, Chapter 1) for the final output class obtained by (1.14) resp. (1.18) in the single model architecture case and by using (1.22) for multi model structures (based on TS fuzzy systems). For the samples where the lowest confidence level is responded (i.e. the lowest certainty from the classifier reported), the expert or operator is asked to provide his/her labelling. This procedure is somewhat intuitive as an unconfident response indicates that the sample either lies around the decision boundary (conflict state) or in the extrapolation region of the input data space (ignorance state) — see also Section 4.5.2. In both cases, it is very welcome to update the classifiers in order to intensify its decision boundary (in case of conflict state) or to extend its structure (in case of ignorance state).

The committee-based selection approach applies a distinct set of classifiers, which are all used for predicting new samples. Those samples whose assigned labels differ most among all classifiers are presented to the experts for annotation. In principle, this is equivalent to the situation when different operators may label a sample differently (and eventually discuss and quarrel about it), and a supervising operator (with more experience and know-how) has to make a final decision (here we have different classifiers and one operator). The drawback of committee-based selection is that it is only applicable in off-line mode, as the diversity of predictions has to be examined for all collected training samples and compared against each other. On the other hand, certainty-based selection is applicable in an off-line as well as on-line learning process. The difference is simply that in the off-line phase some training samples are pre-labelled and, if the classification accuracy is not high enough, additional ones are presented to the expert. This is done in a step-wise iterative procedure. In the on-line phase, feedback from the experts is requested immediately in uncertain prediction cases from the evolved classifiers — for instance for all samples causing classifiers' confidence lower around 0.5 denoting a high degree of conflict or a high ignorance level (see Section 4.5.2), this threshold is finally a matter of a tradeoff between the speed of increase of predictive accuracy and labelling effort of the operators. Furthermore, certainty-based learning is also applicable for identification/approximation problems, as there the (adaptive) local error bars as demonstrated in Section 4.5.1 can be used to detect regions where the model is still quite unsure as having wide error bands. The excitation signal is then designed in a way that these regions are also covered significantly.

Active learning in connection with neural networks and support vector machines is studied in [132] and [426], respectively. An enhanced active learning approach is proposed in [354], where feedback on features in addition to the labelling instances are included and which was successfully applied to text categorization, improving the classifier accuracies significantly. The basic assumption therein is that the expert is able to point to the most important features, based on which a re-weighting of the features can be achieved (and hence curse of dimensionality decrease when some feature are out-weighted). This is only guaranteed for specific application scenarios. Finally, we want to highlight that active learning may even improve the generalization performance as pointed out in [87], according to an appropriate selection of samples. A simple example is that of locating a boundary on the unit line interval. In order to achieve an expected position error of less than ε, one would need to draw $O(\frac{1}{\varepsilon}ln(\frac{1}{\varepsilon}))$ random training examples. If one is allowed to make membership queries sequentially, then binary search is possible and, assuming a uniform distribution, a position error of ε may be reached with $O(ln(\frac{1}{\varepsilon}))$ queries. Another example, the triangle learning problem, is visualized in Figure 5.9 (taken from [87]).

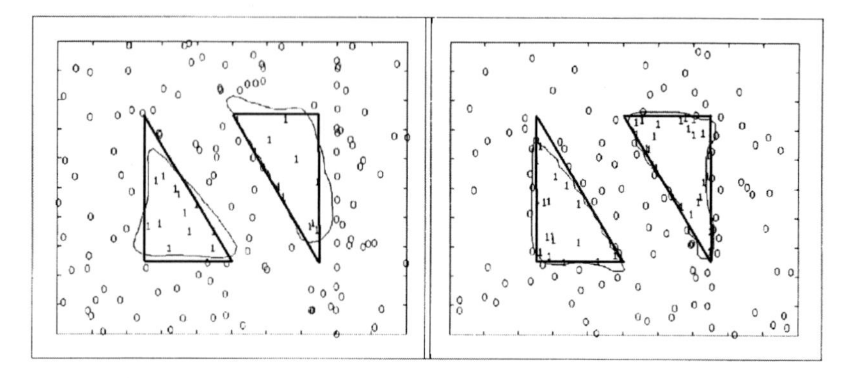

Fig. 5.9 Left image: decision boundaries for the triangle learning problem obtained when using conventional classifier learning; right image: decision boundaries significantly improved (compare solid light with solid dark lines = real decision boundaries) when applying active learning queries (taken from [87]).

Obviously, the decision boundary obtained by using 150 training samples selected from the certainty-based active learning approach (right image) is closer to the real decision boundary (shown as solid triangles) than when using 150 random samples (left image).

5.3.3 *Hybrid Active Learning (HAL)*

(Direct) active learning is a mechanism to reduce the annotation effort or the design of excitation signals for experts significantly. However, it may still require a significant number of samples for the initial model training step. This is because the model has to already have a reasonable accuracy in order to give reliable predictions on new unlabelled training samples and hence in order to be able to decide which samples should be selected for further labelling. However, in real-world application scenarios, it is possible that the number of initial labelled data is small or even zero: for instance there is no expert to carry out labelling or it is simply too time-consuming. For such cases, a novel active learning strategy presented in [269], which is called *hybrid active learning (HAL)* and comes with two stages:

- In the first stage, this strategy tries to select the most informative samples based on unsupervised clustering. Hence, no pre-labelling at all is required, as the selection is fully done based on cluster partitions obtained from the off-line batch data set.
- Once a classifier is set up based on the most informative samples, the certainty-based active learning approach as described in the previous section is applied for new incoming on-line samples (during on-line training mode) → second stage.

The approach is given the name *hybrid active learning (HAL)*, as it combines unsupervised with supervised active learning within an initial off-line and further on-line learning context, where the operator/user interacts with the machine/system in an iterative fashion.

Regarding the first stage, any prototype-based (fuzzy) clustering can be applied. In [269], the *eVQ* approach as described in Section 3.1.1.1 is used, but with an initial number of clusters which is set to the number of possible classes at the system (it can be assumed that this number K can be gained a-priori from expert knowledge). This means, instead of evolving new clusters on demand based on the characteristics of the data, the clusters are kept fixed during the whole learning process, their centers initialized at the beginning (e.g. by using the first K samples) and shifted according to (5.33) with a decreasing learning gain η which decreases with the number of iterations over the whole data set rather than with single samples forming the clusters (as done in original *eVQ*). The whole algorithm is called iterative vector quantization and described in Algorithm 22.

Algorithm 22: Iterative Vector Quantization (for unsupervised active learning)

1. **Input:** Complete data set normalized to $[0,1]$ due to the ranges of variables, number of clusters = number of classes classes K
2. Choose initial values for the K cluster centers $\mathbf{c}_i, i = 1,...,K$, e.g. simply by taking the first K data samples as cluster centers.
3. Fetch out the next data sample of the data set $\mathbf{x}$.
4. Calculate the distance of the selected data sample to all cluster centers by using a predefined distance measure. Euclidean distance measure is the most common choice.
5. Elicit the cluster center which is closest to the data sample by taking the minimum over all calculated distances $\rightarrow$ winning cluster represented by its center c_{win}.
6. Update the p components of the winning cluster by moving it towards the selected sample $\mathbf{x}$:
$$\mathbf{c}_{win}^{(new)} = \mathbf{c}_{win}^{(old)} + \eta(\mathbf{x} - \mathbf{c}_{win}^{(old)}) \tag{5.33}$$
with η a learning gain, decreasing with number of iterations, i.e. $\frac{0.5}{\#iter}$.
7. If the data set contains data samples which were not processed through steps 3 to 6, goto step 3.
8. If any cluster center was moved significantly in the last iteration, say more than ε, reset the pointer to the data buffer at the beginning and goto step 3, otherwise stop.
9. Estimate for each cluster which samples belong to it: yields a set of index vectors $\mathbf{index}_i, i = 1,...,K$.
10. Estimate the variances $\sigma_{i,j}^2$ for each cluster i and each dimension j by applying variance formula over samples of $\mathbf{index}_i$.
11. **Output:** cluster partition with K clusters: centers $\mathbf{c}_1,...\mathbf{c}_K$ and widths $\sigma_1,...,\sigma_K$.

From the cluster partition (representing the class partition), the selection is now performed based on the following two criteria:

- The first criterion is to select those samples which are close to the cluster (=class) centers, also denoted as center-based selection. The reason for this is that these samples characterize the inner core parts of the classes and can be seen as highly informative where the classes have the highest density and their intrinsic location in the feature space. Closeness to the centers can be either expressed by any distance measure d in the normalized feature space or by a membership degree μ how strong samples belong to the clusters — as e.g. achieved in fuzzy clustering approaches such as fuzzy c-means [46]. In the first case, those samples $\mathbf{x}$ are candidates for selection for which

$$\exists \mathbf{c}_i, i = 1, ..., K \;\; d(\mathbf{c}_i, \mathbf{x}) \leq thresh_i \tag{5.34}$$

with $thresh_i$ a small value dependent on the average width of the ith cluster (averaged over all dimensions):

$$thresh_i = 0.1 \frac{1}{p} \sum_{j=1}^{p} \sigma_{i,j} \tag{5.35}$$

In the second case, those samples are selected for which the membership degree μ to any cluster is high, e.g. higher than 0.9.
- The second criterion is to select those samples which are close to the border of the clusters (classes), also denoted as border-based selection. The reason for this concept is that these samples characterize the border line between the classes, i.e. the region where the decision boundary is expected to be drawn by the classifier. Hence, it is necessary to feed the classifier training algorithm with a significant number of these samples. Border samples are characterized by lying near the surfaces of the ellipsoidal clusters; that is the case when the distance calculated by (3.19) with (3.20) when using Euclidean distance measures respectively with (3.24) when using Mahalanobis distance is low. The problem with this approach is that not only border samples inbetween clusters (classes) are selected but also in border regions where only one class is really significant. Furthermore, if two classes strongly overlap, the surface-oriented border samples may lie significantly away from the decision boundary. Hence, in [269] it is suggested calculating membership degrees of each sample to all existing clusters, as to obtaining a membership matrix U where μ_{ij} is the membership degree of the ith sample to the jth cluster (class). Hereby, the membership degree μ_{ij} is calculated in the same manner as done in within the fuzzy c-means training algorithm [46]:

$$\mu_{ij} = \left(\sum_{k=1}^{K} \frac{\|\mathbf{x}_i - \mathbf{c}_j\|^{2/(m-1)}}{\|\mathbf{x}_i - \mathbf{c}_k\|^{2/(m-1)}} \right)^{-1} \tag{5.36}$$

with m a fuzzy factor, often set to 2. Then, for each sample $\mathbf{x}_i, i = 1,...,N$ the two clusters with maximal membership degrees are elicited by:

$$L1 = argmax_{j=1,...,K}\mu_{ij} \quad L2 = argmax_{j=1,...,K\backslash L_1}\mu_{ij} \tag{5.37}$$

and the deviation of membership degree T_i calculated by:

$$T_i = |\mu_{i,L1} - \mu_{i,L2}| \tag{5.38}$$

Finally, the samples with the lowest T_i score are selected for annotation, as these belong to two clusters (classes) with almost equal membership and hence can be seen as border-line samples.

Combining both sample selection strategies is called *CBS* (center- and border-based selection). In sum, k_1 samples from center-based and k_2 samples from border-based selection strategy will be selected, where it is important to select an (almost) equal number of samples from each class (cluster). Otherwise, an imbalanced learning problem would arise, causing weak classification accuracies (see also Section 4.3). For instance, if in sum 20% of the training samples should be selected, then it is quite intuitive to select 10% samples according to center-based selection strategy (again balanced, e.g. in two-class 5% for class #1 and 5% for class #2 represented by the two extracted clusters) and 10% according to border based selection (those 10% with lowest T_i score (5.38)).

A visualization example of the three selection strategies (center-based, border-based in two variants) is presented in Figure 5.10 (a) to (c) for the Wine data set from UCI repository (showing the two most important features). The selected samples by each strategy are surrounded by a circle. From (b) we can clearly recognize that also border line samples in uninteresting regions are selected, whereas in (c) only the border-line samples in the regions between the clusters (indicated as ellipsoids) are selected. Also, it is interesting to see that the second variant of border-based selection automatically selects more samples in the region where the two clusters (for circle and diamond class) overlap, i.e. where the decision boundary is harder to learn for the classifier, and a low number where the boundaries are easier to learn (low class overlap). Experiments on classifiers trained from the selected samples by our unsupervised active learning approach show 1.) that border-based selection variant 2 can outperform the other approaches and 2.) that these classifiers can compete with classifiers trained on all samples. Some detailed results on three well-known benchmark data sets from the Internet are listed in Table 5.1, where RS denotes random selection CBS center- and border-based selection as described above, different rows represents different portions of selected samples. This finally means that similar classifier accuracies are achieved by only using a couple of samples selected from our unsupervised active learning approach as when using all samples. Similar observations could be obtained when using support vector machines and k-NN classifiers instead of CART, see also [269].

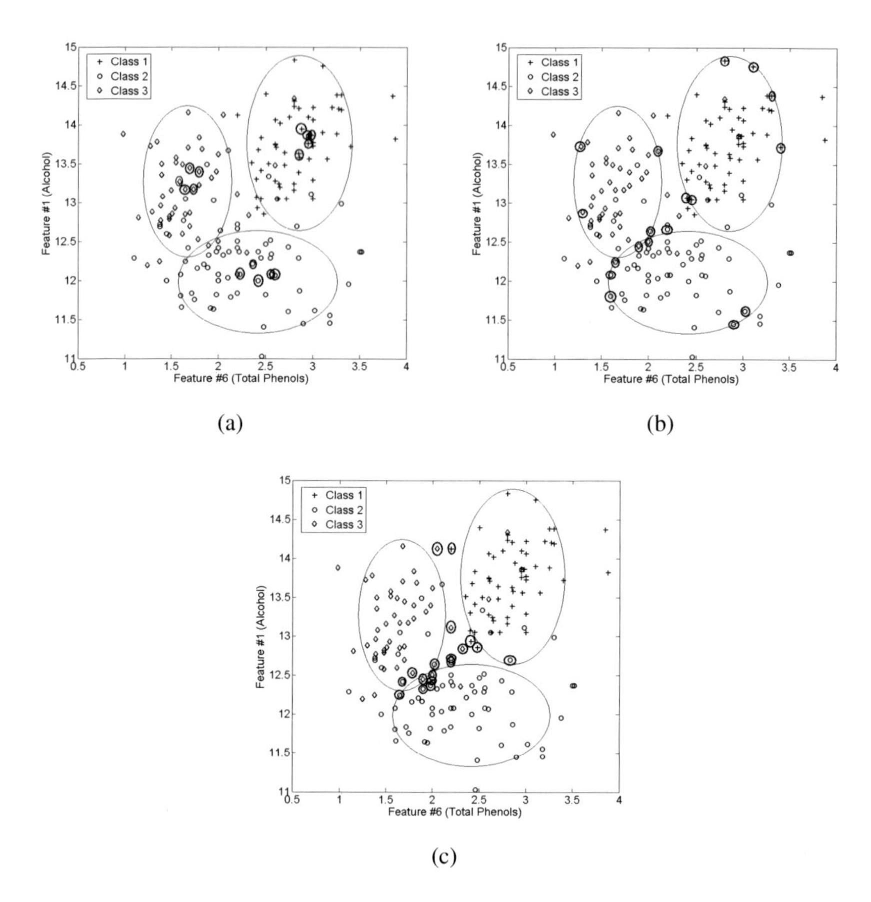

Fig. 5.10 (a): samples selected by center-based selection strategy, (b): samples selected by border-based selection strategy variant 1, (c): samples selected by border-based selection strategy variant 2; the ellipsis indicate the extracted clusters, the selected samples surrounded by black circles

5.3.4 Semi-supervised Learning

Another attempt towards more automatization and lower annotation, labelling and supervision effort for operators, is the so-called semi-supervised learning concept.

Semi-supervised learning is understood as any form of learning in which labelled and unlabelled data are processed synchronously for building reliable models — typically a small amount of labelled data with a large amount of unlabelled data is used.

Table 5.1 Comparison of classification performance (10-fold CV rates ± standard deviation) achieved by CART on several data sets when 1.) p% randomly selected (RS) samples are sent into training of CART, 2.) p% samples selected after center- and border-based selection strategy (CBS) are sent into training of CART and 3.) all training data samples are sent into CART (last row); p is varied from 10% to 50%

% Samples	Meth.	Iris	Yeast	SpamB
10%				
	RS	50±52.50	52.22±15.76	81.82±5.67
	CBS	80±28.71	57.78±11.48	90.00±5.59
20%				
	RS	86.67±23.31	44.74±12.89	83.78±6.11
	CBS	90±16.1	61.67±15.15	89.78±5.56
30%				
	RS	80.00±16.87	47.93±6.18	85.74±7.33
	CBS	87.50±13.18	58.28±11.31	88.24±3.67
50%				
	RS	95.71±9.64	51.84±8.72	88.42±4.09
	CBS	94.29±12.05	54.90±5.04	90.88±3.34
All		94.00±7.34	56.84±4.71	90.48±1.57

Compared to hybrid active learning, which in fact combines unsupervised with supervised learning, however requests concrete class labels from experts for a particular selection of samples in the unsupervised cluster partitions, semi-supervised learning is able to improve models in a completely unsupervised context (i.e. no feedback from experts is required to label a selection of samples). This means that with the help of semi-supervised learning it is possible to fully automatize off-line or on-line enhancements of models with newly collected samples. In fact, many machine-learning researchers have found that unlabelled data, when used in conjunction with a small amount of labelled data (for initial model training), can produce considerable improvement in learning accuracy [77].

Several contributions to semi-supervised learning have been investigated and developed during the last decade, for a comprehensive survey see [472]. One class of semi-supervised learning approaches are those which are linked with a particular learning approach, i.e. which underly strict model assumptions. These are also called generative models, where it is assumed that models can be identified by a complete unsupervised learning problem — e.g. by finding components in form of mixture models [324] [131] with the famous *EM (Expectation-Maximization) algorithm* [101] or clusters [95] [100] with various clustering algorithms. Then, ideally one labelled example per component is required to fully determine the mixture distribution. The problem with this approach is, that the components may widely overlap or do not always follow the natural distinct partition in the unsupervised space. An example of this aspect is demonstrated in Figure 5.11, where the samples in the upper half of the image represent the first class and the samples in the lower half the

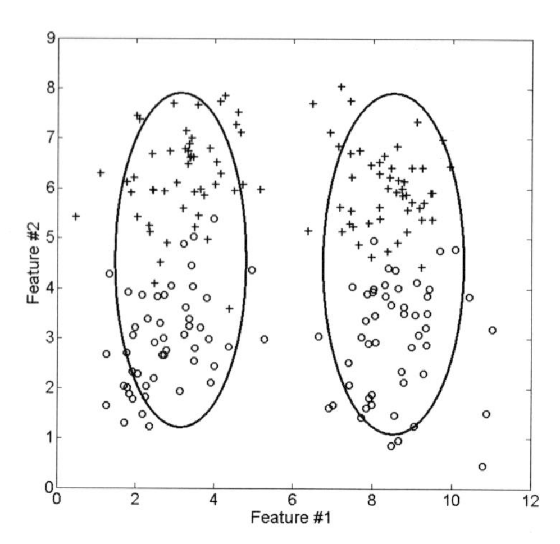

Fig. 5.11 The class distribution does not follow the unsupervised cluster distribution → two Gaussian mixture models include both classes with approximate equal frequency; by labelling just one single sample from each mixture model, a classifier with approx. 50% miss-classification rate would be achieved

second class (indicated with 'o' and '+'). When applying EM algorithm, the Gaussian mixture models end up as shown by the ellipsoids; this is somewhat intuitive as the mixture models try to represent the natural distribution of the data as well as possible: the data gap in the middle along the horizontal axis is by far more distinct than in the middle along the vertical axis. Finally, this would mean that a classifier trained on the classes represented by the mixture models would have around 50% accuracy only. An interesting point now is to inspect how the unsupervised data selection part in the hybrid active learning approach (see previous section) performs for this problem. This is visualized in Figure 5.12, where we realize that due to applying both synchronously, center-based and border-based selection (variant 2), important samples near the natural border between the two classes are suggested to be labelled (surrounded by circles). From this we can conclude, that even in cases when the natural cluster distribution of the samples do not represent the individual classes well, hybrid active learning may still select important samples along the decision boundary and hence be superior to a generative semi-supervised learning technique.

Another semi-supervised learning strategy exploits the idea of avoiding any changes in dense regions by not putting the boundary in high density regions. A well-known approach which uses this concept is the *Transductive Support Vector Machines (TSVM)*, often also called S^3VM *(Semi-Supervised Support Vector Machines)*. There, the goal is to find a labelling of the unlabelled data, so that the linear boundary in the reproducing Kernel Hilbert Space has the maximum margin on both the original labelled data and the (now labelled) unlabelled data. A visualization of the basic idea of this concept is given in Figure 5.13: with unlabelled data the original maximal margin boundary between the two classes ('o' and '□') shown as dotted

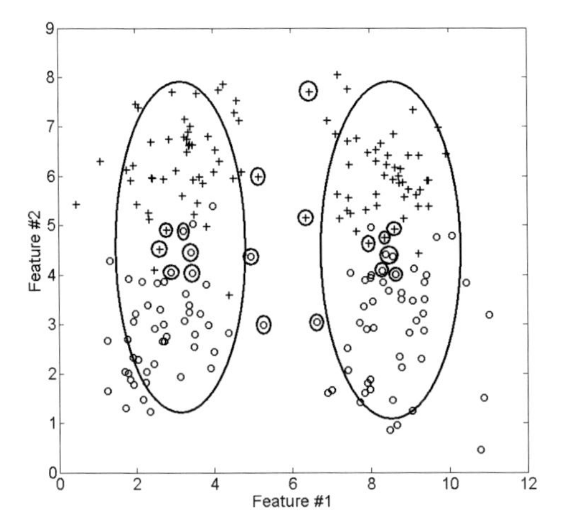

Fig. 5.12 Same as Figure 5.11, where the bigger circles mark the data samples selected by the unsupervised active learning approach for annotation, most of these are in fact along the natural border between the two classes.

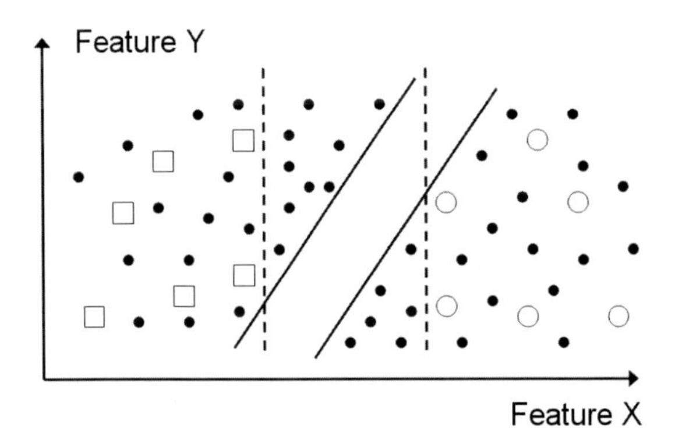

Fig. 5.13 The maximal margin boundary between the two classes ('o' and '□') shown as dotted lines; the updated margin by unlabelled samples (dark dots) shown as solid lines.

line can be significantly refined (→ solid line) and hence improve the significance of the classifier in a previously uncertain region (gap between the two classes). The problem with this approach is that it is NP-hard, so that major effort has focussed on efficient approximation algorithms such as [48] [453]. These are mostly very time-intensive and cannot be applied within an on-line learning context.

Another class of semi-supervised learning approaches can be used independently from the chosen model architecture or learning method, hence are also applicable directly to fuzzy systems. Most of these are able to provide direct class label suggestions sample per sample, hence are also applicable in an incremental on-line learning context for evolving fuzzy systems/classsifiers. One straightforward approach is the so-called *self-learning*, where an initial classifier is first trained with a small number of labelled samples and further used to classify the unlabelled (off-line or on-line) data. Typically, the most confident unlabelled samples (i.e. where the classifier has the highest confidence in its prediction) are added to the training samples (off-line mode) resp. are used for adaptation/evolution of the classifier (on-line mode). In this sense, the classifier uses its own predictions to teach itself. Hence, the problem is that a classification error can reinforce itself. However, for specific classifier training algorithms, there are some analytic results on their convergence when using self-learning [163] [93].

In order to avoid the reinforcement of classification errors, *co-training* [52] is a promising alternative. There, two separate classifiers are trained with the labelled data on two sub-feature sets; usually, a disjoint conditionally independent set of features should be used. Each classifier then classifies the unlabelled data, and teaches the other classifier with some unlabelled samples for which they feel most confident. In off-line mode, each classifier is retrained with the additional training examples whose labels are predicted by the other classifier. In on-line mode, the whole procedure is applied to single samples in alternating fashion: for all odd samples (with an odd sample number) the first classifier makes a prediction based on which the second classifier is updated, for all even samples the second classifier makes a prediction based on which the first classifier is updated (in both cases when the classification was made with high confidence). The condition on an independent split of features into two halves is relaxed in [323], where they show that an artificial split by randomly breaking the feature set into two subsets still helps to improve the accuracy of the classifiers (however, not so much as an analytical conditional independent split).

Another relaxation is presented in [144], where two classifiers of different types are trained on the whole feature set, and those samples used for teaching a classifier, where the other made a prediction with high confidence. This could be further extended to a *multi-training* case (instead of co-training), where multiple classifiers are trained and evolved in parallel on the same set of features and those unlabelled samples used for further adaptation for which the majority of the classifiers agree on their predictions. The higher the number of used classifiers, the more significant the majority decision can be expected, especially when the classifiers are diverse, i.e. applying different learning concepts. Multi-training can be seen as a specific aspect of classifier fusion and ensembling (with majority voting as ensembling method), which will be explained in more detail in Section 5.4.

Similarity-based learning is another attempt of semi-supervised learning of a model. There, the labelled samples most similar to the current unlabelled sample(s) are sought in a reference data base. Similarity is often expressed by a distance metric as also done in nearest neighbors and lazy learning approaches (see Section 5.5.2): the labelled samples nearest to the current unlabelled sample are used and the majority class label over these is sent as input to modify or update the (global) classifier. In fact, it is also possible to use only the similarity-based learning approaches for predicting new samples without feeding these observations into a global classifier update. However, global models often enjoy a great attraction regarding interpretability and system knowledge representation in compact form. Self-learning, multi-training as well as similarity-based learning can be also applied for regression problems, the training outputs are just numerical values of the target concept to be learned instead of class labels.

No matter which semi-supervised learning approach is applied, constraining class proportions on unlabelled data may be important. This is because without any constraints, semi-supervised learning algorithms tend to produce unbalanced output. In the extreme case, all unlabelled data might be classified into one of the classes. The desired class proportions can be estimated from the labelled data set and kept by constraining on the hard labels:

$$\frac{1}{u}\sum_{i=l+1}^{l+u} y_i = \frac{1}{l}\sum_{i=1}^{l} y_i \tag{5.39}$$

where u is the number of unlabelled samples and l the number of labelled samples. This formula can be used in binary classification problems and keeps the distribution strictly. This can be easily extended to continuous predictions. In an on-line learning environment, the situation is a bit different, as there the underlying concept to be learned (i.e. the underlying class distribution) may change over time. Hence, it is quite natural, that the above condition needs to be strongly relaxed (loose label constraints); otherwise, no new system states or operating modes would be included in the models. However, care has to be taken that the class distributions do not get too imbalanced over time, as this usually deteriorates the performance of evolving classifiers [349]. A strategy for preventing such a situation is described in Section 4.3.

5.4 Incremental Classifier Fusion

In case of classification problems in real-world decision processes, a central consideration is often dedicated to the choice of an appropriate classification method or the model architecture in case of global classifiers. This depends on the underlying class distribution implicitly contained in the classification problem resp. the data collected from it. For instance, if the decision boundaries are expected to be linear, the usage of a non-linear classifier would not be recommended as this would cause unnecessary complexity and also increase the likelihood of over-fitting. This means that for different classification problems different classification methods turn out to be the best choice with respect to minimizing the expected prediction error on unseen samples. In other words, there is no unique classification algorithm which is able to outperform the others for a large collection of data sets. This fact is an interpretation of the well-known "No Free Lunch" theorem by Wolpert et al. [448].

As in real-world application scenarios the classification concept to be learned is not known a priori and often cannot be derived directly from the high-dimensional data, it is useful to train and evolve a set of different classifiers on a data set and to combine their predicted outputs on new samples to be classified in order to exploit the diversity of the classifiers and to stabilize the predictive performance. This can be done with so-called *classifier fusion methods* (such as e.g. voting schemes) which merge the outputs of an ensemble of classifiers into a combined decision [237, 345, 103].

Incremental classifier fusion (ICF) can be understood as the merging process of ensembles of classifiers' outputs into one final decision during the on-line learning and evolution process.

Two necessary and sufficient conditions for ensembles of classifiers for achieving a higher accuracy than any of their individual members are [168, 234]:

1. The classifiers are accurate (they perform better than random guessing).
2. The classifiers are diverse (their errors are uncorrelated).

Another advantage of the use of ensembles of classifiers is the increased reliability and robustness of the classification performance [103]. Ensembles reduce the risk of a single classifier going into the "bad" part of the hypothesis space (caused e.g. by selecting a classification method with an inappropriate bias, an inappropriate choice of parameters for the classification method, etc.).

We denote every ensemble member as *base classifier* and any technique combining the outputs of the base classifiers to one single class decision a *fusion method* or an *ensemble classifier*.

Although fixed fusion methods such as voting schemes have interesting properties such as their simplicity and the good results they can produce in certain situations, they are usually suboptimal. In order to optimize the ensemble performance, trainable fusion methods can be used (see e.g. [237, 345]), which take as their input the outputs of the classifiers for some training data and/or the estimated accuracies of the classifiers. Of course, in contrast to the fixed fusion methods, the trainable fusion methods also need to be updated incrementally if they are to be used in an on-line classification system. A number of strategies to update an ensemble of classifiers on-line have been explored, such as

- Dynamic combiners [256] [352] which adapt combination rules on-line, however the ensemble members are trained fully off-line before hand.
- Updating the ensemble members where new data is used to update the ensemble member, but the combination rules are kept fixed. Examples are online *Bagging* and *Boosting* algorithms [330], the *Pasting Small Votes* system [57] and *Learn++* [353].
- Structural changes where the classifiers are re-evaluated or completely removed and replaced by a newly trained one, for instance [435].

In [381] [380] some deficiencies of the above listed approaches are eliminated. There, a generic framework for incremental classifier fusion is proposed, which is

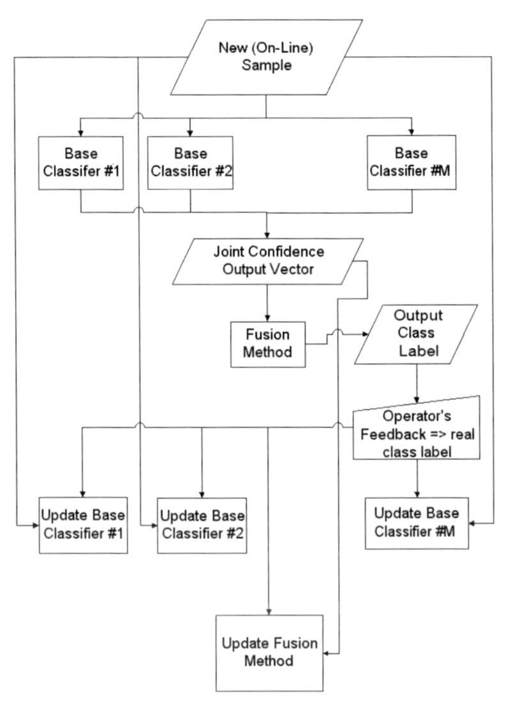

Fig. 5.14 Work-flow of the update strategy in case of updating base classifiers and the fusion method synchronously.

configurable with an arbitrary number and type of base classifiers. Both, the base classifier and the fusion method are updated synchronously based on new incoming samples, the latter uses the joint confidence output vector from the base classifiers to update its parameters and structure. This guarantees improved performance 1.) over static kept ensemble of classifiers which are not updated during on-line mode and 2.) over single incremental base classifiers by exploiting the diversity of incremental based classifiers during on-line mode (this could be empirically shown on real-world on-line data in [381]). The update strategy of this idea is visualized in Figure 5.14. For each new incoming (on-line) data sample, first a classification statement (output class label) is produced using the base classifiers and the fusion method. Then, based on the operator's feedback and the input feature vector of the new on-line sample the base classifiers are updated. The fusion method is updated also based on the operator's feedback and on the joint confidence vector produced by the base classifiers (rather than on the original sample values). For the next incoming data sample, the updated base classifiers and the updated fusion method are used for classification. According to the free choice of the base classifiers and its incremental learning engine, it is possible to include several evolving fuzzy classifiers with different characteristics into this framework aside other classifier methods. For instance, it is promising to include a fuzzy classifier using single- and a fuzzy classifier using multi-model architecture (based on Takagi-Sugeno fuzzy systems) into the framework (such as *eClassA* and *FLEXFIS-Class MM*), as both use completely different training and optimization concepts and architectures, hence the likelihood of a good diversity is high. On the other hand, it would for instance not make much sense to include classifiers with similar architectures and training concept such as for instance the clustering-based classifier *eVQ-Class* and the fuzzy classifier *FLEXFIS-Class SM* — the latter is deduced from the former by projecting the clusters onto the axes and assigning labels in the consequent parts (rather than assigning labels directly to the clusters). This would just increase the computational performance, rather than providing more diversity.

Last but not least, an evolving fuzzy classifier could also be used as a fusion method itself. There, instead of a vector including original features/variables $x_1,...,x_p$, the joint confidence output vectors from the base classifiers are sent as input features into the fusion mechanisms. Assuming that M base classifiers are applied for a classification problem including K classes, a joint confidence output vector for a new sample $\mathbf{x}$ is defined by:

$$\begin{aligned}\mathbf{conf}(\mathbf{x}) &= (D_1(\mathbf{x}),\ldots,D_M(\mathbf{x})) \\ &= (d_{1,1}(\mathbf{x}),\ldots,d_{1,K}(\mathbf{x}),\ldots,d_{M,1}(\mathbf{x}),\ldots,d_{M,K}(\mathbf{x})) \quad . \end{aligned} \tag{5.40}$$

where $d_{i,j}$ is the confidence of the ith base classifiers into the jth class lying in $[0,1]$. If a base classifier does not provide any confidence level (for instance a decision tree), then its output vector will be binary and indicated (one entry 1, the others 0). Some columns in (5.40) may be redundant (due to weak diversity of the base classifiers), which originally was one of the major arguments not to use standard classification methods directly on the confidence output vectors, as maybe causing instabilities in the learning process [238]. However, when applying evolving fuzzy classifiers as incremental fusion technique, these instabilities may only arise during a preliminary off-line step (e.g. when applying matrix inversion steps such as in multi-model and TS architectures for estimating consequent parameters); in these cases, however, the regularization techniques as discussed in Section 4.1 can be applied to improve this situation. Good empirical performance when using *eVQ-Class* as incremental fusion method (called *IDC = Incremental Direct Cluster-based Fusion*) could already be verified in [381].

5.5 Alternatives to Incremental Learning

So far, in the first five chapters of this book we have dealt with various evolving fuzzy systems approaches (Chapter 3), basic required algorithms (Chapter 2) and possible extensions of EFS in order to gain more process safety, user-friendliness and predictive power of the models (Chapters 4 and 5). In all of these concepts, incremental learning served as the basic learning engine, mostly used in a single-pass and sample-wise manner. This means that the algorithms for updating model parameters, model structures, (statistical) help measures, uncertainty measures, feature information criteria etc. were carried out based on a sample per sample basis, ideally without using older data samples (optimizing the virtual memory demand): once an incremental training cycle for one sample is finished, this sample is immediately discarded, afterwards and the next sample (in temporal order) sent into the incremental update process. In this section, we are presenting two alternatives to the sample-wise and temporal incremental learning concept, which are dynamic data mining and lazy learning.

5.5.1 The Concept of Dynamic Data Mining

The dynamic data mining concept [92] [338] deals with the problem that data sets (also called data sites) are available at some data snapshots. The snapshots usually occur as temporal, spatial or spatial-temporal segments, also called chunks of data.

Dynamic data mining (DDM) tries to build models which are inherited through the sequence of chunks in the temporal or spatial domain.

This means that a first (initial) model is built based on the first segment of data, and when the next segment is available (e.g. from a different data source/base or from a further recording in on-line measurement systems), the model training process is run again based on the samples from the new segments starting with the previously obtained model as initial configuration. The big difference to the incremental learning concept dealt with so far in this book is that the data is not presented as single samples to the training algorithm(s), but in blocks and the algorithms in the dynamic data mining concept explicitly take advantage of this fact. This may have some favorable characteristics when trying to optimize parameters, cluster centers etc. in a coherent procedure. For instance, in [340] a collaborative clustering approach based on fuzzy c-means (FCM) is demonstrated, where the centers $\mathbf{c}_1,...,\mathbf{c}_C$ and the membership degree matrix U are optimized based on an augmented objective function including conditions from previous sites weighted according to their importance levels. In particular, in the Ith data site the objective function is defined by [340]:

$$J[I] = \sum_{k=1}^{N[I]} \sum_{i=1}^{C} \mu_{ik}^2[I] \|\mathbf{x}_k - \mathbf{c}_i\| + \beta \sum_{j=1, j \neq i}^{P} \sum_{k=1}^{N[I]} \sum_{i=1}^{C} (\mu_{ik}[I] - \mu_{ik}[I|j])^2 d_{ik}^2 \quad (5.41)$$

with P the number of data sites connected to the Ith data site, $N[I]$ the number of samples in the Ith data site, C the number of clusters, d_{ik}^2 the distance between the kth sample and the ith prototype and μ_{ik} the membership degree of the kth sample to the ith prototype. The symbol β denotes a scaling parameter steering a balance between the optimization guided by the structure in $D[I]$ and the already developed structures at the remaining site. Hence, its value implies a certain level of intensity of the collaboration: the higher its value, the stronger the collaboration. This approach also includes a split-and-merge process for the fuzzy c-means in order to

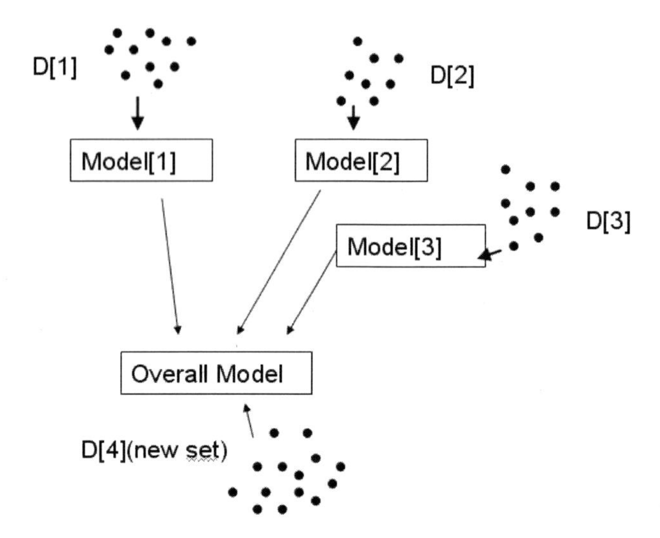

Fig. 5.15 Collaborative modelling by including the models of former data sites D[1], D[2] and D[3] into the new data site D[4]; no interconnections between D[1], D[2] and D[3] is made

accommodate variable changing numbers of clusters over different data segments. Another big difference to (temporal) incremental learning is that the sequence of communication between data segments does not need to be linearly ordered (from one to the other and next), but can have more complex structures. An example is visualized in Figure 5.15, where data site $D[5]$ is linked with all other four data sites, but no other interaction between these four is communicated.

An application of this dynamic data mining concept to evolving fuzzy systems could be in form of a modified least squares optimization problem by including an augmented objective function on the consequent parameters, optimizing on the samples in the new data site with least squares (first term) while bounding the change on the consequent parameters with a scaling factor in the second term (similarly as in (5.41)). Carrying out this for each rule separately (in a local learning context) and for the rules which were also present in the 'former' sites (connected to the new site), would lead to a promising novel learning concept (however, to our best knowledge not investigated so far). The augmented objective function at the Ith site would then look like:

$$J[I]_i = \sum_{k=1}^{N[I]} \Psi_i(\mathbf{x}(k))e_i^2(k) + \sum_{j=1}^{P} \beta_j \sum_{k=1}^{N[I]} \Psi_i(\mathbf{x}(k))\|\mathbf{w}_i[I] - \mathbf{w}_i[j]\| \longrightarrow \min_{\mathbf{w_i}[I]} \quad \forall i \in \{D[1], D[2], ..., D[P]\}$$

$$J[I]_i = \sum_{k=1}^{N[I]} \Psi_i(\mathbf{x}(k))e_i^2(k) \longrightarrow \min_{\mathbf{w_i}[I]} \quad \forall i \notin \{D[1], D[2], ..., D[P]\} \qquad (5.42)$$

where $e_i(k) = y(k) - \hat{y}_i(k)$ represents the error of the ith local linear model (rule) in the kth sample of the Ith data site, P again the number of older data segments (sites): in case of temporal linearly ordered sequences of data blocks, this number will be always equal to 1 (updated from data segment to data segment). The symbol $\mathbf{w}_i[j]$ denotes the consequent parameter vector for the ith rule in the jth data site. The second part in (5.42) follows the conventional least squares optimization for all rules which were newly evolved for the new data site (and therefore not present in the older ones). The symbol β_j denotes the scaling for the jth data site; in case of intending to gain an equally weighted learning situation, this could be set to the proportion of data samples between the jth and the Ith data site. If for any reason an under-weighted data site (a data site with a low number of samples) is equally important with a heavy data site (data site containing a high number of samples), β could be set near to 1 in order to support this in the optimization procedure. Furthermore, forgetting of samples from an older data site is possible by decreasing the corresponding scaling factor β_j. Update of older rule centers and fuzzy sets (from previous sites) as well as evolution of new rules based on the new data site can be autonomously and independently carried out before optimizing (5.42), starting with the initial cluster configuration obtained from the site. This can be simply achieved by using the rule evolution strategies in the EFS approaches mentioned in Chapter 3 by dividing a new data site into single samples. Merging of some rules will be important here, as usually some clusters/rules will strongly overlap when amalgamating the cluster partitions from the different sites.

5.5.2 *Lazy Learning with Fuzzy Systems — Fuzzy Lazy Learning (FLL)*

Opposed to incremental learning which refers to a concept for updating global models in sample-wise manner, lazy learning produces local models in dependency of a query point (to be predicted). *Lazy learning* [6] [400] postpones all the computation cost for training until an explicit request for a prediction is received.

The request is fulfilled by interpolating locally examples considered relevant for the current query according to a distance measures (most frequently Manhattan or Euclidean distances are applied [50]). In this sense, someone may speak about a local modelling procedure, which should not be confused with local learning of consequent parameters in fuzzy systems (there still a global model is trained, but locality refers to the estimation of each local region (rule) separately). For a comprehensive tutorial on local modelling, see also [26]. In lazy learning, there are two basic identification problems:

- Identifying an appropriate number of nearest neighbors of the current query point.
- Estimating the local models from the selected nearest neighbors.

The latter is applied within a linear least squares optimization problem, where a kernel function defines the local neighborhood of query points:

$$J = \sum_{k=1}^{N} (y_k - f_k)^2 K(\mathbf{x}_q, \mathbf{x}_k) \longrightarrow \min_{\mathbf{w}} \tag{5.43}$$

where $f_k = \mathbf{r}_k \mathbf{w}$ with $\mathbf{x}_k$ a regression vector (in the linear case containing the original feature values plus a 1 for the intercept), $\mathbf{x}_q$ the current query point, $\mathbf{w}$ the linear parameters to be estimated and K a kernel function, generally defined as

$$K(\mathbf{x}_q, \mathbf{x}) = D(\frac{\|\mathbf{x}_q - \mathbf{x}\|}{h}) \tag{5.44}$$

with D a distance measure between two points. The solution of the optimization problem in (5.43) is given by the weighted least squares expression in (2.61) (Section 2.3.1), where the weight matrix Q is a diagonal matrix whose entries are $K(\mathbf{x}_q, \mathbf{x}_k), k = 1, ..., N$.

Obviously, this can be extended easily to the case when using a fuzzy system with one rule as f in (5.43) by setting its rule center to the query point $\mathbf{x}_q$, and using the membership degrees $\Psi(\mathbf{x}_k)$ to this rule as kernel function K. We also call this approach *fuzzy lazy learning (FLL).*

An interesting property of lazy learning is that the leave-one-out cross-validation error can be calculated without re-estimating the linear parameter vector $\mathbf{w}$ (which is time intensive and hardly possible during on-line learning mode). This can be achieved by using the PRESS statistic [316] which is for the kth data sample:

$$Press_k = \frac{y_k - \mathbf{x}_k \mathbf{w}}{1 - h_k} \tag{5.45}$$

with h_k the kth diagonal element of the hat matrix $H = QR(R^T QR)^{-1} R^T Q$, R the regression matrix containing all regression vectors $\mathbf{x}_k$ for $k = 1, ..., N$. The overall PRESS statistic $Press$ is then the average of all $Press_k$'s. This finally means that the most sensitive parameter (the bandwidth h resp. σ in Ψ) for the performance of the local models could be tuned by minimizing the leave-one-out error. When applying an indicator function on the kernel, i.e.

$$I(K) = \begin{cases} 1 & D(\|\mathbf{x}_q - \mathbf{x}\|) \leq h \\ 0 & otherwise \end{cases} \tag{5.46}$$

substituting K in (5.44), the optimization of the continuous bandwidth parameter can be conveniently reduced to the optimization of the number k of nearest neighbors to which a weight of 1 is assigned in the lazy learning approach. This means that the problem of bandwidth selection is reduced to a search in the original feature space of k nearest neighbors. Now, we could start with a reasonable lower bound of nearest neighbors (in case of a fuzzy rule $k = 10$ would be such a bound), compute the weighted least squares solution and calculate the PRESS statistic in (5.45) and then repeat this by adding one more sample (the next nearest neighbor). Re-estimation of the whole weighted least squares would take too much computational effort (especially in the on-line case), so we suggest applying the recursive weighted least squares formulas as in (2.62) to (2.64) when adding successively more neighbors (the weights given by $K(\mathbf{x}_q, \mathbf{x}_k)$). This finally ends in a sequence of $k = 10, ..., M$ PRESS statistics (with M a user-defined upper bound on allowed samples in one local modelling procedure), where the one with the lowest error is taken as the optimal choice of nearest neighbors. As both, the PRESS statistics as well as the recursive weighted least squares approach are quite fast to compute, the whole approach is plausible within an on-line modelling scenario. The only remaining problem is an appropriate selection of a reference data base (from which the nearest neighbors are selected) as collecting all samples seen so far to one large data base significantly slows down the selection process over time. This research is still in its infancy, a first attempt tackling it is presented in [44], which is focussed on sample selection for instance-based classification in data streams. Within the context of fuzzy lazy learning for regression as well as classification problems in data streams, an appropriate sample selection strategy has not, to the best of our knowledge, been considered so far.

So far, we have discussed fuzzy lazy learning when applying one fuzzy rule and using its Gaussian multi-dimensional membership function as kernel. However, it is also possible to include more local linearity into the fuzzy lazy learning approach by estimating more rules from the nearest neighbors of one query point. In this case, someone could again apply an increasing sequence of k nearest neighbors and train a fuzzy system by a rule evolution approach (extracting automatically an appropriate number of rules from the data). For each evolved rule the optimization problem in (5.43) is solved separately using multi-dimensional membership functions as kernels, achieving a linear consequent parameter vector $\mathbf{w}_i$ for each rule. The PRESS statistic $Press_i$ is then evaluated for each rule by using those data samples being closest to the corresponding rule. An overall PRESS statistic can be achieved by a weighted average of $Press_i, i = 1, ..., C$, where the weights denote the support of the single rules.

Chapter 6
Interpretability Issues in EFS

Abstract. All the previous chapters were dealing with concepts, methodologies and aspects in precise evolving fuzzy modelling, i.e. fuzzy models were automatically evolved based on incoming data streams in order to keep the models up-to-date with the primary goal to obtain on-line predictions from the models with high precision and quality. No attention was paid to interpretability aspects in the evolved fuzzy models, which are one of the key motivations for the choice of fuzzy systems architecture in an (industrial) on-line learning scenario (as discussed in Chapter 1). This deficit is compensated in this chapter, which will deal with the following aspects in EFS:

- Complexity reduction (Section 6.2)
- Towards interpretable EFS (Section 6) which is divided into
  - Linguistic interpretability (Section 6.3.1) and
  - Visual interpretability aspects (Section 6.3.2)
- Reliability aspects in EFS (Section 6.4)

The first issue (complexity reduction) will describe techniques and ideas on how to perform incremental on-line reduction of complexities, i.e. reducing the number of fuzzy sets and rules in evolved fuzzy models, while still keeping the precision of the models high. The outcome of such techniques can be objectively measured in terms of complexities of the reduced models and their predictive performance. The second issue goes significantly beyond pure complexity reduction steps by describing aspects and methodologies for guiding the evolved fuzzy models to more interpretable power, both in linguistic and in visual form. Of course, it is always a matter of subjective taste, whether one or another fuzzy model is more interpretable, but some principal guidelines for achieving more interpretable models (partially motivated from cognitive sciences) can be given. The fourth issue (reliability) embraces the concept of uncertainty in the predictions of fuzzy models, which can be seen as valuable additional information to a human being operating at the machine learning system.

E. Lughofer: Evolving Fuzzy Systems – Method., Adv. Conc. and Appl., STUDFUZZ 266, pp. 261–291.
springerlink.com

6.1 Motivation

Under the scope of precise evolving fuzzy modelling as discussed in the previous chapters, the major focus was placed on representing the underlying and over time developing relationships within the system (reflected in the supervised data streams) by evolving fuzzy systems as good as possible — in fact, mostly the main goal was to minimize the quadratic error between measured and predicted output values. However, precise modelling usually leads to complex structures in the models (high number of rules, chaotic fuzzy sets etc.) and therefore the interpretability of the evolved fuzzy models may suffer significantly. The reason for this is that in precise modelling the goal is to represent the characteristics of the incoming data in the evolved fuzzy systems in compressed form and in accordance with the natural data distribution in order to achieve a high (predictive) accuracy. This is in conformity with the requirements in many industrial and medical systems to achieve high quality products and standards while saving production time and money in the process. On the other hand, as outlined in the introduction of this book, one key motivation for the choice of fuzzy models (among others such as neural networks, genetic programming, supper vector machines etc.) is their universal approximation capability while at the same time still preserving some insight into their model structures and allowing experts' interpretations of the underlying processes within the system.

Based on our experience working with experts involved in the design of machine learning systems, the following benefits drawn from interpretable (evolved) fuzzy models can be identified:

- Providing insight for operators into the nature of the process: due to the slimmer linguistic and more transparent model description the user may find interesting dependencies between certain process parameters or variables within the system. From this, the experts may gain deeper knowledge about the process (what is really going on), which can be used in the future when designing some hard- or software components for more automatization, productivity and efficiency.
- User interaction in the on-line modelling process: the user may wish to change some structural components of the models as he might think (based on his long-term experience) that they are not completely adequate or correctly placed, especially when the model makes mistakes in its predictions. This also leads to a kind of hybrid modelling scheme, where a fuzzy model is evolved based on two sources, expert knowledge and data and this in an alternating scheme (see Section 10.2.3). Please note that this kind of modelling scheme (also called grey-box modelling) can be done with hardly any other modelling architecture, as models need to be understandable.
- Easiness of on-line supervision purposes: for instance, sometimes a user wants to find the reason for a certain model decision. This can be achieved quite easily with transparent evolved fuzzy models, as those rule(s) can be shown to the user with the highest activation degree. Rules which are easy to follow are necessary in this case, as otherwise the user would not understand, why a certain decision was taken.

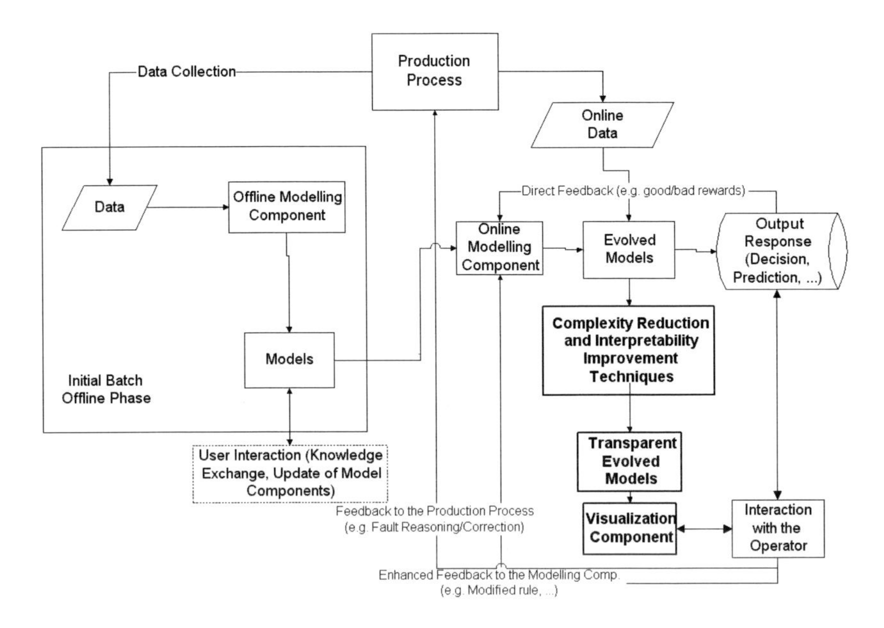

Fig. 6.1 System architecture for on-line modelling and user interaction, the necessary components for improving interpretability and stimulating user interaction are highlighted by boxes with thick lines and containing text in bold font.

- Increasing the motivation of feedback to the modelling process: sometimes the user may overrule the models decisions, and in other cases not. Sometimes he is convinced that his choice is the correct one and the model is wrong, but sometimes he may ask himself, whom he should trust more: the model or himself. In the latter cases transparent rules may support the user for an easier decision and a more consistent feedback. This should also improve the motivation of the user for system appraisal and other types of integration into the whole modelling process.

In an industrial context, for example, an evolving model is used to support decisions made by human experts or, in less critical domains, perhaps even to act and make decisions by itself. To serve this purpose, first of all the model must be transparent, so that a human expert is able to understand it. Approaches not fulfilling this condition will hamper the experts willingness to interact with the system and probably be refused in practice. Figure 6.1 presents an architecture for such an interactive modelling framework. It is worth noting that the user plays an active role in this framework. Instead of only monitoring the model, he can give feedback and interact with the learning system. This feedback can be of different types, ranging from a simple indication of prediction errors (direct feedback) to the manual modification of the model or some of its components (enhanced feedback). A framework of this type was also successfully applied (in a more specific form) within a machine vision systems for surface inspection tasks (classifying images showing surfaces of production items into 'good' and 'bad ones) — see also Section 9.1, Chapter 9.

Interpretation capabilities may be violated in purely precise evolving techniques (as discussed in previous chapters), leading again to (almost) black box models; in this sense, it is an important challenge to improve the interpretability of (precisely) evolved fuzzy models in order to stimulate user interaction. In order to cope with on-line operation modes, the techniques for guiding the models to more interpretable power should therefore be ideally fast and incremental. This is opposed to applying interpretability improvement techniques to conventional fuzzy modelling approaches, which are completely conducted in off-line batch model. For a comprehensive survey on off-line approaches, please refer to [69]. A possible way to tackle the issue of improving understandability of fuzzy models, is to divide the interpretability concept into two parts: low-level and high-level interpretability [469]:

- The low-level interpretability of fuzzy models refers to fuzzy model interpretability achieved by optimizing the membership functions in terms of semantic criteria on fuzzy set/fuzzy partition level, such as moderate number of MFs, distinguishability or complementarity.
- The high-level interpretability refers to fuzzy model interpretability obtained by dealing with the coverage, completeness, and consistency of the rules in terms of the criteria on fuzzy rule level, such as rule-base parsimony and simplicity (number of rules), consistency of rules or transparency/readability of a single rule.

In this book, we describe another way to improve interpretability in fuzzy models, which is manifested in a three-step approach:

- The first step (Section 6.2) includes the reduction of the model complexities by exploiting redundancies in the models evolved from the data over time;
- The second step (Section 6.3.1) exploits some concepts from semantics and cognitive science as well as specific learning mechanisms in order to give the fuzzy sets, the rules and the consequent functions (in case of Takagi-Sugeno fuzzy systems) a clearer, more interpretable and finally understandable meaning.
- The third step (Section 6.3.2) expands the second step by preparing and demonstrating the evolved fuzzy models and their changes over time with new data visually.

The reason why we follow these three stages, especially the distinction between the first stage with the second and third one, lies in the degree of objectivity.

Model complexity is something which can be measured objectively in terms of number of rules, fuzzy sets in the antecedent parts, number of inputs, parameters etc.

In this sense, complexity between two evolved fuzzy models can be simply compared and someone can report which one is the better with respect to this property.

Interpretability and furthermore *understandability* of the fuzzy models is something which is somewhat much more intuitive and therefore subjective. As such, it cannot be measured fully objectively, i.e. there is no objective measure to say that one model has a better interpretable meaning than another one.

In this sense, the topic of interpretability drifts much more into the direction of human cognition and psychology. Section 6.3 we will discuss some possible promising aspects on how to achieve better interpretable power of evolving fuzzy systems.

6.2 Complexity Reduction

As mentioned in the previous section, the reduction of complexity in fuzzy models is a first step to achieving a better interpretability and understandability of the models. We can assume that a complexity reduction step always leads to a more transparent model. Indeed, if comparing two different models obtained from different data sets/sources, it does not necessarily mean that the model with less fuzzy sets and rules yields a better interpretation of the underlying process than the model with a higher number of sets and rules [74]. However, what we can always assume is the fact that reducing the *unnecessary complexity* in a given fuzzy model (learned from one data source) always helps to improve the transparency and furthermore the interpretability of this model. This is a direct consequence of a well-known result from cognitive psychology [310]: a lower number of entities can be more efficiently handled in the short term memory of a human being than a higher one. In terms of fuzzy systems, entities can be seen as structural components within the models, i.e. fuzzy sets and rules.

Unnecessary complexity can be present in form of redundancy in the structural components (strongly over-lapping rules), obsolete rules, unimportant rules (with low support), contradictory rules etc.

Furthermore, a nice side effect is that complexity reduction of fuzzy models may even increase the accuracy and therefore the predictive quality of the models. This is because too many rules may over-fit the training data, while performing badly on separate test data set, i.e. having a low generalization capability (see also Section 2.8 for a discussion on over-fitting and the related bias-variance tradeoff problematic). This is especially the case in evolving fuzzy systems approaches, where, based on a snapshot of new incoming samples, it may happen that new local models (rules) are evolved, which turn out to be unnecessary at a later stage (for instance consider two separate clusters moving together over time, see Section 5.1.1, Figure 5.1).

For off-line batch learning approaches of fuzzy systems, a lot of approaches for reducing the complexity of the models were reported in literature. There are basically two different types:

- Incorporating aspects of model complexity and transparency directly into the learning process (mostly integrated into the optimization problem itself).
- Post-processing techniques, which reduce the complexity of models (once they are trained) based on some information criteria directly extracted from the structure of the models representing their inner characteristics.

The basic idea in the first approach is to reduce complexity and to enforce transparency in the form of constraints, that is, to formalize the learning problem as a constrained optimization problem that does not only seek to maximize predictive accuracy. These constraints can refer to the fuzzy sets and individual fuzzy partitions, for example by limiting the number of fuzzy sets in these partitions [328, 460], or by requiring a certain type of characteristics of the fuzzy partitions (e.g. Ruspini partitions) (as achieved in [178, 98, 196]). An iterative complexity reduction approach connected with evolutionary techniques is presented in [372]. Complexity can also be reduced by reducing the number of input variables, i.e., the dimensionality of the input space within the learning process. To this end, feature selection methods can be used [308]. The basic idea in the second approach is to learn an accurate fuzzy system first, and to reduce its complexity, as far as possible, afterward. Essentially, this comes down to simplifying the rule base, for example by merging similar fuzzy sets and removing redundant or dispensable rules, while preserving as much as possible of its predictive accuracy. Fundamental and widely used techniques in this direction were developed during the 90ties by Setnes et. al, see [393, 394]. Another approach is rule ranking, where the rules, while being learned, are ordered according to their importance [461, 301]. To reduce complexity, such a list of rules can then be cut at any position.

In an on-line learning context, complexity reduction techniques for evolving fuzzy systems are discussed in this book which are fast and can be performed in incremental manner. This prevents us from using time-intensive constrained-based optimization techniques, which cannot be calculated recursively, but usually need an iterative batch learning process. Furthermore, in order to stay independent of the applied evolving fuzzy systems approach, it is beneficial to use techniques which can be applied in a post-processing manner and operate directly on the structure of the evolved models. This finally yields a generic applicability of the techniques to any evolving fuzzy system approach as discussed in Chapter 3.

6.2.1 On-Line Merging of Fuzzy Sets (Distinguishability Assurance)

On-line merging of fuzzy sets is necessary in order to assure *distinguishability* of the membership functions within the fuzzy partitions in the single dimensions. Distinguishability of the fuzzy sets is one of the key features with regard to a high interpretability of the models (see Section 6.3 below). A fuzzy set should represent

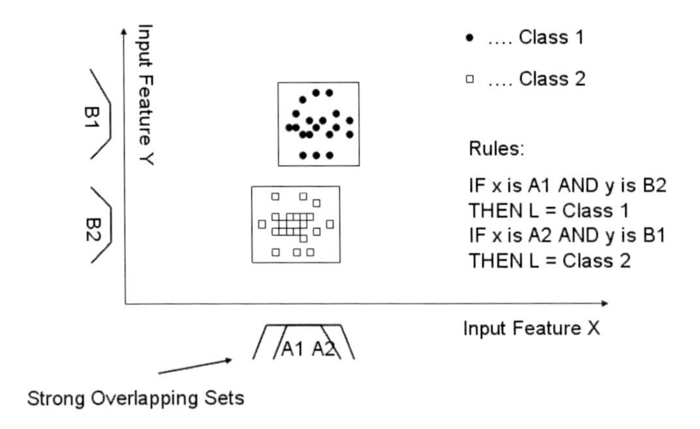

Fig. 6.2 Projection of two rules (shown as two-dimensional boxes) representing two classes onto the axes to obtain the fuzzy sets in partitions of both features; in Feature X two strongly overlapping (redundant) trapezoidal sets occur, as the two rules are lying one over the other in the direction of Feature Y

a linguistic term with a clear semantic meaning. To obviate the subjective establishment of this fuzzy sets/linguistic terms association, the fuzzy sets of each input variable should be distinct from each other. Projecting rules onto the one-dimensional axes which were directly evolved, identified in the high-dimensional feature space as compact regions (spheres, ellipsoids, hyper-boxes) of data clouds (as e.g. done in *FLEXFIS*, *eTS*, *ePL* or *DENFIS* methods with the help of clustering methods), may lead to strongly over-lapping (redundant) representations of the fuzzy sets in the fuzzy partitions. A visualization example for underlining this effect is presented in Figure 6.2, where two evolved rules (shown as two-dimensional boxes) representing two separating classes in classification problem are projected onto the two axes of the input features X and Y. Obviously, the two trapezoidal fuzzy sets in input dimension X are strongly overlapping and hard to distinguish, which leads to a less interpretable fuzzy partition. A merging of these sets into a slightly broader one (as will be demonstrated in the paragraph below) would not lose significant accuracy of the rules.

In *SOFNN* a similar problematic can arise as rules may be added even if the coverage of new incoming samples by already extracted fuzzy sets is high (namely when the model error in the new samples is high — see Case 3 in Section 3.5.2). *SOFNN* only guarantees a good coverage of the samples in the single input features by enlarging the widths of the smallest fuzzy sets, but does not prevent strongly overlapping sets. In *SAFIS*, the overlap degree of fuzzy sets is controlled by an overlap factor, which is embedded in a more complex condition whether a new fuzzy rule is evolved or not (3.62). The likelihood of strongly overlapping fuzzy sets can be decreased in this way, but such fuzzy sets cannot be completely prevented — see Section 3.6. *SONFIN* implicitly contains an own approach for avoiding fuzzy sets with a strong overlapping. This is achieved by aligning clusters which were formed in the input space — see Section 3.7.1. A similar concept is performed in *SEIT2FNN* for evolving type-2 fuzzy systems, see Section 3.7.4.

Merging Based on Similarity Measure

An attempt to overcome the situation of strongly over-lapping and redundant fuzzy sets in evolving fuzzy systems, is demonstrated in [281] for Gaussian membership functions. There, the idea is presented to merge very similar fuzzy sets by exploiting the so-called Jaccard index as similarity measure (following the approach in [393] for off-line trained fuzzy systems), which defines the similarity between two fuzzy sets A and B as follows:

$$S(A,B) = \frac{\int(\mu_A \cap \mu_B)(x)\,dx}{\int(\mu_A \cup \mu_B)(x)\,dx}, \tag{6.1}$$

where the intersection in the nominator is given by the pointwise minimum of the membership functions, and the union in the denominator by the pointwise maximum. This measure belongs to a set-theoretic similarity measure satisfying the following conditions:

$$S(A,B) = 0 \Leftrightarrow \mu_A(x)\mu_B(x) = 0 \;\forall x \in X \tag{6.2}$$
$$S(A,B) > 0 \Leftrightarrow \;\exists x \in X \;\mu_A(x)\mu_B(x) \neq 0 \tag{6.3}$$
$$S(A,B) = 1 \Leftrightarrow \mu_A(x) = \mu_B(x) \;\forall x \in X \tag{6.4}$$
$$S(A*,B*) = S(A,B) \;\; \mu_{A*}(a+bx) = \mu_A(x), \mu_{B*}(a+bx) = \mu_B(x) \;\; a,b \in \mathbb{R}, b > 0 \tag{6.5}$$

The calculation of the integrals in (6.1) can be done by discretization steps within the range of the variables and approximating the integrals by a sum over n discrete points:

$$S(A,B) = \frac{\sum_{q=1}^{n} \min(\mu_A(x_q), \mu_B(x_q))}{\sum_{q=1}^{n} \max(\mu_A(x_q), \mu_B(x_q))} \tag{6.6}$$

The finer the discretization grid is, i.e. the higher n, the more exactly the similarity measure is calculated (however computational effort is increased).

In [281], two Gaussian similar fuzzy sets, i.e. fuzzy sets whose similarity degree S is higher than a pre-defined threshold, are merged into a new Gaussian kernel with the following parameters:

$$c_{new} = (\max(U) + \min(U))/2, \tag{6.7}$$
$$\sigma_{new} = (\max(U) - \min(U))/2, \tag{6.8}$$

where $U = \{c_A \pm \sigma_A, c_B \pm \sigma_B\}$ and $c_.$ the centers and $\sigma_.$ the widths of the fuzzy sets indicated in the indices. The idea underlying this definition is to reduce the *approximate* merging of two Gaussian kernels to the *exact* merging of two of their α-cuts, for a specific value of α. In [281], $\alpha = \exp(-1/2) \approx 0.6$ is chosen, which

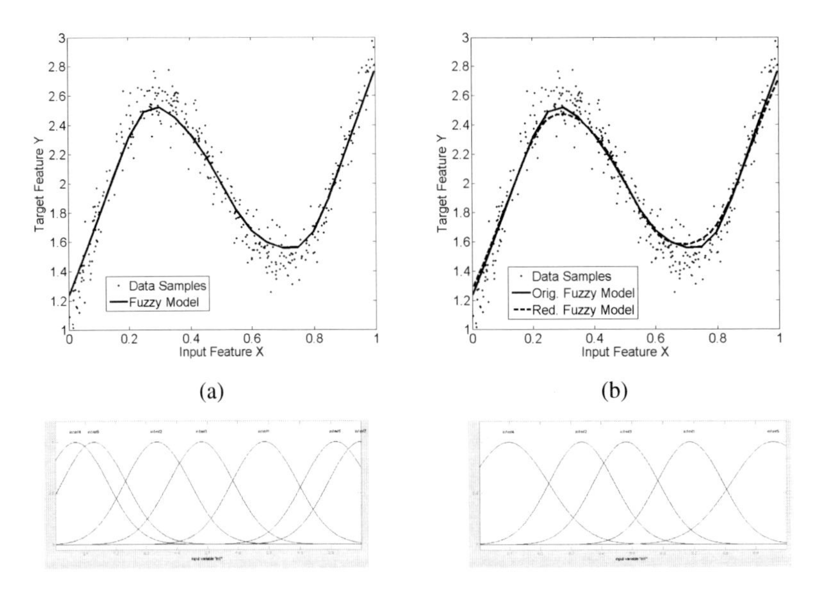

Fig. 6.3 Left: Original model as extracted from the data stream (sin-curve is built up sample-wise); note the two overlapping fuzzy sets on the left and on the right hand side in the fuzzy partition; right: reduced model due to similarity-based merging using a similarity threshold of 0.5 ($\rightarrow$ reduction to 5 fuzzy sets and rules); the dotted line represents the original approximation curves based on the 7 sets = rules, whereas the solid line shows the approx. curves for the reduced model (only tiny differences in the shape can be recognized)

is the membership degree of the inflection points $c \pm \sigma$ of a Gaussian kernel with parameters c and σ.

To see the effect of this merging procedure, Figure 6.3 demonstrates a simple two-dimensional approximation example, where in (a) the original model as extracted from the data stream is visualized containing 7 fuzzy sets = rules, whereas the left and the right two most a significantly overlapping (one of these is obviously superfluous because of the strong linearity in these regions); in (b) a similarity threshold of 0.5 is applied. The upper row compares the approximation quality of the original fuzzy model evolved from the data (solid line) with the one achieved by the reduced model after merging some fuzzy sets (dotted line) (by visualizing the corresponding approximation curves). Clearly, there is no significant difference between solid and dotted line, i.e. the basic tendency of the approximating curve is preserved, while reducing the (unnecessary) complexity of the model in terms of decreasing the number of fuzzy sets (from 7 to 5). The latter are visualized in the lower row for the reduced models. The merging procedure was already successfully applied for various high-dimensional real-world approximation problems such as prediction of NOx emissions at engine test benches, prediction of resistance values at rolling mills (see Chapters 7 and 8) and prediction of bending angles at steel bending machines. For the latter, the number of fuzzy sets could be reduced from

140 to 12, and the number of rules from 28 to 22 without loosing significant accuracy of the models: in fact, the quality (measured in terms of r-squared-adjusted) slightly decreased from 0.9988 to 0.9987 [281].

This merging procedure (applicable to any evolving fuzzy systems approach using Gaussian membership functions as fuzzy sets) can be done in a single-pass incremental learning context as not requiring any past data and just acting on the current fuzzy partitions extracted from the data stream so far. In order to speed up the merging process for fast on-line demands, the similarity measure is not calculated for every fuzzy set pair in each fuzzy partition (for each input feature), but just for those fuzzy sets which appear in the updated rules by the last data block. If performing sample-wise update operations, then maximally only the fuzzy sets occurring in one (=the updated) rule have to be checked whether they become redundant to any other fuzzy set in the same partition due to the rule update.

Reducing Computation Time for Calculating Similarity

However, still the procedure for checking similarities of fuzzy set pairs after model updates with each new incoming sample may be quite time-intensive due to the two sums in (6.6) (including a lot of evaluations of membership functions), especially when n is large in order to achieve a high precision in the calculation of the similarity measure. Instead of using such a set-theoretic similarity measure (such as the Jaccard index), Dourado et al. [360] [359] propose to apply a geometric similarity measure, which represents fuzzy sets as points in a metric space (for instance the centers of the clusters from where the fuzzy sets have been derived). Similarity between the sets is regarded as an inverse of their distance in the metric space in the following way:

$$S(A_1,A_2) = \frac{1}{1+d(A_1,A_2)} \tag{6.9}$$

where the geometric distance between two Gaussian membership functions can be approximated by (also used in [197] for batch trained fuzzy systems):

$$d(A_1,A_2) = \sqrt{(c_1-c_2)^2+(\sigma_1-\sigma_2)^2} \tag{6.10}$$

Instead of using the time-intensive formula in (6.6), equation (6.10) allows an easy and fast calculation of the similarity of two fuzzy sets.

Another possibility for reducing calculation times is to perform a transformation of Gaussian membership functions to trapezoidal ones [79]. In particular, to determine a trapezoidal membership function

$$T(x) = \max\{\min\{(x-a)/(d-a),1,(b-x)/(b-e)\},0\} \tag{6.11}$$

which can approximate a Gaussian function, in [79] the α-cut of a fuzzy set is exploited:

$$A_\alpha = \{x \in X | \mu(x) \geq \alpha\} \tag{6.12}$$

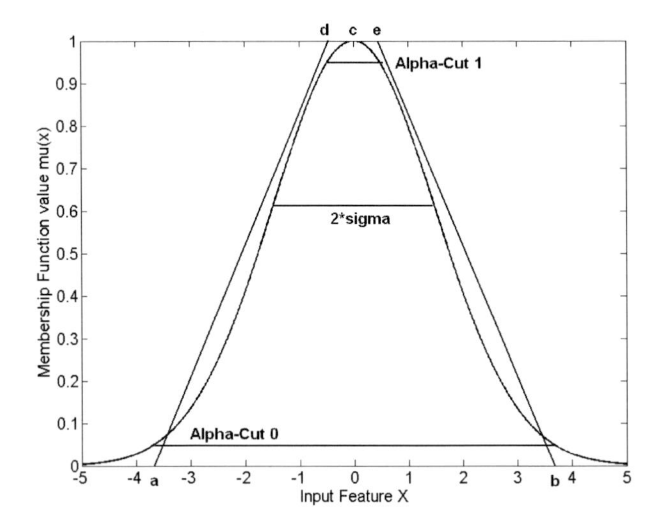

Fig. 6.4 Transferring a Gaussian membership function with parameters c and σ to a trapezoidal function with parameters a,b,d and e by using α-cuts with alpha values of 0.05 and 0.95

where α lies in $[0,1]$. Hence the α-cut of a fuzzy set is a closed interval in $\mathbb{R}$. Now, to identify the parameters of the trapezoidal membership function a,b,d,e on the basis of a given Gaussian function, two special α-cuts of a fuzzy set A are introduced, namely the bottom α-cut A_{α_0} and the top α-cut A_{α_1}. Thus, the parameters a,b,d,e can be decided on the basis of the Gaussian function through $A_{\alpha_0} = [a,b]$ and $A_{\alpha_1} = [d,e]$. Figure 6.4 visualizes the concept of transferring a trapezoidal to a Gaussian membership function for $\alpha_0 = 0.05$ and $\alpha = 0.95$. Expressed in formulas, the parameters a,b,d and e can be calculated from the parameters c and σ of the original Gaussian function as follows:

$$
\begin{aligned}
a &= c - \sigma\sqrt{-2ln(0.05)} \\
b &= c + \sigma\sqrt{-2ln(0.05)} \\
d &= c - \sigma\sqrt{-2ln(0.95)} \\
e &= c + \sigma\sqrt{-2ln(0.95)}
\end{aligned}
\qquad (6.13)
$$

The virtual transfer of the antecedent fuzzy sets included in the updated rule to trapezoidal ones is thus quite fast. Now, the remaining question is how to compare two trapezoidal fuzzy sets, in particular does the transfer to trapezoidal functions really pay off in terms of computational complexity for calculating the similarity measure for these types of functions? - the answer is positive, as similarity can be explicitly calculated in simple formulas based on the parameters a_1,b_1,d_1,e_1 (representing the first set) and a_2,b_2,d_2,e_2 (representing the second set): four cases

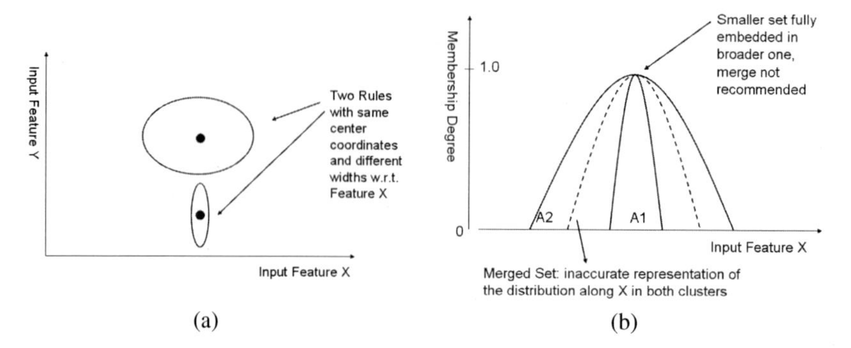

Fig. 6.5 (a): two rules with same center coordinated by significantly differing ranges of influence with respect to Feature X; (b): the projection of the rules onto the X-axis leads to two fuzzy sets, where the smaller is completely embedded in the large one; a merging of these two is not recommended as one single fuzzy set could not reliably represent the data distribution along the X-axis in both rules shown in (a)

needs to be distinguished according to the degree and type of overlap of the two sets, see [79] for further details.

A specific note is dedicated to the situation when one fuzzy set is completely covered by another fuzzy set as shown in Figure 6.5 (b). Here, someone may consider that these two sets can be merged; however, this would either loose the more precise specification represented by the inner set (if the inner set is artificially blown up and merged with the outer set) or cause an over-fitted situation (when the widths of the outer set is reduced). This finally means that, although such a situation is quite nasty from interpretable and fuzzy set-theoretic point of view, not much can be done to improve this situation without losing significant accuracy of the evolved fuzzy models. The reason why such an occurrence can happen is simply because in different parts of the feature space the characteristics of the local regions is different: in some parts, rules are narrower with respect to some dimensions than in others. Now, if two rules with different ranges of influence with respect to a certain dimension happen to be centered around the same coordinates (see Figure 6.5 (a) for a two-dimensional examples of this case), such a situation as presented in Figure 6.5 (b) arises. Clearly, a fuzzy set having (nearly) the same center coordinates and a width which is between the width of the inner and the outer set, would give a weak representation of the data distribution for both clusters with respect to Feature X.

Merging Based on Close Modal Values

The basic idea in this approach is to merge those fuzzy set pairs (A,B) which fulfill the following properties:

1. Their centers (=model values or cores) are close to each other (i.e. closer than a small fraction of the range of the corresponding feature in whose partition they appear).

2. Their spans (widths in the case of Gaussian MFs) are tiny, such that, despite their close modal values, they are not similar or redundant (hence, this situation is not covered by the approach described in the previous section).
3. The corresponding weights (belonging to the dimension where the close and tiny fuzzy sets appear) in the rule consequents of those rule pairs, where the two fuzzy sets appear in their antecedents and their remaining antecedent parts are similar, have similar values. This means that the gradients of local models in two neighboring rules (which are only differing significantly in one antecedent part, one using fuzzy set A, the other using fuzzy set B close to A) with respect to the dimension (variable) where A and B occur in its partition, are nearly the same.

The first point is an idea which was already formulated within the FuZion merging algorithm [121] (there applied for triangular fuzzy sets and iteratively within a batch off-line learning process). The third point guarantees that highly non-linear occurrences in the feature space (along certain dimensions) are still resolved with sufficient accuracy, as in this case the gradients of the two neighboring rules will be completely different (mostly even having different signs), violating this property. On the other hand, when the tendency of the two neighboring rules along the dimension including fuzzy sets A and B is similar, gradients will have similar value, as the non-linearity is weak between the two neighboring local regions.

To underline this aspect, Figure 6.6 demonstrates a small two-dimensional example, where three clusters = rules are extracted from the data, having a very small width along the X-direction, causing tiny fuzzy sets with close modal values at the right border of Feature X (a). There, the tendency of the function to be approximated is the same in all three local regions (no non-linear fluctuating behavior, just going almost like a straight line). In this case, the three fuzzy sets at the right most side can be merged to one bigger one without losing any accuracy of the model, as shown in Figure 6.6 (b) (model curve still approximates the data quite accurately). On the other hand, when the non-linearity is high, the extraction of two or more close fuzzy sets with tiny widths is absolutely necessary and a merging would cause a significant deterioration in the quality of the approximating curve. This is underlined in Figure 6.7 (b) where some fuzzy sets in the left highly non-linear part are merged based on their modal values and widths, as not taking into account the gradients in the corresponding rule consequent functions. Obviously, this deteriorates the approximation quality of the model significantly, hence the third property in above itemization is essential.

The whole approach can be applied easily within an on-line learning context, as parameters of neighboring fuzzy sets are compared, which does not require any past data or re-training phases. The gradients in the rule consequent functions are permanently updated through *recursive weighted least squares*, therefore always representing the latest tendency in the local regions. When using Mamdani fuzzy systems, this concept cannot be applied, as no gradient information in the single rule consequents is provided.

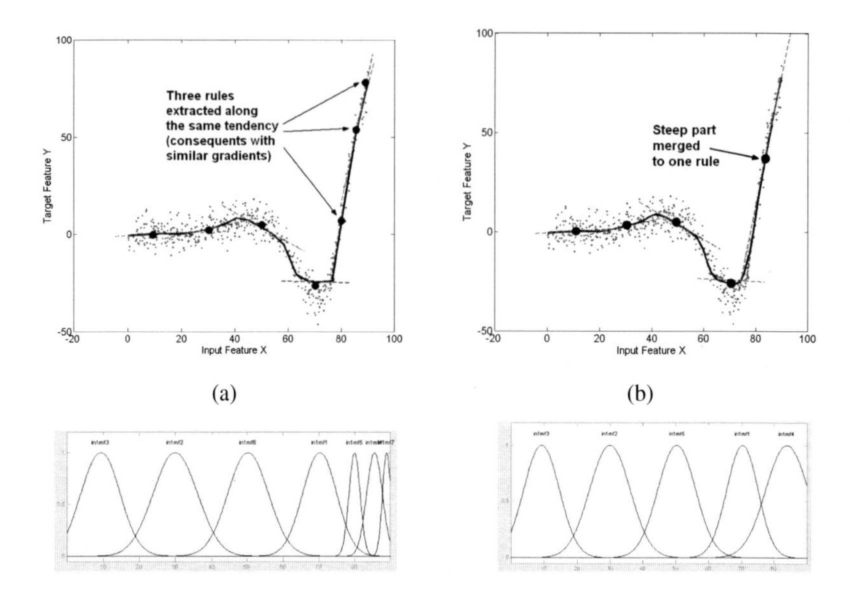

Fig. 6.6 (a): original model as evolved by *FLEXFIS* approach, the rule consequent functions shown as dotted lines, the rule centers indicated by thick dark dots, below the corresponding fuzzy partition of input feature X; (b): improved model by merging three right most fuzzy sets (=rules) to one: approximating curve preserved

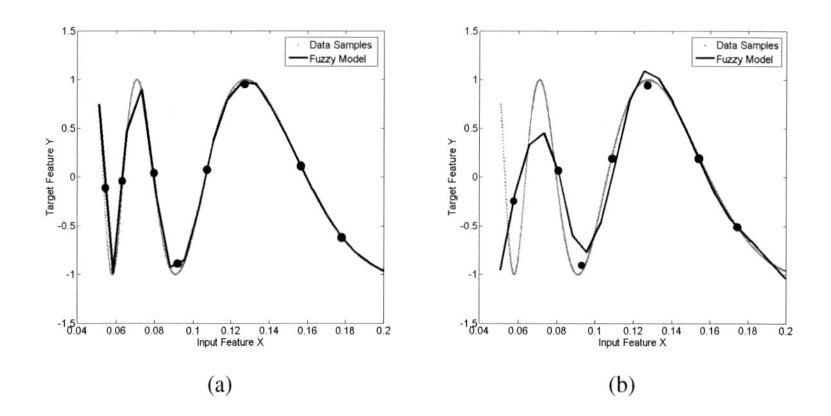

Fig. 6.7 (a): original model as evolved by EFS approach (solid line) for $sin(1/x)$-type relationship, the data samples as grey small dots, the thick dark dots representing the rule/cluster centers, the model is able to approximate also the highly non-linear part at the left hand side accurately; (b): model by merging two left most fuzzy sets with close modal values and small widths at the left side without taking into account the gradients in the corresponding rule consequent functions $\rightarrow$ model accuracy suffer significantly

6.2.2 On-Line Rule Merging – A Generic Concept

In Section 5.1 (Chapter 5) a rule merging concept was demonstrated, which is completely carried out in the original high-dimensional feature space: two ellipsoidal clusters associated with rules are merged whenever certain compatibility criteria between the two clusters are fulfilled. The meaningfulness of this concept is underlined by the fact that clusters = rules, originally representing distinct data clouds, may move together with more and more samples loaded over a certain time frame — see Figure 5.1. This concept, motivated in Section 5.1 for achieving cluster partitions more appropriately reflecting the natural characteristics of the data samples (rather than for interpretability purposes), is applicable for all evolving fuzzy systems approaches which apply an incremental, recursive and/or evolving clustering scheme for doing rule evolution and adaptation of the rules' antecedent parts during incremental learning phase, i.e. for *FLEXFIS*, *eTS*, *DENFIS* and *ePL*. However, this concept is not applicable for other approaches (such as *SAFIS*, *SOFNN* or *SONFIN*), which perform rule evolution and antecedent learning based on completely other concepts than clustering.

Here, we are presenting an alternative rule merging approach, which naturally results from the fuzzy set merging process (according to local redundant information) as demonstrated in the previous section and as such is applicable for any evolving fuzzy systems approach. In the one-dimensional case, when two fuzzy sets are merged in the fuzzy partition of the input features, this automatically leads to a rule merging process, as then the antecedent parts of two rules are falling together. In higher-dimensional cases, the complete antecedent part of two (or more) rules can fall together, whenever fuzzy sets in those premises (=dimensions) are merged, where the rules differ from each other. A visualization of such an occurrence for a three-dimensional example is visualized in Figure 6.8. Now, the remaining question is whether and how to merge the rule consequent parts, if two rule premises turn out to become equal after merging some fuzzy sets. From (Boolean) logic point of view, it is precarious to merge two rules whose antecedent parts are similar (or equal after merging), while having dissimilar consequents. In fact, such rules usually represent inconsistent parts of the rule base as they contradict each other. This concept can be easily extended to fuzzy rule bases by replacing the concept of equivalence between Boolean propositions by the degree of similarity between fuzzy sets.

In this sense, two fuzzy rules given by

IF A THEN C

IF B THEN D

can be seen as *contradictory*, if the similarity between premises A and B is (significantly) greater than the similarity between the consequents C and D.

This crisp condition can be extended to a fuzzy one, delivering a degree of contradiction. For the merging process this means that merging the consequent parts of two rules (and hence finally the two rules together to one new rule) is only valid

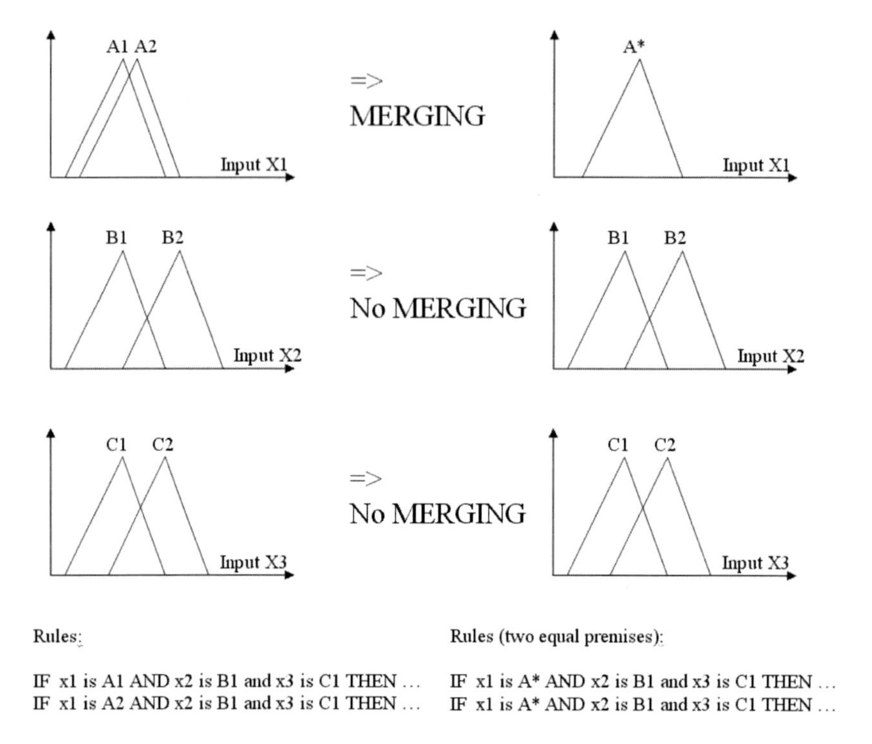

Fig. 6.8 Left: original fuzzy partitions in three input features, two rules; right: two redundant fuzzy sets are merged in input feature $X1$ to a new set $A*$ $\rightarrow$ the antecedent parts of the two rules fall together after merging process

when the original rules' antecedents (i.e. before the fuzzy set merging process) are more dissimilar than the rule consequents (also called dissimilarity balance). Otherwise, the deletion of one of the two rules (after fuzzy set merging) is more appropriate (ideally the one with lower significance, i.e. in the data-driven sense with lower number of samples belonging to/forming it in the past).

To calculate the similarity of complete rule premises between two rules one may exploit the techniques as demonstrated in the previous section, apply it pair-wise for the single antecedent parts (fuzzy sets) and calculate a minimal similarity degree, that is (for the mth and the nth rule)

$$S_{ante}(m,n) = \min_{i=1,\dots,p} S(A_{i,m}, A_{i,n}) \qquad (6.14)$$

with p the dimensionality of the input space = the number of premise parts in each rule and $A_{i,m}$ the fuzzy set in the mth rule belonging to the ith antecedent part (ith dimension). Taking the minimum over all similarity degrees is justified by the possibility that two local regions may be completely torn far apart with respect to already one single dimension. In this case, the rules are far away from having any overlap and therefore redundancy. Another, more enhanced possibility of setting up a similarity measure between rules, is also to include the t-norm used in the rule premise parts in the similarity concept, as is done for instance in [65].

In Mamdani fuzzy systems the similarity between rule consequents can be calculated by using the fuzzy set similarity measures as proposed in the previous section (single fuzzy sets appear in each of the rule consequents there). In Takagi-Sugeno fuzzy systems, rule consequent functions are represented by hyper-planes in the high-dimensional space. When using local learning approach, these hyper-planes snuggle along the real trend of the approximation curve in the corresponding local region, i.e. the hyper-planes represent exactly the tendency of the data cloud in the local regions where they are defined, achieving a good interpretation capability of TS fuzzy systems — see Section 6.3 below. A reasonable measure for similarity is the angle between two hyper-planes, as it measures the difference of the direction the consequent functions follow in the high-dimensional space. The angle between two hyper-planes (corresponding to the mth and mth rule) can be measured by calculating the angle between their normal vectors $a = (w_{m1}\ w_{m2}\ \dots\ w_{mp}\ \ -1)^T$ and $b = (w_{n1}\ w_{n2}\ \dots\ w_{np}\ \ -1)^T$:

$$\phi = \arccos\left(\left|\frac{\mathbf{a}^T\mathbf{b}}{|\mathbf{a}||\mathbf{b}|}\right|\right) \tag{6.15}$$

with $\phi \in [0,\pi]$. The maximal dissimilarity is obtained when the angle between the two normal vectors is $\frac{\pi}{2}$, as the orientation of the vectors does not play a role when using the hyper-planes in function approximation or classification. In the extreme case, when a points into the opposite direction than b, the angle would be π, but from function approximation point of view both hyper-planes would deliver exactly the same response for a data sample to be predicted, hence are equivalent. Figure 6.9 presents two similar consequent functions (for Rules 1 and 3) in case of an angle close to 180 degrees (160 degrees), and two dissimilar consequents (for Rules 1 and 2) in case of an angle of 60 degrees. The predictions are shown by a big dark dot and underline the similarity and dissimilarity of the rules consequent functions. The similarity measure of two hyper-planes y_i and y_j can be defined as:

$$S_{cons}(y_m,y_n) = \begin{cases} 1-\frac{2}{\pi}*\phi & \phi\in[0,\frac{\pi}{2}] \\ \frac{2}{\pi}*(\phi-\frac{\pi}{2}) & \phi\in[\frac{\pi}{2},\pi] \end{cases} \tag{6.16}$$

Comparing $S_{ante}(m,n)$ calculated before the fuzzy set merging process with $S_{cons}(y_m,y_n)$ performs a contradiction check and leads to the decision whether two rules becoming redundant due to the fuzzy set merging (see Figure 6.8) are merged or one (the less significant one) is deleted:

IF $S_{cons}(y_m,y_n) \geq S_{ante}(m,n)$

then merging is performed, otherwise deletion.

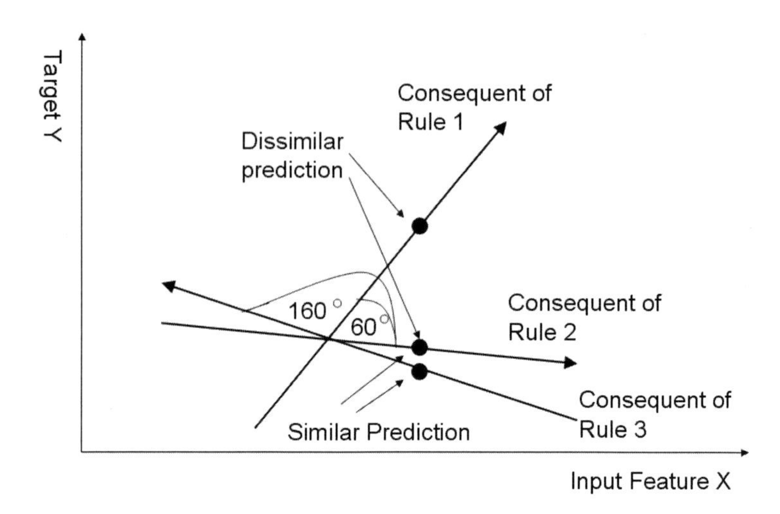

Fig. 6.9 Three rule consequent functions shown as straight lines: an angle of 160 degrees is equivalent to an angle of 20 degrees regarding similarity of predictions (shown as dark dots); 60 degrees between two consequent functions already trigger predictions far away from each other

Significance of a rule can be measured in terms of the number of data samples belonging to the rule, i.e. the number of those samples for which the rule had highest membership degree among all rules.

Based on the significance levels of the two rules and hence their importance degree for the final model output, the merging of two rule consequent functions is done by applying the following formula:

$$w_{(new)j} = \frac{w_{mj}k_m + w_{nj}k_n}{k_m + k_n}, \qquad \forall j = 0, ..., p \tag{6.17}$$

where w_{mj} resp. w_{nj} is the linear parameter in the consequent of the mth resp. nth rule with respect to the jth dimension and k_m resp. k_n is the number of data samples belonging to the cluster corresponding to the mth resp. nth rule (local area). Thus, the parameters of the new rule consequent are defined as a weighted average of the consequent parameters of the two redundant rules, with the weights representing the significance of the rules. This merging strategy can obviously be applied in on-line mode and in fast manner as it does not require any prior training data and just computes a weighted average over p dimensions.

6.2.3 Deletion of Obsolete Rules

During the incremental learning phase, some rules may get out-dated over-time (obsolete rules). This could be either because of a mis-leaded rule evolution process (e.g. due to outliers, high noise levels or failures in the data) or because this partial relation is not important or even valid any longer (e.g. due to changing system states, new operating conditions or drifts in data streams). This can be measured in terms of the time span for which these 'out-dated' rules were not addressed in the past or the relative significance of a rule expressed by $\frac{k_i}{k}$ with k_i the number of samples for which rule was winning rule (i.e. had highest membership degree among all rules) and k the number of all samples seen so far. If this fraction is very low, the ith rule can be deleted. In [12] this concept is proposed and named population-based rule base simplification and is successfully applied to *eTS* approach (as such called *simpl_eTS*). This concept is generic in the sense, that it is applicable for any evolving fuzzy systems approach (k_i can be counted independently from the rule evolution and parameter learning process). An extension of this concept is the utility function as proposed in [20]:

$$U_k^i = \frac{1}{k - I_k^i} \sum_{l=1}^{I_k^i} \Psi^l \tag{6.18}$$

Utility U^i of the ith rule accumulates the weight of the rule contributions to the overall output during the life of the rule (from the current time instant back to the moment when this rule was generated). It is a measure of importance of the respective fuzzy rule comparing to the other rules (comparison is hidden in the relative nature of the basis functions Ψ = membership degrees). If the utility of a fuzzy rule is low, then this rule becomes obsolete or not used very much. In [259] a compatibility index among rule centers is defined in the following way (between rule m and n):

$$\rho(m,n) = 1 - \sum_{j=1}^{p} |\mathbf{c}_{m,j} - \mathbf{c}_{n,j}| \tag{6.19}$$

Two clusters are redundant if ρ is bigger than a predefined threshold near 1. In this case, one rule (the one with lower significance) can be deleted.

6.2.4 Integration Concept for On-Line Fuzzy Set and Rule Merging

This section raises possibilities on how to integrate fuzzy set and rule merging during on-line incremental learning phase. In batch off-line training modes, the situation is quite obvious: the approaches proposed in the previous sections can be applied in a post-processing manner after the training of the fuzzy systems is finished. This may be even done in iterative manner (e.g. iterating over the fuzzy set pairs, consisting of original and already merged set), as there are usually no time restrictions in an off-line setting. After completing the iterations, the improved fuzzy system can be shown to the operator(s), expert(s) and/or user(s). In on-line mode, the situation is different, as fuzzy systems are permanently updated, i.e. their structural components change over time according to the data stream(s) from which they are extracted. Complexity reduction techniques should be ideally applied after each incremental update step in order to be able to present an interpretable fuzzy system which is up-to-date. This can be performed even in single-pass manner with all the approaches aforementioned in the previous sections, as they do not require any past data and contain basically fast operations (also similarity calculation between fuzzy sets could be made fast with specific techniques — see Section 6.2.1). The remaining question is how to proceed further in the on-line learning context, once the merging process was applied to the latest evolved fuzzy system. Here, two basic possibilities can be pursued:

- The improved fuzzy system is further evolved with new incoming data, then again complexity reduction techniques are applied, leading to an alternating scheme (starting with a given fuzzy system): update fuzzy system with new data, reduce complexity of the updated fuzzy system, update reduced (improved) fuzzy system with new data, reduce complexity of the updated fuzzy system (due to the new data, new fuzzy sets and rules may become redundant) etc.
- The improved fuzzy system is only used for visualization purposes to operators/users and the original fuzzy system is further evolved with new incoming data. This leads to a two-layer model building, where one layer represents the evolved fuzzy systems as it is permanently updated from the data and represents its natural characteristics as accurately as possible, and the other layer an improved fuzzy system which is obtained from the latest evolved fuzzy system by applying the complexity reduction techniques after each incremental learning step (hence always representing the latest less complex (more interpretable) version of the evolved fuzzy system).

The work-flows of both concepts are presented in Figure 6.10. The former variant has the advantage of saving virtual memory and in some cases to implicitly decrease the over-fitting effect of the evolved fuzzy system as this is a nice side effect of the complexity reduction techniques (apart from increasing interpretability of the models). Furthermore, it offers the option to perform an enhanced interaction with the user/expert, where the user may change some structural components or parameters of the model which are integrating on-the-fly for the next incremental learning cycle — see Section 10.2.3 for a more detailed description of this strategy. The disadvantage of the former variant is that a back-integration of the modified models (parameters, structures) into the learning context is required: for instance, in case when using recursive clustering algorithms for antecedent learning, a new cluster needs to be associated with the merged rule, also the inverse Hessian of the merged rule needs to be elicited from the two redundant rules (in order to guarantee robust further update of the consequent functions). In the extreme case, merged fuzzy sets represent a 'blurred' situation of local feature distributions and cause a severe drop in model accuracy, whereas the latter variant preserves maximal accuracy in terms of always representing the natural characteristics (local distributions) of the data.

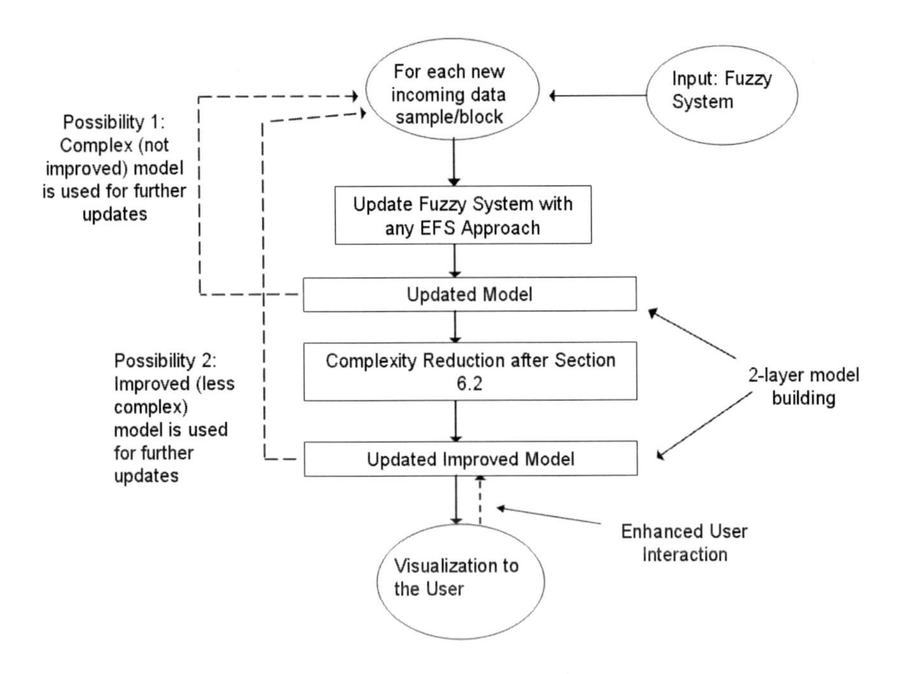

Fig. 6.10 Work-flow for integrating fuzzy set and rule merging into evolving fuzzy systems

6.2.5 On-Line Dimensionality Reduction

The previous sections were dealing with reduction steps affecting the number of fuzzy sets and rules. Another major component in a fuzzy system, significantly responsible for its complexity, is the number of input features/variables. To reduce the dimensionality of the feature space in an off-line setting, feature selection is a feasible and common choice for both, decreasing curse of dimensionality effect while at the same time improving interpretability of the fuzzy models [379] [307] [70]. For on-line learning demands, feature selection coupled appropriately with evolving fuzzy systems is a much more complicated task. In fact, incremental feature selection methods such as the approaches in [254] [459] or [210] could be applied synchronously to the update of the fuzzy systems. This may serve as a by-information about the most essential features, but the problem remains how to integrate this permanently changing information into the evolved fuzzy systems in a smooth way. Then, the problem is how to change the input structure of the models appropriately without causing discontinuities in the learning process. Features can be exchanged dynamically and online in the importance list (e.g. after a new data block, feature #1 replaces feature #2 in the list of five most important features as becoming more important than feature #2), but this requires an exchange in the input structure of the fuzzy models as well. There, however, parameters and rules were learned based on the inputs used in the past. As usually two inputs show different distributions and different relations with other features in the joint feature space, it is not reliable to exchange one input with another one by simply exchanging it in the model definition and continue with the old parameter setting learned for the other input (usually the classifier will immediately produce many wrong classification statements).

A possibility to circumvent this problematic is to apply incremental feature weighting, i.e. assigning feature weights to the various input features, which change smoothly over time. Hence, no disruptions in the input structure is caused; refer to Section 5.2 for the full details about this concept. Although this achieves a smooth reduction of the curse of dimensionality effect, it does not account for any complexity reduction in EFS, as all features are still present in the system, and this with additional weight information. However, this concept may serve as an additional interpretability concept in evolving fuzzy systems (impact of features per rule) — see subsequent section.

6.3 Towards Interpretable EFS

In this section, we describe some further aspects (aside from complexity reduction) which are helpful in guiding the evolved fuzzy systems to more interpretable power. Most of these aspects have not been studied in connection with evolving fuzzy systems so far, so, inspired from concepts used in connection with batch off-line data-driven fuzzy systems, we basically mention criteria and possibilities how to achieve better interpretability of evolving fuzzy systems in the future and do not present complete solutions of these. Basically, someone may distinguish between two views on the interpretability of the models:

- Linguistic interpretability
- Visual interpretability

Linguistic interpretability deals with the semantics of fuzzy systems, for which terms such as readability and understandability of the models are essential in order to open the possibility for a rich interpretation of the models.

This includes answers to questions such as:

- what are the essential meanings of the structural components?
- what do these components stand for and express with respect to the underlying dependencies and relationship hidden in the process? or
- why are certain decisions or predictions made by the models?

The latter question can be an essential one for instance in a fault reasoning process, where the reasons for detected faults are sought-after in order to be able to perform manual fault correction in an early stage, e.g. to switch of some components of the operating system. In fact, the violated rules in the fuzzy models (representing nominal fault-free dependencies at the system) may offer valuable clues about which system variables are affected by the faults and hence restrict the failure(s) to certain components in the operating system — refer also to Chapter 8 for using evolving fuzzy systems within an on-line fault detection process.

Visual interpretability handles the understandability of models on a visual level, i.e. preparing the models in a way, such that they can be understood when graphically shown to the users/experts.

An additional aspect therein is how to show the changes of the components and parameters in the evolved fuzzy systems over time, i.e. how the models develop over time with the recorded data stream(s).

An important partial aspect of both, linguistic and visual interpretability in fuzzy models is transparency.

Model transparency is defined as a property that enables us to understand and analyze the influence of each system parameter on the system output [171] [394].

6.3.1 Linguistic Interpretability

We highlight criteria and concepts which are helpful for a nice interpretability of 1.) fuzzy partitions, 2.) rules, 3.) consequent functions and 4.) input features/variables.

6.3.1.1 Interpretability of Fuzzy Partitions

An attempt is made to ensure a set of properties that the membership functions of the referential fuzzy sets should have in order to yield linguistically interpretable partitions of the inputs and to guarantee the possibility of assigning linguistic labels to each fuzzy set in all rule premises. To accomplish that, five basic semantic properties should be achieved (see [328], [327], also [469]), namely:

- A moderate number of membership functions
- Distinguishability
- Normality and unimodality
- Coverage
- Complimenarity

In the next paragraphs, we will give a summary about each criterion and point out whether and how they can be fulfilled within the scope of evolving fuzzy systems.

Towards a Moderate Number of Membership Functions

A well-known and interesting result from Cognitive Psychology is that the typical number of different entities efficiently handled at the short term memory of a human being is 7 ± 2 [310]. Surprisingly, it is interesting to verify that most of the successful industrial and commercial applications of fuzzy systems do not exceed this limit [412]; however, this statement was made over 20 years ago based on experience from pure knowledge-based fuzzy systems (without using any learning aspects). This number should also be a guideline when extracting fuzzy sets from data, although for specific application cases the number of fuzzy sets have to be set significantly above 9. For instance when the functional behavior between variables possesses a highly fluctuating characteristic, i.e. a strong nonlinear behavior as e.g. shown in Figure 6.7 (especially left part of the curve), a lot of fuzzy sets have to be generated, in order to be able to track every up-and-down swing. When trying to reduce the sets — see right image in Figure 6.7 — a correct approximation is not possible any longer. Hence, a reasonable trade-off between accuracy and a moderate number of membership functions should be achieved. This also depends on the requirements of the predictive quality of the models. i.e. which maximal errors are allowed within an industrial system, such that the models are reliable, reasonable and therefore do not risk being switched off. A possibility for achieving a reasonable trade-off is the inclusion of constraints in the learning process, either by combining an error measure such as for instance least squares with a punishment term including the complexity of the models (as done in [197] for batch learning of fuzzy systems, see also Section 2.8 for alternative objective functions), or by a multi-objective learning approach. In the latter approach, instead of optimizing a single compromise measure between accuracy and complexity, the problem of trading off the first criterion against the second one can also be approached in the sense of Pareto optimality. In [147], for example, a multi-objective evolutionary algorithm is proposed to determine a set of fuzzy systems that are Pareto-optimal with respect to accuracy and complexity as optimization criteria. An on-line approach of multi-objective learning integrated in the concept of evolving fuzzy systems has, to our best knowledge, not been investigated so far and is therefore a promising hot topic in future research.

Distinguishability Assurance

A fuzzy set should represent a linguistic term with a clear semantic meaning. To obviate the subjective establishment of these fuzzy sets/linguistic terms association, the fuzzy sets of each input variable should be distinct from each other. Besides, close or nearly identical fuzzy sets may cause inconsistencies at the processing stage [328]. As a matter of fact, distinguishability assurance algorithms cause a merging of fuzzy sets and therefore directly the reduction of the number of fuzzy sets. From this point of view, distinguishability assurance algorithms stand for the mechanism to bring the fuzzy sets to a moderate number after human-like thinking and classifying. All the concepts and algorithms presented in Section 6.2.1 can be seen as corner stones for assuring distinguishability of fuzzy sets, as these cover the merging of similar, redundant and even close fuzzy sets.

Alternatively, constraints on the movement of knot points in fuzzy sets (as exploited in the batch learning approach *RENO* [62]) to avoid fuzzy sets having large overlaps may be considered as promising alternative option. This has been considered only little in connection with incremental learning and evolving techniques so far.

Normality and Unimodality

Fuzzy sets should be normal, i.e. for each membership function at least one of the datum in the universe of discourse should exhibit full matching, so formally the following condition should be fulfilled:

$$\forall i = 1,...,C \; \exists x \in X_j \quad \mu_i(x) = 1.0 \tag{6.20}$$

where C fuzzy sets are extracted and X_j denotes the universe of discourse for input variable j. From the interpretability point of view, it is guaranteed that each linguistic term has its core value, where it is fully fulfilled. As usually fuzzy sets are used in EFS having the property that at their centers they yield a membership value of 1 (Gaussian, triangular, trapezoidal MFs fulfill this criterion), the normality condition is fulfilled in all EFS approaches discussed in Chapter 3. This is in favor among other fuzzy sets, for instance among n-order B-splines with n greater or equal than 2. Unimodality is also guaranteed in all EFS approaches, as none of these extracts any bimodal membership functions.

Coverage

This property is closely related to completeness. Completeness is a property of deductive systems that has been used in the context of Artificial Intelligence to indicate that the knowledge representation scheme can represent every entity within the intended domain [328]. When applied to fuzzy systems, this states that a fuzzy system should be able to derive a proper action for every input [241]. When the complete universe of discourse is not covered this leads to interpolation holes and to undefined input states, which can cause a breakdown of the whole prediction process or at least a wrong output behavior of the fuzzy inference. So, coverage is an indispensable property in order to ensure process security. In the on-line learning context, as fuzzy sets are permanently shifted, moved through the feature space with new incoming data, it is preferable to use fuzzy sets with infinite support. These always guarantee some (at least tiny) coverage of the input space. Some EFS approaches (such as *SOFNN* approach — see Section 3.5) guarantee ε-completeness of the evolved fuzzy systems, achieving a minimal degree of overlap and hence always significant coverage of the input space.

Complimentarity

For each element of the universe of discourse, the sum of all its membership values should be equal to one. This guarantees uniform distribution of meaning among the elements. Fuzzy partitions following the Ruspini form [376] (with overlap of 1/2) fulfill this criterion. However, the complementarity requirement is only suitable for probabilistic fuzzy systems to guarantee uniform distribution of meaning among the elements so that a sufficient overlapping of MFs is obtained. A possibility fuzzy system (i.e., the sum of all the membership values for every input vector is between 0 and 1) does not consider this requirement. Within the scope of EFS, currently usually possibility fuzzy systems are used and evolved.

On Assigning Linguistic Labels

In order to obtain a complete linguistic description of the several input partitions, linguistic labels have to be assigned to the extracted and/or modified fuzzy sets. A strategy, namely the strategy about so-called *linguistic hedges* [51] [28] may be applied after each incremental learning step. This compares the actual (original or merged) fuzzy sets with reference fuzzy sets and some of their transformations (*linguistic hedges*), standing for the various pre-defined labels such as for instance 'SMALL', 'Very TINY' or 'More or less BIG'. These reference sets and their transformations have a clear mathematical formulation (e.g. for 'More or less BIG' the square-root of the original fuzzy set 'BIG' is taken as reference set). A comparison can be carried out between the extracted fuzzy sets and the pre-defined hedges again based on similarity measures as described in Section 6.2.1. Finally, the linguistic label of the pre-defined reference sets which are most similar to the actual fuzzy sets are assigned to the fuzzy sets.

6.3.1.2 Interpretability of Rules

Regarding the rule base in an evolved fuzzy system, five basic criteria are essential for achieving good interpretability [469]:

- Rule base parsimony and simplicity
- Readability of single rule
- Consistency
- Completeness
- Transparency of rule structure

In the next paragraphs, we will give a summary about each criterion and point out whether and how they can be fulfilled within the scope of evolving fuzzy systems.

Rule Base Parsimony and Simplicity

According to the principle of Occams razor [23] (the best model is the simplest one fitting the system behaviors well), the set of fuzzy rules must be as small as possible under the condition that the model performance is preserved at a satisfied level. A rule base including a large number of rules would lead to a lack of global understanding of the system. Within the concept of EFS, simplicity of rule bases can be forced by specific parameters in the training algorithms (distance thresholds etc.) — see Chapter 3. However, usually the accuracy suffers when doing so, such that it is again a matter of a tradeoff between simplicity and accuracy of the fuzzy models. Rule merging algorithms as proposed in Sections 6.2.2 and 5.1 help to improve the simplicity of the rule bases whenever too many rules have emerged over time (which usually turns out at a later stage in the single-pass incremental learning context).

Readability of single rule

To improve readability, the number of conditions in the premise part of a rule should not exceed the limit of 7 ± 2 distinct conditions, which is the number of conceptual entities a human being can efficiently handle [469] (same consideration as made for fuzzy partitions, see above). This finally means that the fuzzy systems should be restricted to include maximally 7 ± 2 input features. This can be accomplished with feature selection (or weighting) techniques, selecting (or assigning high weights to) the 7 to 9 most important features, which also obliges the reduction of the curse of dimensionality effect (which is often severe in fuzzy systems as they act as partial local approximators in the input feature space). However, due to the considerations presented in Section 6.2.5, this can be only reasonably carried out in an (initial) batch off-line phase or in re-training phases during on-line learning mode, which are usually not terminated in real-time. The concept of incremental feature weights changing smoothly over time (Section 5.2) can be of great help in this regard: features with low importance levels (low weights) could be simply ignored when presenting the fuzzy rules to the experts/users (linguistically or through a graphical user interface), hence achieving a reduction of the size of the premise parts of the rules.

Consistency

Rule base consistency means the absence of contradictory rules in rule base in the sense that rules with similar premise parts should have similar consequent parts [156] [110]. This is guaranteed when applying the rule merging approach (through fuzzy set merging) as demonstrated in Section 6.2.2: rules with similar premises are merged when they have similar consequents or the most significant rule is kept and the others deleted in case of dissimilar consequents.

Completeness

For any possible input vector, at least one rule should fire to prevent the fuzzy system from breaking inference [156]. This is only guaranteed when using membership functions with infinite support (all rules fire for each sample, at least to some small degree). However, in practice an input space partition with very low rule activation (below ε) may deteriorate fuzzy model interpretability. A good choice is to set up a tolerance threshold for rule activation during fuzzy sets and rule extraction from the data streams to prevent rule base from being activated at a very low level.

Transparency of rule structure

The criteria for high-level interpretability of fuzzy models evaluate the structure of fuzzy rules and their constitutions. A fuzzy rule should characterize human knowledge or system behaviors in a clear way [469]. Currently, most of the EFS approaches either use Takagi-Sugeno fuzzy systems or a neuro-fuzzy type architecture similar to TS fuzzy systems, both containing real functions (usually hyper-planes) in the consequent parts of the rules. A consequent variable expressed in terms of a real function does not exhibit a clear linguistic meaning [198], so it seems that Mamdani fuzzy systems (containing linguistic terms in the consequent parts of the rules) offer a more comprehensible way of characterizing system behaviors than the TS system. This could be a motivation for EFS researchers to investigate and develop incremental learning and evolving techniques for Mamdani fuzzy systems in order to account for more interpretability (ideally with similar accuracy as the evolving TS fuzzy systems have) — a first attempt in this direction is performed in [180] where MTS (Mamdani-Takagi-Sugeno) fuzzy models are used in the incremental/evolving learning concept, which are combining the explanatory trait of Mamdani-type fuzzy models with the output accuracy of Takagi-Sugeno fuzzy models. On the other hand, someone may argue that TS fuzzy rules characterize system behaviors in a different way, in which each rule represents a local linear model in nature (see subsequent section). Hence, TS fuzzy model interpretability should be studied from the perspective of the interaction between global model and its local linear models. This is a different view of interpretability (different to the Mamdani concept) and will be handled in the subsequent section.

6.3.1.3 Towards Interpretable Consequent Functions

Indeed, some researchers have proposed to achieve the interpretable Takagi-Sugeno local models in the sense of the following definition [200] [462]: the local linear models (hyper-planes) of a TS model are considered to be interpretable if they fit the global model well in their local regions, and result in fuzzy rule consequents that are local linearizations of the nonlinear system. According to this definition, interpretable partial hyper-planes should snuggle along the actual surface spanned up by the non-linear Takagi-Sugeno fuzzy system they correspond to; in this case, the hyper-planes are actually representing piecewise local linear approximations of the real trend of the variable relationship/dependency to be approximated. Such local approximations may serve as valuable insight into the models' or control behaviors in certain regions of the input/output space. A visualization example is presented in Figure 6.11 (a), where the consequent functions of the five extracted rules (their centers marked as dark dots) are shown as dotted lines. The two left most functions show an almost constant behavior, which can be directly deduced from the values of the linear parameters (weights) in these functions (as representing the partial gradients in the local parts of the input space); therefore, in this part of the input space, the model and its approximation tendency can be considered as nearly constant, opposed to the right most part of the input space, where the model shows a rather steep characteristics. For two- or three-dimensional approximation examples these effects can be directly observed from the model surface, in case of higher dimensional problems the gradients = weights from the hyper-planes are of great help for a better understanding of the model dynamics.

When using local learning approach, i.e. updating the consequent parameters with recursive fuzzily weighted least squares for each rule separately and independently (see Section 2.3.2), the snuggling along the surface of the approximation curve is guaranteed, see [281] [462], also verified in [265]. This is not the case

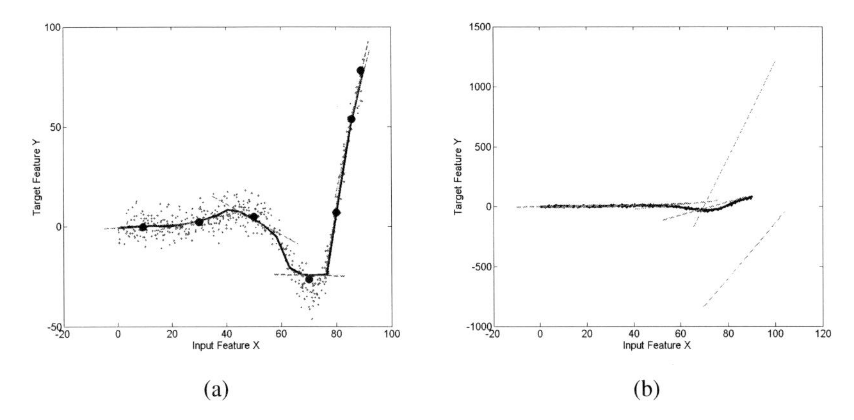

Fig. 6.11 (a): fuzzy model (solid line) and corresponding partial linear functions (dotted line) achieved when using local learning; (b): fuzzy model (solid line) and corresponding partial linear functions (dotted line) achieved when using global learning (note the 'break-out' of the consequent functions in the right hand side, not reliably representing the basic tendency of the functional dependency in this region)

when applying global learning approach, i.e. updating the consequent parameters with conventional recursive least squares (see Section 2.2.2). This is simply because the consequent parameters are trained for all rules together, hence not necessarily delivering piecewise local approximations of the local tendencies — see Figure 6.11 (b) for the resulting consequent functions when applying global learning in case of the approximation problem shown in (a): the functions fall completely out of the original range of the target variable and appear quite chaotic (hence the data samples and the approximation model are visually deformed). An alternative to local learning would be to constrain the recursive learning procedure in order to force rule weights summing up to 1 [128]. This has not been studied in the context of evolving fuzzy systems so far.

6.3.1.4 Interpretability Aspects of Inputs in EFS

Interpretability of inputs in EFS is an important point in order to understand the impact of input features or variables onto the fuzzy systems evolved from the data. This issue is supported by two aspects:

1. a feature importance list which ranks the features according to their importance levels for achieving a proper fuzzy model with high accuracy (most important first, then second most important etc.); incremental feature ranking and subset selection methods can be applied synchronously to the update of the fuzzy systems. However, in this way, the fuzzy systems, in fact the premise parts of the rules are not getting more transparent as features cannot be exchanged permanently during the incremental learning phase, see Section 6.2.5.

2. assigning feature weights (lying in $[0,1]$) as feature importance levels, which can be smoothly updated during incremental learning phase (refer to Section 5.2 for details about the methodologies behind this approach). This aspect brings not only more interpretability about the importance of the features in the evolved fuzzy system themselves, but also for the premise parts of the evolved rules, as unimportant features may be deleted when showing the rules to the experts and users (see previous section).

6.3.2 Visual Interpretability

Describing rule-based models linguistically is one way of representing them in a user-friendly way. An interesting alternative to a linguistic interpretation is a visual representation, i.e., the representation of a model in a graphical form. Visual interpretation aspects have already been studied for batch off-line trained fuzzy systems, see [365] [341] or [318] (NEFCLASS). In the context of EFS, this approach could be especially useful if models evolve quickly, since monitoring a visual representation might then be easier than following a frequently changing linguistic description. Besides, a graphical approach could be especially interesting to represent the changes of an evolving model. In fact, in many applications, not only the model itself is of interest, but also the changes it undergoes in the course of time. Such information may reveal, for example, important tendencies, which could in turn be used for triggering an early warning in a fault detection system. The idea of discovering interesting patterns and regularities in time-dependent data and dynamically changing environments is a topical issue in contemporary data mining research. Hence, it would be an important challenge to investigate data mining methods to discover patterns and trends in evolving fuzzy models. To do so, it will probably be necessary to represent the evolution of a fuzzy model in a suitable way, for example as a trajectory in a suitable parameter space.

6.4 Reliability Aspects in EFS

Another prerequisite for the user acceptance is the reliability of a model. By this, we do not only mean that a model makes accurate predictions. Instead, it is often even more important to offer, apart from the prediction itself, information about how reliable this prediction is, i.e. how certain the model itself is in its prediction. This would serve as the basis of an enhanced interpretation capability of the models, as more or less reliable regions in the feature space can be detected as such. Ideally, a learning algorithm is self-aware in the sense of knowing what it knows and what it does not.

In case of classification scenarios, reliability of classification statements can be expressed by the concepts conflict and ignorance taking into account the class distribution in various local areas of the feature space [185]: higher conflict and ignorance levels indicate regions where the classification statements are less reliable. In case of regression, non-linear system identification and time-series forecasting problems, reliability of the target responses/prediction outputs can be modelled by the concept of adaptive local error bars, incrementally calculated, synchronously to the evolving fuzzy systems: wider local error bars indicate regions where the prediction outputs are less reliable. As both issues are not only an additional basis for a better model interpretation, but also essential parts of enhancing the process safety of evolving fuzzy systems (unreliable predictions should be treated carefully or even eliminated and not processed further), we decided to handle these issues in Chapter 4 (Section 4.5).

Part III
Applications

Chapter 7
Online System Identification and Prediction

Abstract. This chapter is the first one in the sequence of four chapters demonstrating real-world applications of evolving fuzzy systems. Starting with a generic on-line system identification strategy in multi-channel measurement systems (Section 7.1), it deals with concrete on-line identification and prediction scenarios in different industrial processes, namely:

- System identification at engine test benches (Section 7.1.2)
- Prediction of NOx emissions for engines (Section 7.2)
- Prediction of resistance values at rolling mills (Section 7.3)

As such, it should clearly underline the usefulness of on-line modelling with evolving fuzzy systems in practical usage (industrial systems). This is completed by two classical non-linear system identification problems, for which several EFS approaches are compared. Also, there will be a comparison of evolving fuzzy systems 1.) with off-line trained fuzzy models kept static during the whole on-line process and 2.) with an alternative re-training of fuzzy systems by batch modelling approaches during on-line mode. One major result of this comparison will be that re-training is not fast enough in order to cope with real-time demands. Another one is that off-line trained models from some pre-collected data can be significantly improved in terms of accuracy when adapting and evolving them further during on-line mode. In case of predicting NOx emissions, EFS could outperform physical-oriented models in terms of predictive accuracy in model outputs.

E. Lughofer: Evolving Fuzzy Systems – Method., Adv. Conc. and Appl., STUDFUZZ 266, pp. 295–323.
springerlink.com

7.1 On-Line System Identification in Multi-channel Measurement Systems

7.1.1 The Basic Concept

With *online identification* it is meant to identify (high-dimensional, changing, time-varying) system behaviors and dependencies in an on-line setting.

Opposed to off-line identification, where all the collected historic measurement data is sent as so-called training data into the model building algorithms, in online (system) identification measurements are recorded either block-wise or sample-wise with a certain frequency demanding several training steps in order to be up-to-date as soon as possible and hence always including the latest system states. In principle, the online training steps can also be carried out by batch learning algorithms, but depending on the dynamics of the system the computation time can suffer in a way that the application of these methods gets unacceptable as not terminating in real-time or within an appropriate time frame — see also Section 7.1.2 where this claim is underlined based on on-line identification tasks at engine test benches.

The basic idea of an all-coverage system identification framework in a multi-channel measurement system is to model relationships between measurement channels in form of approximation resp. regression models. Assuming that P such channels are recorded in parallel, the goal is to find dependencies between these channels in form of multiple input single output regression models by taking each of the P channels as target variable and a subset of remaining ones as input variables and perform the modelling process. In the static case using stationary data samples for model building (which are usually obtained when averaging dynamic data samples over a fixed time frame) a model for a specific measurement channel x_i is then defined in the following way:

$$\hat{x}_i = \hat{y} = f_k(x_{j_1},...,x_{j_p}), \qquad j_1,...,j_p \in \{1,2,...i-1,i+1,...,P\} \qquad p<P \quad (7.1)$$

with $x_{j_1},...,x_{j_p}$ a subset of p input variables, x_i is also called target variable. The time subscript k indicates that the model was trained based on the first k samples, i.e. in on-line mode f_k represents the model at time instance k. In case of a dynamic model, the time delay of the input channels has to be taken into account, such that a dynamic model for the measurement channel x_i is defined by:

$$\begin{aligned} \hat{x}_i(k) &= \hat{y}(k) = f_k(x_{j_1}(k),x_{j_1}(k-1),...,x_{j_1}(k-l_{max}),..., \\ & x_{j_p}(k),x_{j_p}(k-1),...,x_{j_p}(k-l_{max})) \\ & j_1,...,j_p \in \{1,2,...i-1,i+1,...,P\} \qquad p<P \end{aligned} \quad (7.2)$$

where l_{max} denotes the maximal essential time delay. Finding such models can be important to gain a better understanding of the system (for instance by inspecting which channels have which relationships) during on-line monitoring tasks. Furthermore, such models may also be applied for plausibility check and quality control purposes in order to detect system failures or faulty production items in an early stage — Chapter 8 deals with this topic.

In case of a high number of recorded measurement channels, i.e. P large, it is especially necessary to reduce the dimensionality of the identification problem, i.e. to keep p small (usually $p \ll P$), especially when data is rare because of the curse of dimensionality problems and also due to computational reasons. In this case, we recommend applying a variable selection (for regression) step before the real training of the fuzzy models. This can be either accomplished in an initial (off-line) setup phase (where some data are pre-generated during a simulation phase) or with the first dozens/hundreds of on-line measurement samples. In the former case, it is also feasible to tune the learning parameters within a best parameter grid search process (as e.g. also part of the algorithms in the *FLEXFIS* family, see Section 3.1). A comprehensive survey of variable selection methods for regression is demonstrated in [309]. For engine test benches (see subsequent section) good results with a modified version of forward selection are reported in [153], where regressors are successively added by a dimension(variable)-wise correlation criterion based on R^2 and their joint contribution subtracted from the target variable. This method is quite fast and applicable during on-line mode.

Furthermore, an important point is to estimate and update statistical measures which give rise about the actual model qualities and which are interpretable (i.e. lying in a fixed pre-defined range) and comparable for different models including different target channels. The reason for this step is that some channels in an industrial system are completely independent from each other, hence no useful dependencies (=accurate models) can be extracted for some channels when being selected as target. These models would cause unreliable predictions and therefore could guide the whole industrial process into a wrong direction. Hence, models with bad qualities have to be identified and omitted for further on-line usage. The techniques presented in Section 2.1.4 for estimating and updating the *r-squared-adjusted* quality measure in the case when using Takagi-Sugeno fuzzy systems are a promising option accomplishing this task. This measure lies in $[0,1]$ and indicates bad models when its value is low (near 0.5) and good models when its value is high (near 1).

An alternative to updating model qualities is to apply adaptive local error bars as demonstrated in Section 4.5.1 to estimate the uncertainties in the regression models. The error bars will be automatically wide when there is no useful relationship between a selected set of input channels and the actual target channel and automatically narrow when the opposite is the case. An example for both cases in case of a two-dimensional relationship is visualized in Figure 7.1 for two channel-pairs recorded at engine test benches: (a) shows a pretty accurate correlation between two measurement channels, where (b) demonstrates two channels where no useful model can be built as the data shows now clear tendency over the two-dimensional input/output (x-axis/y-axis) space (as nearly equally distributed inside its range). In the latter case, also the error bars are wide open, automatically indicating an unreliable model over the whole space. Therefore, by providing the bandwidth of the

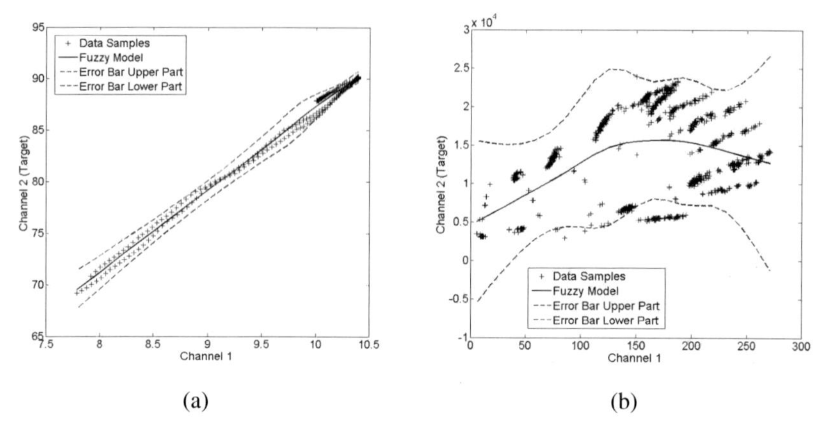

Fig. 7.1 (a): A clear (linear) dependency between two channels, the error bars are surrounding the model quite tightly indicating a reliable model, note that in the right most region the error bars are getting narrower as there more samples are collected than in other parts; (b): two channels having no functional relationship (data samples almost equally distributed), the error bars are surrounding the models quite widely, indicating an unreliable model

error bar together with the actual prediction to the user serves as a good indicator as to whether the prediction is reliable or not. The advantage of this strategy over estimating model qualities is that, instead of omitting low quality models based on a global quality measures, some of these models can still be used whenever they are accurate in some parts of the input/output space. This is because local error bars have the characteristics to change locally according to the reliability of the models in different regions of the input space. On the other hand, the disadvantage of using local error bars is that they provide absolute (and hence un-interpretable) values rather than a clear tendency of the overall global quality (as r-squared-adjusted does by providing values in $[0,1]$).

Combining these aspects together, leads us to the generic on-line identification work-flow as shown in Figure 7.2. The dotted lines (surrounding the boxes or as connectors between boxes) indicate optional functionality:

- The fuzzy models (and its qualities resp. uncertainty regions) may be initially generated based on pre-collected data — whether an initial phase is required or desired depends mostly on whether the experts and operators at the measurement systems blindly trust the purely automatic model building and adaptation from scratch. A fully incrementally built up fuzzy model from scratch is quite often too risky for the experts as it does not allow sufficient insight into the process models as it permanently changes its structure and parameters from the beginning. An initial model with a fixed structure can be first supervised by an expert and upon agreement further evolved, refined etc. to improve its accuracy. Based on our experience with real-world on-line applications, this is a widely-accepted procedure. Furthermore, an initial phase allows an extensive optimization of

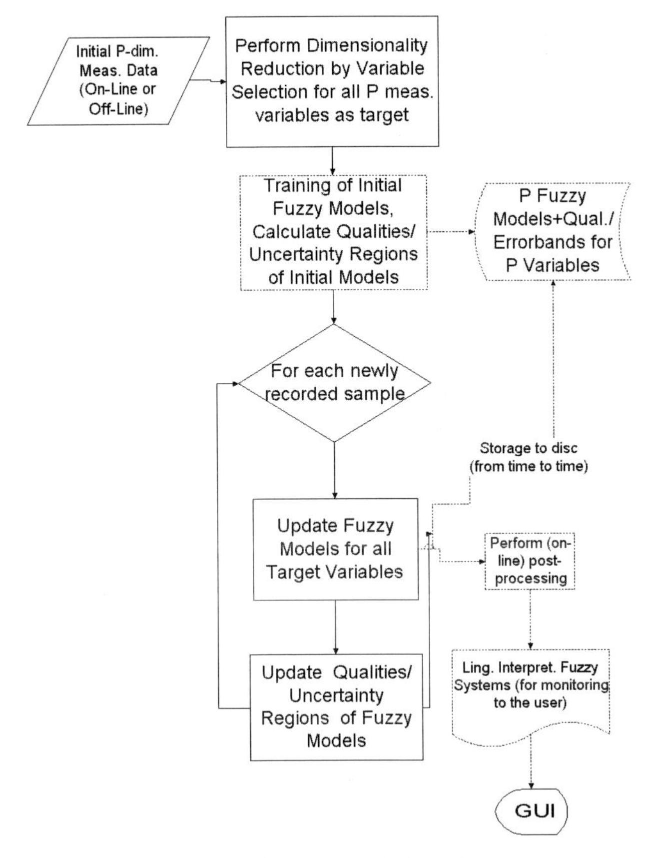

Fig. 7.2 On-line identification work-flow in multi-channel measurement systems (with P variables measured); the dotted lines (surrounding the boxes or as connectors between boxes) indicate optional functionality

parameters used during the on-line learning procedure — see also Section 3.1.1.2 (Paragraph 'Discussion on Incremental Learning from Scratch') for a comprehensive discussion on this problematic.

- The fuzzy models may be improved regarding interpretable power during on-line mode (with the help of (some of) the methodologies discussed in Chapter 6). This may serve as a valuable additional component for an on-line monitoring GUI system in order to get a better understanding of 'what is going on in the system' resp. 'which knowledge can someone extract from the system dependencies modelled by EFS' (for instance by showing most active fuzzy sets, rules and the output behavior in local parts).
- Storage of the evolved models to the hard disc is also an alternative option which can be used. This could be important, when fuzzy models are intended to be updated at a later stage. Then, the stored fuzzy models (together with their qualities and uncertainty regions) can be read in again and further evolved based on new

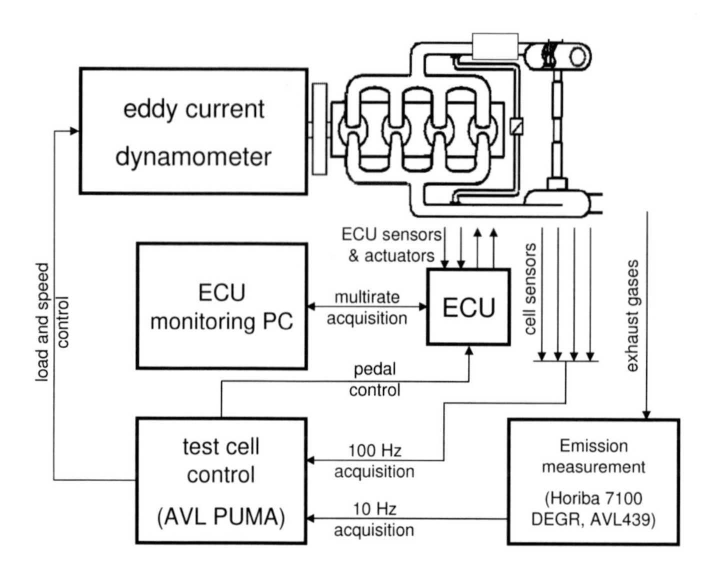

Fig. 7.3 A typical lay-out of an engine test bench (taken from [288])

test phases on the same or also on a similar object with the inclusion of a gradual forgetting of older relations (see Section 7.1.3). Ideally, storage should be performed at the end of an on-line identification run, as storing back the models when updated after each single measurement would usually cause significant computation time.

7.1.2 On-Line Identification at Engine Test Benches

The basic concept demonstrated in the preliminary section is applied to a concrete multi-channel measurement system installed at engine test benches. A visualization of the conventional components of such a system is visualized in Figure 7.3. In this system, up to 1000 different channels may be recorded and sampled in parallel, both in dynamic mode (yielding dynamic measurements) where around each millisecond a new value is sampled and in steady-state mode (yielding static data) averaging the dynamic measurements over 30 seconds. For the latter, one stationary measurement is achieved per minute, and it is necessary to wait 30 seconds before averaging over another 30 seconds in order to allow the system to move from a transient phase to a steady state phase (transient phase data should be not used in the system identification step). One stationary measurement denotes the 'summary' of the system behavior in one operation mode, which is one knot point in a larger grid laid over the two main channels adjustable in the user's frontend panel and controlling the operation modes: rotation speed and torque. An example of such an engine map is shown in Figure 7.4, where the rotation speed is varied from 1200 to 4300 in steps of 50 and the torque varied from -20 to a maximal value in steps of 20 which is different for

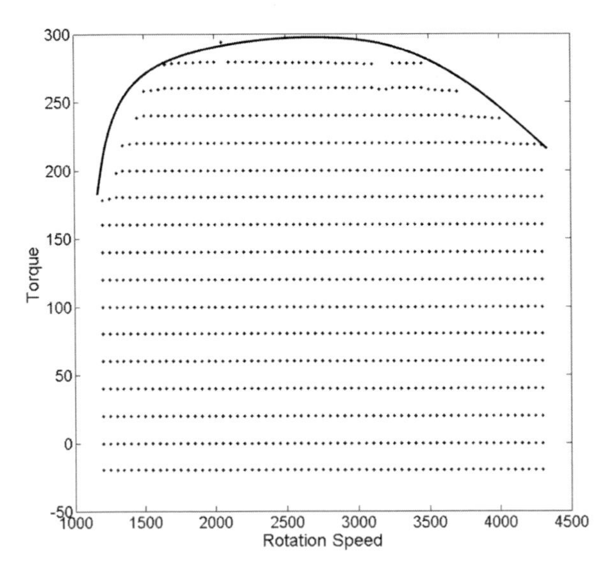

Fig. 7.4 A typical engine characterized with many grid points over rotation speed (x-axis) and torque (y-axis)

different rotation speed values. Usually, the bounding curve across different rotation speed values is a reverse parabola with maximum value at medium rotation speed. Similar operation modes (similar values of rotation speed and torque) are expected to form one operating condition, i.e. one specific region resp. cluster in the sample space of some channel combinations.

Now, our experimental setup for testing the performance and behavior of evolving fuzzy systems for on-line identification and prediction at such a system, was the following:

- For a large diesel engine 1810 stationary measurements were recorded (courtesy of AVL List GmbH), where the knot points in the discrete grid over the engine map were visited in random order; hence, it can be expected that, after having some data available, knot points near the corners of the grid are already included, such that no severe extrapolation situations occur later on.
- For a smaller petrol engine 960 stationary measurements were recorded, where the knot points in the discrete grid over the engine map were visited in ascending order, i.e starting from a minimal rotation speed of 1000revs/min ranging to a maximal one of 5600revs/min and starting from a minimal torque of -20 to a maximal one of 180.
- The most important 80 measurement channels were recorded in the case of the larger diesel engine, for the smaller petrol engine the most important 32 variables.
- In both cases, from the first 200 measurements the most important variables were selected for approximating each measurement channel as target.

- For on-line updating of the fuzzy models, the evolving fuzzy systems approach *FLEXFIS* was applied.
- Additionally on-line adaptation of consequent parameters only (without rule evolution) with recursive fuzzily weighted least squares (for local learning) was applied in order to compare with the impact of rule evolution based on new online samples (also following the comparison made in Section 3.1.1.5 for two-dimensional data).
- Furthermore, batch fuzzy modelling methods such as *FMCLUST* [28], *ANFIS* [194] and *genfis2* [454], *FLEXFIS batch*, were applied in a re-training phase during on-line identification mode. This serves as verification whether a re-training with batch modelling methods is reliable with respect to computation times and model qualities.
- Model qualities in form of an average r-squared-adjusted over the r-squared-adjusted from reliable models, were calculated and updated (see Section 2.1.4). Reliable models are those having a higher r-squared-adjusted value than 0.5.
- As run-time environment for model training and adaptation, a specific software developed by AVL List GmbH, Graz was used (called MAGIC)[1]. In this tool, it was possible to dynamically configure an arbitrary sequence containing various model building algorithms such as statistical models, data-driven fuzzy systems, neural networks and hybrid models (data-driven refinement of parameters in physical formulas) and to load the data either as a whole batch or in single samples into the program. Depending on this buffer-size of the data, batch training (in case when the buffer-size exceed the size of the whole data matrix) or incremental training (with corresponding block-sizes, single samples) were automatically triggered.

From the 80 measurement channels in the larger diesel engine, finally 64 reliable models were obtained (models with r-squared-adjusted higher than 0.5). The average of the r-squared-adjusted values over these models is shown in the second column of Table 7.1 and denote the final model qualities at the end of the whole on-line identification process, calculated on the training data (the 1810 samples which were also used for model update and evolution) — note that r-squared-adjusted already punishes more complex models and avoids some bias towards preferring over-fitted models. The third column and probably the most interesting one shows the model qualities on a separate test data set including 250 measurements. From this, it can be recognized that *FLEXFIS batch* results in the highest generalization performance on new data samples (highest model qualities). The average model quality of 0.904 is a result by a re-training step on the complete data set at the end of the on-line operation phase. *FLEXFIS* as evolving fuzzy training method adapts, evolves, extends the models by only seeing one instance at a time and still reaches a quality of 0.856 in case of sample-wise update (Row # 7) and 0.881 in case of block-wise update (Row # 6) with each block containing 100 samples. This means, that *FLEXFIS* is able to come to the hypothetical batch solution quite closely as indicated and discussed in Section 3.1.1.4. Figure 7.5 shows the tendency of the model qualities with different

[1] https://www.avl.com/avl-magic-automated-data-processing

Table 7.1 Comparison of fuzzy model identification methods with respect to model qualities and computation speed for larger diesel engine at engine test bench, data recorded in shuffled order according to the engine map

Method	Quality Training	Quality Test	Comp. Time online 1 up-to-date every 100 p.	Comp. Time online 2 up-to-date each point
FMCLUST	0.9272	0.902	62m 41s	Not possible
ANFIS	0.9110	0.872	>genfis2	Not possible
genfis2	0.9080	0.893	38m 31s	Not possible
FLEXFIS batch	0.9110	0.904	34m 13s	Not Possible
FLEXFIS conv. adapt ONLY	0.8319	0.818	3m 10s	3m 10s
FLEXFIS full (inc + rule evol.) batch mode 100	0.8712	0.881	4m 36s	Not possible
FLEXFIS full (inc + rule evol.) sample mode	0.8411	0.856	10m 57s	10m 57s

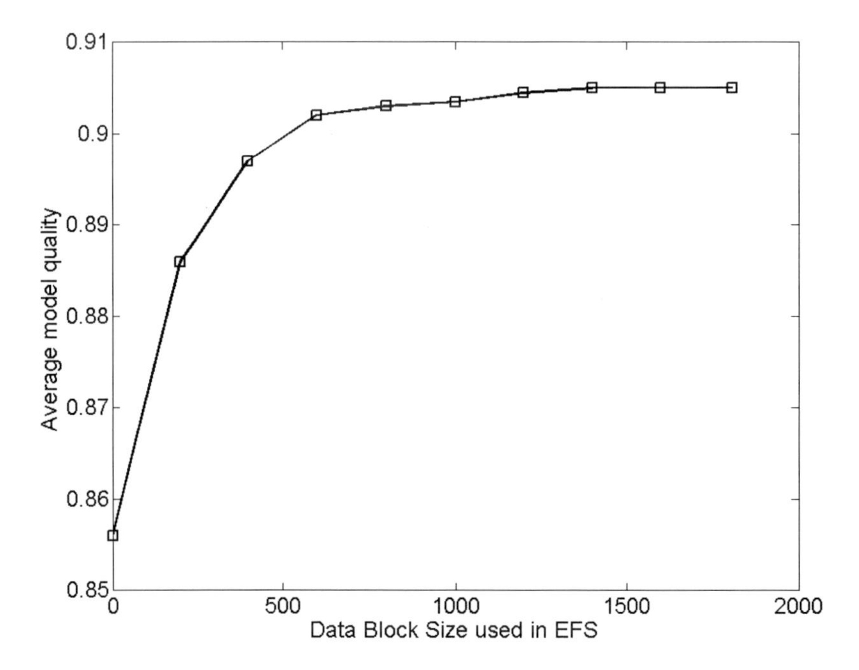

Fig. 7.5 Dependency between block size and average model quality when training with *FLEXFIS*, the rectangles denote the tested block sizes 1, 200, 400, 600, 800, 1000, 1200, 1400, 1600 and 1810 (all samples)

block sizes containing different number of samples. Starting with singe samples as block size (left side) and a model quality of 0.856, the larger the block-sizes get the closer the quality of the batch training case (using all 1810 samples) can be approximated. This is finally an empirical verification of the convergence of *FLEXFIS* to its batch version (obtaining optimal consequent parameters and having the full range available for rule evolution in *eVQ*).

A key reason for underlining the necessity of an evolving fuzzy modelling approach during on-line measuring and identification is presented in Columns #4 and #5. Obviously, when trying to re-train all models with each incoming new sample as indicated in Column #5 (in order to permanently include new or changing operating conditions into the models), the on-line identification process takes several hours or days (hence listed as 'not possible'). Re-training after each data block containing 100 samples is possible, but requires significantly more time than when permanently updating the models with EFS. In fact, *FLEXFIS batch* is the fastest of the batch re-training approaches, but still takes in sum 34 minutes and 13 seconds for the whole training process (18 re-training steps on 1810 samples). This means that at the end of the process one re-training step of all models takes around 115 seconds on average which is not fast enough for real-time demands (as each 60 seconds a new stationary point is measured, see above). This finally means, that the block size has to be further increased in order to cope with on-line real-time demands in case of re-training methods. On the other hand, this further decreases the flexibility of the models to include new system states or operating conditions as fast as possible in order to give reliable predictions in these new states. Compared with the marginal increase in predictive quality over EFS when using re-training steps, EFS seems to be a more reasonable choice.

Another interesting point is the fact that the conventional adaptation of consequent parameters alone (without rule evolution) delivers models whose qualities are not far behind *FLEXFIS* with sample-wise adaptation (0.819 versus 0.856). The reason for this effect lies in the specific measurement plan, i.e. in the shuffled order of knot points over the engine map. This means that the first 100, 200 samples already cover the whole possible range of the measurement variables pretty well and also equally distributed, such that hardly any extrapolation situations occurred and therefore new rules covering these new ranges were hardly ever demanded. This situation changes when tests at the engine test bench are performed based on some other order within the engine map. Is is quite usual to perform the tests and data recordings by starting with the lowest rotation speed, varying torque from the minimal to the maximal value, and successively increasing the rotation speed. In this case, the operating conditions are added 'line per line' (compare the engine map shown in Figure 7.4). Hence, it is quite intuitive that dynamically adding of fuzzy rules is indispensable in order to expand the models to new regions on the engine map. To underline this necessity, conventional adaptation of consequent parameters alone and *FLEXFIS* were applied to the small petrol engine data (including 960 on-line measurements and 32 channels) recorded in the ascending order scheme as mentioned above. The results are presented in Table 7.2. Here, obviously conventional adaptation completely fails, as producing model with much lower qualities as

Table 7.2 Comparison of fuzzy model identification methods with respect to model qualities and computation speed for smaller petrol at engine test bench, data recorded in ascending order according to the engine map

Method	**Quality Training**	**Quality Test**	**Comp. Time**
FLEXFIS with conv. adapt ONLY	0.806	0.679	37 seconds
FLEXFIS full (inc + rule evol.) batch mode 100	0.969	0.963	51 seconds
FLEXFIS full (inc + rule evol.) sample mode	0.958	0.951	131 seconds

in case when using *FLEXFIS* applied to data blocks with 100 and single samples (0.679 versus 0.963/0.951). The conclusion therefore is that, as in the case of an arbitrary online measurement process the distribution of the first k data samples over the input space is usually not known a-priori, the most feasible choice with is to use evolving fuzzy systems. Indeed, in some cases the measurement plan can be elicited in a way that a good initial distribution of the operating channels is ensured, but that does not also automatically ensure that all other measurement channels are affected by this strategy. So, extrapolation on newly recorded data can never be completely avoided.

Figure 7.6 presents the fuzzy partitions obtained for boost pressure (in (a) and (c)) and for engine speed (in (b) and (d)), when evolving a fuzzy systems based for the fill factor of the supercharger as target and using boost pressure, engine speed as well as air-mass flow as input channels. The evolved fuzzy systems achieved a correlation coefficient between predicted and measured output on separate test data set of about 0.97. The top row in Figure 7.6 shows the fuzzy partition directly extracted, evolved from the data according to its distribution in the input/output space, the bottom row shows the partitions improved with respect to complexity and interpretability by using redundancy detection and merging techniques as demonstrated in Chapter 6. These techniques could achieve much better interpretable fuzzy partitions, by losing only a very low fraction in terms of predictive accuracy (0.0124 versus 0.0120 mean absolute error, whereas the range of the fill factor was about 2.4, so the error was below 1%).

7.1.3 Auto-Adaptation of Models to a Similar Test Object

In the previous section, so far we have presented results when identifying system models on-line for one concrete engine. Now, an interesting issue has arisen as to whether it is possible to automatically adapt models generated for one engine (and stored on the hard-disc after the whole on-line experimental/test phase) to channel dependencies triggered by another engine (to be tested in the engine test bed), i.e. to use models elicited for one engine as start configuration for further on-line adaptation when testing a new engine. The motivation of this request is to reduce the requirements on the number of samples for training reliable models and therefore to

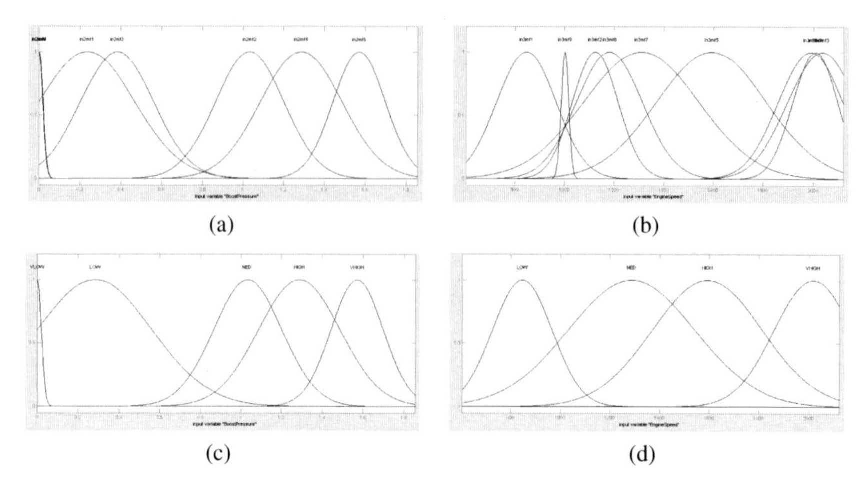

(a) (b)

(c) (d)

Fig. 7.6 Upper row: fuzzy partitions for boost pressure and engine speed without using any interpretability improvement strategies; lower row: fuzzy partitions obtained when applying redundancy detection and merging approaches as demonstrated in Chapter 6

make models for a new engine applicable much earlier (e.g. for quality control purposes) than when re-trained completely from scratch. Similarity is important and means that the expert knows that in principle the already trained models are also somehow relevant for the new engine; the whole idea would not work in the case of two completely different engines showing different behaviors (e.g. a petrol and a diesel engine).

In practice, it turned out that for online identification the number of required samples k could be reduced from around 200, when no model from a similar engine were available, to around 50 in the case of already trained models. These 50 measurements are needed in order to estimate the new ranges of input and output channels occurring in the models and to adapt parameters and structures in the models. A key point for an appropriate adaptation of the models from one engine to the other was the inclusion of a gradual forgetting strategy (as demonstrated in Section 4.2). This was necessary, as otherwise the models from the older engine are usually 'too heavy' (i.e. supported by a large number of measurements) and therefore not adjustable to the new situation quickly enough. A dynamic forgetting factor was used, changing its intensity over time, as the newly learned relationships for the current engine should be not forgotten if they are time-invariant in the outputs. A possible strategy for accomplishing this can be achieved by applying the following formula to the forgetting factor at time instant $m+k$:

$$\lambda_{k+init} = 1 - \frac{1-\lambda_{init}}{k} \tag{7.3}$$

where λ_{init} is the initial forgetting factor for time instance $k=1$. Hence, the difference of 1 to λ is decreased over time (increasing k) converging to a forgetting factor of 1 (=no forgetting).

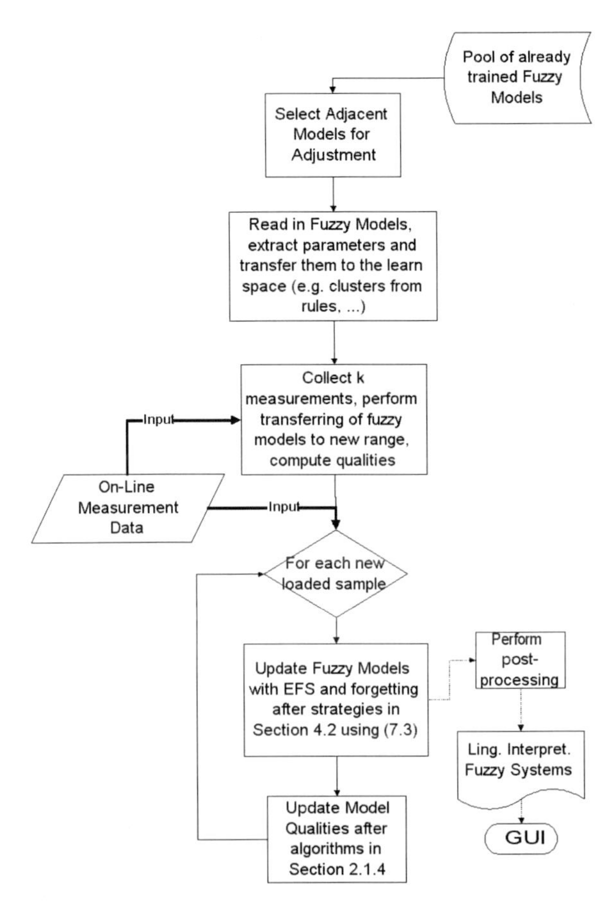

Fig. 7.7 Auto-Adaptation strategy by applying evolving fuzzy models with gradual forgetting after strategies as presented in Section 4.2 using (7.3) for the forgetting factor and updating the qualities of the models by using algorithms from Section 2.1.4; the post-processing component again optional

The auto-adaptation work-flow is shown in Figure 7.7.

The block "Select Adjacent Models for Adjustment" demands expert knowledge about both, the current engine and the preliminary one, for which the models were trained before, so the operator selects those models which can be applied for both engines — in some cases (also depending on the similarity of the engines), he can simply select all the available ones.

The block "Read in Fuzzy Models, Extract Parameters to Form Antecedent and ..." performs the identification of the parameters in the antecedent and consequent parts of the rules. In case of using a cluster-based technique for learning the rules' antecedent parts first a back-projection of the parameters in the fuzzy sets and the rule structure to the high-dimensional cluster space is carried out in order to form the cluster partition. This can be accomplished easily as

- Amount of clusters = amount of rules
- Each rule's premise form one cluster where
- Each premise part corresponds to a fuzzy set whose center can be assigned to the corresponding entry (dimension) in cluster center vector
- The widths of the fuzzy sets can also be extracted and identified with the axis length of the cluster ellipsoids.

The initial setting of the inverse Hessian matrix and of the consequent parameters for the *recursive fuzzily weighted least squares* procedure is also taken from the last setting in the models for the older engine.

In the block "Collect k Measurements, ..." k measurements are collected not only for the initial range estimation of all the read in fuzzy models, but also for the initial computation of model qualities. The initial range estimation is indispensable, as usually for the new engine the ranges of the input channels in the fuzzy models are different (e.g. different rotation speeds for sport cars and trucks). After the range estimation, the following steps are performed for each model:

- Shift centers of all the fuzzy sets to the new range by applying

$$\forall i,j : c_{ij}(new) = \left(\frac{c_{ij}(old) - min_j(old)}{range_j(old)}\right) range_j(new) + min_j(new) \tag{7.4}$$

  where i iterates over all rules and j over all dimensions in the actual fuzzy model, and $min_j(old)$ denotes the minimum of the corresponding input channel estimated from the old engine and $range_j(old)$ the complete range. Both can be extracted easily from the originally trained fuzzy models.
- Adjust widths of fuzzy sets due to the new range (similar to shift of centers) by applying

$$\forall i,j : \sigma_{ij}(new) = \left(\frac{\sigma_{ij}(old)}{range_j(old)}\right) range_j(new) \tag{7.5}$$

- Perform adaptation of the rules consequent parameters by applying *RWLS* with forgetting factor as in (7.3). The forgetting factor ensures a shift in the output as already shown in Figure 4.8 and therefore a movement and re-shaping in the y-direction.

7.1.4 Two Further Examples in Non-linear Dynamic System Identification

The previous two subsection basically dealt with on-line identification scenarios for stationary data samples, i.e. measurement data were averaged over a fixed time horizon (mostly 30 seconds) to deliver one data sample. In this sense, the time delays of the variables for prediction were not really relevant, therefore we were dealing with time-independent models as defined in (7.1). In this section, we provide two application examples for on-line non-linear dynamic system identification, i.e. the time delays of input variables do have an impact on the target, and inspect and compare the performance of several different EFS approaches (all of these discussed in Chapter 3). This yields dynamic models in the form (7.2); of course, the maximal possible time delay depends on the idleness of the whole system: changing rotation speed in cars will have an much more immediate influence on the emission intensity than the influence on the water level when changing the throughput at a water power plant.

The first non-linear dynamic system to be identified is described by Wang and Yen in [440]:

$$y(n+1) = \frac{y(n) * y(n-1) * (y(n) - 0.5)}{1 + y(n)^2 + y(n-1)^2} + u(n) \tag{7.6}$$

where $u(n)$ is uniformly selected in the range $[-1.5, 1.5]$ and the test input is given by $u(n) = \sin(\frac{2\pi n}{25})$, $y(0) = 0$ and $y(1) = 0$. For the incremental and evolving training procedure, 5000 samples were created starting with $y(0) = 0$ and further 200 test samples were created for eliciting the RMSE (root mean squared error) on these samples as reliable estimator for the generalized prediction error. Table 7.3 shows the results and underlines the strength of all EFS variants with respect to accuracy (compare RMSE with the range of $[-1.5, 1.5]$). Some methods could achieve this quite fine performance by keeping the number of rules quite low. Parameter optimization was carried out by manual tuning in all approaches.

Table 7.3 Comparison of the EFS approaches *FLEXFIS*, *eTS*, *Simp_eTS*, *SAFIS*, *MRAN* and *RANEKF* based on a non-linear dynamic system identification problem [268]

Method	RMSE on Test Samp.	# of rules/neurons
eTS	0.212	49
Simp_eTS	0.0225	22
SAFIS	0.0221	17
MRAN	0.0271	22
RANEKF	0.0297	35
FLEXFIS	0.0176	5

The second non-linear dynamic system identification example is the famous MackeyGlass chaotic time series problem, widely used as a benchmark example in the areas of neural networks, fuzzy systems and hybrid systems and described in detail in [294]. This is given by:

$$\frac{dx(t)}{dt} = \frac{0.2x(t-\tau)}{1 + x^{10}(t-\tau)} - 0.1x(t) \tag{7.7}$$

where $x(0) = 1.2$, $\tau = 17$ and $x(t) = 0$ for $t < 0$. The equation displays a dynamical behavior including limit cycle oscillations, with a variety of wave forms, and apparently aperiodic or "chaotic" solution. The task is to predict $x(t+85)$ from the input vectors $[x(t-18)x(t-12)x(t-6)x(t)]$ for any value of t. To do so, 3000 training samples were collected for t in the interval $[201, 3200]$. 500 test samples for $t \in [5001, 5500]$ were collected in order to elicit the NDEI (Non-Dimensional Error Index) on unseen samples, which is the root mean squared error divided by the standard deviation of the target series.

Table 7.4 Comparison of *FLEXFIS*, *DENFIS*, *eTS* and *eTS+* based on chaotic time series data [10]

Method	NDEI on Test Samples	# of rules/neurons
DENFIS	0.404	27
eTS	0.373	9
exTS	0.320	12
eTS+	0.438	8
FLEXFIS	0.206	69

7.2 Prediction of NOx Emissions

7.2.1 Motivation

NOx (nitrogen oxides) is one of the more important pollutants in compression ignited engines. NOx formation is mostly due to the oxidation of the atmospheric nitrogen during the combustion process at high local temperatures. There exists legal guidelines on the upper bound of allowed NOx concentrations absorbed by engines during run-time. Hence, during development of new engines, this upper bound must not be exceeded, however approached as close as possible in order to maximize the efficiency of the engine. One possibility to supervise the NOx content within the full gas emission of an engine, is to measure it with a sensor at an engine test bench (as done for other channels, such as pressures, temperatures etc, see preliminary section). Although this is the ideal means of measuring because it is the only way that fully addresses the diagnosis function, the technology in order to be able to produce low cost, precise and drift-free sensors, however, is still under development and depends on the considered pollutant [312]. Hence, NOx emission models are of great interest, which are able to predict the NOx content based on some other variables in the systems and are included in the engine control system and the on-board diagnostic system.

Currently, two basic approaches for setting up emission models are being pursued by the experts [24]:

- A direct mapping of the pollutant emitted by a reference engine as a function of rotation speed and torque can be used. This method, usually implemented as a series of look-up tables, is straightforward because it has exactly the same structure as many other maps already available in the ECU, and hence calibration engineers can easily calibrate them.

- A physical-based model developed by engine experts (based on the knowledge of the inner physics), based on some engine operating parameters continuously registered by the ECU can be used.

Direct engine maps are usually unable to compensate for production variations and variations in the operating conditions (e.g., warming-up of the engine, altitude, external temperature, etc.) in the engine during the vehicle lifetime. Hence, they are usually not flexible enough to predict the NOx content with sufficient accuracy. Physical-based models compensate this weakness of direct engine maps by including a deeper knowledge of experts about the emission behavior of an engine. Usually, a large set of differential equations are set up, requiring a deep study and a high development effort in terms of expert man power. Furthermore, complex differential equations require a huge computational power to predict the NOx contents during on-line operation modes.

The evolving fuzzy modelling approach tries to find a compromise between a physical-oriented and a pure mapping approach by extracting automatically high-dimensional non-linear fuzzy models from steady-state as well as transient measurements recorded during the test phases of an engine. These measurements reflect the emission behavior of the corresponding engine and hence provide a representation of the intrinsic relations between some physical measurement channels (such as temperatures, pressures, rotation speed, torque etc.) and the NOX content in the emission. The methodology of a machine-learning driven building up of fuzzy models is able to recognize this relation and hence to map input values (from a subset of measurement channels) onto the NOX content (used as target) appropriately and with high precision. A major advantage of such a data-driven model extraction over a physical-oriented model is that it requires a very low manpower for developing one model for a specific engine. The complete model validation and final training phase can be performed completely automatically (without any parameter tuning phases as optimal parameters are elicited over a pre-defined grid), just the data needs to be recorded before hand. Considering, that for each new engine an engine developer has to develop an accurate physical NOX model (almost) from scratch, one can imagine the benefit of such a 'plug-and-play' modelling method in terms of manpower and money. Another advantage is that one single model for steady and transient states cannot be obtained by a physical approach, and the application of one model (steady or dynamic) for the two possible states is usually risky as typically extrapolation situations may occur. Finally, the evolving modelling approach provides the possibility of model adaptation and refinement during on-line operation mode. That could be used for further improving the models during on demand, for instance the inclusion of new operating conditions or system states, not present in the original measurements. This also helps to reduce the effort for measurement recordings and data collection during the initial off-line experiment phase.

7.2.2 Experimental Setup

For verification purposes, the following data sets were collected during on-line operation phase (engine tests and engine test bench):

- Steady-state data including 363 measurements: tests to produce this data ranging from full load to idle operation, and different repetitions varying EGR rate (i.e. oxygen concentration in the intake gas), boost pressure, air charge temperature and coolant temperature were done. Test procedure for each one of the steady tests was as follows:

  1. Operation point is fixed, and stability of the signals is checked. This last issue is specially critical because of the slow thermal transients in the engine operation.
  2. Data is acquired during 30 s.
  3. Data is averaged for the full test.
  4. Data is checked to detect errors, which are corrected when possible.

- A dynamic data including the 42 independent tests delivering 217550 measurements in sum: this data set was down-sampled to 21755 measurements by taking every 10th sample from the original data matrix. 16936 of these down-sampled measurements were used for final training of the fuzzy models, the remaining ones for testing its generalization performance. The dynamic data set was obtained by transient tests which covered European MVEG homologation cycle and several driving conditions. These include the engine speed and torque profiles during an MVEG cycle (top left plot), a sportive driving profile in a mountain road (top right plot) and two different synthetic profiles (bottom plots). Several repetitions of the last two tests were done varying EGR (exhaust gas recirculation) and VGT (variable geometric turbocharger) control references, in a way that EGR rate and boost pressure are varied from one test to another. During dynamical operation the engine reaches states that are not reachable in steady operation. For example, during a cold start the engine coolant temperature is always lower than the nominal coolant temperature, which needs several minutes to be reached. That means that transient tests are needed for the system excitation, since not all system states can be tested in steady tests, nor in the full operation range. In Figure 7.8 boost and exhaust pressures are represented for the steady tests and for a dynamical driving cycle, note that the range of the variation during the transient operation clearly exceeds that of the steady operation. Furthermore, steady tests do not show the dynamical (i.e. temporal) effects.
- Mixed data which appends the steady-state data to the dynamic measurements to form one data set where the fuzzy models are trained from. This is a specific novelty in our approach where one unique model is generated including dynamic and static data. In the physical approach two models are required which have to be switched automatically during on-line engine operation mode (which is not always possible).

The input set consisted of nine essential input channels used in the physical-oriented model (which were selected as a result of a sensitivity analysis with a higher order physical-based model) which were extended by a set of additional 30 measurement channels, used as intermediate variables in the physical model (such as EGR rate, intake manifold oxygen concentration, etc.). For the dynamic and mixed data

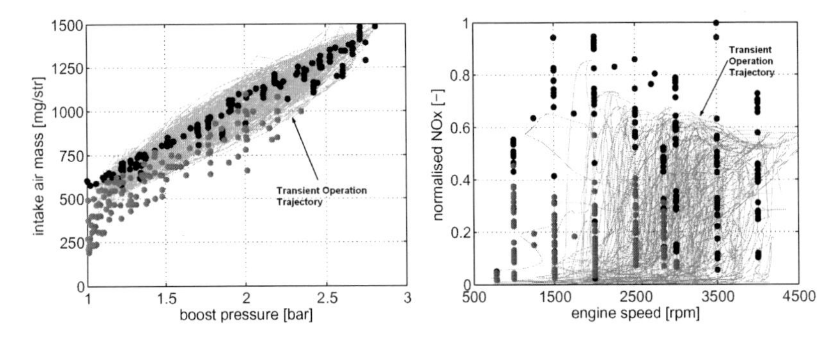

Fig. 7.8 Comparison of the range of several operating variables during the steady tests without EGR (black points), those with EGR (grey points) and during the transient test (light grey line).

set all of these were delayed up to 10 samples. This together with the down-sampling rate of 10 finally means that models are affected by samples up to 10 seconds ago.

The measurements are split into two data sets, one for model evaluation and final training and one for model testing (final validation). Model evaluation is performed within a 10-fold cross-validation procedure [411] coupled with a best parameter grid search scenario. For the latter a parameter grid for our fuzzy modelling component is defined consisting of two dimensions. The first dimension iterates over the number of inputs in order to reduce the curse of dimensionality (if present). The second dimension iterates over the sensitive parameter(s) in EFS (e.g. in case of *FLEXFIS* this vigilance parameter is taken from $0.1\frac{\sqrt{p}}{\sqrt{2}}$ to $0.9\frac{\sqrt{p}}{\sqrt{2}}$ with p the dimensionality of the input/output space).

Once the 10-fold cross-validation is finished for all the defined grid points (in the two-dimensional space over the number of inputs and the vigilance parameter), that parameter setting is elicited for which the CV error, measured in terms of the mean absolute error between predicted and real measured target (NOx), is minimal; this is also called optimal parameter setting as minimizing the expected error on new samples (note that the CV error is a good approximation of the real expected prediction error, according to [174]). For this optimal parameter setting, a final model is generated using all training data and perform an evaluation on the separate test data set.

As evolving fuzzy modelling method, *FLEXFIS* (in batch mode) together with the heuristic regularization (as demonstrated in Section 4.1) and rule base procrastination for omitting outliers (described in Section 4.4) was applied and compared with physical-based emission models with respect to predictive power on separate test data sets. Evolving here refers more to the automatic elicitation of a reasonable cluster = rule partition based on all the measurement samples at hand rather than updating the models during on-line mode. In fact, implicitly in the (off-line) CV procedures (i.e. in the training run for each fold), the training of fuzzy models was done fully incrementally and in evolving manner. A visualization of this concept is

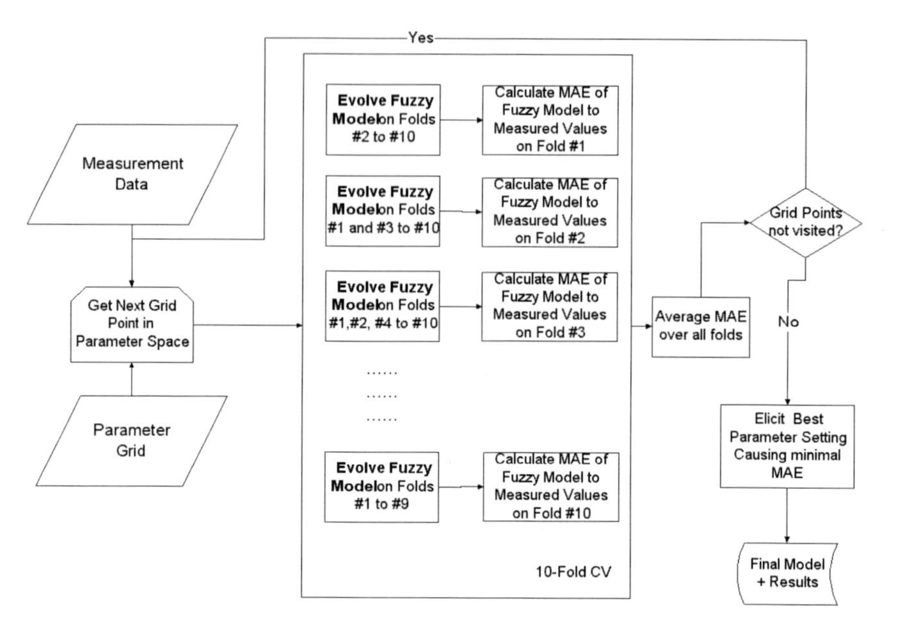

Fig. 7.9 Evolving fuzzy modelling approach applied within a 10-fold cross-validation and best parameter grid search scenario

given in Figure 7.9. As the data was not shuffled and hence each fold represents (a snapshot of) the natural on-line order of the data samples, the results presented below can also be seen as representative for performance verification of on-line trained fuzzy models.

7.2.3 *Some Results*

Figure 7.10 visualizes the results obtained for the static data set. Figures included show the correlation between predicted and measured values (left plot), the absolute error over all samples (middle plot) and the histogram of the errors normalized to the unit interval (right plot). From the left plot, it is easy to realize that the samples are concentrated around the first median, which indicates a model with reasonable predictive accuracy: the closer the distances of these samples to the first median are, the more accurate the model. This is underlined in the right plot, where a major portion of the mean absolute errors (MAE) are lying around 0, indicating that for most of the measurements a high predictive quality can be achieved. For comparison purposes, the results obtained with the analytical physical oriented model on the same data set are shown in Figure 7.11. From this, it can be realized that the error performance is slightly worse as in case of evolving fuzzy modelling (see also Table 7.5 below for concrete numbers: the normalized MAE is around 20% worse in case of physical models). A clear improvement of the fuzzy modelling approach over the physical-based model can be realized when comparing the two right most

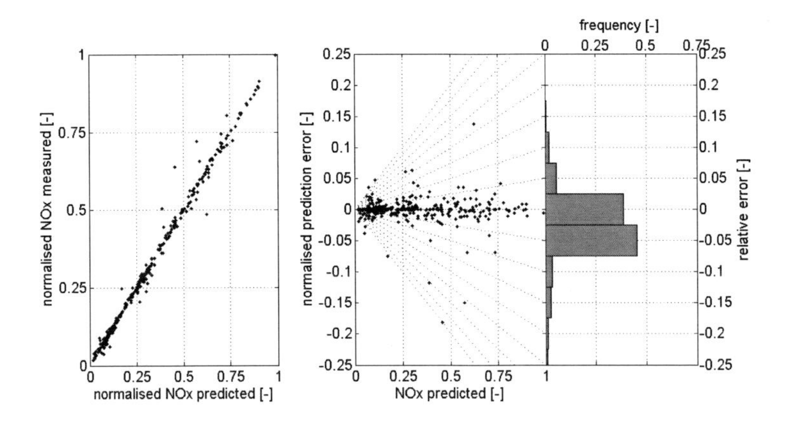

Fig. 7.10 Prediction results from machine-learning based NOx emission model obtained by evolving fuzzy modelling when using static data set; left: correlation between predicted and measured values, middle: absolute error over all samples, right: distribution of the normalized error

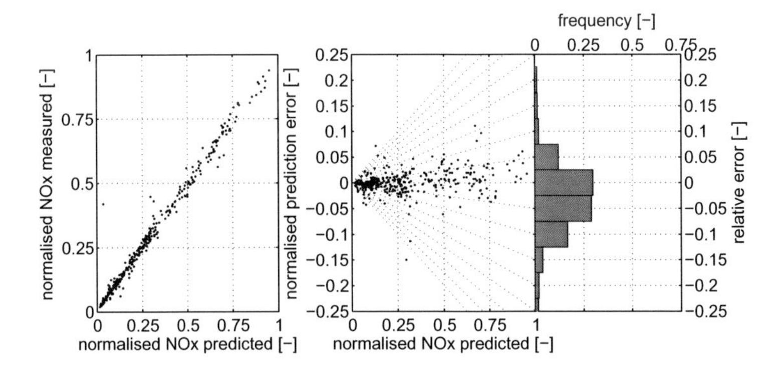

Fig. 7.11 Physical-based model results when applied to steady tests.

plots in Figures 7.10 and 7.11: significantly more samples are distributed around 0 error, also the number of samples causing an absolute deviation of 0.05 (so samples causing 0.05 or -0.05 error) is higher when using the evolving fuzzy modelling component.

For dynamic data, having a time dependent component, it is interesting to see how the predicted NOX content behaves over time compared to the real measured value. Figure 7.12 plots the results for four of the eight dynamic test data sets. Obviously, the EFS model is able to follow the highly fluctuating trend of the measured NOX content during dynamic system states quite well (compare lines in dark = measured values with light = predicted value).

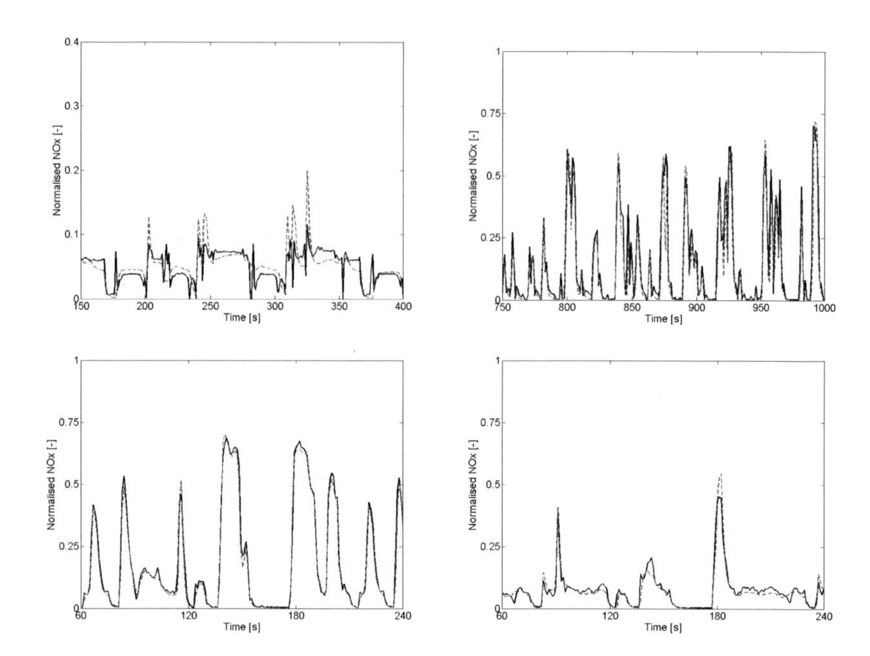

Fig. 7.12 Measured (dark line) versus predicted (light line) NOx content for four portions of dynamic test data using EFS

One major problem in the physical modelling approach is that a separate model for static and dynamic data has to be applied when predicting new samples. Checking whether a steady state or a dynamic measurement often requires a too time intensive feedback loop in the online system, so that it would be nice to have one joint model at hand. In fact, someone may simply use the dynamic model for static measurements or vice versa. This option was verified in [288]. The conclusion could be drawn that this procedure is not reliable in the sense that any reasonable models could be produced: MAEs were very high compared to the static and dynamic models, also see Table 7.5 below. Hence, it is a big challenge to have one single model for transient and steady states available. This can be accomplished by appending the static data set at the end of the dynamic data matrix (including shifts of the variables according to their time delays), by copying the same (static) value of the variables to all of their time delays applied in the dynamic data matrix. When doing so, similar model accuracies could be achieved as when building and testing models separately for static and dynamic states — see Table 7.5, which reports the normalized MAEs for all modelling variants: note the small errors when training a fuzzy model on the mixed data set and using static and dynamic test data sets, compared to the large errors when training fuzzy models on dynamic resp. static data alone and then testing these models with static resp. dynamic data. There, also a comparison with other data-driven (batch) modelling methods is performed: EFS can clearly outperform these (best results reported in bold font), which is a

Table 7.5 Comparison of the prediction error (normalized MAE) of our fuzzy modelling component on various data sets, with physical model and with other data-driven modelling techniques (second part of table)

Method	MAE Static	MAE Dyn.	MAE Mixed / Static	MAE Mixed / Dyn.
EFS	**1.32**	**2.04**	**1.61**	**2.46**
Physical	1.57	2.23	NA	NA
Ridge Regr.	2.91	3.04	5.76	2.76
SVR	3.61	3.44	4.94	4.61
ANFIS	2.37	3.26	4.04	4.74
NN	1.49	2.65	7.06	3.49

remarkable performance as these are seeing all the data at once, while EFS always builds up the models in incremental fashion (seeing one sample at a time).

7.2.4 Further NOx Prediction Results

Another dynamic data set containing 6700 samples and including NOx measurements was recorded for another engine at another engine test bench. At this system, the following measurement channels were elicited to be the most important ones (after down-sampling by a factor of 10 and inclusion of time delays):

Te = Engine Output Torque
P2offset = Pressure in Cylinder number 2
N = Engine Speed
Nd = Speed of the Dynanometer
Alpha = Accelerator Pedal
Tgas = Exhaust Temperature
P1offset = Pressure in Cylinder number 1
COL = CO value

Using only the first four measurement channels as inputs were sufficient for producing models with high qualities on separate test data (greater than 0.9 r-squared), see also [268] for further details; the remaining ones did not really improve the qualities further.

Model complexity is an essential point in this application, as the out-coming model should replace the sensor for NOX completely in future (also called soft sensor). For this purpose, the model has to be interpretable and understandable for the experts, as they have to verify if the model is reliable and save, so that no incorrect output feedback from the model will appear. Figure 7.13 visualizes the extracted fuzzy sets when using the five input channels $\mathrm{N}(k-4), \mathrm{P2offset}(k-5), \mathrm{Te}(k-5), \mathrm{Nd}(k-6), \mathrm{N}(k-6)$. These transparent partitions (fuzzy sets quite evenly distributed over the input space) could be obtained by applying the complexity reduction methods for deleting redundancies in the fuzzy sets as described in Chapter 6 after the complete training process.

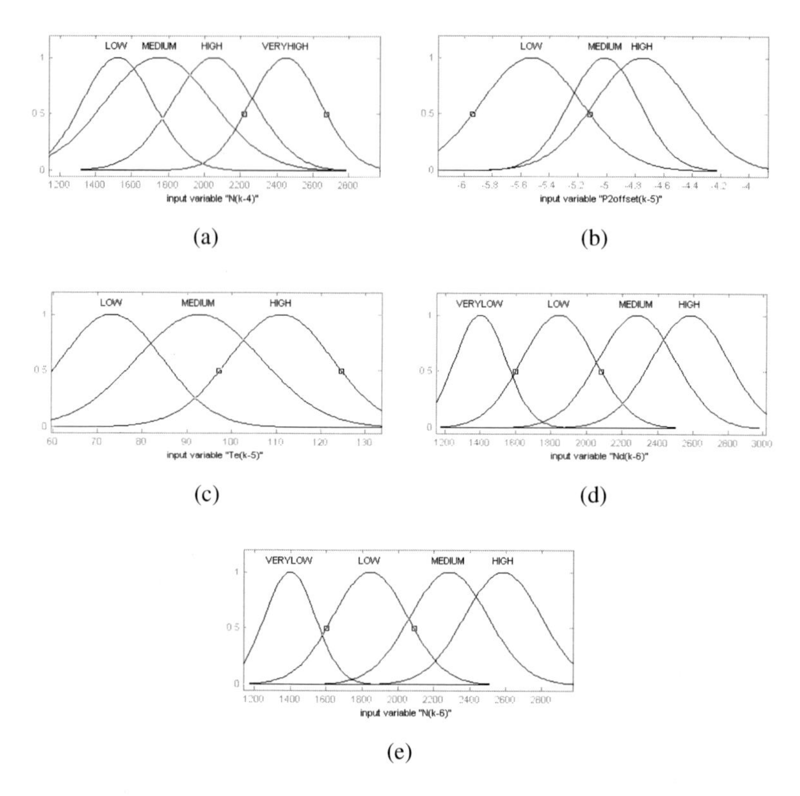

Fig. 7.13 Transparent fuzzy sets for the four input channels N(k-4), P2offset(k-5) and Te(k-5), Nd(k-6) and N(k-6) of a fuzzy prediction model for emission channel NOX (taken from [268])

Furthermore, EFS (using *FLEXFIS*) produced a quite acceptable performance of seven rules, which are listed in Figure 7.14. From these fuzzy sets and rules someone can now obtain the insight, that P2offset(k-5) and Te(k-5) have a significant impact according to their range (seen from Figure 7.13) on the rule consequents for rules #1 to #5, but almost no influence for rule #6 and a medium influence for rule #7. This could be important information for a fault isolation approach (as e.g. [115]) taking into account influence of variables in partial local regions of the input space. This is a valid observation here, as in *FLEXFIS* the local approach is applied, which learns the linear consequent parameters of each rule separately and independently by including the weights of the rule fulfillment degrees in the various data samples. This means, only data samples around the center of a rule affect the learning mechanism of this rule significantly and data samples lying further away have almost no influence. Consequently, the linear consequent hyper-plane of one rule in fact represents a kind of linear regression model through the data samples close to the rule centers. In this sense, a linear consequent hyper-plane represents the local trend of the model

Rule 1: If $N(k-4)$ is HIGH and $P2off(k-5)$ is LOW and $Te(k-5)$ is HIGH and $Nd(k-6)$ is LOW and $N(k-6)$ is LOW
Then $NOX(k) = 0.04 * N(k-4) - 61.83 * P2off(k-5) + 1.54 * Te(k-5) + 0.48 * Nd(k-6) - 0.52 * N(k-6) - 251.5$
Rule 2: If $N(k-4)$ is HIGH and $P2offset(k-5)$ is MED and $Te(k-5)$ is MED and $Nd(k-6)$ is MED and $N(k-6)$ is MED
Then $NOX(k) = 0.04 * N(k-4) - 53.36 * P2off(k-5) + 1.2 * Te(k-5) + 0.35 * Nd(k-6) - 0.37 * N(k-6) - 207.3$
Rule 3: If $N(k-4)$ is LOW and $P2offset(k-5)$ is LOW and $Te(k-5)$ is MED and $Nd(k-6)$ is LOW and $N(k-6)$ is LOW
Then $NOX(k) = 0.02 * N(k-4) - 38.02 * P2off(k-5) + 1.59 * Te(k-5) + 0.42 * Nd(k-6) - 0.54 * N(k-6) + 40.55$
Rule 4: If $N(k-4)$ is HIGH and $P2offset(k-5)$ is LOW and $Te(k-5)$ is HIGH and $Nd(k-6)$ is VLOW and $N(k-6)$ is VLOW
Then $NOX(k) = 0.04 * N(k-4) - 36.86 * P2off(k-5) + 1.77 * Te(k-5) + 0.14 * Nd(k-6) - 0.13 * N(k-6) - 274.4$
Rule 5: If $N(k-4)$ is VHIGH and $P2offset(k-5)$ is HIGH and $Te(k-5)$ is HIGH and $Nd(k-6)$ is MED and $N(k-6)$ is MED
Then $NOX(k) = 0.05 * N(k-4) - 5.3 * P2off(k-5) + 0.87 * Te(k-5) + 0.07 * Nd(k-6) - 0.08 * N(k-6) - 0.23$
Rule 6: If $N(k-4)$ is HIGH and $P2offset(k-5)$ is HIGH and $Te(k-5)$ is HIGH and $Nd(k-6)$ is LOW and $N(k-6)$ is LOW
Then $NOX(k) = -0.02 * N(k-4) - 0.0004 * P2off(k-5) + 0.008 * Te(k-5)$
$+ 0.09 * Nd(k-6) + 0.09 * N(k-6) + 0.0001$
Rule 7: If $N(k-4)$ is MED and $P2offset(k-5)$ is HIGH and $Te(k-5)$ is LOW and $Nd(k-6)$ is HIGH and $N(k-6)$ is HIGH
Then $NOX(k) = 0.02 * N(k-4) + 0.26 * P2off(k-5) + 0.99 * Te(k-5) + 0.29 * Nd(k-6) - 0.25 * N(k-6) + 0.004$

Fig. 7.14 7 rules obtained with *FLEXFIS* for the NOX data [268]

around the center of the corresponding rule (cluster) — see also Chapter 6 for a verification of this aspect.

7.3 Prediction of Resistance Values at Rolling Mills

7.3.1 Motivation

Cold rolling is a metalworking process in which metal is deformed by passing it through rollers at a temperature below its recrystallization temperature. Cold rolling increases the yield strength and hardness of a metal by introducing defects into the metal's crystal structure. These defects prevent further slip and can reduce the grain size of the metal, resulting in Hall-Petch hardening. Cold rolling is most often used to decrease the thickness of plate and sheet metal. A schematic picture of the principal process is visualized in Figure 7.15. One important task is to predict the yield strength of a steel plate in order to guarantee a smooth rolling process, i.e finally a steel plate with the intended a priori defined thickness. The predictions have to be carried out during the on-line rolling process in order to compensate deviations from intended values quickly.

Before the idea of a data-driven identification for setting up prediction models had been created, the situation at the company (VOEST Alpine Stahl GmbH) was as follows: An analytical physical-based model was already available, motivated by material laws which go back to Spittel and Hensel [176]: in modified form the resistance value can be estimated by an exponential function of the temperature, speed and thickness of the steel plate (by taking the logarithm and estimated the parameters through a conventional least squares regression approach). However, the basic problem of this physical model was that it included only static behavior, so there was no inclusion of the history of certain system parameters in order to be able to follow time-dynamic system behaviors. In this sense, the accuracy of the models was

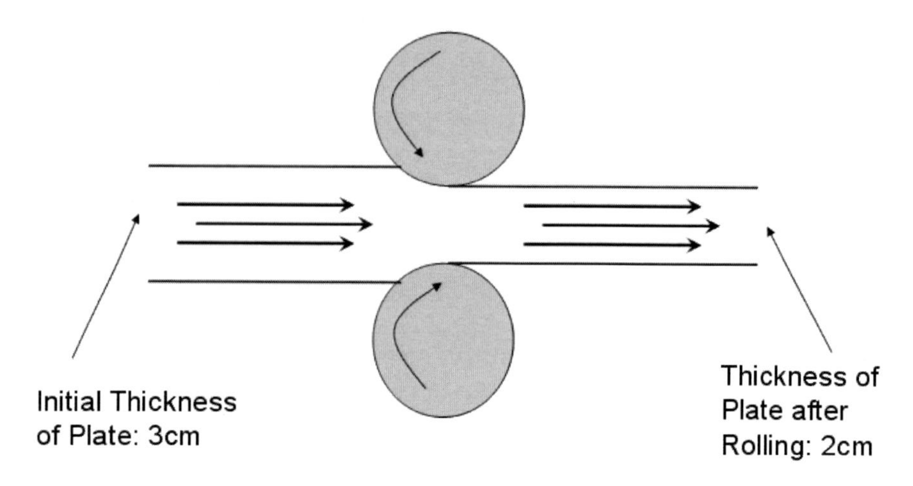

Fig. 7.15 Typical rolling mill process

not always satisfactory. Due to the high physical complexity of the whole rolling mill process, it was not possible to deduce time-variant differential equations to improve the old model by Spittel and Hensel. Therefore, it was an important challenge trying to improve the model with a data-driven approach, i.e. prediction models extracted from measurement data recorded in the past for different types of steel. An additional demand was to refine the model permanently with on-line measurements.

7.3.2 *Experimental Setup*

For this purpose the evolving fuzzy modelling component *FLEXFIS* was used, which can be applied for building up an off-line trained models which is further refined, evolved during the on-line process. Approximately 6000 training data points and approximately 6600 on-line data samples including 13 system variables (12 inputs, 1 output) such as thickness of the steel plate before and after the rolling mill, average temperature, grain size and others, were collected during on-line production. Time delays up to 10 were added for each input variable, leading to 121 input channels. The maximal delay was set to 10, as from the experience of the operators it is known that the yield strength is affected by influence values recorded maximally 8 to 10 passes ago (where one pass delivers one measurement). A 10-fold cross-validation procedure [411] with the training data was conducted for different parameter settings, i.e. varying the number of rules in the fuzzy model and varying the input dimension from 1 to 20, whereas the most important inputs were selected by using a modified version of forward selection [153]. It turned out that the maximal delay in any of the selected input variables was 8. The best performing fuzzy models obtained from the cross-validation step (Algorithm 5 Steps 1 and 2), were compared on the independent on-line data with the analytical model in terms of mean absolute error, the maximal errors from the upper and the lower side (estimated value too high or too low) and the number of values estimated 20 units too low than the measured one (it was important to keep this error very low; generally, higher estimations were not such a big problem as lower ones).

Based on the best-performing parameter setting (elicited during the initial off-line modelling step), the fuzzy model was further adapted and evolved with the 6600 on-line data samples, where for each sample first a prediction was given, compared with real measured value for updating the MAE error and then adapted with the real measured value; this was done because of mainly two purposes:

- In order to verify whether an on-line adaptation improved the accuracy of the model — note that this step is in fact possible as at the system first a prediction for the resistance is given, influencing the whole process at the rolling mill, whereas a few seconds later (after the steel plate is passed), the real value for the resistance is measured, which can then be incorporated into the model adaptation process.
- In order to be able to gradually forget older samples for reacting onto drifts which may come up during the on-line rolling process. A justification of this point is that the operation process at rolling mills is divided into different "stitches". One stitch represents one closed cycle in the rolling process. During the on-line mode, the measurements come in continuously from stitch to stitch. However, for the current processed stitch, the previous stitch should play only little or even no role. However, the measurements from the previous stitch are already included in the fuzzy models as updated by their samples. Thus, this means that older samples from the previous stitch should be forgotten when including samples from the current stitch. Another aspect is that here no drift/shift detection is required, as the drift/shift is indicated by the beginning of a new stitch. This start signal is transferred to the computer and was also included in the stored measurement data.

7.3.3 Some Results

The results are demonstrated in Table 7.6. The rows represent the different methods applied, whereas the row 'Static fuzzy models' denotes the fuzzy models obtained in the off-line CV step and not further updated with the separate on-line data set (so kept fixed during the whole on-line prediction process), opposed to Row 'Evolving fuzzy models' where the fuzzy models are further evolved with *FLEXFIS* approach. The columns represent the different validation measures. From this table, it can be realized that the results could be significantly improved when using adaptation/evolution of the models based on new measurements in parallel to the predictions instead of keeping them static. In sum, four additional rules were evolved as shown in Figure 7.16: based on 21 rules at the start after initial training, 4 rules were further evolved. It is also remarkable, that the static fuzzy models could outperform the physical model in off-line mode. The last row represents the case, when no regularization is included when training the static fuzzy models: the accuracy

Table 7.6 Comparison of evolving fuzzy prediction models obtained by EFS (using *FLEXFIS* as learning engine) with an analytical and a static fuzzy model for the resistance value at rolling mills, also with the inclusion of drift reaction

Method	**MAE**	**Max Error Too High / Max Error Too Low / # Errors > 20 Too Low**
Physical	7.84	63.47 / 87.37 / 259
Static fuzzy models	6.76	45.05 / 81.65 / 176
Evolving fuzzy models	5.41	38.05 / 78.88 / 159
Evolving fuzzy models with forget.	4.65	31.99 / 74 / 68
Evolving fuzzy models no reg.	9.88	85.23 / 151.21 / 501

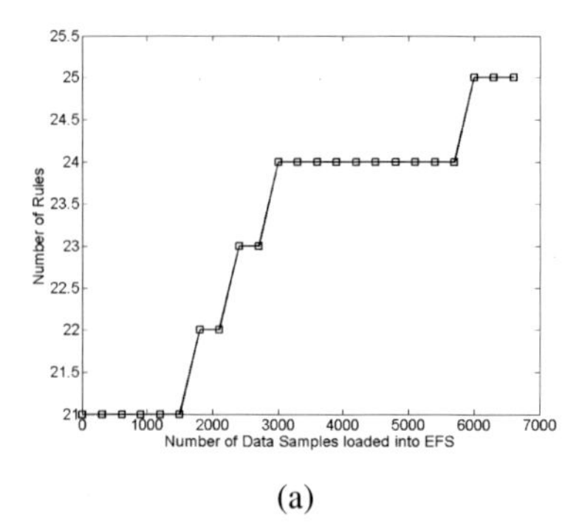

(a)

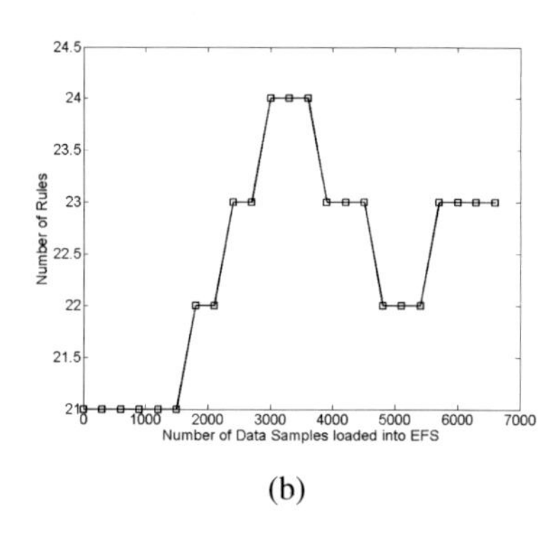

(b)

Fig. 7.16 (a): evolution of the number of rules over the number of loaded data samples into EFS; (b) the same as in (a), but applying rule merging in the cluster space whenever clusters are moving together (as described in Section 5.1), hence decreasing the final complexity of the evolved fuzzy model

suffers drastically, which proves that regularization is necessary to achieve any kind of reasonable results.

Another essential row in Table 7.6 is the last but one which demonstrates the impact of the gradual forgetting when triggered at each new stitch — compare with the evolving fuzzy models where no forgetting takes place (4th row): the number of predictions for which the errors were too high could be reduced by about over one half and the maximal error lowered to almost 31.99 which was quite beneficial. Another interesting aspect is that the error on the single measurements starts to drift over time when gradual forgetting is not applied. This is underlined in the left image of Figure 7.17 which shows the single errors over the 6600 on-line samples: note the drift of the main error area away from the zero line at the end of the data stream. The right hand side plot demonstrates the single errors when applying gradual forgetting, the *drift* occurrence at the end of the plot could be eliminated.

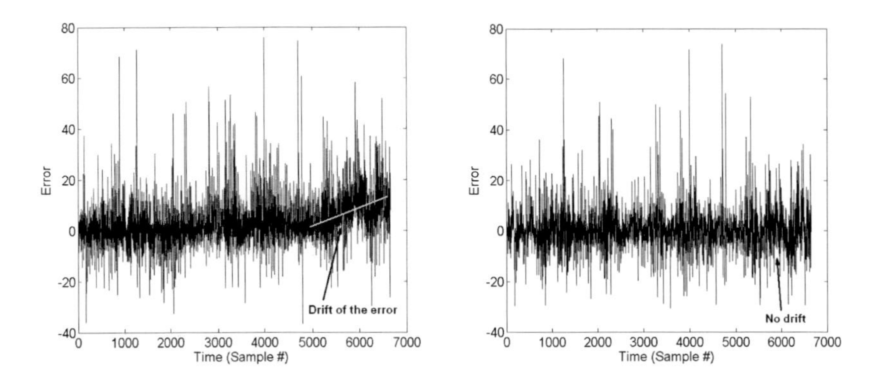

Fig. 7.17 Left: The error curve for the 6600 on-line samples when no forgetting is applied: at the end the error starts drifting away from the zero error line as the body area around zero increases to a value around 5 to 10; right: no drift in case of gradual forgetting as applied at the beginning of each stitch.

Chapter 8
On-Line Fault and Anomaly Detection

Abstract. This chapter deals with on-line quality control systems where measurements and (audio) signals are processed and examined whether they show untypical occurrences, significantly deviation from the normal operation process. The first part of this chapter (Section 8.1) is a natural successor of the previous one, as it deals with the application of on-line identified models at multi-channel measurement systems for early detection of failures in industrial systems. Therefore, a novel on-line fault detection strategy coupled with the usage of evolving fuzzy systems will be presented and its performance demonstrated based on on-line measurements from engine test benches. Section 8.2 deals with the detection of any types of untypical occurrences, also called anomalies (not necessarily faults, but also transient phases), in time series data. Univariate adaptive modelling strategies are applied whose (univariate) responses are used by an integration framework in order to obtain a final statement about the current state and behavior of the production process. Section 8.3 concludes with an application from the signal processing area, dealing with audio signals coming from analogue tapes which should be digitized onto hard disc. There, the detection of a broad band of noises is a central aspect in order to estimate the quality of a tape and to communicate to an operator whether a feasible digitization is possible or not.

E. Lughofer: Evolving Fuzzy Systems – Method., Adv. Conc. and Appl., STUDFUZZ 266, pp. 325–355.
springerlink.com

8.1 On-Line Fault Detection in Multi-channel Measurement Systems

8.1.1 Motivation

During the last two decades a significant growth of the size and complexity of the technological installations in the automotive, power, chemical, and food industries has been observed [82]. A side-effect of this growth is the increase in the concentration of measuring, processing, and control devices. The likelihood of appearance of a fault that may lead to a breakdown of a component or the whole system increases with the complexity of the system [78]. Faults can affect the system itself, the measuring and monitoring devices, or the control system (which also modifies the system behavior). The IFAC Technical Committee SAFEPROCESS has defined a fault formally in the following way (also referenced in [189] and [82]):

> A *fault* is an unpermitted deviation of at least one characteristic property or variable of the system from acceptable/usual/standard behavior.

Since faults affect system performance and product quality, and they can also risk the system integrity and become a safety risk for human beings (for instance a broken pipe where gas effuses), fault detection (FD) algorithms are necessary which allow an automated detection on incipient faults, coping with a huge amount of (measurement) variables and high-frequented dynamic data. Indeed, humans are able to classify sensor signals by inspecting by-passing data, but these classifications are very time-consuming and also have deficiencies because of underlying vague expert knowledge consisting of low-dimensional mostly linguistic relationships. Industry requirements for the application of FD approaches include high rate of detection of faults, while keeping a low level of false alarms. In addition, parametrization effort must be kept as low as possible, in order to shorten start-up delay and cost.

Hence, many automatic fault detection approaches have been developed during the last decade (see [229], [82] and [142] for comprehensive surveys), which can be loosely divided into:

1. Pattern recognition and classification: techniques such as principal component analysis (PCA), Fisher discriminant analysis (FDA), partial least squares [82] or any other type of classification models [111] are exploited in order to directly train patterns of faulty and fault-free cases from supervised (measurement) data. New incoming data can then be discriminated into faulty versus non-faulty points by processing them through the inference of the classification models.
2. Structured (statistical) hypothesis testing [42], where two hypotheses testing [326] or multiple hypotheses testing [34] directly based on the (measured) data samples can be carried out. In the two hypotheses testing approach the two hypotheses can be formulated as 'some fault mode can explain the measured data' and 'no fault mode can explain the measured data'.

3. Signal analysis: signal processing and filtering techniques are used, for example, in so called intelligent sensors [414] to detect sensor faults such as peaks, (mean) drifts [278] or other anomalies in both time and frequency domains of the measurement signals [38] (an alternative approach in this direction will be presented in Section 8.2).
4. Residual-based FD approach: it is based on the idea to have models that represent the ideal, fault-free process at a system and to compare the real measurements with those reference models. The resulting deviations (residuals) can then be observed (in off-line or on-line mode) and used as a trigger to a fault detection mechanism [78, 399].
5. Fault detection approaches which apply fuzzy models such as [251, 252] or neural networks [378, 45] as fault-free reference models.
6. On-line model-based fault detection approaches, such as [442, 8, 452] using recursive parameter estimation during on-line operation.

8.1.2 The Basic Concept

Although some of the fault detection approaches cited in the section above are able to deal with time-evolving systems or to self-tune, these abilities are based on the global recursive learning techniques for updating model parameters during on-line operation. That means that the model structure is not changed, thus significantly limiting the extension of the models to non-conventional operation conditions. On the other hand, even when the model is able to adapt itself to the evolution of the system, the fault detection logic is usually based on pure threshold criteria, which do not consider the confidence level of the evolving model.

In this book, we present a fault detection approach which overcomes these deficiencies as including the following concepts (see also [280] [279] for full details):

- Non-static fuzzy modelling with the help of evolving fuzzy systems, which are able to dynamically change its structure (not just a parameter update) with newly recorded measurements and hence able to include flexibly and quickly dynamic changes or extensions in the system behavior. In this sense, it is more flexible than static analytical or knowledge-based models and also than recursive parameter estimation methods.
- A residual based algorithm which considers the local confidence levels of the evolving fuzzy models (as reported in Section 4.5.1), taking into account confidence deterioration due to extrapolation and interpolation of the models.
- Statistical analysis of the residual signals over time, which can be seen as a self-automatic adaptive thresholding concept.

These concepts are a part of a larger on-line fault detection framework, dealing with multi-channel measurement systems and hence a pool of models automatically identified based on on-line recorded measurements with the concepts as demonstrated in Section 7.1. In particular, it is assumed that most of the possibly occurring faults are reflected in the measurements, which are recorded from P system variables, usually describing some physical, chemical or biological processes inside the

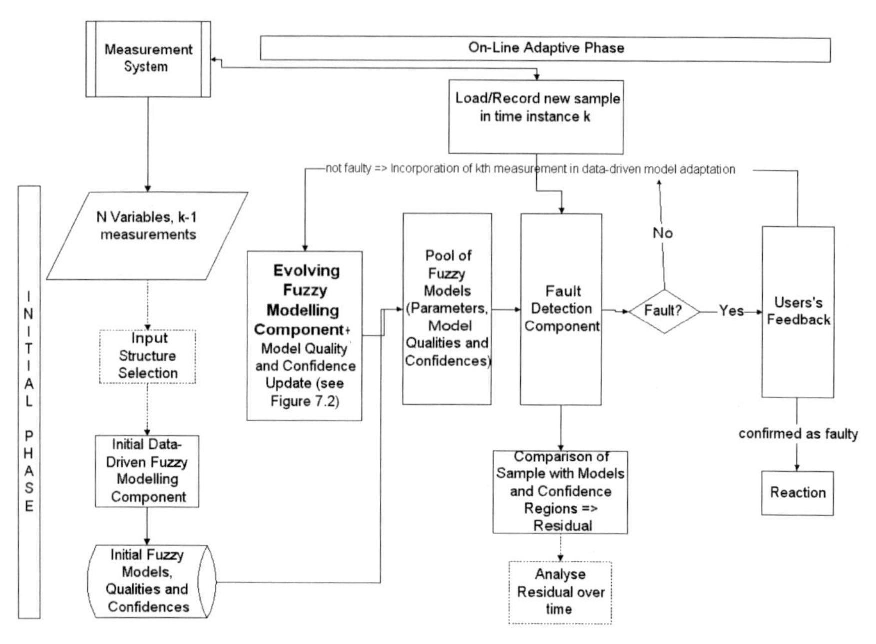

Fig. 8.1 On-line Fault Detection Framework in an Online Measurement System using **Evolving Fuzzy Systems**

system. In order to cover the detection of as many faults as possible in all P variables, dependencies between some variables are identified based on (online) measurement data in form of approximation/regression models — see Section 7.1 for a detailed description. The complete fault detection framework including the on-line identification component is shown in Figure 8.1, where the left path is entered only for the first dozen/hundred of measurement points to generate initial fuzzy models (afterwards measurement data is loaded online, sample-wise and directly processed through the fault detection component). The models identified are taken as a fault-free reference in order to generate residuals between estimated and real measured values and to analyze these residuals (in component 'Fault Detection' in Fig. 8.1). The algorithms in this component will be explained in detail in the subsequent section. Each measurement classified as faulty and confirmed as faulty by an operator is not incorporated into the incremental learning process of the models' parameters. If the measurement is not confirmed as faulty, at least one model's decision was wrong (usually due to a new operating condition or due to a badly generalizing model) and the measurement is incorporated into the evolution process in order to improve and extend the fuzzy models to new system states (e.g. by incorporating new fuzzy sets and/or rules). In this sense, a human-machine interaction takes place for online refinement, improvement and evolution of the data-driven fuzzy models.

8.1.3 *On-Line Fault Detection Strategy with EFS*

8.1.3.1 Residual Calculation and Fault Conditions

Assuming that P models are identified with evolving fuzzy modelling techniques as demonstrated in Section 7.1, and assuming that m of these P models have a quality (measured by r-squared-adjusted) greater than a pre-defined threshold (commonly, 0.6 was used [280]) at current (on-line) time instance (note that the qualities of all P models are permanently updated and hence may yield different numbers m of high quality models at different time instances) and furthermore represent a fault-free nominal operating situation, hence they are called reference models. Then m absolute residuals from m models for the current measurement (say the kth) can be calculated by:

$$res_{k,m} = \|\hat{x}_{k,m} - x_{k,m}\| \tag{8.1}$$

where $\hat{x}_{k,m}$ is the estimate value of the mth model using the kth measurement. Of course, this residual calculation is applicable sample-wise in online processes, as no prior data is required. In fact, such reference (fault-free) models can be obtained for instance in a preliminary setup phase from simulated clean data or by applying filtering and outlier deletion techniques for pre-recorded measurement data.

However, a comparison of the residual with a maximal allowed percentual deviation is only valid in the case of perfect models from noise-free data. Noise-free data is usually quite rarely the case in measurement systems when data is recorded from sensors. Hence, the model errors have to be integrated into the residual calculation in order to obtain correct and stable results. The whole model error is usually formed by bias and variance error as $\sqrt{bias\_error_m^2 + var\_error_m}$ (see Section 2.8, also [174]) and taken into account into the values estimated from the fuzzy models:

$$\hat{x}_{k,m} = \hat{f}_{fuz,k,m} \pm model\_error_m$$

where $\hat{f}_{fuz,k,m}$ denotes the output of the mth fuzzy model as defined in (1.9) for the kth measurement. The model error can be for instance well estimated by the in-sample error [174] — as defined in (2.96), which is mean-squared error plus a punishment terms with the help of the covariance between estimated and measured values. Calculating it only by the mean-squared error or the absolute error would omit the variance part of the model error (see Section 2.8 for a deeper discussion on this aspect).

Using this notation and $\tilde{x}_{k,m} = x_{k,m} \pm \varepsilon_{x_m}$, where ε_{x_m} denotes the inaccuracy level of those sensor which sample the model's output variable x_m

$$\begin{aligned} \hat{x}_{k,m} - \tilde{x}_{k,m} = \\ \hat{f}_{fuz,k,m} \pm model\_error_m - x_{k,m} \mp \varepsilon_{x_m} \end{aligned} \tag{8.2}$$

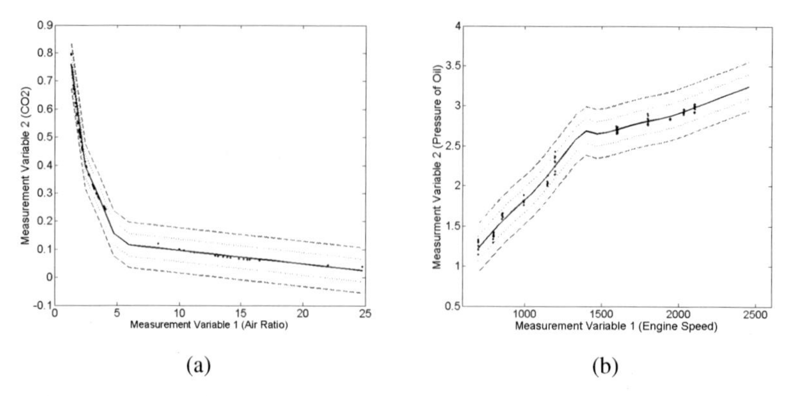

(a) (b)

Fig. 8.2 Constant confidence regions surrounding the models by applying two different thresholds in (8.4) (dotted and slashed lines), the dark dots denoting data describing a two-dimensional relationship in (a) between air ratio and CO2, in (b) between engine speed and pressure of the oil

leading to the fault condition

$$\begin{aligned} \exists m: & \; \hat{f}_{fuz,k,m} - x_{k,m} \mp \varepsilon_{x_m} - model\_error_m > t \\ \vee & \; \hat{f}_{fuz,k,m} - x_{k,m} \mp \varepsilon_{x_m} + model\_error_m < -t \end{aligned} \tag{8.3}$$

The existence operator is used over all m models because a significant violation of one model should be sufficient in order to classify the measurement as faulty. Let us choose (without the loss of generality) t to be a positive integer factor of $model\_error_m$, denoted by thr; note that ε_{x_m} and $model\_error_m$ are always positive. This finally leads to the basic fault condition [287]

$$\begin{aligned} \exists m: & \; \frac{\hat{f}_{fuz,k,m} - x_{k,m} - \varepsilon_{x_m}}{model\_error_m} > thr \\ \vee & \; \frac{\hat{f}_{fuz,k,m} - x_{k,m} + \varepsilon_{x_m}}{model\_error_m} < -thr \end{aligned} \tag{8.4}$$

If this condition is fulfilled, the actual measurement (the kth) is classified as faulty, otherwise as non-faulty. With thr the size of the confidence region around the model can be steered, see Figure 8.2, where different values of thr (2 and 4) trigger confidence regions with different width (obviously a threshold of 4 is too generous and samples affected by faults may be not detected). In this sense, the choice of this parameter is essential to control the performance of the fault detection framework, which is a difficult task. In fact, it is expected that this threshold decreases with the number of samples already included in the evolving fuzzy modelling procedure, i.e. decreasing over-fitting and the variance error: this is a direct consequence of the bias-variance tradeoff — see Section 2.8 (Chapter 2). The problem is that the relationship between the number of samples already forming the model and the threshold value cannot be formulated within a mathematical operation function for an arbitrary data set (manual tuning is required to see the effect of the number of samples: this varies from application to application, i.e. data set to data set) $\rightarrow$ in Section 8.1.3.2 we will present an application-independent and automatic adaptive thresholding concept based on the behavior of the residual signals over time.

When inspecting the confidence regions produced by (8.4) in Fig. 8.2 it can be recognized that a constant error band is triggered, possessing the same width throughout the whole range of the input variable. In this sense, it does not take into account holes in the data (as between 1200 and 1600 in (b)) or an extrapolation region where no data occurred for training the model (see the most right part reaching from 2200 to 2500 in (b)). This is a precarious thing, as in the case of large holes or regions far outside the prior covered input range the correctness of the model's approximation is quite unreliable. For instance, in Figure 8.2 it can be not guaranteed that the model is keeping the shape and tendency as shown in the extrapolation region (right part in the image). This means that if a new point comes in lying in the extrapolation region of the model, a wrong decision produced by (8.4) is very likely. Tendentially, the further the point is away from the training samples, the more likely a wrong decision gets.

Hence, a more reliable representation of a confidence region would lie in the calculation of a more flexible error band, which takes into account data holes, extrapolation regions as well as regions where fewer data points were available for training. This can be achieved with so-called *local error bars* as demonstrated in Section 4.5.1, Chapter 4. Applying the local error bar information for evolving fuzzy systems as given in (4.47) for fault detection, the fault condition in (8.4) is substituted by [280] (advanced fault condition):

$$\exists m: \frac{\hat{y}_{fuz,k,m} - x_{k,m} - \varepsilon_{x_m}}{\sqrt{cov\{\hat{y}_{fuz,k,m}\}}} > thr \qquad (8.5)$$
$$\vee \quad \frac{\hat{y}_{fuz,k,m} - x_{k,m} + \varepsilon_{x_m}}{\sqrt{cov\{\hat{y}_{fuz,k,m}\}}} < -thr$$

i.e. the constant band width $model\_error_m$ is substituted by the locally changing error bar $\sqrt{cov\{\hat{y}_{fuz,k,m}\}}$ in dependency of the actual (the kth) measurement, where $\hat{y}_{fuz,k,m}$ is the estimated output value for the kth measurement of the mth fuzzy model (recall: approximating the mth (measurement) variable x_m).

In Figure 8.3 (a) the error bar regions are shown for the same data set as in Figure 8.2 (here only the result for $thr = 2$ is visualized). From this figure, the impact of the error bars (opposed to the global error bands in the previous section) gets clear, as within the bigger data hole in between 1200 and 1600 in (b) resp. between 4 and 7 in (a) the error bars significantly widen and hence give a larger and more inaccurate confidence region for the model. The same is true in the extrapolation region on the right hand side of both images in (a) and (b). The farer away the curve is from the data samples, the wider error bars get, which is the desired effect, as there the model gets more and more uncertain. In regions where the data samples are more dense (left parts in (a) and (b)), the error bars are getting narrower as the prediction in these

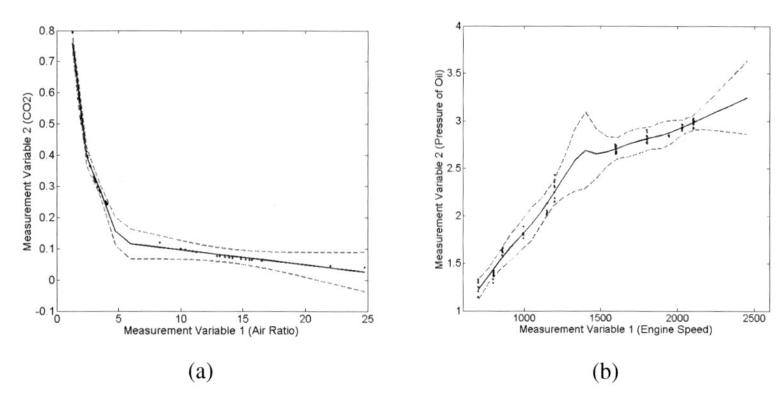

(a) (b)

Fig. 8.3 Varying confidence regions (slashed lines) surrounding the model (solid lines) by applying $thr = 2$ in (8.5), in (a) for the air ratio-CO2 relation, in (b) for the engine speed-pressure of oil relation

regions are quite certain. Finally, for the fault detection performance this means that not only the false detection rates can be decreased for samples lying in areas where the model's prediction is not so confident, but also the detection rates can be increased, as in very confident regions the error bars will get very narrow, such that smaller fault levels can be detected there. This will be empirically confirmed in Section 8.1.5, where a comparison of the performance of both fault detection conditions on high-dimensional real-recorded data from engine test benches will be given.

8.1.3.2 Statistical Analysis of Residual Signals

Even though the local model errors are integrated into the fault condition (8.5) and give a kind of normalization of the residuals, an optimal threshold thr may vary from data set to data set. This could be observed in various test cases, especially between simulated, noise-free data and real-recorded noisy data at a measurement system. However, a manual tuning of the threshold is only possible in the off-line (test) case by sending the data set more than one time through the whole fault detection process. In the online case, where a-priori the nature of the recorded data is not known or can even change during normal operation, the threshold should be dynamically adapted.

In order to cope with this problem, a statistical analysis of the normalized residual signals is reported in [280]: each (normalized) residual

$$res_{i=1,\dots,k-1;m} = \frac{\min(|\hat{y}_{fuz,i,m} - x_{i,m} - \varepsilon_{x_m}|, |\hat{y}_{fuz,i,m} - x_{i,m} + \varepsilon_{x_m}|)}{\sqrt{cov\{\hat{y}_{fuz,i,m}\}}} \tag{8.6}$$

denotes one sample in the residual signal for the mth model. Now, for the kth data sample (measurement), the residuals $res_{k;m}$ are computed for all m models with (8.6) and are not checked versus a fixed defined threshold as in (8.5), but are

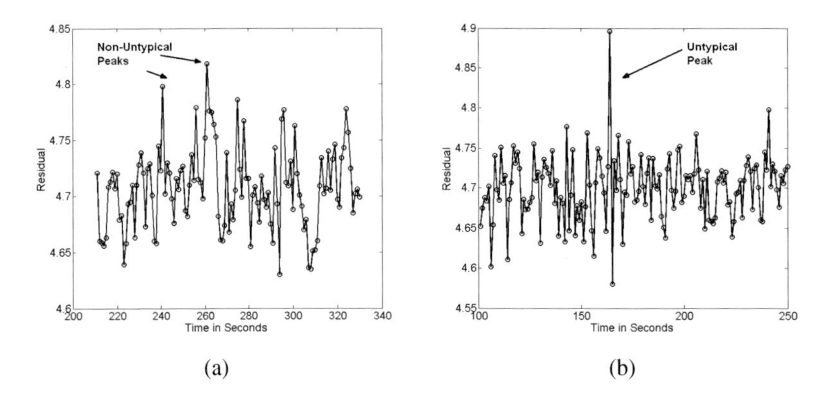

Fig. 8.4 (a): residual signal example for fault-free case, (b): residual signal example for faulty case, the fault is indicated by the distinct peak in the middle of the signal

compared with the previous residuals from the corresponding models. This comparison includes an analysis, if they trigger an untypical peak, jump or another unusual behavior in the signal, which usually indicates faults. To visualize this, in Fig. 8.4 two different residual signals are presented. In the left image a high fluctuation with several peaks seems to be quite normal and it is even hard to say if the slightly more distinct peak in the middle part of the image indicates a fault or not. In the right image it is much clearer that the peak there is really distinct and hence indicates a fault.

In [280] polynomial regression models with linear parameters in form of

$$f(t) = \beta_0 + \beta_1 t + \beta_2 t^2 + \ldots + \beta_n t^n \tag{8.7}$$

are exploited to track the basic tendency of the residual signals over time t and to predict the actual (say the kth) residual values. In this sense, t is regressed on the residual values and the batch solution for the linear parameters in (8.7) is

$$\hat{\beta} = (R^T R)^{-1} R^T \mathbf{res} \tag{8.8}$$

where R the regression matrix containing the regressor values (=rows) obtained by the polynomial regressors $t, t^2, \ldots, t^n$ (=columns) and the 1-vector as first column (for estimating the intercept β_0) and **res** containing the previous residual values. In an on-line fault detection context, the regression models have to be incrementally trained and should represent the latest local trend and not the global trend of the whole residual signal seen so far. This can be achieved by the *recursive weighted least squares* approach [264] including a forgetting factor λ, where the parameters at time instance k can be elicited from the parameters at time instant $k-1$ (also see Section 4.2.3):

$$\hat{\beta}(k) = \hat{\beta}(k-1) + \gamma(k-1)(res(k) - \mathbf{r}^T(k)\hat{\beta}(k-1)) \tag{8.9}$$

$$\gamma(k-1) = P(k)\mathbf{r}(k) = \frac{P(k-1)\mathbf{r}(k)}{\lambda + \mathbf{r}^T(k)P(k)\mathbf{r}(k)} \tag{8.10}$$

$$P(k) = (I - \gamma(k-1)\mathbf{r}^T(k))P(k-1)\frac{1}{\lambda} \tag{8.11}$$

with $P(k) = (R^T R)^{-1}(k)$ and $\mathbf{r}(k)$ the regressor vector at time instance k (which is $[1kk^2...k^n]$). The symbol λ is a forgetting factor lying in the interval $[0,1]$, which steers the locality of the regression models. The further away λ is from 1, the faster older residuals are forgotten and hence the more local the models behave, achieving a sliding regression modelling approach in the residual domain. Instead of the linear regression defined in (8.7), an evolving fuzzy model could be applied as well by integrating the forgetting concept as demonstrated in Section 4.2.3 in order to account for more non-linear dynamics.

As pointed out in Section 4.5.1, the confidence region $conf\_region_k$ of a linear sliding regression model at time instant k can be calculated through (4.44) by updating $\hat{\sigma}^2$ with (4.48). If a newly calculated residual for the actual residual res_k lies significantly outside the confidence region of $f_k(t)$ (which is obtained by using $\hat{\beta}(k)$ in (8.7)), a change in the nature of the system is very likely. In particular, if the actual residual res_k calculated by (8.6) is significantly higher than its estimation $\hat{res}_k = f_k(t_k)$ from the tendency function f_k, i.e. condition

$$res_k > \hat{res}_k + conf\_region_k \tag{8.12}$$

is fulfilled, a fault alarm is triggered. An example is shown in Fig. 8.5, where the new residual lies significantly outside the region of the confidence band of the regression model built from the past local samples. If the fault alarm is not confirmed as correct, the measurement can be assumed as fault-free and is taken into account for the evolution of the fuzzy models, see Fig. 8.1. In this case, a new operating condition is more likely and the sliding regression models updated with a decreased λ in order to adapt faster to the new residual signal behavior and therefore to reduce false alarms. On the other hand, if the fault alarm is confirmed as correct, the corresponding residual res_k is not sent into the parameter update (8.9) to (8.11) and the measurement is not taken into account for the evolution of the fuzzy models. In this sense, the sliding regression model on the residual signals together with its confidence region always represents the last fault-free behavior of the residuals.

This strategy is sufficient for detecting so-called abrupt faults causing abrupt changes of the residual signal behavior; this covers a wide range of faults such as interface defects, breakdowns of whole components or leakages in pipes etc. In the case of incipient faults, e.g. sensor drifts, the behavior of the residual signal will not cause abrupt deviations, but more a transient phase of slowly increasing residual values, see Fig. 8.6. This is because such faulty occasions become worse as time progresses.

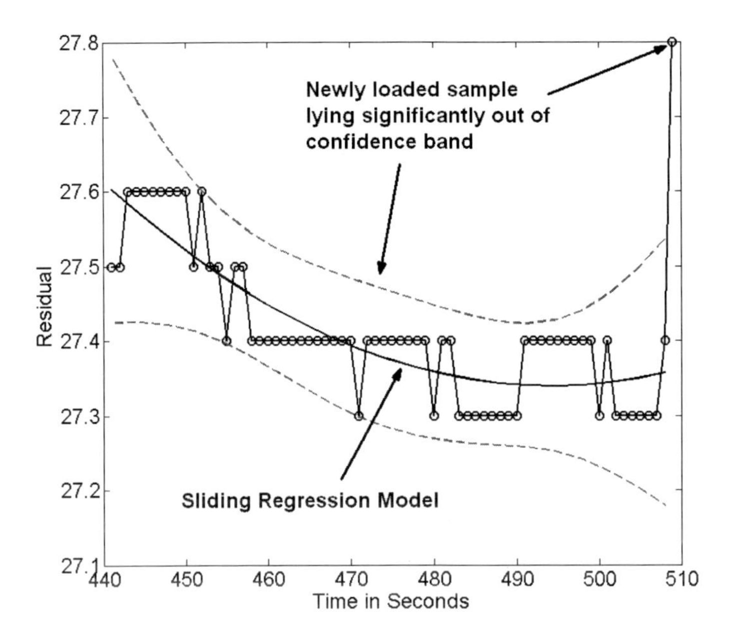

Fig. 8.5 Sliding regression model with its confident region; the new sample lies significantly outside this region, hence a fault alarm is triggered

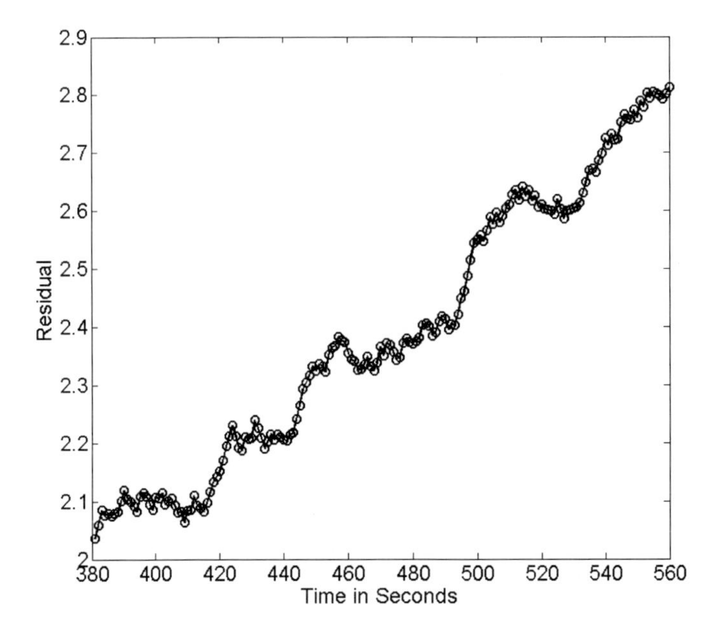

Fig. 8.6 A typical residual signal in the case of a (slow) sensor drift

A possibility to detect such faults is to estimate a linear regression model (in the same way as above the polynomial model), which then serves as a kind of low pass filter. This is important, as the residuals will not always be increasing in a strict monotonic way sample per sample, rather the whole tendency of the residual signal will show a monotonic behavior (with some noisy sample-wise fluctuations - see Fig. 8.6). When doing so, only two parameters of (8.7) (β_0 and β_1) are estimated and the forgetting factor λ has to be set to a quite low value in order to achieve locality of the regression model. The check, if now a transient phase in the residual signal is present or not, is based on the comparison with the normalized gradient from the linear curve fit β_1 with a predefined threshold:

$$\left|\hat{\beta}_1(\max(\mathbf{res}) - \min(\mathbf{res})\right| > s, \tag{8.13}$$

where $\max(\mathbf{res})$ respectively $\min(\mathbf{res})$ denote the maximal and minimal residual signal value over the last M samples and s denotes a threshold. The normalization of the gradient is indispensable to obtain a consistent threshold value. As soon as condition (8.13) is fulfilled multiple times in a row, a transient condition is detected. If now the current residual value res_k exceeds a $n - \sigma$ area around the mean value of those residual signal samples lying before the whole transient phase, a fault alarm is triggered. In this case, the same procedure upon operator's feedback is carried out as for the abrupt changes, see above. If the threshold s is pessimistic, drifts will be detected quite early and the evolving fuzzy models will not be affected much. However, the over-detection rate increases then.

Note: the mean value can be trivially updated in an incremental way, whereas σ_{res}, the variance of the residual data (outside the transient phase), is updated through the recursive variance formula [350] (see Section 2.1):

$$k\sigma_{res}(k)^2 = (k-1)\sigma_{res}(k-1)^2 + k\Delta\mu(k)^2 + (\mu(k) - res_k)^2 \tag{8.14}$$

where $\Delta\mu(k)^2$ is the difference between the old mean value $\mu(k-1)$ and the new one $\mu(k)$.

8.1.4 Fault Isolation and Correction

8.1.4.1 Fault Isolation

Once a fault is detected, it is often a challenge to further analyze the origin of the fault (which is also referred to as fault reasoning in literature, see [229]). This is necessary in order to be able to react properly and within a reasonable time frame after the fault alarm is triggered, in this way prohibiting any major system failures or even breakdown of components. An important step towards supporting the operators in this direction is fault isolation.

Fault isolation (FI) addresses the problem of isolating the fault, once it is recognized by the fault detection system. Isolation is here understood as finding those measurement channels which were affected by the fault most significantly.

Finding those measurement channels serves as valuable information as to where to search for the problem in the system, i.e. serving as support for an early and fast failure localization. Fault isolation is obvious in fault detection system, analyzing measurement channels separately in a univariate way. In our case, the fault isolation scenario gets more complicated, as we are dealing with residual signals coming from m high-dimensional models. Now, if for instance in $m_1 < m$ (out of m) residual signals an untypical behavior could be observed (e.g. by using (8.12)), indicating a fault in the system, all measurement channels occurring in these models are potential candidates to be affected by the fault. This means in our problem setting, fault isolation reduces to rank all the involved measurement channels in the m_1 models according to the likelihood that they were affected by the fault. A fault isolation value in $[0,1]$ indicates this likelihood (1 = very likely, 0 = not likely). Figure 8.7 presents the reason why a significant deviation from a model does not necessarily mean that the target channel is affected by the fault (as someone might expect from the figures in the previous sections), but it can be also the input channel — note that the deviation of the new measurement (dark dot) may be because of a disturbance in the target channel (vertical disturbance), but also because of a disturbance in the input channel (horizontal disturbance). To find out which one is more likely, is the task of fault isolation.

A fault isolation approach which solves this task within our problem setting (m high-dimensional fuzzy regression models) is demonstrated in [115], which exploits the idea of calculating a sensitivity vector as proposed in [125]. The basic idea is to calculate the normalized partial derivative at the current measurement (for which a fault alarm was triggered) with respect to each variable in m_1 models (all those violating the fault condition). The normalized partial derivative serves as indicator of how strongly a certain variable influences the model. In case of fuzzy models, the gradients are calculated numerically through Newton's difference quotient, i.e. for the mth model after the ith variable by

$$\frac{\partial f_{fuz,m}}{\partial x_i} = \lim_{\Delta x_i \to 0} \frac{f_{fuz,m}(x_i + \Delta x_i) + f_{fuz,m}(x_i)}{\Delta x_i} \tag{8.15}$$

Intuitively, the higher the sum of the normalized partial derivatives of a variable (normalized by the range of this variable) over all m_1 models (where it is occurring) gets, the more important this variable in the m_1 violated models. Hence, the likelihood that this variable is affected by the fault increases. Furthermore, also the (normalized) degree of deviation of the sample in each of the m_1 violated models and the qualities of these models (measured in terms of r-squared-adjusted) also play a crucial role when calculating the final fault affectance likelihood for each

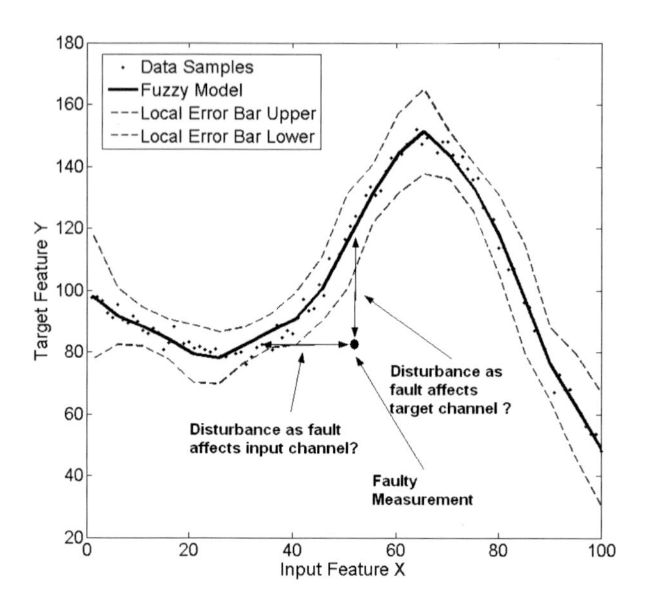

Fig. 8.7 A two dimensional visualization of the fault isolation problematic: a new measurement affected by a fault (dark dot) appears significantly outside the error bar of a two-dimensional relationship (modelled by a fuzzy model, shown as solid line); it is not obvious whether this 'disturbance' occurs because the input channel or because the target channel was affected by the fault (or even both)

channel. In [115] this is achieved with the help of Taylor series expansion, which is further developed in [114]. The normalized degree of deviation in each model is calculated by applying a transformation function mapping the deviation to $[0,1]$ based on the local error bands: the threshold value in (8.5) is mapped to 0.5, two times the threshold value is mapped to 1 and zero deviation to the model is mapped to 0.

8.1.4.2 Fault Correction

Whenever faults in the system are detected, an alarm should be triggered, telling the operator(s) that something in the system is wrong (ideally supported by a fault reasoning process or at least further information about the fault — see previous section). In some cases, it may be important to correct the measurements recorded from the affected channels in order to prevent a repetition of the recordings and so to save time and money for development and test efforts.

Fault correction (FC) is a technique which addresses the problem of delivering correct measurement values during faulty system states.

Based on the fault detection and isolation approaches discussed in the previous section, the following fault correction strategy can be obtained:

- If the faulty channel isolated using fault isolation is not a target channel in any model, then no corrected value can be delivered for this channel. This may happen when no useful functional relationship to at least one other measurement channel in the system exists for the isolated channel.
- If the faulty channel isolated using fault isolation is a target channel in only one model, take the estimated value from this model (= the target value) and deliver this as expected = corrected value (+confidence).
- If the faulty channel isolated by fault isolation is a target channel in more than one model, take the estimated value from the model which is most reliable, i.e. with the highest model quality — the other models for that channel can obviously be considered obsolete for fault correction as they possess a lower quality due to the less precise representation or lower flexibility (e.g. if input channels are missing etc.).

8.1.5 Application to Plausibility Analysis and Failure Detection at Engine Test Benches

Coming back to engine test benches (see also Section 7.1.2, previous chapter), the task was to recognize upcoming failures during the test and development phase of an engine. Categories of faults are:

- Sensor over-heatings resulting in sensor drifts
- Broken interfaces
- Pipe leakages
- Erratic behavior of the engine
- Undesired environmental conditions

The idea was to apply the generic fault detection framework shown in Figure 8.1 by using the fault detection strategies demonstrated in Section 8.1.3 and to further evolve and adapt the models based on the response of the fault detection component (i.e. to adapt the models only in fault-free cases).

8.1.5.1 Results on Real Faults

Eight different types of possible occurring faults were simulated at an engine test bench by switching off some components or reproducing leaks in pipes etc. In sum, 1180 data points were recorded for 32 different measurement channels. Hence, variable selection as described in [153] (also already used in Chapter 7 for on-line identification purposes) was carried out preliminary to the incremental on-line training of fuzzy models as shown in Figure 8.1 by the dotted box at the upper left side. It turned out that 30 measurement channels could be reasonably approximated with maximal five others, i.e. a high-qualitative approximation with up-to five-dimensional fuzzy models was possible. 80 of the 1180 data points were affected by the eight faults

Table 8.1 Comparison of basic and advanced fault detection logic on simulated faults at engine test bench

Test Case	Method	Detection Rates
Half on-line with batch trained fuzzy models	*basic*	71.59% / 6 out of 8
Half-online with batch trained fuzzy models	*advanced*	87.50% / 7 out of 8
On-line with EFS	*basic*	61.36 / 5 out of 8
On-line with EFS	*advanced*	75% / 7 out of 8

(so 10 points for each fault in row, whereas the faults are spread over the complete data set). The on-line incremental training and fault detection process as demonstrated in Figure 8.1 were simulated in off-line mode by sending point-per-point into a fault detection and data analysis framework. In order to simulate operators' feedback on classified measurements, a rating column was appended at the end of the data set, whose entry was either 0 (=non-faulty measurement) or 1 (=faulty measurement). This information was taken into account for confirmation, whenever a measurement was classified as a fault. This is necessary for a fault-free training of fuzzy models as well as for a correct analysis of the residual signals (see Section 8.1.3.2).

In Table 8.1 the results obtained by the conventional fault detection logic (using constant MAE error bands) are compared with those obtained by the advanced fault detection logic (applying adaptive local error bars as described in Chapter 4). Here, we report only the detection rates as the over-detection rates could be kept at a very low value near 0 in all cases. The detection rate is the relative proportion of measurements which were correctly classified as faults to the total number of faulty measurements (in ideal case 100%, so no fault is missed). The over-detection rate is the relative proportion of measurements which are wrongly classified as faults to the number of fault-free measurements in sum (in ideal case 0%, so no fault-free case is detected as fault). This comparison includes the detection rates on measurement basis (entry before the slash in the third column) i.e. all measurements affected by faults are counted (in sum 80) as well as on the fault basis (entry after the slash in the third column), i.e. all different kind of faults are counted (in sum eight). It also contains the half on-line case, where fuzzy models are trained completely in off-line mode (with the complete training data set at once) and are then used for on-line fault detection. It can be recognized that both fault detection logics could achieve quite good results as significantly more than 50% of the faults as well as faulty measurements could be detected. Moreover, advanced fault detection logic can significantly outperform the basic one, as seven instead of five of the eight faults could be detected, where one of these two additional detected faults was a major one and hence mandatory to be detected. It is also remarkable, that the on-line case (modelling procedure obtaining sample per sample) comes close to the performance of the batch case, which is in accordance to the identification results as presented in Table 7.1 (Section 7.1.2).

As we are proposing an on-line fault detection approach here, the computational speed is also an interesting issue. For both fault detection logics the checking of a point by sending it through the fault conditions in (8.4) respectively (8.5) is negligible small (below 1 millisecond on a PC with 2.8GHz Processor). The incremental/evolving step on the 32 fuzzy models for each new incoming data point is about 0.01 seconds, such that data loaded with a frequency of about 100Hz can be handled.

8.1.5.2 Achievable Bounds of FD Approach

Simulated data (for another car engine) were generated and slightly disturbed by some artificial faults with fault levels of 5% and 10%. A fault level of 10% here means that for instance a value of 100 was re-set artificially to a value of 90, representing a (rather small) fault causing 10% disturbance in the data. This gives a reasonable elicitation of the achievable bounds of the fault detection approaches, i.e. the answer to the question which fault levels can be detected at all. Moreover, this data set was used as validation how the manually tuned threshold for the data set in the previous section performs on this data set without any re-tuning phase. 2050 points were generated in sum, where half of it were disturbed with 5% and 10% fault levels in all of the six measurement channels. As it is a quite low-dimensional data set, no variable selection was carried out before starting the incremental learning process of the fuzzy models. All other things were applied in the same way as for the data set in the previous section.

First, it turned out that when using the same threshold as manually tuned and optimal for the data set in the previous section, no fault could be detected at all. A reason for this could be that the fault levels of the real-simulated faults at an engine test bench were all significantly higher than for this simulated (noise-free) data set. Furthermore, the threshold for the data set in the previous subsection could not be set lower as quite a lot of over-detections occurred then. Hence, the adaptive analysis of the residual signals as pointed out in Section 8.1.3.2 was applied for this data set in order to simulate on-line fault detection from scratch without using any manual tuning phase of a threshold or other parameters. When doing so, the results in Table 8.2 could be achieved, there again the basic and the advanced fault detection logic are compared against each other. This table shows us the achievable bounds in on-line mode quite drastically: while for the 10% fault levels the results are quite good and comparable with those in the previous section, for fault levels of 5% the detection rate goes down between 20% and 26%. The false detection rates stayed between 0% and 1% for all test cases, which is usually quite acceptable. According to the experience reported by several engine test specialists, the usual critical failures affect the signal value in one or several measurement channels by more than 10%. In this sense, the fault detection rate achieved by physical-based models (mathematically deduced by physical knowledge about the car engine) could be almost doubled by applying evolving fuzzy models with a flexible structure for incorporating new system states (represented in newly recorded data). Table 8.2 also shows that fuzzy models coupled with the FD conditions together with the analysis of residual signals can significantly out-perform other types of regression models such as local

Table 8.2 Comparison of fault detection approaches on simulated car engine data with small built in errors (10% and 5% fault levels): basic versus improved fault detection strategy = global versus local error bars coupled with EFS, analytical, linear regression and local linear models are always coupled with global error bars (no local ones exist).

Fault Level	Method	Detection Rates
10%	*EFS, basic*	53.36%
10%	*EFS, advanced*	66.23%
10%	*analytical models*	35.13%
10%	*linear regression models*	20.67%
10%	*local linear (lazy learning)*	49.08%
5%	*EFS, basic*	20.45%
5%	*EFS, advanced*	26.23%
5%	*analytical models*	23.92%
5%	*linear regression models*	12.88%
5%	*local correlation (lazy learning)*	20.04%

correlation (based on the lazy learning [6] and LOESS data weighting [85] concepts) and linear high-dimensional regression, both using non-adaptive conventional MAE error bands. Lazy learning is described in detail in Section 5.5.2, whereas the LOESS algorithm exploits this concept, but integrates a specific scheme for data weighting, introducing neighborhood parameter α, which inducts the smoothness of the regression prediction in the current query sample $\mathbf{x}$:

- If $\alpha \leq 1$: the weight of the ith data sample is $w_i(\mathbf{x}) = T(\Delta_i(\mathbf{x}), \Delta_{(q)}(\mathbf{x}))$ with $q = \alpha n$ truncated to an integer.
- If $\alpha > 1$: $w_i(\mathbf{x}) = T(\Delta_i(\mathbf{x}), \Delta_n(\mathbf{x}) * \alpha)$

where $\Delta_i(\mathbf{x})$ the distance of the query point to the ith sample in the reference data base, $\Delta_{(q)}(\mathbf{x})$ the qth distance in an ordered sequence of distances (from smallest to largest) to the query point $\mathbf{x}$, and T defined by a tri-cube function:

$$T(u,v) = \begin{cases} 0 & u \geq v \\ (1-(\frac{u}{v})^3)^3 & 0 \leq u < v \end{cases} \tag{8.16}$$

Based on the comparison in Table 8.2, we can conclude the following:

- There is obviously significant non-linearity in the system dependencies between measurement channels, as the (global) linear regression models significantly fall behind the non-linear models (fuzzy, analytical and local linear) in terms of detection rate.
- The lazy learning concept for local linear models brings in some non-linearity and improves the performance of the global linear regression models; however, its partial (step-wise) linear model characteristics is not able to perform as well as the smooth non-linear surfaces functions triggered by the evolving fuzzy systems.

- For both, 5% and 10% fault levels, the fuzzy models are achieving a higher detection rate than all the other methods, whereas the advanced fault detection strategy (based on adaptive local error bars rather than constant error bands) is better than the basic fault detection strategy (based on constant model error bands).

8.1.6 An Alternative Concept for On-Line Fault Detection and Diagnosis

An alternative concept for on-line fault detection was recently introduced in [242], where an evolving fuzzy classifier trained based on the *ePL* (*evolving participatory learning*) approach is applied for including different operation modes of the monitored process. The key feature of this approach is that it is not only able to detect faults, but also to classify the faults into different fault classes. One fault class corresponds to one operation mode from a set of various operation modes which are also including different fault-free system states/operating conditions/system behaviors. From this point of view, it can be seen as a generalization of the concept proposed in the previous section (distinction between different failure and non-failure modes is possible). On the other hand, this approach is not fully automatic, as the operation modes have to be specified by an operator during the on-line process on demand: whenever a new cluster is created and the current operation mode estimation is below a threshold (i.e. the membership degree to any of the already available fuzzy rules, each one representing a specific operation mode, is low), then an information from the operator is requested whether the event is a new operation mode or a mode already identified. In the former case, the operator has to identify the mode. In case when a new cluster is created and the current operation mode estimation is above a threshold, the consequent label (=operation mode) of the new rule is set to that one proposed by the output of the classifier (after the inference scheme). Figure 8.8 presents a work-flow of the fault detection and diagnosis approach.

The conventional *ePL* approach is extended to deal with updating ellipsoidal clusters in arbitrary shape (covariance update after (2.7) and update of the cluster centers by (3.41)) and to integrate an automatic statistical-oriented monitoring of the compatibility index and threshold violations. According to that, the threshold for the arousal index (responsible for the creation of new clusters) can be adopted to a given significance level, preventing the *ePL* algorithm to generate new clusters in case of outliers — for details about the statistics behind see [242]. The evolving fuzzy classifier architecture is deduced from the clustering approach evolving ellipsoidal clusters in arbitrary shape and therefore includes multivariate Gaussian mixture models in the high-dimensional space in its antecedent parts:

$$\mu(\mathbf{x}) = e^{-\frac{1}{2}(\mathbf{x}-\mathbf{c})\Sigma^{-1}(\mathbf{x}-\mathbf{c})^T} \tag{8.17}$$

with Σ the co-variance matrix and $\mathbf{c}$ the center of the Gaussian. The fuzzy rules are presented directly in high-dimensional form by (so no projection to the one-dimensional axes is carried out):

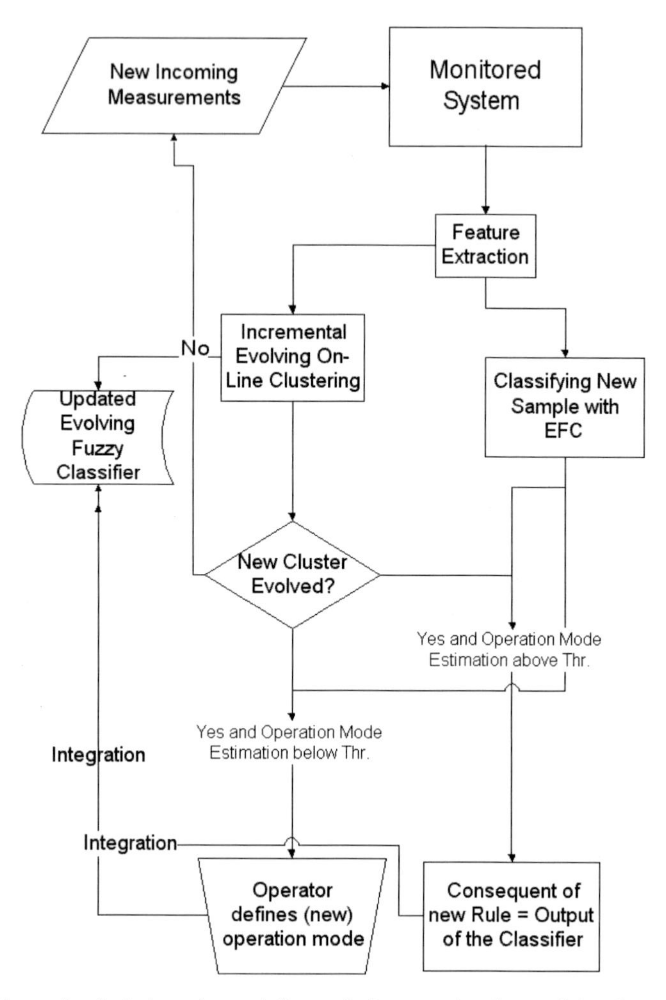

Fig. 8.8 Alternative fault detection and diagnosis framework using evolving fuzzy classifiers and including arbitrary operation modes (failure and non-failure classes)

$$\text{Rule}_i : \text{IF } \mathbf{x} \text{ IS } A_i \text{ THEN } l_i \text{ IS } O_j$$

with A_i a multi-dimensional Gaussian fuzzy set and O_j the jth operation mode which may appear in more than one rule consequent. This looses some interpretability capabilities of evolving fuzzy classifiers (as high-dimensional rules are hardly linguistically understandable — see also Chapter 6), but on the other hand may gain some classification accuracy, as a more exact representation of the natural characteristics of local data clouds is possible (especially, partial local correlations between variables are better represented). In [242], this approach was successfully applied for monitoring the conditions of induction motors, commonly used in electrical drives:

all operation modes could be detected correctly immediately upon their appearance during the on-line process.

8.2 On-Line Anomaly Detection in Time Series Data

8.2.1 Motivation and State of the Art

Following the previous section about general on-line fault detection concepts, strategies and algorithms in measurement systems, we are dealing here with a related problem, i.e. the detection of anomalies in time series data, which occur as successive measurements of the process parameters recorded over time. On the one hand, this problem setting is a special case of that one in the previous section, as it deals with univariate time series data of process parameters, which are completely independent from each other (hence no interactions in form of high-dimensional models between the variables are observed); on the other hand, from the operators' point of view, it is a general case of that one in the previous section, as anomalies are not only necessarily stemming from faults in the system, but also may represent some transient phases of process parameters during on-line operation mode and production. This also means, that different types of anomalies require, once they are detected, a different intervention of the operator(s): some may only trigger an information or a warning, others may yield a fault alarm (as also triggered by the methods presented in the previous section).

In this regard, an *anomaly* can be defined as an untypical occurrence of a process parameter/system variable which is not necessarily unpermitted, but may also represent special cases during the production process.

Usually, an anomaly in a system causes also an untypical pattern in the time series data of the corresponding process parameter(s), for instance a significant change in the amplitude or in the frequency domain or simply a shape or sample sequence in the time series curve which never occurred before during usual operation phase. Two anomaly examples are visualized in Figure 8.9, the upper figure demonstrating an untypical sudden jump/outlier in the signal stream, the lower figure visualizing three untypical patterns in an almost sinusoidal time series (indicated by ellipsoids).

Different ways on how to tackle the detection of anomalies were studied in the past, such as [38], where models are learned for windows of the data and the model parameters for new data compared with those of a reference data set; or [5], using distance measures to find clusters of the normal data and then classifying new points occurring outside the clusters as anomalies; or TARZAN [215], estimating the frequencies of patterns in the discretized time series and indicating those patterns occurring more or less frequently than expected. The approach in [216] has the advantage in that it allows a parameter free detection of anomalies, by evaluating the compression ratios of sequences of normal and test data. In [214] the determination

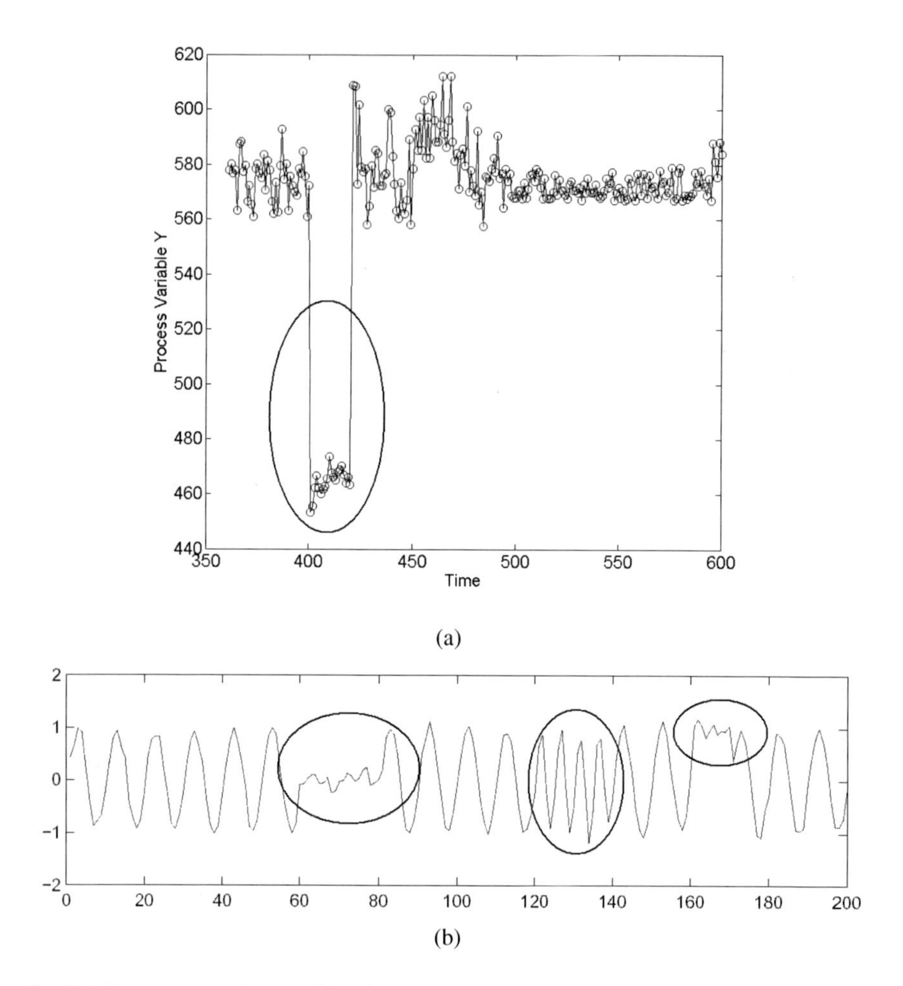

Fig. 8.9 Upper: anomaly as sudden jump, lower: anomaly as untypical patterns in terms of abnormal amplitudes and frequencies

of discords is described, the most unusual part of a time series. All these approaches have in common that they are off-line method, hence usually are detecting anomalies in a post-processing step, once the whole time series data is available. This is nice, but has the significant drawback that the whole production process can be only retouched in a post-mortem manner and therefore bad production items (stemming from 'malicious' anomalies) not really prevented at the time when they are actually produced. Hence, in order to save production waste and money, it is an essential challenge to directly detect anomalies during the on-line production process, such that an intervention in the early phase of an upcoming error (influencing the quality of the production items) can be made possible. In this context, ideally each newly recorded sample is checked based on the past information (representing the normal state), and immediately an information whether an anomaly is present or not shown to the operator.

8.2.2 Methodologies and Concept

Three different approaches are proposed for handling on-line anomaly detection for each new incoming sample (see [388] for details):

- Detection of sudden untypical patterns, occurrences in form of peaks, jumps or drifts in the time series, by exploiting the trend of the signal curves in the near past. These types of anomalies (which can be more or less also seen as outliers) usually represent typical fault cases at the system.
- Detection of more complex upcoming untypical patterns in the time series curves by exploiting relations in sample sequences. These types of anomalies sometimes represent fault cases, but may also arise because of intended changing process parameters.
- Detection of transients in the signal curves denoting changes in the environmental settings or operating conditions, mostly not indicating a fault, but a start-up phase at a system.

8.2.2.1 Detection of Untypical Patterns

The first issue can be tackled with the same techniques as used for analyzing the residual signals in Section 8.1.3.2. The only difference is that here more complex models are used to follow the basic trend of a time series in the near past: evolving fuzzy systems are applied instead of polynomial regression models with linear parameters. In order to guarantee a localization effect of the evolving fuzzy models, i.e. the models only represent the basic trend over the past 10, 20, 30, ... samples, a permanent gradual forgetting in the antecedent and consequent parameters is applied (see Section 4.2 for details about this methodology). An appropriate setting of the forgetting factor λ is essential and depends on the window size specifying the locality of the model and mostly application-dependent. Assuming to have a window size M given and assuming that the sample at the border of this window, i.e. the sample which lies M time steps back, should have only a low weight ε when included into the fuzzy model training, then the wished forgetting factor λ can be automatically elicited by:

$$\lambda = e^{\frac{\log \varepsilon}{M}} \tag{8.18}$$

As with this forgetting strategy, the fuzzy models slide over time to track the basic trend of the time series, we call this concept *sliding fuzzy regression models*.

For the detection of more complex untypical patterns in the process signals, the concept of auto-regressive models (AR) which predict current system states based on past ones is a feasible option. Formally, such models are defined by:

$$y_t = f(y_{t-1}, y_{t-2}, \ldots, y_{t-m}) + \varepsilon_0 \tag{8.19}$$

in extended form (ARMA) defined by:

$$y_t = f_1(y_{t-1}, y_{t-2}, \ldots, y_{t-m}) + f_2(\varepsilon_{t-1}, \varepsilon_{t-2}, \ldots, \varepsilon_{t-m}) \tag{8.20}$$

with y_t the value of the current sample in a time series sequence and y_{t-l} denoting past samples in this sequence and ε_t a noise term at time instance t. In the classical case [56], the function f explicitly consists of linear parameters. However, in today's complex industrial systems and processes linearities only play a minor role. So, applying evolving fuzzy systems as function of f and letting the rule base evolve over time with more and more samples recorded is necessary to handle non-linearities properly (auto-regressive evolving fuzzy models). For each newly recorded sample, first a prediction response from the fuzzy models is provided (based on the m past samples in the time series) and then compared with the real measured value by exploiting the same techniques as in Section 8.1.3 (i.e. calculating the degree of deviation of the measured value to the adaptive local error bars surrounding the fuzzy models $\rightarrow$ residuals). Gradual forgetting is also possible (in the same manner as accomplished for detecting untypical peaks, jumps), but not recommended, as the fuzzy model should include all the past patterns of a time series in normal operation mode.

8.2.2.2 Detection of Transients

Detection of transients is accomplished by a gradient information measure strategy. To calculate the gradient information, it is not feasible to calculate the gradient based on two consecutive samples (by subtraction and normalization with the duration between two time instances). This is because usually the time series signals are affected with noise significantly and every small down-and-up swing would influence the gradient information. A visualization of this problem is given in Figure 8.10, where the left plot shows the global view onto the time series (a long term transient over 160 seconds can be recognized), and the right plot shows a zoom into the left part of the signal (first 14 samples): the basic trend over the window of 14 samples of a decreasing gradient is almost diminished, and calculating the gradient for each sequence of consecutive sample would lead to an even more misguided ('too local and fluctuating') gradient information. Hence, (fuzzy) regression models are extracted from samples located in a time window of size M, providing the basic tendency in a given time window (same concept as for sliding regression models discussed above) and achieving a filtering and smoothing effect over the window. Note that these regression models can be in principle newly extracted from the samples lying in the current window, however this requires significantly more computation time as when updating the fuzzy models sample-wise and with a gradual forgetting strategy. The gradient β from the models is simply calculated by applying Newton's difference quotient as defined in (8.15) and taking the absolute value (decreasing and increasing transients are not distinguished). Using the determined gradient directly would require the selection of a threshold, to which it should be compared, for determination of unusual high (or low) values. To make the selection of the threshold

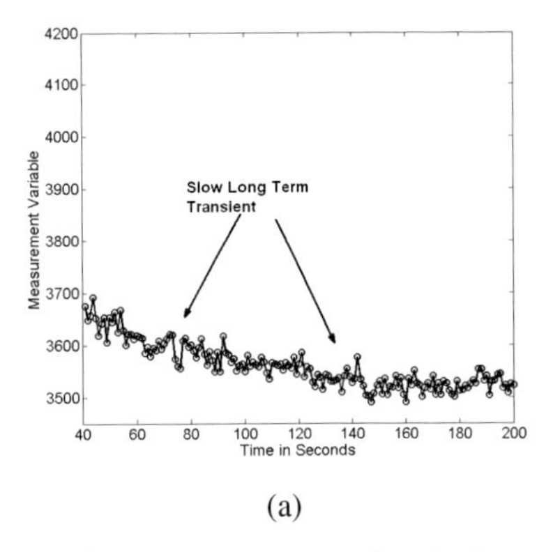

(a)

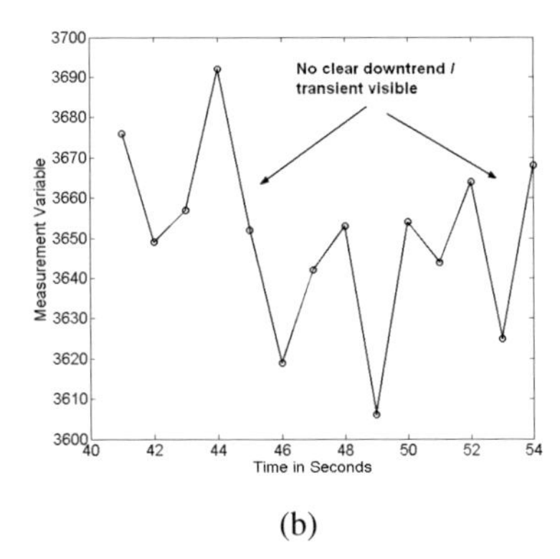

(b)

Fig. 8.10 (a): long term transient lasting over 160 seconds, clearly visible from the global point of view (inspecting the time range as a whole); (b): by laying a (too small) local window over the first 14 samples, the downtrend is not really visible any longer; when taking gradient between two consecutive samples, the effect of localization is even worse

easier and to achieve better generalization, normalization of the gradient has to be performed. The following can be seen as a collection of possible normalization and pre-processing steps, some of which can be selected, depending on the problem at hand — for further details see [388]:

- Window size M: should be sufficiently large for balancing out random fluctuations over small time horizons.
- Continuation of long transients: a measure n_{high} is used which is the number of preceding time steps, in which transients were detected without interruption.
- Variance σ_N: as measure of the average deviation of new channel values from the mean of recent values: for channels with highly fluctuating values, an observed small gradient is less probable to be significant, than for a channel with very little noise.
- Fit error e_N: measure used for normalizing the gradient, as higher error of the linear regression fit indicates a higher noise level and therefore a higher uncertainty in its gradient.

Integrating all the normalization methods explained above, the following computations are performed to obtain the normalized transient measure value β_1' from the original gradient β_1:

$$\beta_1' = d \; \frac{\left(|\beta_1| \,/\, \sigma_N\right)^2}{1 + \left(e_N \,/\, \sigma_N\right)^w}, \text{with} \tag{8.21}$$

$$d = \sqrt{M}\sqrt{1 + n_{\text{high}}}.$$

where $\beta_1'(t)$ denotes the β_1' computed for the tth data point (for $t < 1$ these are initialized with a configured constant β_{def}').

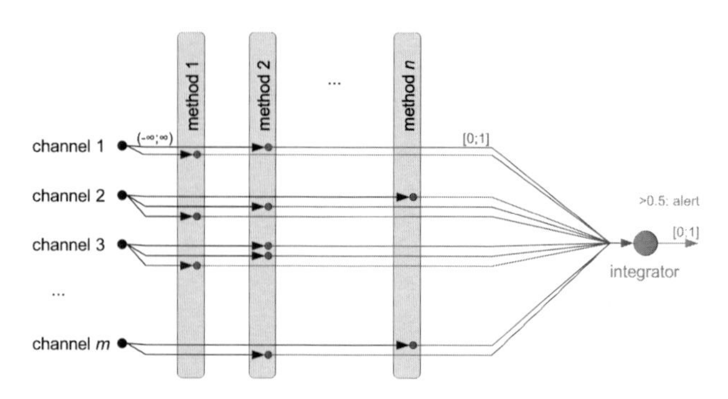

Fig. 8.11 Anomaly detection framework consisting of different methods flexibly configurable for any of the available channels. The channels themselves can potentially assume any value, while the output of each method is normalized into the interval $[0,1]$. The individual predictions are finally integrated by a given integrator to obtain a single instability prediction value (taken from [388]).

To allow the transient detection to work for a diverse range of channels, the threshold s for comparison with the normalized transient measure as defined in (8.21) is defined adaptively:

$$s = c_{\text{thresh}} \ \text{quantile}\big(B,q\big) \tag{8.22}$$

$$B = \big(\beta_1'(N-M_B),\ldots,\beta_1'(N-1)\big) \tag{8.23}$$

with M_B a list of the last observed gradients, $q = 50\%$ and c_{thresh} a pre-configured factor [388].

8.2.2.3 Integration Concept

Usually, many channels (process parameters) are recorded and supervised in parallel. When dealing with univariate methods for untypical pattern and transient recognition as shown above for each time series data (corresponding to each one of the channels) separately, it is necessary to find some amalgamated statement about the system state (transient phase, abnormal phase, faulty state etc.). Together with the aspect that applying several methods for each process parameter time series, an integration concept as shown in Figure 8.11 is required. It can be configured, which method with which parameter settings should be applied to any given channel, assuming that the outputs of the single anomaly detection methods are all normalized to $[0,1]$ and hence comparable among each other. Normalization can be achieved by a transformation function, mapping the residual onto $[0,1]$ (maximal residual mapped onto 1, 0 residual mapped onto 0, residual at the edge between normal and anormal state mapped onto 0.5 and a smooth transition inbetween, for instance achievable by a double parabola function — see [388]).

Each model, after application of the normalization, returns a value between 0 and 1. All the values have to be integrated into a single instability prediction. An obvious way is to return the maximum of the predictions $\rho_{m,l}$, any of the methods $m \in \mathcal{M}$ returns on any of the channels $l \in C(m)$ configured for m, as the integrated or total instability index ρ:

$$\rho = \max_{m \in \mathcal{M}, l \in C(m)} \rho_{m,l} \tag{8.24}$$

This seems reasonable, as the final prediction honors the largest instability detected by any of the methods in any of the channels.

More sophisticated integration methods are possible. It could be useful to take into account dependencies between the model responses, introduced because they work on the same or related channels or because they use the same detection method. This could help to avoid over-detections (because more than one model must detect anomalies), or to detect weak anomalies by allowing predictions of different models to amplify each other. An idea taking this into account is

$$\rho = 1 - \prod_{m \in \mathcal{M}} \prod_{l \in C(m)} \left(1 - \rho_{m,l}\right). \tag{8.25}$$

The motivation for this formula is, that the global prediction is larger than or equal to the largest single prediction; each further detection increases the global value, only slightly for small detections and more for stronger ones.

8.2.3 Application to Injection Moulding Machines

The methods described in the previous section were developed for anomaly detection in a variety of injection moulding machines. Data is sampled from the available channels once for each part produced, and contains information about configuration settings, timing, forces and pressures, temperatures, speed, and dimensions, among others. Fig. 8.12 shows examples for data from some of the more important channels, and the detections by different instances of the presented methods: while interpreting these plots, it should be considered, that the methods had only data up to the prediction point available, not the whole time series.

Several channels contain mainly instabilities manifesting themselves as more or less obvious jumps (panels 3 and 4 in Fig. 8.12); these can be detected quite well using the sliding regression and sometimes the ARMA method. Panel 4 shows, e.g. at samples 70 and 165, some smaller detection peaks, which reflects the fact that it is also hard for humans to decide about a possible instability. Depending on the detection sensitivity desired, either the threshold (here set to 0.5) or the configuration of the method can be adapted to allow detection of these as well.

Furthermore, the really interesting question would be, whether the produced part was acceptable, or not. As it is sometimes not possible to tell for sure, the consequence is, that these channels and the detections should be seen as hints, when it would make sense to check for the quality of the produced parts. A guess from the on-line production process is that the framework can detect around 80%-90% of anomalies by having less than 5% over-detection rate.

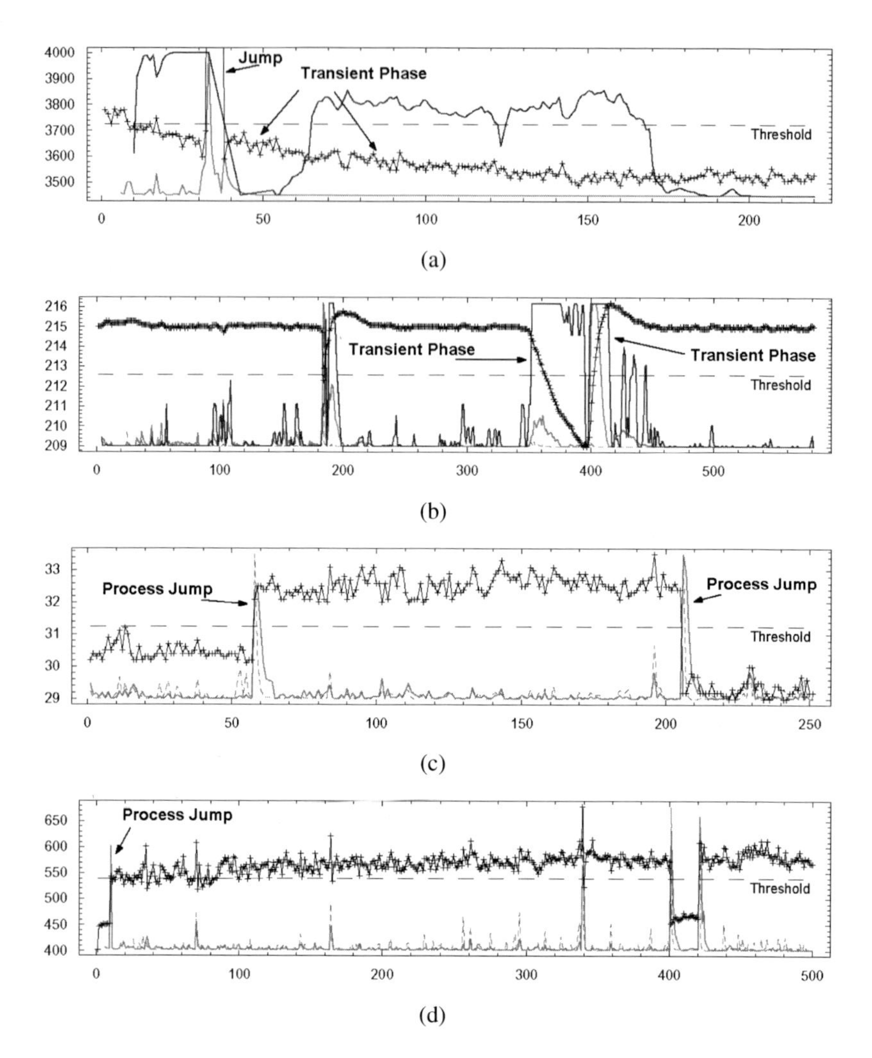

Fig. 8.12 Examples of method performances on injection moulding data. The samples are placed along the horizontal axis, the vertical axis shows the channel values. Instability predictions by the different methods are overlaid, with 0 (no instability) at the bottom and 1 (almost certain instability) at the top of the plot; an example for an alert threshold of 0.5 is also shown. Black solid lines with "+" marker: channel data, Dark solid lines: transient detection, Light grey lines: Sliding evolving fuzzy regression method, Light grey dashed lines: ARMA method with EFS. Not all methods are used on all channels.

8.3 Noise Detection on Analogue Tapes

Analogue tapes enjoyed a great attraction during the 50s to 80s, not only for private usage, but also especially in tone studios, where original recordings were caught by broadband tapes being able to record four tracks or even more in parallel. Over the last two decades, the digitalization of music on CDs, mini discs, DVDs have become more and more popular and replaced the old style of analogue tapes, music cassettes and long players. Nowadays, a lot of tone studios (especially in TV and broadcast stations but also partly in music production companies) still possess tonnes of analogue tapes in their archives. For them, it is a big challenge to get all these analogue sources digitized onto hard discs with tera-bytes storage size. This can be achieved with modern devices such as analogue digital converters combined with up-sampling techniques [329] (in order to reach the desired frequency rate: CDs resp. DVDs have a frequency rate of 44100 Hz resp. 48000Hz, whereas older tapes are often recorded with 22000 Hz or even less — this usually depends on the quality and age of the tapes) and with converging the sample resolution by discrete amplitude transformations (often from 8-bit to 16-bit, but also from 8-bit or 16-bit to 24-bit).

One intrinsic problem during the digitizing of analogue tapes is the checking whether the data source on the tape is recyclable at all, or whether the tape contains too much noise, such that an archiving is pointless and the tape should just be dropped into the bin without any re-usage in digital form. In this regard, noise detection (ND) algorithms come into the game and play a major role for classifying analogue tapes into good ones, mediocre ones and really bad ones (after the traffic light scheme: green-yellow-red). In fact, such a classifier is constructed by machine learning tools based upon the frequency and intensity of noises on an analogue tape. A summary characteristics in form of numerical features can be extracted from the joint noise collections found on the tapes (such as number of noise types of noise, average and maximal intensity over all noises etc.), and a fuzzy classifier trained upon these features plus some annotated samples. This fuzzy classifier can be tuned on-line based on operator's feedback about the classifier's decision on new tapes to be digitized: whether the classifier was true or not $\rightarrow$ update the classifier with incremental, evolving mechanisms in the latter case.

Regarding the noise detection component(s), various signal processing- and statistical-based algorithms have been developed for detecting a broad range of noises such as drop-out based noises (cut begins and ends, holds and mutes, analogue drop-out), frequency distortion based noises (clicks, ticks, crackles, pops, glitches), bad edit, print thru, flutter, azimuth, phase correlation, wow, hum, DC offset and many more. The detectors for some of these noises (especially drop-out based and frequency-distortion based noises) were developed based on collected typical patterns from the raw wave length signals, either on a sample-per-sample basis or on a global basis, where samples are compressed to and inspected as hull curves (and then the noise patterns gets visible). Usually, the noise patterns differ significantly from the noise-free environment in shape, behavior of the sample sequence, outlook. Based on the visualization of these patterns, strategies for detecting the start and end points of such anomalies were developed

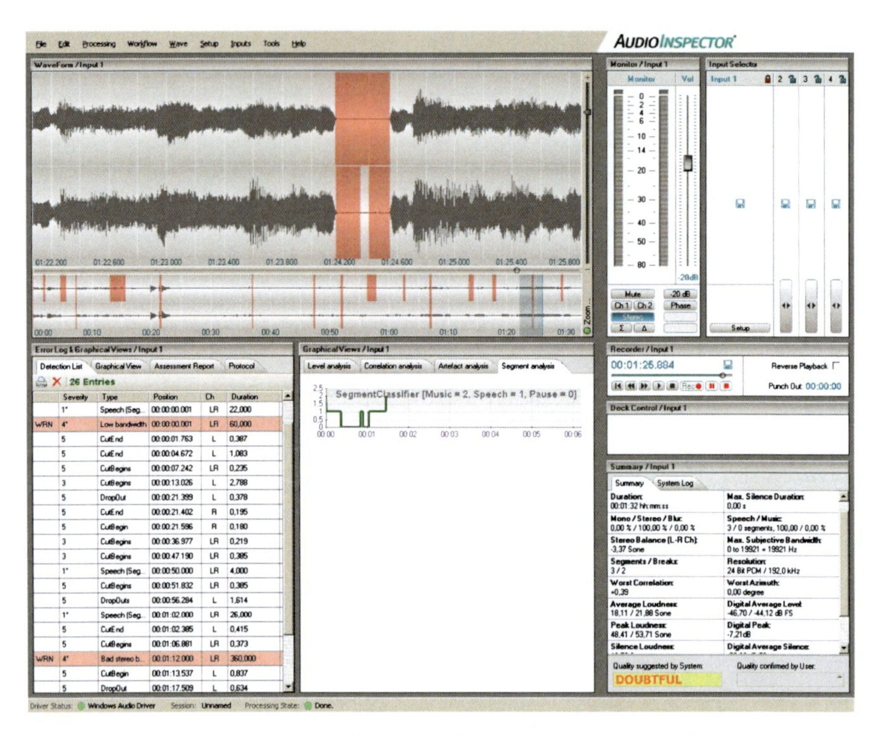

Fig. 8.13 Audio Inspector Software Tool including noise detection algorithms, music-speech discriminator, overall quality assessment tool, hull curve visualization, error logging box.

(e.g. untypical gradients or untypical behaviors in the high-frequency part of the Fourier spectrum or sudden changes of gradients to 0 without swinging out of the amplitude) in order to generate noise hints. Once, the start and end points of such potential fault candidates were identified, features were extracted characterizing the amplitude inbetween these points, but also in the nearby lying environment of these points ($\approx \pm$ 1 millisecond). Based on a set of training samples, where noise occurrences were marked throughout the whole tapes, fuzzy classifiers were trained from the extracted features of these samples plus from extracted features from some noise-free parts on the tapes. For all noise types, an accuracy of at least 80% could be achieved (for some noise types even above 90%), where the over-detection rate could be kept at 5% or lower.

All the algorithms were integrated into a joint software tool with a comprehensive GUI and hardware devices to the audio tapes, the final frontend to the user is visualized in Figure 8.13. A special feature of the software is the visualization of the music hull curves as they come in from the tapes and are digitized. In Figure 8.13 this visualization component is represented by the big upper left window, where the big area marked in different grey shading denotes a (correctly) detected drop-out part in the signal. The bottom left window shows some over-all statistics of the currently processed tape and an overall quality assessment statement in form of 'OK' (green),

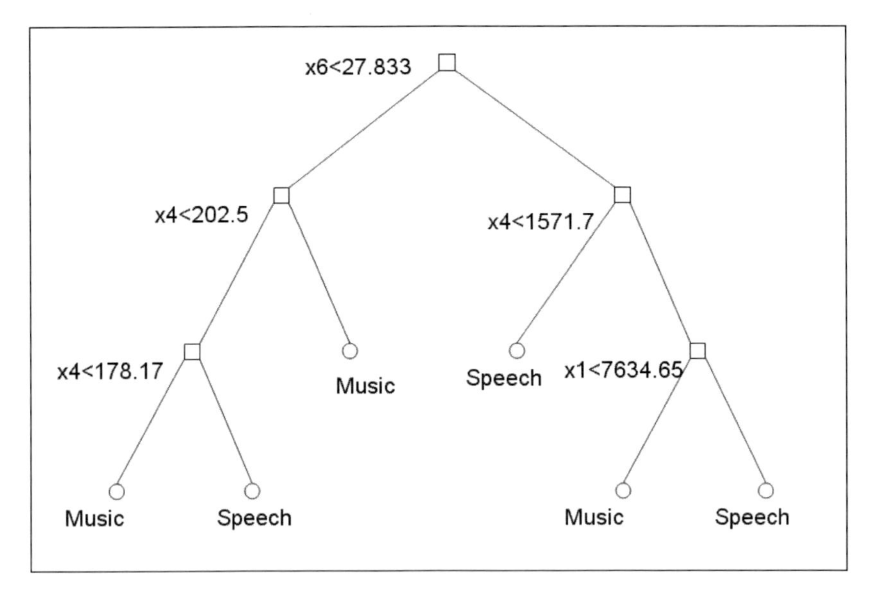

Fig. 8.14 Decision tree for discriminating between speech and music.

'Doubtful' (yellow), 'Noisy' (red). Another feature is an integrated music-speech discriminator, showing the operator whether currently music or speech signals are transferred to the hard disk (middle-bottom graphics window). The middle lower window in Figure 8.13 shows a typical output in this window (0 denotes 'Pause', 1 denotes 'Speech' and 2 denotes 'Music'). The discriminator works on a second-per-second basis, for each second six features are extracted (some of these also found useful and used by Scheier and Slaney [386]):

- Zero-crossing rate (x1)
- Entropy (x2)
- Spectral flux (x3)
- The frequency with the maximal amplitude entry in the fourier spectrum (x4)
- The proportional factor between the low frequency entries and the overall frequency entries (up to 10khZ) (x5)
- The longest consecutive slot below a certain threshold in the energy buffer, obtained from the fourier spectrum of small size pieces in the one-second frame: one entry in the energy buffer contains the sum of the amplitudes in the fourier spectrum of one small piece (=energy of the signal in this small piece) (x6).

A rule-based classifier in form of a decision tree (using the CART approach [58]) could be trained from selected speech and music samples containing different quality of the speech (interview, broadcast, old live recordings from talks) and different sort of music (classic, rock, pop, jazz, ...) — see Figure 8.14. The tree only uses three of the six input features (automatically selected in the embedded feature selection process) and could achieve a classification accuracy above 95%.

For a more detailed description of the audio inspector tool see www.audioinspector.com.

Chapter 9
Visual Inspection Systems

Abstract. Following the previous chapter, the first part of this chapter again deals with quality control and fault detection aspects. However, instead of supervising raw process data (measurement signals), here we deal with two-dimensional visual context information in form of images. In particular, in Section 9.1 we demonstrate how images showing the surface of production items will be analyzed if they show any untypical occurrences which may indicate faults on single items, usually caused by failures in the production systems. The second part of this chapter deals with visual context information in form of textures, which can be seen as special types of images containing repeating patterns over their complete range. Human perception modelling for visual textures will be the core aspect of the second part. This means that models will be built for the purpose to associate human perception and emotions with given visual textures. These association models require a visual texture as input and are able to respond to a certain emotion the input texture may trigger for customers.

E. Lughofer: Evolving Fuzzy Systems – Method., Adv. Conc. and Appl., STUDFUZZ 266, pp. 357–391.
springerlink.com

Fig. 9.1 Visual inspection of die casts parts performed by an operator manually.

9.1 On-Line Image Classification in Surface Inspection Systems

9.1.1 Motivation

Nowadays, in many industrial manufacturing and assembly systems the inspection of items or parts produced during on-line operation mode plays a major role for quality control. Opposed to fault detection in measurement signals (as dealt with in the previous chapter), the inspection is basically performed within an a posteriori process and mostly conducted visually based on images showing the produced items or parts. This means that failures in the production systems, once occurring in the system, can be recognized when inspecting the produced items. Depending on the problem setting (especially the size of possible faults on the items and the speed of production), cameras with different properties (high-resolution, fast snapshots etc.) are installed to record the images showing surfaces of the production items. An example of a typical fault on an obstacle would be a yolk on an egg (in an egg production process): eggs are transported from the mother hens and coming along a conveyor belt portion per portion which need to be supervised to make sure they are ok or not before they are delivered in boxes.

In order to circumvent time-intensive manual checks and to save significant effort for the experts/operators (as shown in Figure 9.1), the whole inspection process has to be automated within a machine vision system. In this case, one has to solve the problem of how to implement a decision-making process (good part vs. bad part) in the software and to reproduce human cognitive abilities. Usually, this requires a step-by-step reprogramming, manual adaptation or parametrization of the software, which may last for several months until satisfying results are obtained: for instance, in a CD imprint production process, a manually configured fuzzy classifier for discriminating between good and bad discs was developed based on expert knowledge and specifically dedicated to the characteristics of the CD imprints, requiring time-intensive manual tuning phases. Other classification approaches have strong focus on classical image processing and segmentation techniques and were specifically

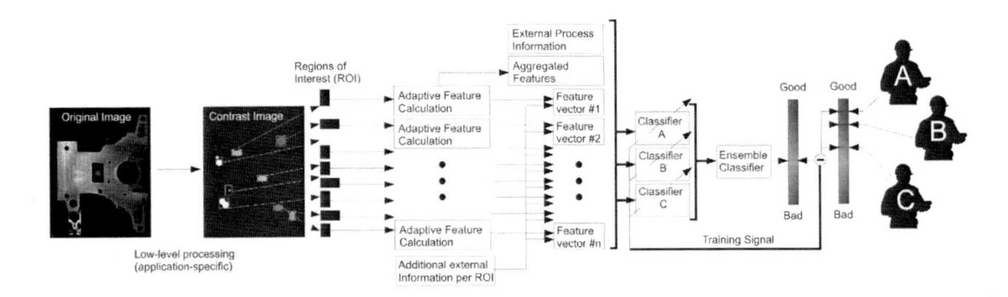

Fig. 9.2 On-line image classification framework at surface inspection systems.

developed for certain applications and their very special needs, for instance in [86], [64], [331], [384] for textile products, in [217] for wood, in [188] for paper, in [392] for leather and in [89] for optical media.

In these cases, we may speak about *application-dependent visual quality control* approaches, which usually are not applicable to a broader range of inspection systems without any time intensive re-development and re-programming phases.

9.1.2 The Framework and Its Components

In order to circumvent these deficiencies of manual tuning, long-time developments and application-dependent visual inspection systems, a framework was developed which is applicable to a wider range of applications [382]. This is achieved by removing application-dependent elements and applying machine vision and learning approaches based on image descriptors (also called features) extracted fully automatically from the images. In this way, the problem of a step-wise deduction of classification rules (based on analytical and expert knowledge) is shifted to the problem of gaining sufficient training samples. The latter can be done fully automatically during production phases, the annotation of the training samples is made as easy as possible by a user-friendly GUI frontend integrated in an annotation wizard (see also [289]). Furthermore, the framework contains a lot of other aspects for extending the applicability, improving the user-friendliness and the accuracy of visual inspection systems and providing a coherent technology for joining various components and responding a unique accept/reject decision.

In the following, we describe the components of our framework, which is shown in Figure 9.2 in detail (see also [382], [118]), from methodological point of view.

- Low-level processing on the images for removing the application dependent elements (contrast image): hereby, the basic assumption is that a fault-free master image is available; for newly recorded images during the production process the deviation to these master images (deviation image) is calculated by subtraction ($\pm$ a threshold for an upper and lower allowed bound). The pixels in a deviation image represent potential fault candidates, but need not indicate necessarily a failure in the production item. This depends on the structure, density and shape of the distinct pixel clouds (called regions of interest). The degree of deviation is reflected in the brightness of the pixels. For color images the color-wise (RGB) deviation from the master are calculated and averaged over the absolute values of the deviations to produce one gray level image.

- Recognition of regions of interest (objects) in the contrast image: the deviation pixels belonging to the same regions are grouped together; therefore, various clustering techniques were exploited which can deal with arbitrary shape of objects and arbitrary number of objects. The applied clustering approaches are iterative prototype-based clustering achieving ellipsoidal clusters (the iteration is over different parameter settings used in *eVQ* [267] (Section 3.1.1.1), hierarchical clustering [191], reduced delaunay graph (graph-based clustering) [335], DBSCAN (density-based clustering) [123] and normalized cut (pretty slow, but mathematically funded) [396]. For a sensitivity analysis among these methods w.r.t classification accuracy we refer to [355].
- Extraction of features with a fixed and adaptive feature calculation component: object features characterizing single objects (potential fault candidates) and aggregated features characterizing images as a whole are extracted. Aggregated features are important to come to a unique final accept/reject decision and are for example the number of objects, the maximal local density of objects in an image or the average brightness of objects — see Table 9.2 for a complete list of aggregated features. Adaptive techniques were developed [117] [118] for guiding the parameters in the feature calculation component (e.g. radius defining local density or thresholds during binarization) to values such that the between-class spread is maximized for each feature separately, i.e. the single features achieve a higher discriminatory power. A key aspect was the extraction of as many features as possible in order to cover a wider range of different application scenarios and problem settings. Hence, a feature selection step during classifier training is included.
- Building of high-dimensional classifiers based on the extracted features and label information on the images (or even single objects) provided by one or more operators. In some cases, the label information contained not only a good/bad label for an image, but also additional information like for instance the uncertainty in his/her decision (during annotation). The training process consisted of three steps:

  – Dimension reduction (deletion of redundancies and filter feature selection approaches with a specific tabu search method [416]) in order to reduce the high initial set of features.
  – Best parameter grid search coupled with 10-fold cross-validation (in order to find the best performing classifiers with the best performing parameter setting).

  - Training of the final classifier with all training samples and the optimal parameter setting achieved in the previous step.

- A feedback loop to the image classifiers based on the operator's feedback upon the classifiers decisions. This is a central aspect for improving the performance of the classifiers set up during the off-line phase (which will be underlined in Section 9.1.5) and reducing the workload of operators for annotating sufficient images (i.e. images sufficiently covering the possible data space). It requires incremental learning steps during the on-line operation mode as a re-building of the classifiers usually does not terminate in real-time. The EFC approach *FLEXFIS-Class MM* and the evolving clustering-based classifier *eVQ-Class* were used to achieve this goal.
- Classifier fusion methods for resolving contradictory input among different operators: various operators may label the same training examples or give feedback during on-line production mode on new samples. Hence, some contradicting feedback may arise, especially when the skills of the operators vary. Fusion methods are able to resolve these contradictions by performing a voting, democratic decision or more complex decision templates (obtained in a training step where the confidence outputs from the single base classifiers serve as features). This classifier type was also applied for boosting performance of single base classifiers: the only assumption thereby is that the classifiers are diverse. The concept of incremental on-line classifier fusion (as described in Section 5.4) was applied for updating fusion methods synchronously to the 'base classifiers', some results will be reported in Section 9.1.7 (see also [381]).
- Early prediction of success or failure of a classifier: it was examined how classification accuracies behave with an increasing number of samples and at which point of time the classifier cannot be improved further by feeding more samples into the training algorithm (see [401] for more information).

From a human-machine interaction point of view, the framework is dealing with five important issues for enhanced operator communication required at the surface inspection systems for guaranteeing user-friendliness and guiding the framework to a higher performance. These are marked by red ellipses in Figure 9.3 named 'HMI 1' to 'HMI 5' and can be summarized as:

1. HMI 1: The incorporation of operator's feedback on classifier(s) decisions during on-line mode: the operator may wish to bring in his knowledge and experience into the (automatically trained) image classifiers, for instance by overruling some classifier decisions. This usually leads to a refinement and improvement of the classifiers accuracy, especially when changing operating conditions or system behaviors arise during on-line mode, which were not included in the original training data set and hence not covered by the trained classifier(s) $\rightarrow$ a fixed kept static classifier would be outdated and deliver wrong classification results after a while. This issue demands an evolving image classifier learning mechanism and will be handled in more detail in Section 9.1.5.

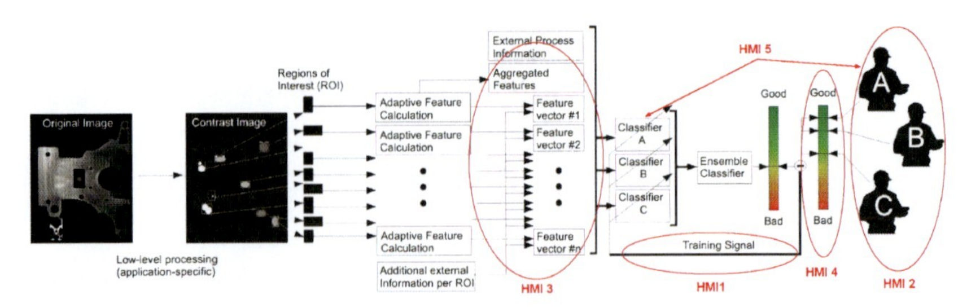

Fig. 9.3 On-line image classification framework at surface inspection systems, the human-machine interaction issues highlighted by ellipses.

2. HMI 2: Feedback or labels from different experts (operating at the same system) may be contradictory, hence this should be resolved by fusion mechanisms of different classification outputs. In connection with the previous demand, it is advantageous that incremental update of the classifier fusion is supported.
3. HMI 3: Labels may be provided at different levels of detail (i.e. either only on images or also on single objects in the images) according to the speed of on-line processing and according to the effort for labelling off-line data.
4. HMI 4: The operator(s) may provide additional information in the form of a confidence level, as not being completely confident in his/her (their) decision(s) due to low experience or occurring similar patterns between faulty and non-faulty regions. This is a first step towards an enhanced feedback, instead of just responding a good/bad reward.
5. HMI 5: The operator(s) may wish to have more insight into the classifier structures in order to obtain a better understanding of the classifier decisions respectively the characteristics of the faulty situations at the systems. In special cases, he may wish to interact with the structural components of the classifiers (changing or adding some rules). Such an interaction also demands an improvement of the interpretability of (fuzzy) classifiers (refer to Chapter 6).

For detailed description and improved results when integrating these HMI issues see also [289].

9.1.3 Experimental Setup on Three Different Application Scenarios

9.1.3.1 The Surface Inspection Problems

The whole framework and its image classification components were applied to four real-world surface inspection problems:

- CD imprint inspection
- Egg inspection
- Inspection of bearings
- Inspection of metal rotor parts

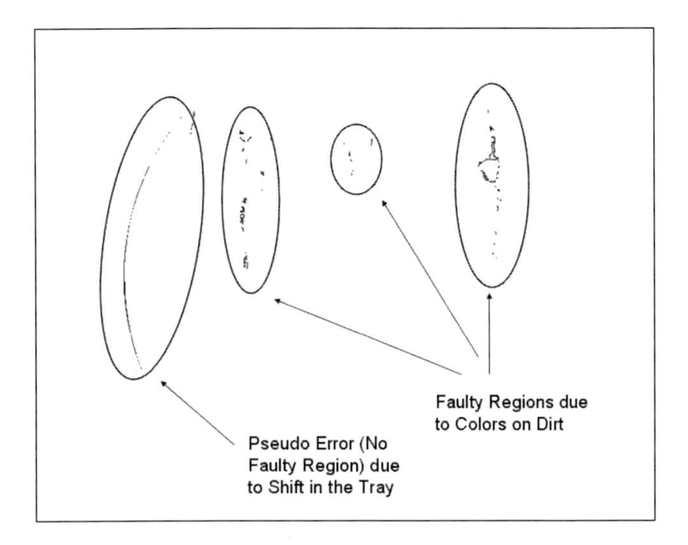

Fig. 9.4 Deviation image from the CD imprint production process, different grey-levels and pixel groups represent different ROIs; note that most of the image is white which means that the deviation to the fault-free master is concentrated in small regions; the faulty and non-faulty parts are exclusively marked as such.

In the CD imprint inspection system the task was to identify faults on the Compact Discs caused by the imprint system, e.g. a color drift during offset print, a pinhole caused by a dirty sieve (→ color cannot go through), occurrence of colors on dirt, palettes running out of ink, and distinguish them between so-called pseudo-faults, like for instance shifts of CDs in the tray (disc not centered correctly), causing longer arc-type objects or masking problems at the edges of the image or illumination problems (→ causing reflections). Figure 9.4 shows a typical deviation image (after taking the difference to the master) from this production process, containing three faults caused by colors on dirt and one pseudo-error because of a shift of the CD in the tray.

In the egg inspection system, the main task was to identify broken eggs on a conveyor belt by recognizing some spoors of yolk and scratches on them. Here, the main problem was to distinguish these spoors from dirt on the eggs, which do not indicate any fault. An example of a deviation image for an egg is shown in Figure 9.5, where the dirt part of the egg has a different outlook than the yolk part, basically with respect to its brightness and shape.

Another real-world data set is taken from an inspection problem of bearings. The main challenge in this application is whether the trainable system is able to learn how to distinguish between different types of faults with only a good/bad label for the whole image. This is of particular importance as quite often the quality expert can only spend very short time on each part and is not able to provide more detailed training input.

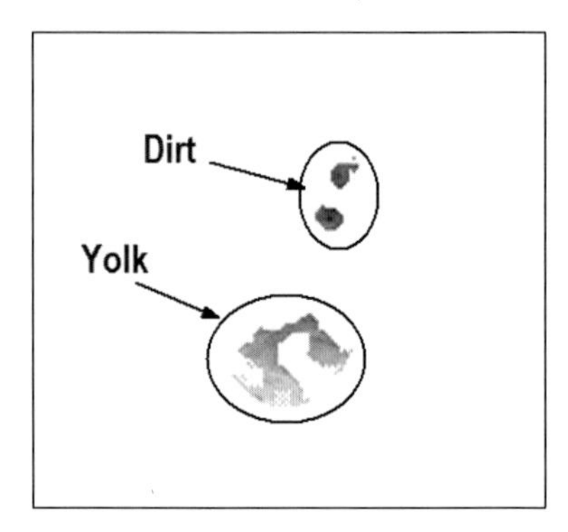

Fig. 9.5 Deviation image of an egg, showing a dirt and a yolk occurrence

Table 9.1 Image data sets from surface inspection scenarios and their characteristics

	# Images	# Tr. Samples	# Feat.	Obj. Labels	# Cl.	Cl. Dist. in %	# Op.	Uncert.
CD Imprint	1687	1534	29/74	Yes	2 (bad/good) + 12 (6/6)	15-20/80-85	5	Yes
Eggs	4341	11312	57	Yes	2 (yolk/dirt)	51.4/48.6	1	No
Rotor	225	225	74	No	2 (bad/good)	66.6/33.3	1	No
Bearing	1500	1500	74	No	2 (bad/good)	69.8/30.2	1	No
Artificial	5x20000	5x20000	74	No	2 (bad/good)	50/50	1	No

In the compressor inspection system, rotor parts of the compressors which are die-cast parts that are afterwards milled or grinded are examined with respect to faults. The main problems are gas bubbles in the die cast part (so called link-holes or pores). As the excess material is cut away during milling or grinding, the gas bubbles become visible on the surface. They usually appear as dark spots on the surface with a mostly circular shape. They are 3D deformations of the surfaces and cause problems particularly on sealing or bearing areas, since casings will not be tight any more. This is a specific problem with the high pressures found in compressors.

9.1.3.2 Data Collection

In each of the aforementioned applications data were collected during on-line operation and stored in the same order onto a hard disc as they were recorded (with ascending numeration). This is essential for a correct processing in the on-line simulation framework for testing the impact of the incremental training and evolution of the image classifiers during on-line production mode (see Sections 9.1.5 and 9.1.7 below). The number of collected image samples varied from application to application and are summarized in Table 9.1. These image samples were annotated by a single operator (in case of egg, bearing and metal rotor part inspection) or by five different system experts (in case of CD imprint inspection). For bearing and metal rotor part data only labels on the whole image ('good' or 'bad'), for egg and CD

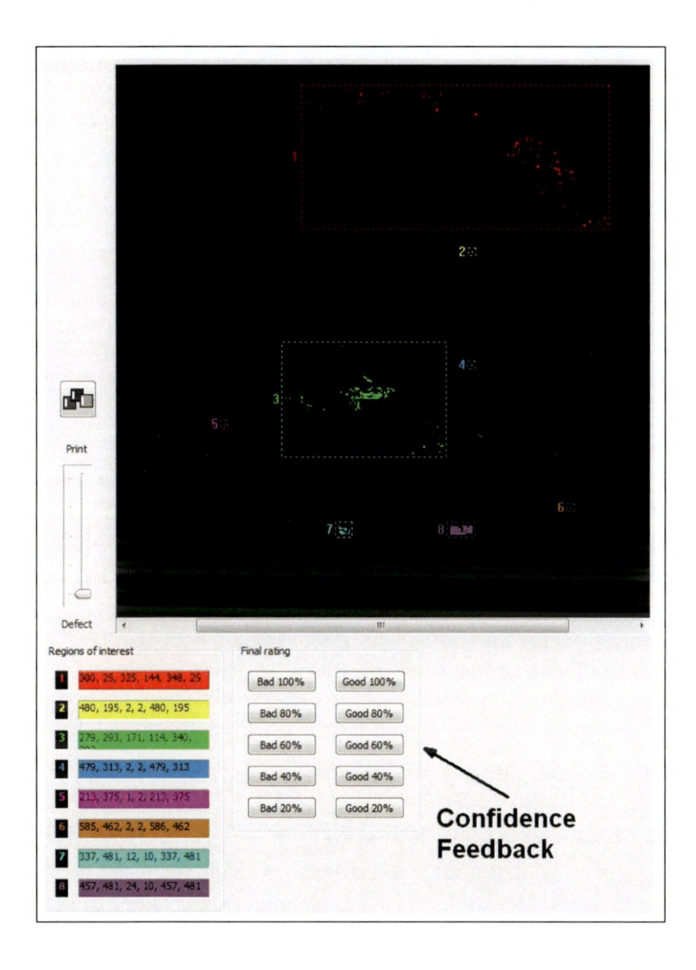

Fig. 9.6 GUI of image annotation tool, here 8 regions of interest are found which can be all labelled by a drop-down control element (lower left), image labels can be assigned together with a confidence level representing the uncertainty of the operator in her/his annotation (buttons in the lower middle part)

imprint data also the object labels were provided, differentiating between 'dirt' = no fault and yolk = 'fault' in the case of egg data and between 6 fault and 6 non-fault classes in the case of CD imprint data. For the latter, the operators also provided the uncertainty in their annotations for each single image as indicated in the last column of Table 9.1. This could be achieved quite comfortably and rapidly by pushing a single button in the GUI frontend (5 buttons for bad and 5 buttons for good were used, indicating confidence levels of 20%, 40%, 60%, 80% or 100%), which is shown in Figure 9.6.

Table 9.2 List of aggregated features used. These features have been chosen because of their relevance for a very wide range of surface inspection applications.

No.	Description	No.	Description
1	Number of objects	10	Max. grey value in the image
2	Average minimal distance between two objects	11	Average grey value of the image
3	Minimal distance between any two objects	12	Total area of all objects
4	Size of largest objects	13	Sum of grey values in all objects
5	Center Position of largest objects, x-coord	14	Maximal local density of objects
6	Center Position of largest objects, y-coord	15	Average local density of objects
7	Maximal intensity of objects	16	Average area of the objects
8	Center Position of object with max. intensity, x-coord	17	Variance of the area of the objects
9	Center Position of object with max. intensity, y-coord		

Additionally, 5 artificial data sets were produced containing a large number of samples (20000) and an automatic label process by fixed pre-defined rules, successively increasing their complexity from the first to the fifth set. The rules were based on descriptions that are regularly found in quality control instructions, such as "part is bad, if there is a fault with size $>$ 1.5 mm". The rules also included more complicated combinations, such as "part is bad, if there is a cluster of 4 faults, each with a size $>$ 0.75 mm". Three to five such rules were logically combined for each set of images.

9.1.3.3 Feature Extraction and Classifier Training

From the collected image samples a wide range of features are extracted characterizing both, single objects (denoted as object features) and images as a whole (denoted as aggregated features). The objects can be recognized as regions of interest by various clustering techniques (see [355] for a complete survey); as default setting, the agglomerative hierarchical clustering procedure [191] was used. In sum, 17 aggregated features and 57 object features are extracted, which were a priori defined by image processing experts at the surface inspection systems mentioned above and also from other industrial partners employing experts with a deeper knowledge in visual quality control. The basic idea was to generate a large set of features and to let the automatic classification process decide which features are important for which application scenario (resp. data set collected at this). This can be done in a kind of automatic feature selection process before the real classifier training starts; in some cases, feature selection is even embedded and closely linked to the classifier training process itself. The 17 aggregated features and their meaning are listed in Table 9.2. The following formulas define how these features were calculated; the symbol $featX$ belongs to the Xth feature in Table 9.2, C denotes the number of objects (clusters), P_m the number of pixels belonging to the mth object, $p_{i,m}$ the ith pixel in the mth object, $g(p_{i,m})$ its grey level (intensity) lying in $[0,255]$ $p_{i,m}(x)$ respectively $p_{i,m}(y)$ the x- respectively y-coordinate of the ith pixel in the mth object:

$$feat1 = C$$

$$feat2 = \frac{2}{C(C-1)} \sum_{i=1}^{C} \sum_{j=i+1}^{C} \min_{m,n} |p_{i,m} - p_{j,n}|, \quad m = 1,...,P_i \;\; n = 1,...,P_j$$

$$feat3 = \min_{i,j=1,...,C} |p_{i,m} - p_{j,n}| \quad m = 1,...,P_i \quad n = 1,...,P_j$$

$$feat4 = P_m, \quad m = argmax(P_1,...,P_C)$$

$$feat5 = \bar{O}_m(x) = \frac{1}{P_m} \sum_{i=1}^{P_m} p_{i,m}(x) \quad m = argmax(P_1,...,P_C)$$

$$feat6 = \bar{O}_m(y) = \frac{1}{P_m} \sum_{i=1}^{P_m} p_{i,m}(y) \quad m = argmax(P_1,...,P_C)$$

$$feat7 = \max_{j=1,...,C} \left((\frac{1}{P_j} \sum_{i=1}^{P_j} g(p_{i,j})) * P_j \right)$$

$$feat8 = \frac{1}{P_m} \sum_{i=1}^{P_m} p_{i,m}(x) \quad m = argmax_{j=1,...,C} \left((\frac{1}{P_j} \sum_{i=1}^{P_j} g(p_{i,j})) * P_j \right)$$

$$feat9 = \frac{1}{P_m} \sum_{i=1}^{P_m} p_{i,m}(y) \quad m = argmax_{j=1,...,C} \left((\frac{1}{P_j} \sum_{i=1}^{P_j} g(p_{i,j})) * P_j \right)$$

$$feat10 = \max_{j=1,...,C} (\max_{i=1,...,P_j} g(p_{i,j})) \qquad feat11 = \frac{1}{\sum_{j=1}^{C} P_j} \sum_{j=1}^{C} \sum_{i=1}^{P_j} g(p_{i,j})$$

$$feat12 = \sum_{j=1}^{C} P_j \qquad feat13 = \sum_{j=1}^{C} \sum_{i=1}^{P_j} g(p_{i,j})$$

$$feat14 = \max_{j=1,...,C} |\{O_i | i \neq j \wedge \exists p \in O_i : (\bar{O}_j(x) - p(x))^2 + (\bar{O}_j(y) - p(y))^2 < r^2\}|$$

$$feat15 = \frac{1}{C} \sum_{i=1}^{C} |\{O_i | i \neq j \wedge \exists p \in O_i : (\bar{O}_j(x) - p(x))^2 + (\bar{O}_j(y) - p(y))^2 < r^2\}|$$

$$feat16 = \bar{P} = \frac{1}{C} \sum_{i=1}^{C} P_i \qquad feat17 = \frac{1}{C} \sum_{i=1}^{C} (P_i - \bar{P})^2$$

with

$$r = \frac{\#imageRows + \#imageColumns}{30}, \tag{9.1}$$

where *#imageRows* denotes the vertical and *#imageColumns* the horizontal size (in pixel) of the image. Some examples of object features are the size and the roundness of an object or statistical values characterizing the grey level distribution over an object, such as histogram, kurtosis, skew, quantiles over the grey levels.

Once, the features are extracted, a feature pre-processing step is initiated based on the label information the operator has provided:

- If the operator only provides labels for the whole images (as was the case for metal rotor parts and bearings), then aggregated information from object feature vectors (extracted from the single objects per image) is extracted and appended at the end of the aggregated features, obtaining 74-dimensional image feature vectors — standard aggregation was the maximum operator, so taking the feature-wise maximum over all objects in one image. The whole features vectors are called extended unsupervised aggregated features.
- If the operator provides also labels on the single objects, a classifier on the object features is trained, which is used 1.) for discriminating between types of faults and 2.) for generating features which are used additionally to the aggregated ones: the number of feature vectors per image falling into each class is counted and appended to the aggregated features ($\rightarrow$ extended supervised aggregated features). In case of CD imprint inspection containing 12 classes, this leads to feature 29-dimensional image feature vectors. In case of egg data, directly training a final good/bad decision on the object labels is performed, as dirt automatically indicates a good egg and yolk automatically a bad one (achieving a 57-dimensional feature space).
- If the operator also provides certainty levels for his image labels, (pre-processed) image feature vectors are duplicated according to these levels: once in case of 20%, twice in case of 40% and so on up to 5 times in case of 100% certainty.

From the pre-processed features, the classifier training starts including a feature selection, 10-fold cross-validation and final training phase (based on optimal parameter setting eliciting during cross-validation phase) in the off-line case. For classifier training, several data mining methods were applied which are referred to as top-10 data mining methods over the past century [451] such as k-nearest neighbor algorithm, *CART (Classification and Regression Trees)* [58], *Support Vector Machines (SVMs)* [434] [389], *AdaBoost* [88], *Bayesian* classifiers, *discriminant analysis* [111] or *possibilistic neural networks* [444], and also fuzzy classifiers (using multi-model architecture) resp. the purely clustering-based approach *eVQ-Class* approach (Section 3.1.3).

9.1.4 *Off-Line Results*

Each of the classification method has 1 or 2 significant parameters which are sensitive in regard to the final performance of the trained classifier. These parameters are varied over a pre-defined parameter grid and for each knot point a 10-fold CV procedure is initiated. The accuracies for all runs are compared and the parameter setting belonging to the highest accuracy can be taken for final training of the classifier using the whole data set. The achieved (maximal) CV accuracies on the four real-world and the one artificial application scenarios are presented in Table 9.3. Heuristic regularization (see Section 4.1.2) was used in all partial TS models when applying multi model structure by using *FLEXFIS-Class MM*, otherwise no results could be obtained at all due to (nearly) singular Hessian matrices. The sensitivity with respect to the folds in the 10-fold CV is nearly negligible, as laid below $\pm 1\%$

Table 9.3 Cross-validation accuracies achieved on pre-processed features for different classifiers

Data set	CART	SVM	NN	Bagg	AdaBoost	eVQ-Cl.	Fuzzy MM (FLEXFIS-Class)
Art. #1	92.05	91.53	90.90	92.35	91.50	82.20	90.00
Art. #2	95.68	94.90	92.55	96.30	96.30	91.90	94.40
Art. #3	94.82	93.40	94.05	94.60	94.20	88.73	92.47
Art. #4	90.50	90.33	91.54	91.70	90.60	85.37	89.10
Art. #5	92.11	91.37	91.87	92.70	91.90	86.53	89.40
Print #1	94.39	95.10	94.01	92.83	96.30	93.40	91.02
Print #2	97.71	98.22	95.88	95.63	96.35	96.67	95.29
Print #3	97.95	96.54	94.13	94.78	95.31	94.12	94.09
Print #4	96.99	96.93	95.66	95.63	95.63	95.56	93.40
Print #5	95.21	95.71	94.13	95.63	96.15	91.74	89.39
Bearing	72.73	95.15	78.23	73.02	70.71	82.73	77.07
Egg	96.08	96.15	95.50	95.90	96.00	85.18	94.94
Metal	87.27	89.09	91.4	89.78	91.56	91.36	84.55

for most of the classifiers and data sets. From this table, we can conclude that both, *eVQ-Class* as well as *FLEXFIS-Class*, can nearly compete with the other (predominantly) top-10 data mining methods, falling behind at most 3% to 4% in worst cases (although these training algorithms were designed for the purpose of on-line training recieving the training data in sample-wise manner rather than the whole set at once). The necessity of further evolving classifiers, initially trained on off-line data, during on-line operation mode for significantly improving performance will be demonstrated in the subsequent section. In case of CD imprint data set #2 some further improvement of accuracies (around 1.2%) could be obtained when using Morozov discrepancy principle instead of heuristic regularization (see also [282]).

Also, the performance of the classifiers on the original image feature vectors (lying in the 17-dimensional aggregated feature space) is examined: the performance deteriorated by 22% in one case (bearing) and by 4% in two other cases (artificial data set #1 and CD imprint data set # 5). In all cases, the feature pre-processing step helped to improve the results (for details see also [118]). In all cases, the image data samples represented critical cases on the edge between showing real faults or pseudo errors. Taking into account that usually around 96% are non-critical images, which can be easily classified as good (as appearing as black images showing no deviations from the master), a final classification rate of over 99.8% is obtained in all cases, which was a satisfactory number for the industrial partners and their customers (as 99% was required in fact).

Another examination was dedicated to the sensitivity of *FLEXFIS-Class MM* with respect to disturbances in the feature vectors. With disturbances in the feature vectors it is meant that, according to the applied technique for finding the regions of interests (simply called objects), the feature vectors extracted from these objects may get different. For instance, when the object recognition part is not able to find the correct regions of interest, the feature vectors do not represent the real faulty and non-faulty objects correctly, hence are 'disturbed' by the object recognition part. The impact on classification accuracy of classifiers trained from the (more or less) disturbed features can be up to 6-7% (for the real-world data sets DBSCAN clustering algorithm [123] performed best, for the artificial data sets clustering based on normalized cut technique [396] could outperform the others, where in all cases simple connected components algorithm [409] was the worst performing one). An interesting point was that no matter which classification technique was applied, the impact stayed in the same intensity range. From this it can be concluded that 1.) the applied object recognition approach in fact does have an impact on the final classification accuracy and 2.) for all the classification methods the impact had a similar tendency, hence no favor for any classification method in terms of robustness against feature disturbance can be made. For further details on this analysis see [355].

9.1.5 Impact of On-Line Evolution of (Fuzzy) Image Classifiers

To show the impact of evolving classifiers, the following experimental setup was used to simulate real on-line operation mode for classifier update:

- Each of the data set was split into a training and test data set (as indicated in Table 9.1).
- The first half of the training set was used for initial model training (+ cross-validation coupled with best parameter grid search) as done in the off-line case.
- The second half of the training set was used for further on-line adaptation and evolution of the classifiers (keeping the same optimal parameter setting elicited during initial phase). This portion of the data set is sent sample-wise into the memory and the classifiers update process in order to simulate on-line operation/production mode (where items also come in one by one).
- The test set serves as separate data, temporally seen actually occurring after the training data, as samples and hence feature vectors in the feature matrices were stored in the same order as recorded.
- This test set served as basis for comparing the accuracies of evolved classifiers during the adaptation phase with 1.) re-trained classifiers (using all the data samples seen so far) and 2.) initially trained classifiers kept fixed during the whole on-line process. Both variants (re-trained and keeping static classifiers) are potential possible alternatives to incremental adaptation and evolution, even though re-training may have some downtrends in computational time, once more and more samples are loaded into the memory (refer also to Section 1.1.4). Therefore, it is interesting to see 1.) whether the accuracy of on-line updated classifiers can converge to the accuracy of re-trained classifiers (always seeing all data at once) $\rightarrow$ stability aspect as discussed in Sections 1.2.2.2, 3.1.1.2 and 3.1.1.4 and also examined in Section 7.1.2 for identification and regression problems; and 2.) whether the accuracy of the on-line updated classifiers outperforms the initial classifiers kept fixed $\rightarrow$ to answer the question whether an update is meaningful at all or not.
- EFC in connection with single-model and multi-model architectures (using *FLEXFIS-Class SM* and *MM*) as well as *eVQ-Class* in both variants were

applied for sample-wise on-line evolution of the image classifiers, and *k-NN* and *CART* to re-train the image classifiers during on-line phase.

- Only in case of CD imprints, eggs and rotor parts, the order of the data reflected the recording order during on-line production process. Hence, a fair report is possible only on these three scenarios.

In Table 9.4 the accuracies on the separate test data set are shown. The table is split into two parts, the first part denotes the accuracies on the static classifiers, i.e. the image classifiers are trained on the first half of the training period and then kept fixed during the whole incremental learning case, the second part shows the accuracies on the further evolved image classifiers.

From these values, it can be clearly seen that an on-line evolution of the image classifiers is strongly recommended, sometimes even mandatory (as increasing accuracy about 20%) in order to guarantee high-performance classifier and hence an overall increase of the quality of the production items.

eVQ-Class variant B can outperform *eVQ-Class variant A* for CD imprint and egg data and is equal to *eVQ-Class variant A* for rotor data, hence it is recommended to use *variant B* as there is no difference in the classifier structures to *variant A*, only the classification is done in another way (incorporating the position of the sample with respect to the decision boundary). The evolving fuzzy classifier (EFC) trained with single-model architecture has a weaker performance than EFC using multi-model architecture, which can clearly outperform the clustering-based classifiers for egg data. It is also remarkable, that this method has quite weak accuracy for CD imprint and rotor data after initial training and then can improve this more drastically due to evolution than any other method (compare also Figure 9.7 below). This is quite intuitive, as this method trains internally two (high-dimensional) fuzzy models (for the two classes) and hence requires more data than the other methods for a reasonable performance at the beginning. The last two lines in both parts of Table 9.4 mention the performance of two well-known batch classifiers, *k-NN* [174] and *CART* [58], which are completely re-trained after the whole incremental learning phase. In this sense, they serve as reliable benchmark, which accuracies for the different inspection problems can be achieved when using all the training data at once. Here, we can conclude that a re-trained batch classifier is not really better than EFC approaches (and their spin-offs).

Figure 9.7 shows the accuracies on the separate test data as they evolve during on-line adaptation phase. The left column demonstrates the evolution when using sample-wise adapted, evolved classifiers, the right column the evolution when using the re-trained classifiers. From top to bottom, this includes CD imprint data, egg data and rotor data. We can also realize that *CART* has some severe problems when doing re-training on CD imprint data (upper right image), as the accuracy decreases after 700 samples.

Table 9.4 Comparison of the accuracies (in %) between static image classifiers built on the first half of the training data and sample-wise evolved image classifiers with the second half of the training data for the three surface inspection problems of CD imprints, eggs and rotor parts.

	CD imprints	*Eggs*	*Rotor*
Static Image Classifiers			
eVQ-Class variant A	75.69	91.55	66.67
eVQ-Class variant B	88.82	90.11	66.67
EFC SM	78.82	95.20	66.67
EFC MM	73.53	95.89	54.67
k-NN	79.61	91.51	53.33
CART	78.82	91.78	52.00
Evolved Image Classifiers			
eVQ-Class variant A	89.22	91.12	86.67
eVQ-Class variant B	90.39	93.33	86.67
EFC SM	78.82	96.21	64.00
EFC MM	87.65	97.19	78.67
k-NN (re-trained)	90.98	96.06	74.67
CART (re-trained)	90.59	97.02	52.00

9.1.6 *Impact of Active Learning and Feature Weighting during Classifier Evolution*

A further test was made for simulating the case that an operator does not give a feedback for each single sample, but only for those ones for which the image classifiers were not quite confident in their decisions. This procedure saves significant amount of work effort for the operators and therefore helps the on-line machine vision system to provide more user-friendliness. The evolution of the test accuracies when applying this active learning approach (described in more detail in Section 5.3) are shown in Figure 9.8 for all data samples and applying a threshold of 0.9 on the confidence level (all samples below are used for adaptation). When comparing the top image (CD imprint data) with the top image in Figure 9.7, then we can realize that the *eVQ-Class* approaches perform similarly, whereas *EFC MM* has a better performance boost in the beginning but a weaker one later. This occurrence can be explained by the fact that at the beginning there were a lot of unconfident decisions belonging to the under-represented class and all the others which belonged to the

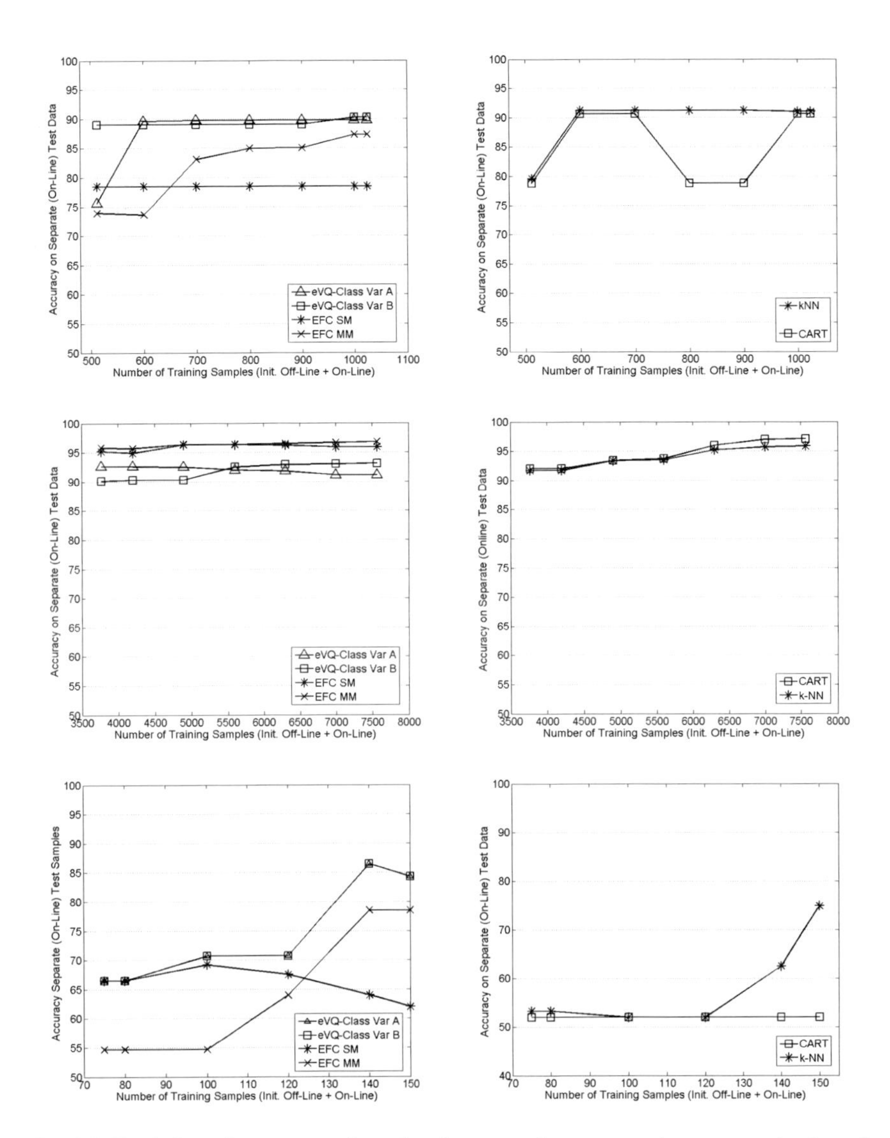

Fig. 9.7 Evolution of test accuracies when incorporating more and more samples into the image classifiers with the incremental/evolving methods: from top to bottom for CD imprint data, egg data and rotor data; left: for on-line sample-wise adaptive approaches (eVQ-Class and Evolving Fuzzy Classifiers = EFC), right for re-trained batch approaches (CART and k-NN)

over-represented class were classified with a high confidence. For the egg data set, the results of the active learning approach are very similar to the results without using active learning, except for the *EFC SM* approach, which performs a bit weaker in case of active learning. For the rotor data set, the active learning approach leads

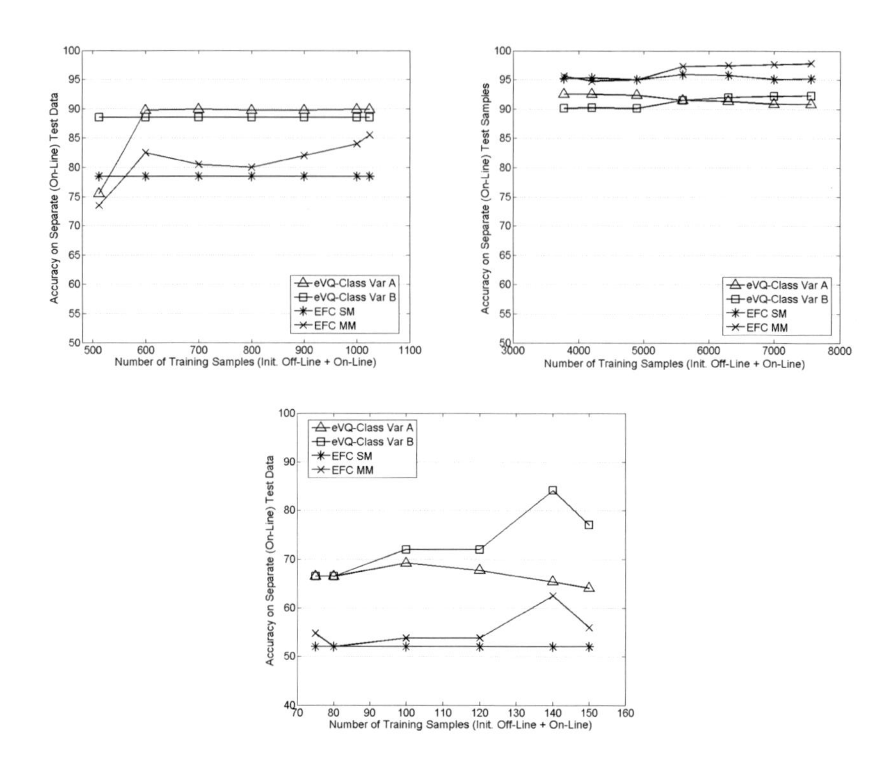

Fig. 9.8 Evolution of test accuracies when incorporating more and more samples into the image classifiers by the active learning approach (only samples with a low confident decision are incorporated): from top to bottom for CD imprint data, egg data and rotor data

to a severe drop in accuracy for all methods. The reason for this unpleasant effect is obviously the low number of samples (just 75 for adaptation), such that an omission of a portion of these for updating can be not afforded (in fact, finally only up to 15 samples were used for updating).

Furthermore, it was investigated how the incremental feature weighting concept demonstrated in Section 5.2.2 (Chapter 5) performs in the CD imprint inspection scenario. Figure 9.9 presents the evolution of the accuracies during on-line update of the fuzzy classifiers with *EFC MM* on CD imprint data (however labelled by a different operator, hence not directly comparable with the evolution in Figure 9.7, upper images). The first figure (a) visualizes those obtained without applying any feature weighting strategy implicitly. Here, we can recognize that in fact the accuracy increases with more samples included into the incremental update, however this increase is not as stable as in case when applying leave-one-feature-out feature weighing (shown in (b)) or dimension-wise feature weighting (shown in (c)) (both described in Section 5.2.2), as showing some down trends, especially for the first part of on-line samples (up to sample #800). Furthermore, the levels of accuracy are significantly higher in both incremental feature weighting variants, especially at

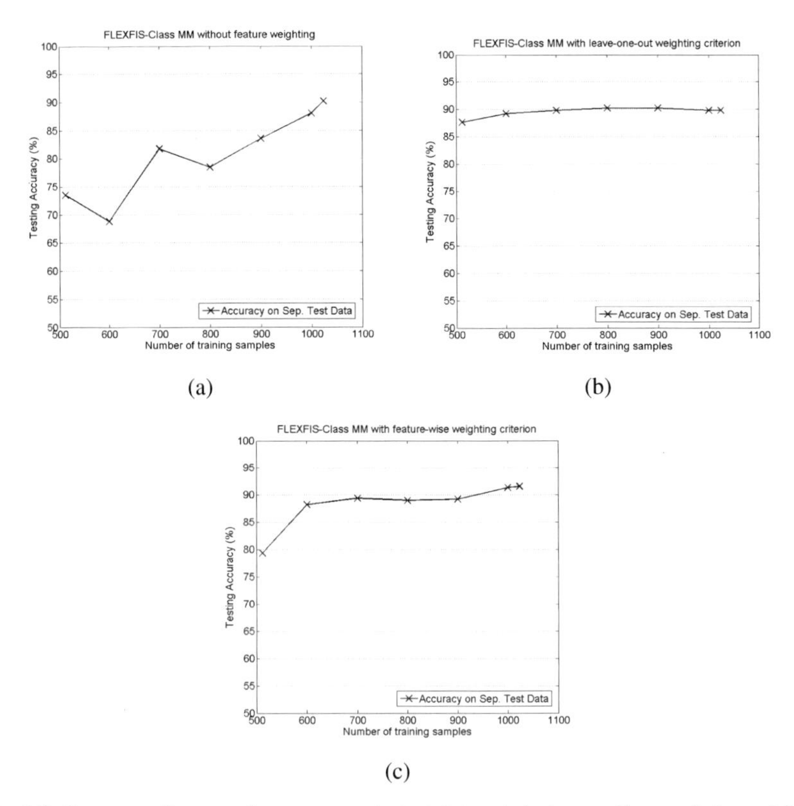

Fig. 9.9 Progress of accuracies on separate test data set during on-line evolution of fuzzy classifiers with *EFC MM* for CD imprint data set, (a): without feature weighting, (b): with leave-one-out criterion and (c): with feature-wise separability criterion.

the beginning of the whole update process, i.e. from the 512th sample on — note that the first 512 samples are used for initial training and feature weights are also included there in (b) and (c), the first box in the accuracy trends denotes the accuracy after the initial training. The bigger discrepancy in predictive performance at the beginning is no surprise, as curse of dimensionality (74-dimensional problem) is more severe when having less samples included into the evolving fuzzy classifiers. When more and more samples are loaded into the update of the fuzzy classifiers, the curse of dimensionality effect usually gets weaker, hence the impact of feature weighting (or selection) diminishes (but is still present).

9.1.7 Impact of Incremental Classifier Fusion

Finally, we study the impact of incremental classifier fusion on the predictive accuracy on the separate test data set. Therefore, three incremental base classifiers *eVQ-Class variant A* are applied, an incremental version of the *Naive Bayes*

Table 9.5 Accuracies (in %) of base classifiers and ensembles after the initial batch training using the first half of the training data

	Op. 1	*Op. 2*	*Op. 3*	*Op. 4*	*Op. 5*
Base Cl.					
eVQ-Class	**85.88**	**89.80**	75.29	75.69	**70.00**
NB	82.55	82.35	**81.18**	**78.63**	69.22
k-NN	68.04	79.61	79.41	77.06	64.31
Ensemble Cl.					
FI	**85.88**	**90.59**	**85.29**	**86.08**	**72.35**
DT	78.63	82.35	80.39	78.43	64.71
DS	82.55	82.35	80.39	78.63	64.31
DDS	78.63	82.35	80.39	78.63	64.31
DC	**85.88**	**90.59**	84.51	85.88	64.31

Table 9.6 Accuracies (in %) of base classifiers and ensembles after the incremental adaptation of the initial models using the second half of the training data — see improved accuracies among those in Table 9.5

	Op. 1	*Op. 2*	*Op. 3*	*Op. 4*	*Op. 5*
Base Cl.					
eVQ-Class	**82.94**	90.78	**88.82**	**89.22**	65.88
NB	78.63	80.98	81.76	80.00	61.96
k-NN	81.76	**90.98**	84.12	88.43	**83.33**
Ensemble Cl.					
IFI	**85.69**	**90.78**	89.02	**89.41**	70.00
IDT	**85.69**	**90.78**	89.02	**89.41**	82.94
IDS	**85.69**	**90.78**	89.02	**89.41**	76.27
IDDS	**85.69**	**90.78**	**89.80**	**89.41**	82.94
IDC	82.35	**90.78**	84.12	**89.41**	**83.33**

classifier and *k-NN* (with permanently updating the reference data base), and the incremental (trainable) fusion methods *Fuzzy Integral (FI)*, *Decision Templates (DT)*, *Dempster-Shafer combination (DS)*, *Discounted Dempster-Shafer (DDS)* and *Incremental Direct Cluster-based Fusion (IDC)* using *eVQ-Class* as learning engine (for details on these approaches see Section 5.4 and [381] [380]).

The results on the five CD imprint data sets (labelled by five different operators) are presented in Tables 9.5 to 9.7: for initial classifiers and fusion methods, updated classifiers and fusion methods and for re-trained classifiers and fusion methods on all training samples. The interpretation of these results can be summarized as follows:

1. At any stage of the learning process the performance of the best ensemble method is in general equal to or better than the best base classifier, except for Op. 2 (if the classifiers are updated incrementally – see Table 9.6), for which the result of the best base classifier is 0.2% better than the best ensemble method. The best

Table 9.7 Accuracies (in %) of base classifiers and ensembles after batch training using all training data — compare with Table 9.6

	Op. 1	*Op. 2*	*Op. 3*	*Op. 4*	*Op. 5*
Base Cl.					
eVQ-Class	81.76	90.39	**87.25**	87.25	64.90
NB	**83.53**	87.65	86.86	87.06	64.71
k-NN	82.15	**91.18**	84.31	**88.43**	**82.94**
Ensemble Cl.					
FI	**84.71**	90.98	**87.84**	88.24	69.41
DT	80.78	**91.18**	82.55	**88.63**	81.57
DS	**84.71**	90.98	**87.84**	**88.63**	74.71
DDS	82.16	90.98	82.55	**88.63**	82.55
DC	82.16	**91.18**	82.16	88.04	**82.94**

improvement achieved is 7.45% for Op. 4 if both classifiers and ensembles are kept static. In this sense, we can conclude that incremental classifier fusion helps to improve the performance of single incremental (base) classifiers.

2. The accuracy of the incremental classifier fusion methods is in general equal to or better than the accuracy of the classifier fusion methods which are trained in batch mode for the entire training data (comparing Tables 9.6 and 9.7). This shows the incremental learning capabilities of the proposed concept in Section 5.4 and the fusion methods therein. This is quite an important result, as the batch ensembles see the data set all together, rather than one data item at a time. In this sense, it opens the question whether time-intensive retraining in the on-line data streams is meaningful at all.
3. It is a necessity to perform an incremental adaptation of the base classifiers and the ensemble methods during the on-line phase, as the accuracies of the static ensembles may be significantly worse (up to 10.98%) than the updated ensembles, as can be seen by comparing Tables 9.5 and 9.6.
4. Even though one of the incremental base classifiers (*incremental Naive Bayes*) does not always perform as well as the other two (*k-NN* and *eVQ-Class*), the incremental ensemble methods are not affected by this as their accuracies remain at a high level. This underlines another benefit of the use of the (incremental) classifier fusion methods: their robustness with respect to poorly performing classifiers. In fact, this diversity in the decisions of the classifiers helps to build a diverse ensemble, enabling it to increase their performance.
5. From the experiments presented here it is not clear which incremental fusion method is the most suited for this kind of application. A slight preference could be given to the *Incremental Discounted Dempster-Shafer Combination (IDDS)* method, as it is always amongst the best performing algorithms.

We also studied a larger variety of base classifiers and obtained similar or generally even better results for incremental classifier fusion techniques. The larger the diversity of base classifiers, the higher the accuracy obtained. Figure 9.10 presents

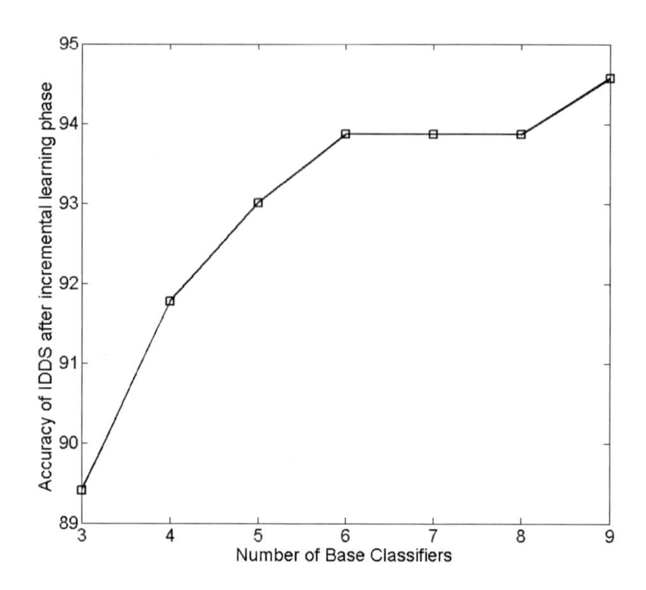

Fig. 9.10 Accuracy (y-axis) versus diversity of the base classifiers (x-axis) on Op.4 labels of CD imprint data when using *IDDS* as classifier fusion method

a plot of accuracies achieved for Op. 6 data (y-axis) when using *Incremental Discounted Dempster-Shafer Combination (IDDS)* on different varieties of classifiers (x-axis) starting from 3 (*k-NN*, *Naive Bayes* and *eVQ-Class*) to 9 base classifiers, increasing the number by one classifier at a time according to the order in the following itemization:

- Incremental *Naive Bayes*
- *k-NN* [111] updating reference data-base
- *eVQ-Class*
- Incremental *SVMs* [102]
- *Ada-boosted* weak classifiers [88] (retrained)
- Evolving fuzzy classifiers using
  - *FLEXFIS-Class MM* and
  - *FLEXFIS-Class SM*
- statistical-oriented (quadratic) discriminant analysis function [174] (updated by updating the covariance matrices for the single classes and for the whole data — see Section 2.1.2 for covariance update)
- Decision trees with *CART* approach [58]) (re-trained)

From this figure, we can realize that adding *incremental SVMs*, *Ada-boost* and *FLEXFIS-Class MM* brings significant step-wise increase in classification accuracy. Then, a sort of saturation is reached, especially adding *FLEXFIS-Class SM* and statistical discriminant analysis classifiers does not bring any further improvement.

The reason for this is that *FLEXFIS-Class SM* is in large parts correlated to *eVQ-Class*, see Sections 3.1.2.1 and 3.1.3 (the only difference is a projection of clusters to axis (→ rules) and including the width of the rules in *FLEXFIS-Class SM*, whereas in *eVQ-Class* only the distance to the centers is measured when classifying new samples), whereas statistical discriminant analysis classifiers were among the worst performers of all base classifiers (together with Naive Bayes). Finally, the adding of *CART* (in re-trained mode) helped to improve the accuracy a little more.

9.2 Emotion-Based Texture Classification

9.2.1 Motivation

The second part of this chapter is dedicated to emotion-based visual texture classification.

With *emotion-based visual texture classification* it is understood that for a given texture visually shown to a human being, the 'affected feeling', or in psychological sense the 'cognitive perception and emotion' this texture triggers, should be measured.

Based on this measuring, textures can be classified to different groups according to the cognitive emotions/perceptions they trigger. A visual texture is a concept that is present in our every-day lives [390] and contains information about an object's surface. It is considered as a specific case of an image, which

1. contains significant variation in intensity levels between nearby pixels [390] and
2. is stationary, i.e. under a proper window size, observable sub-images always appear similar. [445]

Six examples of visual textures are presented in Figure 9.11.

The motivation for a psychological-oriented texture classification is motivated by the fact that nowadays visual textures play a significant role for comfort, certain feelings, perceptions, finally causing emotions in our every day life. This is because textures are present in living rooms (carpets, wallpapers etc.), shop designs, or on surfaces of products sold to customers. It is therefore an essential issue to give people/customers a comfortable feeling resp. to inspire them to buy certain products. For instance a design of wallpapers and storage racks in a heavy metal shop usually has a completely different appearance than in a jewellery store. Therefore, to find appropriate textures addressing specific expectations of regular and potential new customers by triggering the adequate perception or even emotional reaction is an important issue.

To do so, the first step to take is to examine and understand human perception and emotion associated with visual textures. The result of this process should be a kind of model which explains to a user which texture will cause (probably) which emotion to a certain degree (Section 9.2.2). Based on such a model, a reverse process can

Fig. 9.11 Some visual texture examples

be initiated, which is able to fully generate and synthesize new textures evoking an a priori defined (target) emotion. Thereby, an emotion is usually defined by a numerical vector in the aesthetic space containing the most important adjectives and their antonyms defining an emotion from psychological side (for instance 'rough-smooth' or 'artificial-natural'). For the reverse process, genetic algorithms can be exploited (as optimization process in the low level feature space — for details see [154]).

9.2.2 *Human Perception Modelling (State of the Art)*

While in the past a lot of research has been dedicated to the characterization of texture segregation, comparatively little effort has been spent on the evaluation of aesthetic texture properties. In particular, the question which perceived textural properties are the essential building blocks of aesthetic texture content remains unanswered [419]. A lot of attempts have been made to set up models for associating human perceptions with visual textures. For instance, in [218] a textile indexing system is proposed which is able to classify textile images based on human emotions. This is done by extracting features of textile images such as colors, texture and patterns which were used as input for feed-forward neural network models. In [405], a mapping-function is derived by exploiting neural networks for mapping features extracted from clothing fabric images to several human Kansei factors such as (strange,familiar), (bright, dark) or (clear,indistinct).

The majority of research related to human visual texture perception, conducted in the last decades, was focused on describing and modelling texture segregation and classification phenomena [204], [41]. Comparably little research has been dedicated to understand or measure higher level texture perception, such as naturalness, roughness, feeling (like/dislike) or beauty [190].

9.2.3 *Machine-Learning Based Modelling*

9.2.3.1 Layered Model Structure

A novel attempt for visual texture perception modelling is presented in [419], where a hierarchical multi-layer model with data-driven machine learning techniques is generated. This layered model

1. is capable to predict human aesthetic judgements with high accuracy and
2. is psychologically interpretable and therefore allows for analysis of the relationships between low level texture features and aesthetic properties.

For interpretability reasons, a model structure is defined, consisting of three different layers (inspired by Norman [325]):

1. *affective layer*: descriptive words which relate to physical properties of the object; "how can the object be described?"
2. *judgement layer*: words which describe higher level properties of the object; "what the object says about itself?".
3. *emotional layer*: words which describe emotional properties or feelings; "how people feel when interacting with the object?".

Each layer contains a number of aesthetic properties (termed intermediate adjectives from now on), which were categorized based on inter-subject variability of judgements and three assignment questions. These questions determine whether the property represents affective, judgemental or emotional aspects of human texture perception.

A visualization of the layered model structure is presented in Figure 9.12. In all three layers, low level features are used as input which are directly extracted from the textures by image processing techniques. These low level features yield a kind of visual characterization of textures, representing all the texture properties which are measurable and therefore completely objective. Hence, they can be seen as additional valuable input for setting up models which are able to associate human emotions with visual textures. Using the aesthetic properties alone, would yield models with higher psychological interpretation power, but on the other hand lose some ground in association accuracy.

9.2.3.2 Definition of Core Adjectives

27 aesthetic antonyms were defined, such as *rough-smooth, elegant-not elegant* or *like-dislike* based on a series of preliminary focus group sessions, as listed in Table 9.8. A subset of these antonyms is related to the 13 aesthetic dimensions defined in [213], which have been successfully applied in other modelling approaches such as [218]. The remaining antonyms were chosen based on extensive discussions with psychological experts, who recommended using them for a better understanding of the cognitive textures-perception association process. Six of these properties

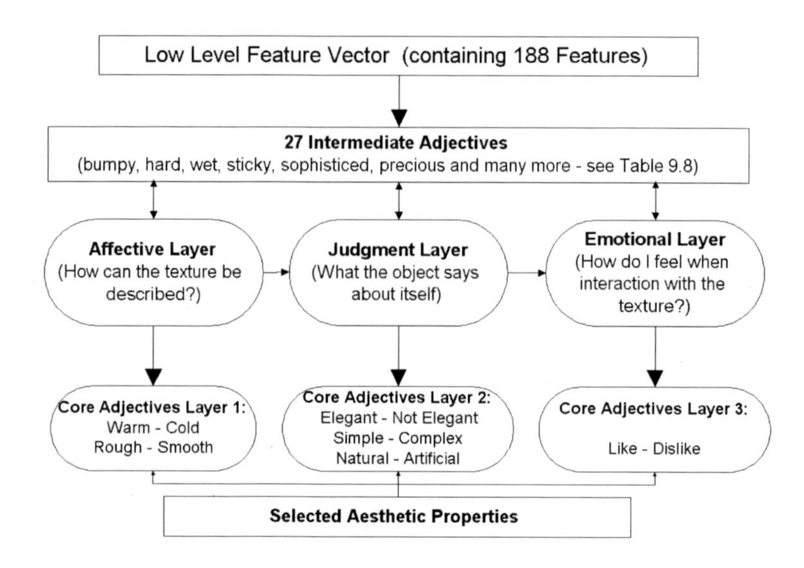

Fig. 9.12 Structure of the hierarchical feed-forward model of aesthetic texture perception [419]. The model consists of 3 layers, containing 27 intermediate aesthetic properties. From these properties 6 are selected (2 from the affective layer, 3 from the judgement layer and 1 from the emotional layer) as core set, representing the final output of the texture↔perception association models; note that adjectives in layer #1 are used as inputs for predicting adjectives in layers #2 and #3, adjectives in layer #2 are used as input for predicting adjectives in layer #3

Table 9.8 27 aesthetic antonyms divided into three layers, with *C* marking members of the core set [419].

Property	Layer #	Property	Layer #
Wood/Non-Wood	1	Simulated/Original	2
Stone/Non-Stone	1	Indulgent/Resistant	2
Glossy/Beamless	1	Precious/Worthless	2
Bumpy/Flat	1	Premium/Inferior	2
Silky/Non-Silky	1	Practical/Non-Practical	2
Grainy/Non-Grainy	1	Cheap/Expensive	2
Harsh/Hollow	1	Distinguished/Non-Dist.	2
Wet/Dry	1	Sophisticated/Non-Sophisticated	2
Warm/Cold *C*	1	Modern/Old-fashioned	2
Rough/Smooth *C*	1	Simple/Complex *C*	2
Hard/Soft	2	Elegant/Non-Elegant *C*	2
Pretentious/Plain	2	Natural/Artificial *C*	2
Sticky/Non-Sticky	2	Like/Dislike *C*	3
Delicate/Non-Delicate	2		

(together with their antonyms) were selected as core properties (also called core adjectives), responsible for defining the emotional aesthetic space and final outcome of the association models, namely:

- (warm,cold)
- (rough,smooth)
- (simple,complex)
- (elegant,non-elegant)
- (natural,artificial)
- (like,dislike)

The values of each of the properties listed in Table 9.8 are defined to lie in the range $[-100, 100]$ (100 means the property (e.g. 'wood') is fully the case, -100 means the antonym (e.g. non-wood) is fully the case). The range $[-100, 100]$ represents a natural support of human cognition: human beings tend to annotate textures regarding their perceptions/emotions with the help of percentual fulfillment degrees (100 would mean that a specific perception is caused by a given texture to 100%); in order to be able to capture also the opposite perception within one value (instead of two values, e.g. one for 'warm', another separate for its antonym 'cold'), the range was expanded from [0,100] to [-100,100]. A value of 0 then reflects perfectly a neutral feeling (e.g. neither 'warm', nor 'cold'). For instance, a six-dimensional core adjective vector such as $[-50, 100, 100, 0, -100, 50]$ would represent a texture (according to the same order as in above itemization) which is 'cold' with a degree of 50, very 'rough', very 'simple', neither 'elegant' nor 'non-elegant' (0 always represents a neutral feeling), completely 'artificial' and to 50% liked.

9.2.3.3 Modelling Scenario

Based on interview data, i.e. a texture set plus judgements on this texture set for each property obtained from various human beings, the goal was to design a single prediction model for each core property (following the layer structure illustrated in Figure 9.12, e.g. using properties from layer #1 as inputs for properties from layer #2 and #3). The focus was placed on a single model, instead of distinct models for single interviewed persons, in order to investigate general relationships, applicable to the population as a whole. The resulting model is applicable to different people without re-adjustments. Dividing the interview data according to groups of people would most likely increase the accuracy of the models (as getting more specific), at the price of reduced generalization. As the judgements were made based on a continuous scale, the task was to build regression problems when extracting models for the different layers from the data.

On the Tradeoff between Accuracy and Interpretability

For the model setup process, there are two central issues to be dealt with:

- accuracy of the model and
- interpretability of the model.

Accuracy is important in order to have a model with high predictive power, i.e. mimicking human aesthetic perception for given visual textures. Obviously, in case of an inaccurate model, the predictions are worthless (e.g. to guide a product designer who has to select a texture for an object surface). Even more important is, that the conclusions, based on the relationships and structure of the model are of lesser relevance if the model predictions are too different from human judgements. Interpretability on the other hand is important, too, for gaining insights into the relationships between different aesthetic properties. This enables an investigation whether models are reasonable from a psychological point of view or not. Interpretability of the models is closely linked to their complexity. Hence, it is obvious that accuracy and interpretability somehow contradict each other: a perfectly interpretable model will just use a single term or structural component, but usually increases the model error as it is too simple to approximate complex, non-linear relations between texture and aesthetic perceptions. Finally, it is a matter of a good tradeoff between interpretability and accuracy.

An essential issue for assuring a good tradeoff is the choice of an appropriate model architecture. For the training purposes, linear regression models and fuzzy systems were applied, as both are appropriate choices regarding interpretability of the models (in contrast to neural networks, applied in [218] [240]). Linear regression models are providing a flat model structure, with linear weights (when multiplied by the ranges of the corresponding features or aesthetic properties from lower layers) indicating the importance of the variable. The order of the variables in the model also gives rise about their importance, according to a filter feature selection procedure after [153], where the most important feature or aesthetic property is selected first, then the second most important one etc. (see also Section 7.1). Linear regression models usually require less training samples than fuzzy systems in order to build reliable models.

Extended Optimization Measure

In order to achieve a good tradeoff in the model training process, it is important to avoid optimization using a standard error measure that takes into account the accuracy of the models exclusively. Therefore, a punishment term was incorporated for measuring the model complexity into the error measure. This resulted an extension of the mean absolute error (MAE) (measuring the absolute deviation of the interview values and the model responses) by adding a punishment term (representing model complexity, normalized with the maximum model complexity) and performing a convex combination with the normalized MAE ($\rightarrow$ MAE punished).
The following formula for the MAE punished was defined by [419]:

$$MAE\_punished = \alpha \frac{MAE}{range(MAE)} + \beta \frac{ModelComplexity}{max.ModelComplexity} \tag{9.2}$$

with α and β linear weights summing up to 1. These are adjustable and control the ratio of accuracy versus interpretability (interpretability is emphasized with increasing β). The effect of the new error measure on the final model is shown in

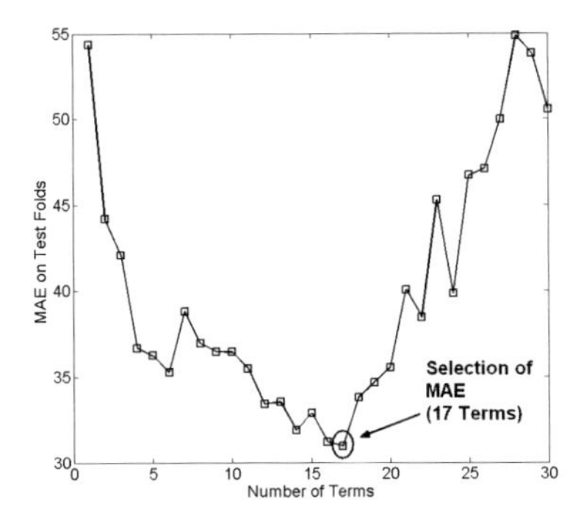

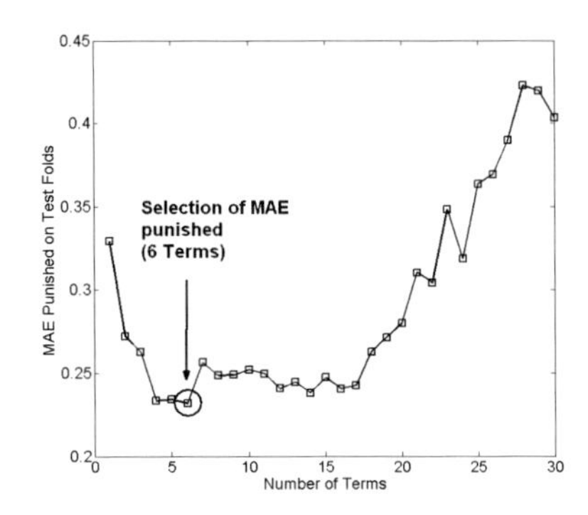

Fig. 9.13 Left: selection of the final model complexity based on the MAE curve (17 input terms would be selected as optimal, leading to a complex model). Right: selection of the final model complexity based on the new MAE punished measure, the first distinct local minimum is selected, suggesting only 6 terms, resulting in an increased original MAE of only 5 (i.e. a relative MAE of 2.5%).

Figure 9.13. Note the difference between the left image (using standard MAE) and the right one (using MAE punished). In the left image, the global minimum of the MAE is reached for a linear regression model with 17 terms, resulting in a highly complex model. The MAE punished forces the error curve to increase with the number of terms (note the monotonically increasing tendency over the number of terms), achieving a global minimum for a model with only 7 terms (which is at the first distinct local minimum in the conventional MAE curve). The model complexity for a linear regression model is simply defined by the number of inputs (=number of regressor terms), for a fuzzy system by the number of rules times the number of inputs (as each rule has the whole number of inputs in its antecedent parts).

The Training Process

In a first round of the training process each of the intermediate adjectives contained in the first model layer is used as target adjective in a regression modelling step, whereas the whole set of low level features is used as input. The second round uses the whole set of low level features plus all intermediate adjectives in the first layer (the values obtained from the interviews) as input to the regression modelling step for the target adjectives from layer #2. The third round uses the whole set of low level features plus all intermediate adjectives in the first layer and second layer (the values obtained from the interviews) as input to the regression modelling step for the target adjectives from layer #3. Prior to each round a filter forward variable selection procedure is applied, selecting the most appropriate variables (low level features + intermediate adjectives in a joined set of inputs) to explain the target adjectives. In this sense, the order of the selected variables gives rise to their importance degree in the models and serves as valuable information for psychologists. In Section 9.2.5, we will see that there are even some synergies about the importance levels of our machine-learning based models with psychological-based models.

The regression model building step for both model types (linear regression and fuzzy systems) is based on a 4-fold cross-validation procedure [411], using the MAE punished as error measure. Cross-validation, estimating the expected model error, is repeated several times, varying the parameter responsible for steering the complexity of the obtained models (number of inputs in case of linear regression, number of rules and inputs in case of fuzzy sets). The parameter setting leading to the optimal models in terms of the best average error measure (averaged over all 4 folds) is used for the final model training step. As error measure for the models for the intermediate adjectives two variants are applied: 1.) the average MAE punished measure as defined in (9.2) and 2.) the average punished classification accuracy, which is obtained in the same way as MAE punished, but using classification accuracy instead of MAE in (9.2). The classification accuracy is obtained by mapping the regression outputs, lying in $[-100,100]$ onto $\{-1,1\}$ (0 is mapped onto 1 by definition) and comparing with the corresponding interview values also mapped to $\{-1,1\}$: the percentual accordance of these values denotes the classification accuracy. Finally, that parameter setting is selected for which

$$\min(\min_{1,\dots,params} MAE\_punished, \min_{1,\dots,params} Class\_punished) \tag{9.3}$$

is achieved. The reason for this procedure is that for some intermediate adjectives, it is expected to obtain a very high classification accuracy (so a lot of textures are judged into the same direction by our models and by the human beings), whereas the spread in the $[-100,0[$ resp. $[0,100]$ range and therefore MAE punished is quite high (this was the case for 'stone' and 'wood'). In this case, by using the classification mappings of these (intermediate) adjectives as inputs into the models (and the model training step) for the adjectives in the subsequent layer(s), the accuracy of these models can be increased (as the preliminary layer models are more accurate).

For generating fuzzy systems from the experimental data (within the cross-validation step), *FLEXFIS* (in its batch version) is applied (including rule evolution and consequent learning).

9.2.4 Experimental Setup

A texture set containing 69 various real-world textures was used, including patterns from stone, wood, carpets, wall-papers, ground floors, nylons etc. From these 69 textures a set of 188 low level features was extracted, characterizing these textures by objective image-processing-like properties. According to this quite huge number of low level features, it is possible to represent a huge variety of important texture characteristics. Due to space restrictions, it is not possible to list all of the 188 features and to explain their meanings and impact on perception modelling in detail. Loosely, they can be divided into six big groups, namely:

- Gray level co-occurrence matrix features (88 in sum)
- Neighborhood gray-tone difference matrix features (10 in sum)

Fig. 9.14 Visual texture perception annotation tool: the main panel shows the current visual texture; in the bottom the rating area for an aesthetic properties, including the antonyms, can be found; the rating is done by a single mouse-click on the rating area.

- Tamura features (6 in sum)
- Fourier power spectrum energy (44 in sum)
- Color features (9 in sum)
- Gabor energy map features (31 in sum)

For further details refer to [420] and [419].

Ratings of subjects for all aesthetic properties defined in Section 9.2.3 were collected. In total, 19 subjects (of different gender, age, education, nationality, etc.) rated 69 visual textures against 27 aesthetic antonyms. A software tool supported the human beings by displaying the texture images and collecting the ratings. The tool is composed of 2 major components, as depicted in Figure 9.14. Centered, the visual texture is displayed. The user rates the texture by placing a mouse click on the continuous rating area in the bottom of the tool. The rating area is labelled with the current aesthetic antonyms. Instead of using a 7 point rating scale as [84], a continuous rating scale within the interval $[-100, 100]$ was used. This is particularly useful as regression instead of classification models were built. This circumvents the

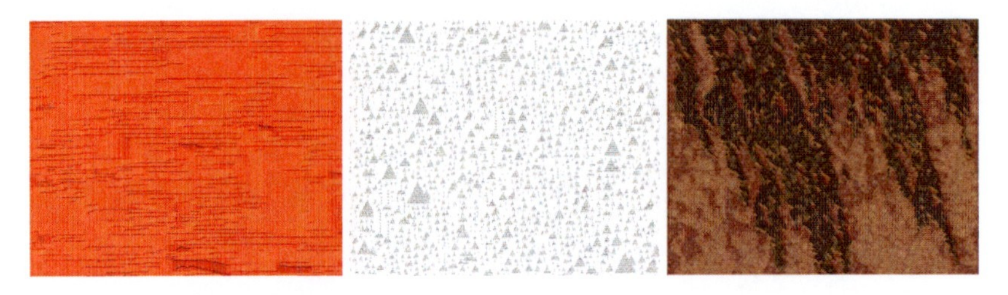

Fig. 9.15 Three extreme texture samples synthesized, using a texture mixing approach, close to the corner points in the aesthetic space, defined by the 6 core properties.

problem of an imbalanced data set (i.e. classes are not equally distributed), usually causing models with weak predictive performance for the under-represented classes [349]. The visual textures are presented in pseudo-random order for each aesthetic property. The total experiment takes about 1 to 1.5 hours. Both, the ordering of the aesthetic properties and the polarity of a single aesthetic property was randomized per subject. The subjects were able to choose between an English or German translation of the software tool. As someone wants to obtain general insights into the relevant factor of human texture perception, it is not feasible to deal with individual experimental data (e.g. to build an individual model for every subject). On the contrary, the mean over (subjects) experimental results was applied for the model building in order to use the average impression how people feel when getting these 69 textures shown. In case of a higher number of interviewed people, it is plausible to provide a more enhanced pre-processing based on statistical examinations. The outcome of this annotation process (numerical values in $[-100,100]$ for all properties) together with the extracted low level features were used as input (69 training samples, 188-dimensional feature space) for the modelling process as described in the previous section.

Additionally, a set of 64 textures were synthesized, according to the definition of $2^6 = 64$ corner points in the six-dimensional core adjective hyper-cube $[-100,100]^6$. These 64 textures, representing a kind of 'extreme' scenario from human perception side (hence also called 'extreme' textures), were annotated by 9 English speaking and 9 German speaking people, again by using the tool as shown in Figure 9.14. These textures were used as separate test data set for final verification of the models built on the 69 textures, also to verify the intensity of error propagation through the model layers. The goal was to elicit the achievable bounds of the models, i.e. which increase of the model error (MAE) can be expected in case of extrapolation (=low level features lying significantly out of the ranges of the low level features extracted from the training texture set). Three texture examples of this extreme set are shown in Figure 9.15.

9.2.5 *Some Results*

Table 9.9 shows the results for the core properties defining the final human aesthetic perception when applying linear regression models for each property in the layer

Table 9.9 Comparison of the modelling results, using linear regression models. The column *MAE*, *MAE pun.* and *Complexity* illustrate the effects of our optimized error measure. The rightmost column contains the MAE results for error propagation (sending low level features to the top layer and taking the *predicted* values from previous layers as input for the subsequent layer).

Core property	MAE	MAE pun.	Complexity	MAE whole layer
Warm/Cold	12.55	0.0997 (16.00)	3 (38)	15.52
Rough/Smooth	12.02	0.1096 (16.16)	6 (7)	19.97
Simple/Complex	8.68	0.1041 (15.73)	5 (46)	15.67
Elegant/Non-Elegant	10.83	0.0996 (11.23)	12 (13)	16.22
Natural/Artificial	10.55	0.136 (15.87)	6 (7)	21.41
Like/Dislike	6.27	0.072 (9.57)	6 (30)	14.49

Table 9.10 Prediction error and complexity for the aesthetic prediction model built from evolved fuzzy models, following the structure outlined in Figure 9.12.

Core property	MAE	MAE pun.	Complexity	MAE whole layer
Warm/Cold	14.33	0.1000 (15.96)	24	17.16
Rough/Smooth	16.89	0.1451 (18.72)	20	29.98
Simple/Complex	12.31	0.1122 (15.83)	42	20.03
Elegant/Non-Elegant	14.44	0.057 (14.44)	50	24.77
Natural/Artificial	15.81	0.0647 (15.81)	30	25.55
Like/Dislike	10.47	0.0632 (10.47)	30	22.41

structure. The second column shows the minimal mean absolute error for 1 to 50 inputs (= number of regressor terms). The third column shows the minimal MAE punished value and in braces the corresponding MAE values (in order to compare the loss in the accuracy when gaining less complex and hence easier interpretable models). The fourth column contains two values of model complexity: the first one is the final model complexity by using MAE punished, the second number (in braces) is the original model complexity. The fifth column shows the MAE when using only low level texture features as input for the model (omitting the ratings of the intermediate properties), to elicit the effect of error propagation and error summation through the layers. When comparing these numbers with those in the second column, the reader can observe the intensity of this effect.

Table 9.10 contains results (i.e. prediction error measures and complexity), utilizing evolving fuzzy models instead of linear regression models (Table 9.9) in order to be able to track also non-linear relationships between low level features, intermediate and core adjectives. However, in this case we can see that more non-linearity in the modelling does not really pay off in terms of accuracy. One explanation is the low number of available data samples compared to the input dimensionality of the problem (69 samples versus 188 + some intermediate adjectives input features). Furthermore, the model complexity is higher, as every local region (= number of inputs times the number of rules) needs to be counted. On the other hand, more

Table 9.11 Results (MAEs) for the prediction of human perception on a set of extreme textures synthesized based on corner points in the aesthetic space; these define a kind of worst case scenario on the expected errors (MAEs).

Core property	MAE English	MAE German
Warm/Cold	18.07	25.35
Rough/Smooth	37.73	31.62
Simple/Complex	34.32	30.94
Elegant/Non-Elegant	32.61	28.89
Natural/Artificial	31.07	46.16
Like/Dislike	24.25	22.38

detailed interpretation can be obtained, as different aesthetic properties are usually important with different intensity in different local regions. For instance, it could be found out that the influence of 'beamless' for the core property 'warm' was much higher for non-wood-type textures than for wood-type textures. Finally, we may realize the achievable bounds of (evolving) fuzzy systems according to the number of samples used and the dimensionality of the problem setting.

Regarding the extreme data set, Table 9.11 shows the results in terms of MAE, computed for German and English speaking subjects. We can see that for 5 out of 6 core properties, the MAE is around 30 (i.e. 15% difference to the ranges of the core properties, $[-100, 100]$). When taking into account, that these textures are among the set of the most extreme ones, we can conclude that the models are able to predict human aesthetic perceptions for new textures with an accuracy of 15% to 20% in a worst case scenario.

Another major outcome of our examination was that the fuzzy model built for 'natural' with *FLEXFIS* had some synergies with the psychological models obtained through mediator analysis coupled with principal judgement findings [419], underlining the psychological plausibility of the machine-learning based models; see [190] for further details on mediator analysis. In fact, the final evolved fuzzy model for natural contained the following inputs (order matters: most important ones first):

- 'fourier energy' (low level feature)
- 'wood' (layer #1 adjective)
- 'stone' (layer #1 adjective)
- 'GLCM feature measuring the texture randomness' (low level feature)
- 'Gabor filter (vertical orientation, secondary finest) response' (low level feature) and
- 'bumpy' (layer #1 adjective).

with highest average effects on the output of the model caused by 'wood' and 'Gabor filter (vertical orientation, secondary finest) response'. The effect of a variable can be interpreted as a comparison measure whether a change of variable i with X percentage affects the outcome of the model equal/more/less than the change of variable j with X percentage. For instance when variable i has an effect of 30 and variable j has an effect of -50, then someone can conclude that small changes in j influence the output of the model more than small changes in i and this in the opposite direction (because of the opposite sign). As this depends on the position of the current input feature vector (a non-linear fuzzy model has a changing, non-constant gradient), the average effect over a sample of grid points defined in the multi-dimensional space was taken.

Chapter 10
Further (Potential) Application Fields

Abstract. The first section in this chapter deals with a summary of various application fields, where evolving fuzzy systems approaches have been successfully applied and installed, covering *evolving inferential sensors*, tracking of objects in video streams, adaptive knowledge-discovery in bio-informatics and dynamic forecasting tasks in financial markets. The second part of this chapter (Sections 10.2.1 to 10.2.3) is dedicated to application fields, where evolving fuzzy systems and classifiers have not been applied so far (to the best of our knowledge, status: 30th of June 2010), but may serve as potential components in the near future in order to enhance adaptability, flexibility, degree of automatization and also the level of human-machine interaction in industrial systems. These enhancements can be seen as basic building blocks for a better productivity and higher process safety and user-friendliness of these systems during on-line operation mode. The potential fields of application for evolving fuzzy systems handled in this chapter are the following:

- Adaptive open-loop control (Section 10.2.1): evolving fuzzy systems are applied for automatic generation of non-linear controllers from data, which can be further updated during on-line control mode to permanently improve the control rules/strategies, e.g. adapting to new control system states.
- Chemometric modelling (Section 10.2.2): deals with quantification of the composition of chemical substances in materials; here, it is a big challenge to automatize the quantification process by (non-linear) models extracted from measured spectral data and updated during on-line production process.
- Interactive machine learning (Section 10.2.3): a generic concept for enhancing interaction and increasing knowledge exchange between humans and machine learning systems; evolving fuzzy systems could play a central role in such a concept, as, when combined with the techniques for improving interpretability of EFS (Chapter 6), providing linguistic and also visual insight into the behavior and underlying dependencies of the process.

E. Lughofer: Evolving Fuzzy Systems – Method., Adv. Conc. and Appl., STUDFUZZ 266, pp. 393–410.
springerlink.com

10.1 Further Applications

10.1.1 eSensors

Soft sensors play a crucial role in various industrial systems, as they are able to predict target variables from a source of input variables by providing accurate real-time estimates. With the help of soft sensors it is possible to reduce the costs for expensive hardware sensors and even to circumvent (sensor) installations in dangerous environments (e.g. consider acid gases or liquids in chemical laboratories and environments). An example for reducing costs for sensors was already demonstrated in Section 7.2, when building data-driven fuzzy models for the purpose to predict NOx emissions from a bunch of other by-measured input variables and therefore to omit a cost-intensive (and often also error-prone) hardware sensor for measuring the NOx portions in engine emissions. In the past, many data-driven methods have been used in soft sensors such as neural networks [356], partial least squares [129], support vector machines [457] or genetic programming [230]. However, in all these approaches an initial model is built from some historic measurement data and then kept fixed during the whole on-line process. In fact, in historical data, usually the broadest possible ranges from the inputs and outputs are selected; however, due to various system behaviors, new operating conditions and regime fluctuations, these ranges are usually exceeded by more than 20% of the collected on-line data [231]. Some enhanced concepts for dealing with extrapolation situations in (evolved) fuzzy systems are demonstrated in Section 4.6. These may assure well-tempered extrapolation beyond the left and right hand side of the fuzzy partition boundaries, preventing reactivation of inner fuzzy sets and hence of relations which are valid in the core part of the variables ranges. However, there is no guarantee that the outer most fuzzy sets are sufficiently describing the extrapolation region when constantly (zero-order) or linearly (first-order) [342] continued in it (in fact, the model will usually continue linearly and is not able to track new non-linear dependencies — also compare with Figures 2.9 and 2.10 in Section 2.5).

In this sense, it is a challenge to adapt and evolve soft sensors during on-line operation modes in order to follow new data patterns, range extensions, drift and shift cases by performing automatic self re-calibration steps in form of parameter adaptations and evolution of structural components. Soft sensors with such capabilities are called *eSensors*.

In principle, any evolving fuzzy systems approach demonstrated throughout this book can serve as basic methodology in an *eSensor* as fulfilling above mentioned requirements. To give concrete application cases, in [15] *eSensors* are successfully applied for product composition estimation in a distillation tower, in [292] they are applied to the oil and gas industry, both using *eTS* and *eTS+* approaches

Table 10.1 Results on estimating product composition using *eSensor* and all features [15]

Criterion	Case1	Case2	Case3	Case4
Correlation pred. vs. meas.	0.95	0.919	0.832	0.948
# of Rules	2	4	3	4
# of Features	6	47	47	23

Table 10.2 Results on estimating product composition using *eSensor* and feature selection [15]

Criterion	Case1	Case2	Case3	Case4
Correlation pred. vs. meas.	0.944	0.921	0.847	0.989
# of Rules	3	3	3	6
# of Features	2	2	4	2

(see Sections 3.2.1 and 3.2.1.5) for flexible model updates. The first application includes the following four cases

- Case 1: Estimation of product composition in a distillation tower
- Case 2: Estimation of product composition in the bottom of a distillation tower
- Case 3: Estimation of product composition in the top of a distillation tower
- Case 4: Evaluation of propylene in the top of a distillation tower

for which *eTS* could produce reliable results with a reasonable number of rules and correlations above 0.9 in three of the four cases. The concrete results are listed in Tables 10.1 (using all features) and 10.2 (using only a selection of features). Improved results by about 3-5% by including the detection of and reaction onto a drift in *eTS* approach (in Case 1) are described in [275] [274].

The second application deals with monitoring the quality of oil products on-line, where the crude oil distillation tower is the first and most important part. Hence, an accurate prediction of crude oil distillations is necessary. This was achieved by four different *eSensors* (one for each of the variables of interest) and for four different operation modes. In all 16 cases, *eTS* could produce a quite low number of rules (2 to maximal 5) with a reasonable accuracy between real outputs and the predictions — for further details see Table 17.2 in [292].

10.1.2 Autonomous Tracking of Objects in Video Streams

Autonomous tracking of objects in video streams is important in visual surveillance and security systems, for instance for recognizing human beings, passengers, vehicles not being allowed to enter certain building, rooms, facilities etc. Usually, the tracking is performed on a three-step basis:

- In a first step, foreground pixels are discriminated from background pixels; a conventional way for achieving this is background substraction by estimating the density of the pixels along N image frames [119].

- In a second step, the pixel with maximal density value among the identified foreground pixels is chosen to represent the novel object ($\rightarrow$ object detection); in order to be able to include the detection of multiple objects, it is checked whether the selected pixel = novel object is close to an already existing one: if no, a new object is introduced, otherwise the pixel belongs to an already existing object.
- In a third step, all the identified objects are tracked over time on a frame per frame basis; the aim is to predict the position of the objects for the next frame based on the information of the current and the previous frames.

In [377] an approach for autonomous tracking of objects in video streams is presented which is based on recursive estimations of pixels density distributions (over image frames) with Cauchy-type functions omitting time-intensive re-estimations for sliding windows. This is leaned on the recursive calculation of potentials (3.30) as used in *eTS* approach (Section 3.2.1). Based on the recursive density estimation, a new condition for foreground (and furthermore novel object) identification is demonstrated which do not require any threshold parameter, but relies on a comparison with pixel densities in all previous frames (hereby the pixels lying in the same (i, j)th image coordinates over all frames are compared: if the density in a new frame reaches its minimum, a foreground pixel is identified). For object tracking purposes, the *eTS* approach as demonstrated in Section 3.2.1 is applied and able to out-perform conventional Kalman filter (as linear recursive estimator) [205] in terms predicting the locations of the objects by means of both, vertical and horizontal coordinates: the root mean squared error between predicted and actual coordinates was around 10% better in case of *eTS* than when using Kalman filter.

The recursive spatial density estimation with Cauchy type functions was also applied for landmark detection in [470] and self-localization in robotics in [471].

10.1.3 Adaptive Knowledge Discovery in Bio-informatics

Bioinformatics is the application of statistics and information sciences for the analysis, modelling and knowledge discovery of biological processes in living organisms [33].

A major attraction in Bioinformatics is dedicated to the observation and analysis of genes, which are contained in specific segments of the DNA (deoxyribonucleic acid) of living organisms and are used in the cells to produce proteins. Genes are complex chemical structures and they cause dynamic transformation of one substance to another during the whole lifetime of an individual as well as the life of the human population over many generations. When genes are activated, the dynamics of the processes in which a single gene is involved are very complex, as there is a high interaction level with many other genes and proteins in the living cells, which is influenced by many environmental and developmental factors. Modelling such interactions, learning about them and extracting important knowledge and information is one of the major goals of bioinformatics.

In bioinformatics, usually someone is tackled with small data sets containing only a few hundreds of biological samples, where at the same time the samples a very high-dimensional (one gene in a DNA sequence corresponds to one single dimension) and are presented in form of bit-strings also containing void entries. Often, these data sets are static and many traditional machine learning and data mining methodologies (such as principal component analysis [463], self-organizing maps [296], support vector machines [61]) were used during the last decades for feature extraction, clustering, information expression and classification purposes. On the other hand, there are many problems in bioinformatics that require their solutions in the form of dynamic adaptive learning. In [208], the *EFuNN* (*Evolving Fuzzy Neural Networks*) approach (which is a predecessor of the *DENFIS* approach) as described in Section 3.4) is applied to several adaptive tasks such as [208]:

- Discovering patterns (features) from DNA and RNA sequences.
- Analysis of gene expression data and gene profile creation of diseases.
- Protein discovery and protein function analysis.
- Modelling the full development (metabolic processes) of a cell.

Regarding the first issue, usually a sliding window along the sequences is applied and data is submitted to a machine-learning based or neural network classifier, which identifies whether of one already known pattern is contained in the actual window. If there is a continuous flow of new labelled data, this can be integrated on-the-fly into the pattern analysis process by an adaptive evolving scheme. In [208], the *EFuNN* approach is successfully applied to ribosome binding site (RBS) identification, where the task is to decide whether in a new 33 base long sequence an RBS is contained or not, i.e. a two-class classification problem (also dealt in [130]). The adaptive networks is able to predict this with a high accuracy by just evolving nine fuzzy rules in sum. Another application of *EFuNN* to pattern discovery is splice junction identification.

Regarding gene profiling, a classification model with high accuracy could be evolved with *EFuNN*, which discriminates between colon cancer and normal genes. An interesting examination was that the pre-processing of the data by 1.) normalization, 2.) log transformation and 3.) filtering for eliminating high noise and outliers could boost the classification accuracy of the evolved fuzzy neural network by about 20% (from 60% to 80%). The accuracy for the gene profiling of two classes of Leukaemia was even higher (about 94% to 97%, depending on the parameterization of the *EFuNN* method), where k-NN with specific setting (as batch method seeing all data at once) could not achieve more than 97% accuracy.

10.1.4 Dynamic Forecasting in Financial Domains

In nowadays financial domains, it is an important task to analyze and forecast financial markets. Analysis makes it possible to learn from past bad decisions or severe mistakes regarding investments, credits or cutting/raising interest rates. For instance, in case of real financial crises as have been present at the beginning of the 30ties or in the recent past (years 2008 and 2009), important conclusions could be drawn how to prevent such cases in the future, based on the analysis of the financial markets and financial global situation in general. Forecasting of various financial issues can be a very useful tool for supporting fund managers in their decisions about buying and selling different stocks, bonds or borrowings. Forecasting of oil prices, currency exchange rates or the prices of important big indices may be also very helpful for influential politicians guiding them in their political plans for their countries.

As financial domains, especially financial markets, often possess a very short-dated behavior which may suddenly completely change by just some extraordinary events on the world (terrorism attacks, catastrophes in nature etc.), it is a central challenge to build forecast models which are adaptable, extendable on-the-fly with new incoming financial data. In this sense, evolving fuzzy systems are an appropriate methodology for handling this issue. The forecast models usually have the appearance of k-step-ahead prediction models, where k stands for the time horizon of the prediction: for day traders in stock markets, k is usually kept at a much lower time frame (minutes or even only seconds), whereas for fund managers looking for long-term investments k may be large horizon of months or years. Usually, the larger the horizon, the more uncertain the predictions will get (the number of extraordinary events or political changes increases). This requires the usage of dynamic evolving models including time delays of certain variables (as also applied in Section 7.2 for predicting the NOx emission content in engines' dynamic states). Figure 10.1 visualizes the prediction of the major stock market index *Dow Jones* for the years 2004 to 2010, based on an evolving fuzzy system trained on the prices during the period 1930 to 2003 (linear scale, monthly basis). Figure (a) shows the prediction as dotted line in the right part of the figure), Figure (b) the estimated versus real values in a zoom in snapshot; important historical facts marked as such.

In [346], the *SOFNN* approach (described in Section 3.5) is successfully applied to forecasting of currency exchange rates, where a 6-step ahead autoregressive prediction model in the following form was applied:

$$\hat{y}(t+6) = f(y(t), y(t-1), y(t-2), y(t-3), y(t-4), y(t-5)) \tag{10.1}$$

so just using the past values of the exchange rate for predicting rates in the future and not using any other environmental influences. The self-organizing fuzzy neural network evolved on 3000 training samples could achieve low root mean squared error of 0.0183 on separate test data, which were defined as the last 1281 observations (out of 4281 in sum) in a sequence of currency exchange data recorded between 3rd of January 1986 and 31st of May 2002. The data reflected the daily average exchange rates between the UK pound and the US dollar. A specific characteristics of this type of data was that it represented a highly time-variant problem, where the effect of older data on the models parameters/structures should be decayed over time. Therefore, a sliding window approach was applied with different window sizes and a re-evolving step of fuzzy neural network with *SOFNN*. The obtained model complexities and accuracies with different window sizes are listed in Table 10.3.

Another possibility how to tackle evolving regression over a sliding window is by using a forgetting factor λ in the model updates whose methodology is described

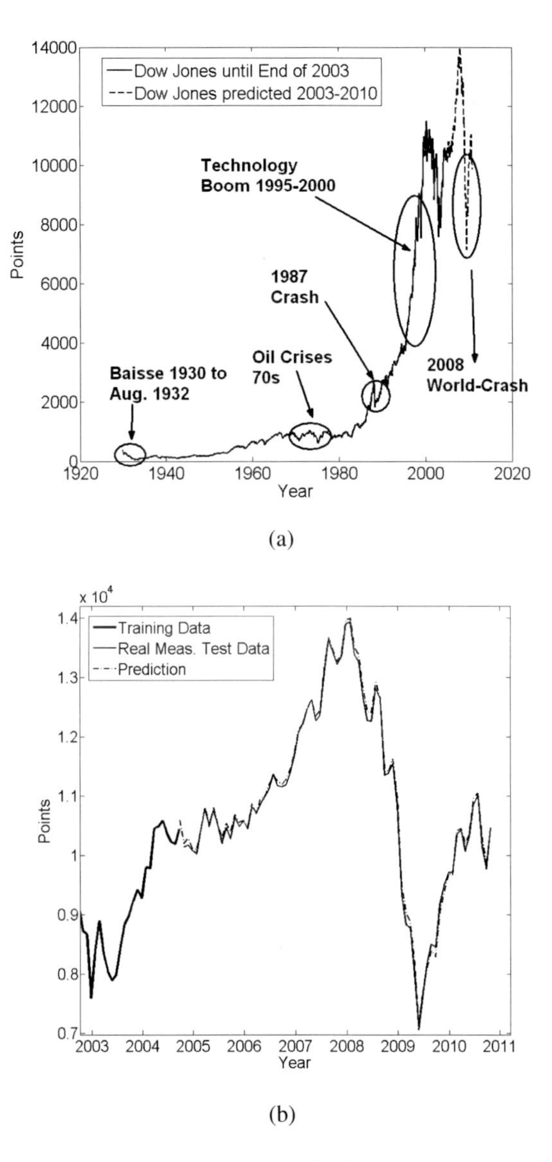

Fig. 10.1 (a): Dow Jones index from 1930 to 2010, predicted values from 2004 to 2010 indicated by dotted line; predicted (dotted line) versus real measured (solid line) index values from Fall 2004 to Fall 2010

in detail in Section 4.2 (recursive weighted least squares estimator and more intense movements of rules' antecedent parts) and applied on a concrete application for detecting untypical patterns in time-series in Section 8.2.

Table 10.3 Results on forecasting the currency exchange rates between UK pound and US dollar in the period between 3rd of January 1986 and 31st of May 2002 using *SOFNN* as EFS approach [346]

Window Width	# of Neurons	RMSE (Testing)	Corr. Coeff.
60	2	0.0480	0.9518
120	18	0.0221	0.9804
360	21	0.0309	0.9722
No window	25	0.0183	0.9824

10.2 Potential (Future) Application Fields – An Outlook

10.2.1 Adaptive Open-Loop Control

In classical control engineering, the behavior of a system is described by means of analytical approaches such as systems of differential equations derived from physical laws. Thereby, usually in a first step a model of the system is identified (also called control path) and then in a second step on the basis of this model an appropriate controller is designed analytically. In reality the system behavior quite often depends on a huge amount of concomitant factors increasing the complexity in a way, such that an analytical description of the model and the controller is impossible. However, for complex systems sometimes expert knowledge based on the experience of some operators is available in a linguistic form describing the principal behavior of the whole system in an intuitive manner.

This circumstance suggests the usage of fuzzy logic [464, 467] to build up a linguistic fuzzy controller [306], which had for example a significant success in application areas such as robotic [157, 333], automotive industry [40, 81], cement kilns [313, 181], waste water treatments [425] or water level control [own project experience]: a purely linguistic fuzzy controller including about 400 rules for the water level of two power plants along the Danube was designed and finally successfully implemented in online mode. One major problem was that extracting the expert knowledge was very time-consuming; a particular problem was the resolution of contradictions in the opinion of different experts.

Sometimes no expert knowledge is available at all, but measurement data has been recorded in the past or can be recorded easily on demand. The basic idea is to exploit measurement data in order to identify a model of the system and derive the fuzzy controller from this model automatically, leading to a purely data-driven controller design. In addition, the on-line adaptation of the model with the help of evolving fuzzy systems techniques makes it then possible also to update the fuzzy controller with new data during on-line operation mode. This means a controller designed in this way can be updated with new control strategies and rules, which is necessary in case of system states, not included in the initial measurements. From the view of the evolution process it makes no difference, how the fuzzy controller was developed, it makes only a difference how the inner structure of the underlying

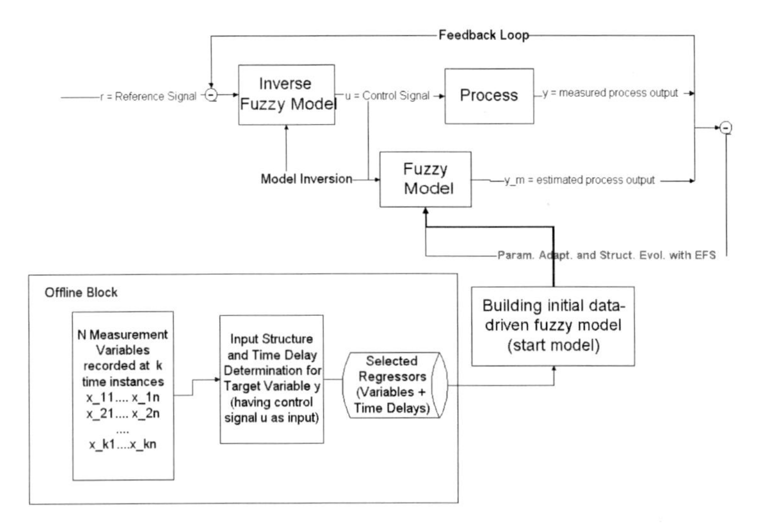

Fig. 10.2 Design of a feedback controller using an evolving fuzzy model and adaptive fuzzy controller (=inverse fuzzy model)

fuzzy model looks. Hence, if some linguistic expert knowledge exists only for some special operating cases, a hybrid fuzzy controller design by refining and extending the linguistic fuzzy controller on the basis of measurements is also possible (see Section 10.2.3 below). In principle, a data-driven design with neural networks or other types of regression models is possible, too, but this usually results in uninterpretable black box models and additionally inhibits the fusion of data-driven models with expert knowledge.

Concluding, the design of an adaptive fuzzy controller by means of (off-line and on-line) measurement data is a good option for the following application cases:

- No analytical description and no expert knowledge are available at all
- Expert knowledge is available only for some special operating cases
- Initial measurements do not cover all operating cases $\rightarrow$ adaptation makes the controller more accurate and process-save.

The final concept of a controller design using fuzzy approach with the option to adapt the controller further is visualized in Figure 10.2. In this approach, fuzzy models or rule bases are not used as batch building components for conventional P, PI or PID controllers as proposed in [206] or incorporated into an objective function which characterizes the impact of future control signal values onto values of state variables and whose minimization is intended — see [127, 257]. However, besides the online adaptation concept of fuzzy models, it includes the possibility of generating the fuzzy models from high dimensional dynamic data — as a structure and time delay identification procedure follows — and to perform control operations directly by inverted fuzzy models. The last point omits the need of a complex objective function to be minimized and triggers a linguistically interpretable and understandable fuzzy controller. Moreover, it is well known that the use of an inverse model possesses the advantages of open-loop control, i.e. inherent stability and 'perfect' control with zero error both in dynamic transients and in steady states [113]. In the case, when linguistic expert knowledge is available, the off-line block in Figure 10.2 can be omitted by coding directly this expert knowledge into a fuzzy model, which can be refined through the adaptation and evolution process with new incoming measurements.

Furthermore, it is also possible to apply the structure and time delay determination as well as the fuzzy model adaptation components for other control approaches (not necessarily inverse model control), such as for instance for model-based predictive control: there the deviation between predicted reference values (with a pre-defined time horizon of seconds, minutes into the future, which depends on the idleness of the system) and the current position of the (to be controlled) reference variable is minimized in an optimization procedure — see [63] [137] for details.

10.2.2 Chemometric Modelling

The key challenges in the chemical industry are to improve product quality, increase productivity and to minimize energy consumption as well as waste production. There is an ongoing effort to utilize existing potentials and to identify new potentials for the continued optimization of chemical production processes. To achieve this goal and to be able to judge the improvements made, in depth characterization of the process under investigation is required. In general terms this implies the ability to measure key parameters from which a sound judgment on the process status and product quality can be made. Very often these key parameters are concentrations of one or several chemical compounds in a given process or product, or are directly related to such chemical information. Therefore, in the field of analytical chemistry - the scientific discipline which develops and applies instruments capable of obtaining information on the chemical composition of matter in space and time - a sub-discipline, termed Process Analytical Chemistry (PAC), has emerged in recent years [224] [232].

About 85% of the production processes are chemical batch processes, which are only controlled by indirect physical parameters. Safety margins should ensure constant product quality and process stability. Chemical information is usually acquired by off line laboratory analysis, mainly to verify the quality of the final product after the production process has been finished. In current state-of-the art systems, time-intensive gathering and analysis of probes from the alloys are collected and measured manually whether they lie in an allowed band-width or not. This is essential for guaranteeing process-save operation and functionality at these systems and to recognize system faults and failures early. Furthermore, due to the inherent time delay of the arrival of the chemical information and the dynamics of batch processes, this information is often not feasible for production control. Hence, during

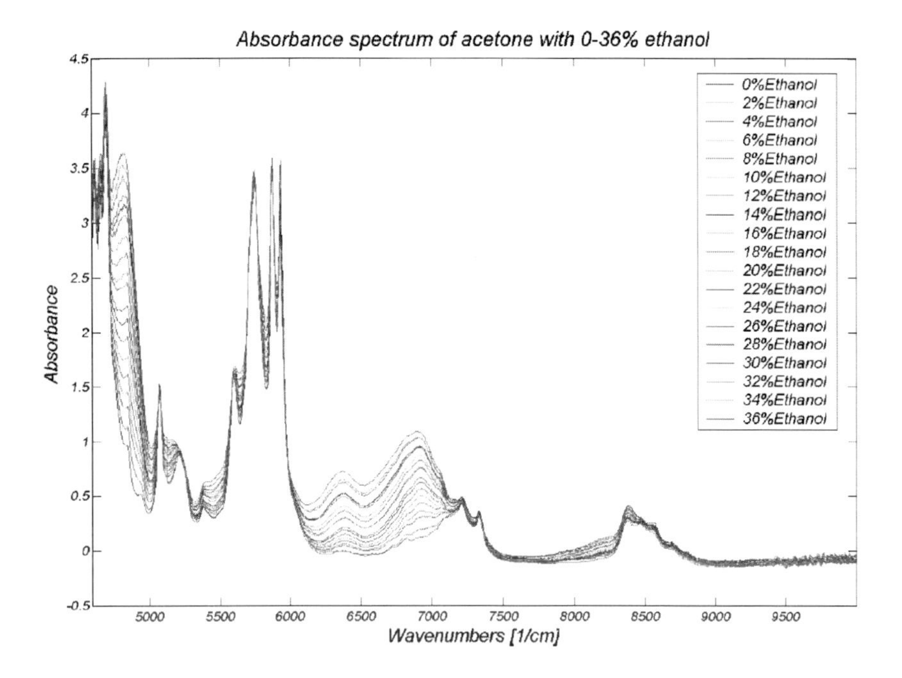

Fig. 10.3 Example of an absorbance spectrum of acetone with 0-36% ethanol

the last decade, the field of *chemometric modelling* (or short *Chemometrics*) [59] [300] [433] emerged to an important field of research for the community.

The central idea of *chemometric modelling* is to automatize the time-intensive measuring and analysis process of the chemical constitution of materials and substances. This is achieved by quantification models which are calibrated/identified based on past experience, often represented by historic data (recordings).

The type of the historic data depends on the application and the type of the data acquisition method: often spectral data are recorded with NIR methodologies [427] or also newer Raman [138] or QCL techniques [183], yielding two-dimensional signal waves, containing up to several thousand wave-lengths.

An example of (absorbance) spectral data in acetone for various ethanol contents is shown in Figure 10.3: one single wave signal represents one specific ethanol concentration. In some wave length regions, the spectra look quite similar, no matter which ethanol contents is measured, in some other parts, they are different. The task in chemometric modelling is now to extract the relevant information from the spectra data in order to build a quantification model, which is able to quantify the concentration of a substance based on a new measurement sample (= newly recorded spectra) as accurate as possible. In literature, such a model is also called *calibration model*, as historic measurements (=spectra) are used for calibrating the model. Thereby, a multivariate calibration process consists of the following steps:

1. Selection of calibration samples
2. Determination of properties for calibration samples (target, output variables)
3. Collection of spectra from calibration samples (input variables)
4. Identification and validation of calibration model
5. Application of model for the analysis of unknowns
6. Monitoring of the output of calibration models

Steps 1 to 3 are pre-processing phases prior to the actual data analysis and data-driven modeling phase in steps 4 and 5. Step 2 is necessary for quantification applications where a reference measurement of the desired target quantity is needed (typically done in the laboratory). The values are usually continuous values (as also in the case of our example), hence mostly regression problems are the case in chemometric modelling. In Steps 4 and 5, each sample spectrum is typically considered as a sequence of discrete data points and one tries to build up a computational model based on a set of training samples, gathered a priori. Step 6 is important in order to monitor the model output on-line and to react appropriately if something goes wrong. In general, Steps 1 to 6 are more or less independent from the chosen data source (NIR, MIR, QCL, Raman).

A central challenge in the model identification part is the handling of high-dimensional input data. This is because each discrete wave length (along the x-axis of a spectrum) denotes one dimension in the feature space. In conventional (old-fashioned) systems, several thousands of such wavelength are recorded within few hundreds of signal waves. This means that in usual cases the dimension of the feature vectors exceeds significantly the number of observations. Hence linear approaches such as partial least squares (PLS) [162] or principal component regression (PCR) [201] were applied to multivariate calibration during the last two decades to overcome the problems of correlated inputs and dimension reduction (i.e. feature selection) at the same time; see for instance [364], [415], [432]. As long as there are basically linear relations between spectra waves and target content and the number of historic spectra data (from which the models are learned) is low, linear models are a sufficient tool to calibrate the models. With modern methodologies and devices for faster data sampling and substances in more complex chemical scenarios to be measured and analyzed, the necessity for non-linear models is becoming more and more important; first attempts for non-linear models in chemometrics have been reported recently, for instance with the help of support vector regression (SVR) in [177]. Furthermore, all the applied (mostly statistical) techniques are static approaches, i.e. once the off-line calibration phase is finished, the identified models are kept fixed during the on-line operation. Hence in the near future, it will be an important challenge to develop non-linear models which are able to refine, update and extend themselves on-the-fly during online operation mode, ideally combined with an implicit and adaptive feature selection or extraction approach in order to cope with the high-dimensionality of the data. Evolving fuzzy systems coupled with the incremental feature weighting approach as discussed in Section 5.2 could be a powerful tool to solve these problems.

Similar considerations can be achieved in the case of the usage of three-dimensional hyper-spectral images [75] [149], which has enjoyed upcoming great attention in the field of chemometrics during the last years. A hyper-spectral image includes conventional spectral data (appearing as one-dimensional single curves over several wave lengths) for each pixel in the image. This can be visualized as a bunch of spectra in a three-dimensional plot, see http://en.wikipedia.org/wiki/Hyperspectral_imaging for an example of a hyper-spectral image cube. Someone may produce models and quantify the content of new samples pixel per pixel, obtaining a quantification matrix, from which an overall or local quantification statement can be extracted. Another possibility is to segment the multidimensional spectral data by averaging, clustering or other pre-processing steps [408] and then to apply the (off-line, on-line, linear vs. non-linear) techniques as used for one-dimensional spectra (for each cluster or for the averaged information).

10.2.3 Enhanced Interactive Machine Learning (IML)

The previous Chapters 7, 8 and 9 and the previous sections in this chapter were dealing with applications falling into one of the two categories regarding incremental and evolving learning strategies:

- Fuzzy systems are updated fully in-line and automatically based on newly recorded measurements without any interaction without the expert or operator (Chapters 7 and 8, Section 10.2.1); this is generally possible for regression/approximation problems, whenever the (system) target variable in the models is measured synchronously to the (system) input variables (hence for the evolving approach the correct responses as 'learn direction' are known).
- Fuzzy classifiers are updated based on operator's feedback (Chapters 9 and Section 10.2.2); the interaction in these cases to the user is present, but is reduced to a good/bad reward on the classifier's decision (see Figure 9.2 for a visualization of this strategy in an image classification system).

In the first case, we speak about a *fully automatic update*, in the second about a *half-automatic update and evolution* of fuzzy models.

In both cases, the human-machine interaction is restricted to passive supervision, responses to models' outputs, good/bad rewards or switching off some components of the system when something goes wrong (in case of fault detection scenarios). However, none of these application cases implies a more detailed and funded interaction between the machine learning components and the operators/experts working with these components resp. at the plants where these components are installed. The components try to build and evolve models from data permanently collected during on-line operation mode, but do not incorporate the knowledge of the operators about the system (e.g. based on past long-term experience) directly into the concepts (structures, parameters) of the learned models.

Hence, we see it as a big challenge for the future to develop evolving machine learning components which are able to go beyond this in order to achieve a deeper, evidence-based understanding of how people may interact to machine learning systems in different contexts, finally to deliver improved interfaces between *Natural and Artificial Intelligence*. We call this bridge *natural-inspired artificial intelligence.*

Important question when building such components will be:

- what are the elements that constrain level and nature of feedback?
- how can environmental and process related information be incorporated to augment user input?
- how can someone elicit levels of confidence, and of engagement with the system?
- and how can the interaction be made intuitive from the user's perspective?

The answer to these questions will help to stimulate and motivate the users for an enhanced interaction with the systems, in ideal case to bring in his own knowledge (from past experience) in order to manually modify some structural components of the models (for instance to correct bad relations reflected in the models) — whether by inputting rules, by directly manipulating the decision model boundaries, or by requesting data items/synthesised exemplars in certain regions.

All these issues tend towards 'letting the user build classifiers' [443] respectively to have a significant participation in the model evolution and update phase during on-line processes. The whole concept is called *human-inspired evolving machines/models (HIEM).*

Figure 10.4 visualizes the basic components of a flexible machine learning framework that facilitates multi-modal, bi-directional human-machine interaction. One essential aspect in such an interactive learning framework is to prepare the models purely trained from data to be interpretable and understandable for the users, otherwise they may get distracted, uninterested etc. This is the point where evolving fuzzy systems come into the game as a potential powerful tool for delivering user-friendly machine learning models. This is because they offer capabilities for fast and accurate on-line update of the models based on new incoming single samples including parameter adaptation, structure evolution and (range) extension to new system states (Section 3) while at the same time

1. offering robust and process-save operation and prediction qualities (achieved by the techniques described in Chapter 4)

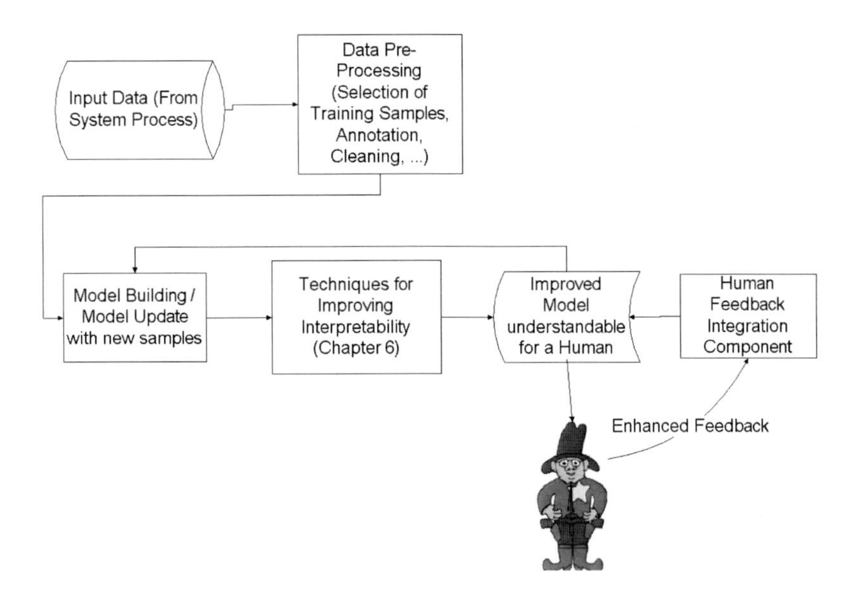

Fig. 10.4 Components of an enhanced interactive machine learning framework

2. offering enhanced aspects for dealing with high-dimensional data, more user-friendliness and spatially divided data sources (Chapter 5)
3. and, most importantly, by combining these issues with the techniques for improving interpretability of the models as described in Chapter 6, yielding models which are linguistically as well as visually understandable and interpretable for users.

The last point also includes the aspect of responding uncertainties in their decision as additional output aside the predictive statement (output class etc.), together with a reason in case for low confidences = high uncertainties (see Section 4.5). This is the main motivation for applying evolving fuzzy systems within an enhanced interactive machine learning framework.

Finally, this may lead us to the *next generation evolving fuzzy systems* (NextGenEFS).

Other types of models such as neural networks, support vector machines or evolutionary approaches allow no or only a little interpretation of intrinsic dependencies and relations in the system. Decision trees, which can be represented in rule form (where one path from the root to a terminal node can be read as one rule [351]), are a powerful alternative (as also used in the visualization approaches [186] and [436]), but 1.) their incremental learning capabilities are strongly restricted, as sub-trees and leaves have to be re-structured and traversed regularly after some new incoming samples [431] [91], usually requiring high computation time and 2.) they do not offer any vague statements and confidence levels in their decision (leave/rule fulfillment is either 0 or 1), which often results in too 'black' and 'white' interpretation of the model's decisions.

A slim version of the bi-directional human-machine interaction as shown in Figure 10.4, is the so-called generation of gray box models.

Gray box models can be seen as an uni-directional interactive machine learning approach, where models are initially set up based on experience of the operators and are further refined with measurement data recorded at the system.

Operator experience is usually in a form such as 'if this and that happens, then the reaction should be in this and that manner' and can be linguistically encoded in (fuzzy) rule bases. As usually such a linguistic based (fuzzy) model (also referred as weak white box model — see Chapter 1) has not sufficient accuracy and does not cover the full range of possible operation modes, it is beneficial to refine such a model with measurement data. Sometimes it is sufficient to adapt the model parameters alone, which has the favorable effect that the linguistic interpretable input partitions (as defined originally from experts) remain. However, most of the time, the experts do not have all the possible system conditions together with its correct reaction in their minds, so an adjoining of new structural components (fuzzy rules) for incorporating new conditions is beneficial. A flowchart for the generation of grey box models is demonstrated in Figure 10.5, which should give the basic idea about this uni-directional interaction mode. The first two blocks, i.e. collecting expert knowledge and coding this expert knowledge into a fuzzy system, denote the off-line part of this flowchart and is usually characterized by high development effort, as collecting expert knowledge requires a lot of discussion rounds in several meetings etc. Also contradictory opinions about different situations at the system (e.g. control strategies in various cases) have to be clarified and resolved at this stage. The coding itself can be done in straightforward manner, as expert knowledge usually can be transferred into rules with clearly defined premises and conclusions in the if-then-parts (usually, the operator knows how the response should look or how to react when this and that happens at a system).

Generally, every type of fuzzy system (see Section 1.2.1) can be applied as model architecture in which the expert knowledge is coded. Mostly preferred are Mamdani-type fuzzy systems, as they contain also linguistic formulations in the conclusion of the rules. When applying EFS for further adaptation and evolution of the knowledge-based fuzzy models, the Mamdani-type fuzzy systems have to be transferred to an equivalent (Takagi-)Sugeno or even neuro-fuzzy system: this depends on the application context and especially which EFS approach is applied to evolve the models further (*FLEXFIS*, *eTS* and *ePL* require Takagi-Sugeno fuzzy systems, where most of the others, i.e. *DENFIS*, *SAFIS*, *SOFNN* etc., require neuro-fuzzy type architectures, also refer to Chapter 3). However, in some of these approaches the extension to the Mamdani case is quite straightforward: for instance

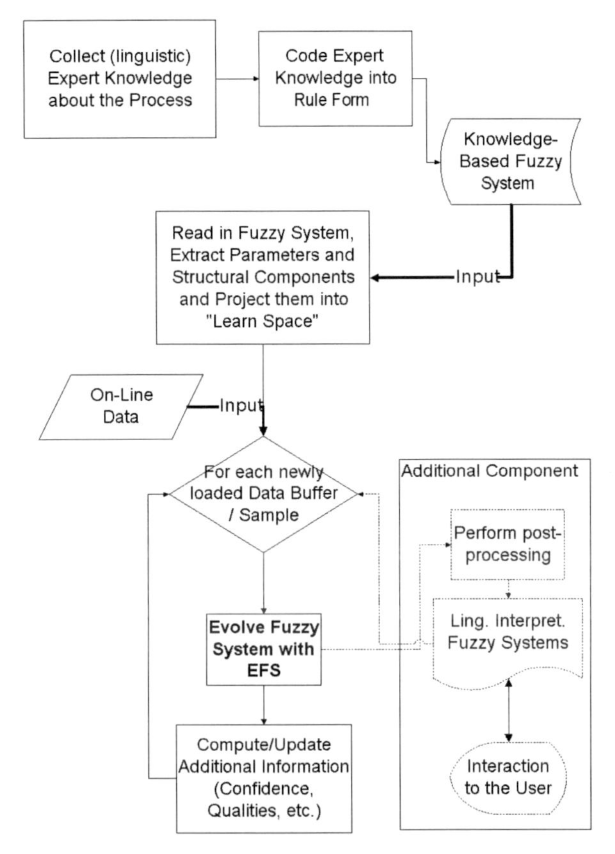

Fig. 10.5 Gray box fuzzy modelling strategy with applying EFS as refining and update mechanism of an initial knowledge-based model

in *FLEXFIS* and *eTS*, both applying an incremental clustering concept to learn the antecedent parts of the rules, clusters could also be projected to the output (and not only onto the input axes) in order to obtain a fuzzy partition for the target variable. A drawback of this variant is that usually Mamdani fuzzy systems are not able to compete with TS fuzzy systems in terms of accuracy and predictive power. Also possible is an which use a subset of variables and expand the fuzzy models on an on demand basis in a tree-like structure as long as the error does not fall below a threshold. This approach is also called *fuzzy pattern trees* [391]. Finally, it is a matter of a tradeoff between user-friendliness (and the associated interpretability and understandability) and accuracy and predictive performance to decide whether to use this or that fuzzy system architecture.

An essential aspect in the whole grey box generation work-flow is the question on how to transfer the parameters and structural components identified during the initial model setup (based on expert knowledge) to the training space of the corresponding applied EFS approach. For instance, in clustering-based EFS approaches, this means identifying the cluster centers and surfaces from the given fuzzy partitions — this can be directly achieved by the following considerations and assignments:

- Number of clusters = number of rules.
- Each rule's premise form one cluster where.
- Each premise part corresponds to a fuzzy set whose center can be assigned to the corresponding entry in cluster center vector.
- The width of the fuzzy sets can also be extracted and stored together with cluster centers, denoting the ellipsoidal surface of the cluster in the feature/variable space.
- The impact of each cluster (= the number of data samples forming each cluster in the purely data-driven case) can be assigned to the significance level of the rule (which the operator has to define according to his confidence in the formulated rule).

In case of TS fuzzy systems, the start parameters of the consequent functions for *RWLS* can be directly read from the linear parameters in the user-defined hyperplane. In case of Mamdani fuzzy systems, the parameters of the consequent functions for *RWLS* are set to the center values of the fuzzy sets in the corresponding rule conclusions. In both cases, the inverse Hessian matrix, needed in *RWLS* to guarantee safe and optimal solutions, is set to αI with α a big positive integer as this has some favorable properties regarding parameter convergence, see Section 2.2.2.3. In case of classification tasks, usually the responding class labels are defined by the experts per rule. This can be directly used in the single model architecture approach. For the multi-model architecture, each class label can be transferred to a singleton (intercept) parameter with value 0 (not belonging to the corresponding class model) or 1 (belonging to the corresponding class model). The other linear parameters (for the variable weights in the consequent hyper-planes) are all set to 0 according to Section 2.2.2.3. In the case of using *eTS*, the remaining question there is to estimate the potential of already existing cluster centers from the knowledge-based models. A possibility here could be to use the significance level of a rule, depending on the operator's confidence in it.

The block "Evolve Fuzzy System with EFS" incorporates the evolution strategy as implicitly applied by the chosen EFS approach. The block "Compute/Update Additional Information" first computes the qualities on the first k measurement data, if no model quality is given by the expert linguistically, and then updates it in the same way as for the identification framework shown in Figure 7.2 and described in Section 2.1.4. Additionally, it calculates confidence bands, error bars and confidence levels in classification statements with the help of techniques described in Section 4.5.

An additional component could be again an enhanced user interaction on structural level (bottom right in Figure 10.5) in order to achieve an iterative and alternating fuzzy modelling approach (expert knowledge for initial setup, refinement with data, correction by experts, inclusion of new data, again correction by experts and so on, ...).

Epilog – Achievements, Open Problems and New Challenges in EFS

This book deals with many aspects in the field of evolving fuzzy systems which have emerged rapidly during the last decade and meanwhile have become well established in the fuzzy community. The establishment is underlined by a huge amount of publications in international fuzzy conferences and journals (such as Fuzzy Sets and Systems (Elsevier), IEEE Transactions on Fuzzy Systems (IEEE Press), IEEE Transactions on Systems, Man and Cybernetics part B (IEEE Press), Information Sciences (Elsevier), International Journal of Approximate Reasoning (Elsevier), Applied Soft Computing (Elsevier), Evolving Systems (Springer) and others) and by the organization of several special sessions and workshops during the recent years at international fuzzy conferences and symposia. This book covers not only a comprehensive survey of the most important evolving modelling approaches and the most important incremental learning algorithms they contain, but also presenting significant extensions for guiding evolving fuzzy systems to higher process safety, predictive quality, user-friendliness and interpretability as well as understandability of its components and finally the processes which they are modelling. Applications from different industrial fields including multi-channel measurement systems (identification, prediction and plausibility analysis), visual and audio inspection tasks, psychological aspects in texture analysis, applications in bio-informatics, eSensors or financial market forecasting should underline the necessity, applicability and reliability of evolving fuzzy systems in these fields. Therefore, and as the evolving concepts presented in this book allow the machines to permanently learn from changing environments and enrich their knowledge, the book may serve as another cornerstone for a step towards computational intelligence. Even though, based on all these facets, the reader may get a nice impression about the range of applicability and the methodological richness of evolving fuzzy systems, there are still some open problems in EFS which have not been handled with sufficient care and necessary detail so far (although discussed in some sections in this book): [1]

- **Avoidance of over-fitting**: currently most of the evolving fuzzy system approaches use the least squares error optimization problem for optimizing linear parameters (mostly in rule consequent functions of TS-type fuzzy systems): this leads to a recursive least squares (in case of global learning) or recursive fuzzily weighted least squares (in case of local learning) algorithms, which may overfit significantly on the (permanently loaded) training data. This is because, the model complexity is not included in the formulation of the optimization problem; contrary in an off-line setting, the learning based on least squares error often is performed through N-fold cross-validation which can resolve the problem of over-fitting by eliciting error on separate test samples (folds) — see also Section 2.8 for a more extensive discussion on this topic. In fact, an appropriate rule merging approach as discussed in Section 5.1 may overcome some deficiencies in this regard, however this denotes a more heuristic approach rather than relying on analytical incremental optimization procedures with a more theoretic basis.
- **Fault handling**: although there exist some concepts (rule base procrastination, incremental quantiles [428], recursive potentials) for dealing with (unique, couple of) outliers in various EFS approaches, currently, there exists no enhanced concept for dealing with systematic faults, appearing in new regions of the feature space; in fact, the distinction to the appearance of new system states or operating conditions also extending the feature space has not been studied so far.
- **Drift handling**: although we presented a concept for the automatic detection and treatment of drifts and shifts in data streams in Section 4.2, this is only a first attempt in this direction, especially designed for *FLEXFIS* and *eTS* and this in a quite heuristic fashion, and therefore far from being fully investigated. In this regard, also the concept of multi-task learning [68] in connection with EFS has not been studied so far.
- **On-line curse of dimensionality reduction**: Section 5.2 indeed demonstrates a reasonable concept for a soft dimension reduction in in EFS by using feature weights whose values are changed slightly with new incoming samples, therefore discontinuities in the learning process (as would be forced by crisp on-line feature selection) can be avoided. However, this approach is far from being optimal, as the feature weighting is performed by a linear filter approach and as such not related/embedded with the underlying fuzzy model architecture. Indeed, important features in a linear sense are also important ones for non-linear fuzzy models, but some other unimportant features may turn out to be important when

[1] At this point, the author wants to emphasize that this book covers all approaches, methods and algorithms which were published before 30th of June 2010 as crisp deadline; the author apologizes for eventually overlooking some important novel concepts and aspects which were published before this date and also for not being able (due to time restrictions) to take into account any novel concepts and aspects published after this date — in both cases the author of this book would be grateful and invites the reader for any valuable comments, critics, claims, suggestions for improvement.

inspected through a non-linear approach (ideally coupled with the fuzzy model architecture as embedded approach).
- **Extrapolation problematic**: it is not fully understood how the left and right most fuzzy sets in fuzzy partitions should be optimally extended in the extrapolation region in order to give reliable predictions and classification statements; first attempts are presented in Section 4.6, especially dealing with the avoidance of re-activating inner membership functions with infinite support.

New challenges for EFS include the following aspects:

- **Active learning scenarios**: So far, the assumption in most of the EFS approaches is that the response data (labels, target values) are provided together with the input feature/variable vectors and incremental training and evolution of the models continues in a life-long learning mode. However, the concept of active learning in connection with EFS has not been sufficiently studied so far (a first attempt is presented in Section 5.3): active learning may be very important in order to decrease the annotation effort and response times of operators during on-line mode, especially for classification problems, as active learning makes explicit selection of important samples possible which may help the models to improve their qualities, and synchronously neglecting unimportant samples. This makes the application of evolving models and especially evolving fuzzy systems much more attractive.
- **Linguistic and visual interpretability**: A large part of Chapter 6 deals with aspects of complexity reduction in EFS, which can be performed in a fast on-line matter; this is far from providing interpretable and for operators, users, experts understandable EFS, only some first ideas in this direction are provided in Chapter 6. Basically, we can say that all the current EFS approaches are precise modelling approaches neglecting completely the interpretable aspect of fuzzy systems, even though including some aspects regarding deletion of obsolete rules or rule merging processes. An exception is a recent publication [180], where more interpretability is achieved by applying a kind of mixed Mamdani and Takagi-Sugeno type model architecture in order to gain more interpretability while still preserving the good approximation accuracy of TS fuzzy models. Visualization aspects and visual interpretation of (high-dimensional) EFS are completely missing in literature, but could be very important for the operators to get a better understanding of what is going on in the on-line processes of the their systems, respectively motivating the users for an enriched human-machine interaction and communication, see subsequent point.
- **Enhanced human-machine interaction**: Combining aspects of active learning with EFS and interpretable and understandable EFS together in an enriched user-interaction concepts, can be one of the key aspects in the near future such that the models benefit from both parties, data samples on the one hand and expert knowledge on the other hand in an alternating exchanging context, leading to the concept of *human-inspired evolving machine/models/fuzzy systems* (see also Section 10.2.3).

- **Dynamic data mining**: the application of EFS approaches was so far mainly concentrated on temporally changing environments. Evolving fuzzy systems have not as yet been applied in a dynamic data mining context (see also Section 5.5.1), where spatially distributed data is used for knowledge exchange via dynamic models.
- **Transfer learning**: transfer learning addresses a similar learning problem as dynamic data mining, where already trained models are updated based on samples loaded from a new data source. The difference to dynamic data mining is basically that models are adjusted to new test objects, devices or environments in order to include the new information — this can also stem from temporal changes and is not necessarily triggered by spatially distributed data sources. Data samples may be given different weights for different devices, objects etc.
- **Combination of EFS with other machine learning techniques**: currently, EFS have hardly been studied in the context of incremental learning in combination with other machine learning methods. We see this as an important and useful future challenge, as both parties, the (evolving) fuzzy systems as well as the machine learning community may benefit from such an exchange w.r.t. learning aspects, model components or parameter adaptation and tuning, enriching the diversity of data-driven methodologies in today's real-world systems.

References

1. Abonyi, J.: Fuzzy Model Identification for Control. Birkhäuser, Boston (2003)
2. Abraham, W., Robins, A.: Memory retention – the synaptic stability versus plasticity dilemma. Trends in Neurosciences 28(2), 73–78 (2005)
3. Abramovic, M., Stegun, I.: Handbook of Mathematical Functions with Formulas, Graphs, and Mathematical Tables. Dover Publications, New York (1970)
4. Adams, D.E.: Health Monitoring of Structural Materials and Components. John Wiley & Sons, Chichester (2007)
5. Aggarwal, C., Yu, P.: Outlier detection for high dimensional data. In: Proceedings of the 2001 ACM SIGMOD International Conference on Management of Data, Santa Barbara, California, pp. 37–46 (2001)
6. Aha, D.: Lazy Learning. Kluwer Academic Publishers, Norwell (1997)
7. Aha, D., Kibler, D., Albert, M.: Instance-based learning algorithms. Machine Learning 6(1), 37–66 (1991)
8. Albertos, P., Sala, A.: Fault detection via continuous-time parameter estimation. In: Proceedings of IFAC Symposium on Fault Detection, Supervision and Safety for Technical Processes, SAFEPROCESS, pp. 87–92. Helsinki Univ. Technol., Espoo, Finland (1994)
9. Alsabti, K., Ranka, S., Singh, V.: An efficient k-means clustering algorithm. In: Proceedings of IPPS/SPDP Workshop on High Performance Data Mining, Orlando, Florida, pp. 556–560 (1998)
10. Angelov, P.: Evolving takagi-sugeno fuzzy systems from streaming data, eTS+. In: Angelov, P., Filev, D., Kasabov, N. (eds.) Evolving Intelligent Systems: Methodology and Applications, pp. 21–50. John Wiley & Sons, New York (2010)
11. Angelov, P., Filev, D.: An approach to online identification of Takagi-Sugeno fuzzy models. IEEE Transactions on Systems, Man and Cybernetics, Part B: Cybernetics 34(1), 484–498 (2004)
12. Angelov, P., Filev, D.: Simpl_eTS: A simplified method for learning evolving Takagi-Sugeno fuzzy models. In: Proceedings of FUZZ-IEEE 2005, Reno, Nevada, U.S.A., pp. 1068–1073 (2005)
13. Angelov, P., Giglio, V., Guardiola, C., Lughofer, E., Luján, J.: An approach to model-based fault detection in industrial measurement systems with application to engine test benches. Measurement Science and Technology 17(7), 1809–1818 (2006)

14. Angelov, P., Kasabov, N.: Evolving computational intelligence systems. In: Proceedings of the 1st International Workshop on Genetic Fuzzy Systems, Granada, Spain, pp. 76–82 (2005)
15. Angelov, P., Kordon, A.: Evolving inferential sensors in the chemical process industry. In: Angelov, P., Filev, D., Kasabov, N. (eds.) Evolving Intelligent Systems: Methodology and Applications, pp. 313–336. John Wiley & Sons, New York (2010)
16. Angelov, P., Lughofer, E.: Data-driven evolving fuzzy systems using eTS and FLEXFIS: Comparative analysis. International Journal of General Systems 37(1), 45–67 (2008)
17. Angelov, P., Lughofer, E., Klement, E.: Two approaches to data-driven design of evolving fuzzy systems: eTS and FLEXFIS. In: Proceedings of NAFIPS 2005, Ann Arbor, Michigan, U.S.A., pp. 31–35 (2005)
18. Angelov, P., Lughofer, E., Zhou, X.: Evolving fuzzy classifiers using different model architectures. Fuzzy Sets and Systems 159(23), 3160–3182 (2008)
19. Angelov, P., Xydeas, C., Filev, D.: Online identification of MIMO evolving Takagi-Sugeno fuzzy models. In: Proceedings of IJCNN-FUZZ-IEEE 2004, Budapest, Hungary, pp. 55–60 (2004)
20. Angelov, P., Zhou, X.: Evolving fuzzy-rule-based classifiers from data streams. IEEE Transactions on Fuzzy Systems 16(6), 1462–1475 (2008)
21. Angelov, P., Zhou, X., Filev, D., Lughofer, E.: Architectures for evolving fuzzy rule-based classifiers. In: Proceedings of SMC 2007, Montreal, Canada, pp. 2050–2055 (2007)
22. Angelov, P., Zhou, X.W.: Evolving fuzzy systems from data streams in real-time. In: 2006 International Symposium on Evolving Fuzzy Systems (EFS 2006), Ambleside, Lake District, UK, pp. 29–35 (2006)
23. Ariew, R.: Ockham's Razor: A Historical and Philosophical Analysis of Ockham's Principle of Parsimony. Champaign-Urbana, University of Illinois, Urbana (1976)
24. Arrègle, J., López, J., Guardiola, C., Monin, C.: Sensitivity study of a NOx estimation model for on-board applications. SAE paper 2008-01-0640 (2008)
25. Aström, K., Wittenmark, B.: Adaptive Control, 2nd edn. Addison-Wesley Longman Publishing Co., Inc., Boston (1994)
26. Atkeson, C., Moore, A., Schaal, S.: Locally weighted learning. Artificial Intelligence Review 11(1-5), 11–73 (1997)
27. Avriel, M.: Nonlinear Programming: Analysis and Methods. Dover Publishing, New York (2003)
28. Babuska, R.: Fuzzy Modeling for Control. Kluwer Academic Publishers, Norwell (1998)
29. Babuska, R., Verbruggen, H.: Constructing fuzzy models by product space clustering. In: Hellendoorn, H., Driankov, D. (eds.) Fuzzy Model Identification: Selected Approaches, pp. 53–90. Springer, Berlin (1997)
30. Backer, S.D., Scheunders, P.: Texture segmentation by frequency-sensitive elliptical competitive learning. Image and Vision Computing 19(9-10), 639–648 (2001)
31. Baczynski, M., Jayaram, B.: Fuzzy Implications. Springer, Heidelberg (2008)
32. Bakushinskii, A.: The problem of the convergence of the iteratively regularized gauss–newton method. Comput. Math. Phys. 32(9), 1353–1359
33. Baldi, P., Brunak, S.: Bioinformatics - A Machine Learning Approach. MIT Press, Cambridge (2001)
34. Basseville, M., Nikiforov, I.: Detection of Abrupt Changes. Prentice Hall Inc., Englewood Cliffs (1993)

35. Bauer, F.: Some considerations concerning regularization and parameter choice algorithms. Inverse Problems 23(2), 837–858 (2007)
36. Bauer, F., Kindermann, S.: The quasi-optimality criterion for classical inverse problems. Inverse Problems 24(3), 35,002–35,021 (2008)
37. Bauer, F., Lukas, M.: Comparing parameter choice methods for regularization of ill-posed problems. Inverse Problems (2010), http://bmath.de/Docs/mainLowRes.pdf
38. Bay, S., Saito, K., Ueda, N., Langley, P.: A framework for discovering anomalous regimes in multivariate time-series data with local models. In: Symposium on Machine Learning for Anomaly Detection, Stanford, U.S.A. (2004)
39. Bellman, R.: Dynamic Programming. Princeton University Press, Princeton (1957)
40. Berenji, H., Ruspini, E.: Experiments in multiobjective fuzzy control of hybrid automotive engines. In: Proceedings of the Fifth IEEE International Conference on Fuzzy Systems FUZZ-IEEE 1996, New York, U.S.A., pp. 681–686 (1996)
41. Bergen, J.R., Landy, M.S.: Computational modeling of visual texture segregation. In: Landy, M.S., Movshon, J.A. (eds.) Computational Models of Visual Processing, pp. 253–271. MIT Press, Cambridge (1991)
42. Berger, J.: Statistical Decision Theory and Bayesian Analysis. Springer, Heidelberg (1985)
43. Beringer, J., Hüllermeier, E.: Online clustering of parallel data streams. Data & Knowledge Engineering 58(2), 180–204 (2006)
44. Beringer, J., Hüllermeier, E.: Efficient instance-based learning on data streams. Intelligent Data Analysis 11(6), 627–650 (2007)
45. Bernieri, A., Betta, G., Liguori, C.: On-line fault detection and diagnosis obtained by implementing neural algorithms on a digital signal processor. IEEE Transactions on Instrumentation and Measurement 45(5), 894–899 (1996)
46. Bezdek, J.: Pattern Recognition with Fuzzy Objective Function Algorithms. Kluwer Academic/Plenum Publishers, U.S.A. (1981)
47. Bharitkar, S., Filev, D.: An online learning vector quantization algorithm. In: Proc. of Sixth International Symposium on Signal Processing and its Applications, vol. 2, pp. 394–397 (2001)
48. Bie, T.D., Cristianini, N.: Semi-supervised learning using semi-definite programming. In: Chapelle, O., Schoelkopf, B., Zien, A. (eds.) Semi-Supervised Learning, pp. 113–131. MIT Press, Cambridge (2006)
49. Biehl, M., Gosh, A., Hammer, B.: Dynamics and generalization ability of LVQ algorithms. Journal of Machine Learning Research 8, 323–360 (2007)
50. Birattari, M., Bontempi, G.: The lazy learning toolbox (1999), ftp://iridia.ulb.ac.be/pub/lazy/papers/IridiaTr1999-07.ps.gz
51. Blazquez, J.M., Shen, Q.: Regaining comprehensibility of approximative fuzzy models via the use of linguistic hedges. In: Casillas, J., Cordón, O., Herrera, F., Magdalena, L. (eds.) Interpretability Issues in Fuzzy Modeling, pp. 25–53. Springer, Berlin (2003)
52. Blum, A., Mitchell, T.: Combining labelled and unlabelled data with co-training. In: Proceedings of the Workshop on Computational Learning Theory (COLT), Madison, Wisconsin, pp. 92–100 (1998)
53. Botzheim, J., Cabrita, C., Kóczy, L., Ruano, A.: Estimating fuzzy membership functions parameters by the Levenberg-Marquardt algorithm. In: Proceedings of the IEEE International Conference on Fuzzy Systems, FUZZ-IEEE 2004, Budapest, Hungary, pp. 1667–1672 (2004)

54. Botzheim, J., Lughofer, E., Klement, E., Kóczy, L., Gedeon, T.: Separated antecedent and consequent learning for Takagi-Sugeno fuzzy systems. In: Proceedings of FUZZ-IEEE 2006, Vancouver, Canada, pp. 2263–2269 (2006)
55. Bouchachia, A.: Incremental induction of classification fuzzy rules. In: IEEE Workshop on Evolving and Self-Developing Intelligent Systems (ESDIS) 2009, Nashville, U.S.A., pp. 32–39 (2009)
56. Box, G., Jenkins, G., Reinsel, G.: Time Series Analysis, Forecasting and Control. Prentice Hall, Englewood Cliffs (1994)
57. Breiman, L.: Pasting small votes for classification in large databases and on-line. Machine Learning 36(1-2), 85–103 (1999)
58. Breiman, L., Friedman, J., Stone, C., Olshen, R.: Classification and Regression Trees. Chapman and Hall, Boca Raton (1993)
59. Brereton, R.: Chemometrics: Data Analysis for the Laboratory and Chemical Plant. John Wiley & Sons, Hoboken (2003)
60. Brown, K.: Voronoi diagrams from convex hulls. Information Processing Letters 9(5), 223–228 (1979)
61. Brown, M.P.S., Grundy, W.N., Lin, D., Cristianini, N., Sugnet, C.W., Furey, T.S., Ares, M., Haussler, D.: Knowledge-based analysis of microarray gene expression data by using support vector machines. Proceedings National Academic Sciences 97, 262–267 (2006)
62. Burger, M., Haslinger, J., Bodenhofer, U., Engl, H.W.: Regularized data-driven construction of fuzzy controllers. Journal of Inverse and Ill-Posed Problems 10(4), 319–344 (2002)
63. Camacho, E., Bordons, C.: Model Predictive Control. Springer, London (2004)
64. Campbell, J., Hashim, A., Murtagh, F.: Flaw detection in woven textiles using space-dependent fourier transform. In: Proc. of ISSC 1997, Irish Signals and Systems Conference, pp. 241–252. University of Ulster, Londonderry (1997)
65. Carmona, P., Castro, J., Zurita, J.: Contradiction sensitive fuzzy model-based adaptive control. Approximate Reasoning 30(2), 107–129 (2001)
66. Carpenter, G.A., Grossberg, S.: Adaptive resonance theory (ART). In: Arbib, M.A. (ed.) The Handbook of Brain Theory and Neural Networks, pp. 79–82. MIT Press, Cambridge (1995)
67. Carreira-Perpinan, M.: A review of dimension reduction techniques. Tech. Rep. CS-96-09, Dept. of Computer Science. University of Sheffield, Sheffield, U.K (1997)
68. Caruana, R.: Multitask learning: A knowledge-based source of inductive bias. Machine Learning 28(1), 41–75 (1997)
69. Casillas, J., Cordon, O., Herrera, F., Magdalena, L.: Interpretability Issues in Fuzzy Modeling. Springer, Heidelberg (2003)
70. Casillas, J., Cordon, O., Jesus, M.D., Herrera, F.: Genetic feature selection in a fuzzy rule-based classification system learning process for high-dimensional problems. Information Sciences 136(1-4), 135–157 (2001)
71. Castillo, E., Alvarez, E.: Expert Systems: Uncertainty and Learning. Computational Mechanics Publications, Southampton Boston (1991)
72. Castro, J., Delgado, M.: Fuzzy systems with defuzzification are universal approximators. IEEE Transactions on Systems, Man and Cybernetics, Part B: Cybernetics 26(1), 149–152 (1996)
73. Celikyilmaz, A., Türksen, I.: Modeling Uncertainty with Fuzzy Logic: With Recent Theory and Applications. Springer, Berlin (2009)
74. Cernuda, C.: Experimental analysis on assessing interpretability of fuzzy rule-based systems. Universidad de Oviedo, Spain (2010)

75. Chang, C.I.: Hyperspectral Data Exploration: Theory and Applications. John Wiley & Sons, Hoboken (2007)
76. Chao, C., Chen, Y., Teng, C.: Simplification of fuzzy-neural systems using similarity analysis. IEEE Transactions on Systems, Man and Cybernetics, Part B: Cybernetics 26(2), 344–354 (1996)
77. Chapelle, O., Schoelkopf, B., Zien, A.: Semi-Supervised Learning. MIT Press, Cambridge (2006)
78. Chen, J., Patton, R.: Robust Model-Based Fault Diagnosis for Dynamic Systems. Kluwer Academic Publishers, Norwell (1999)
79. Chen, M., Linkens, D.: Rule-base self-generation and simplification for data-driven fuzzy models. Fuzzy Sets and Systems 142(2), 243–265 (2004)
80. Chen, S., Donoho, D., Saunders, M.: Atomic decomposition by basis pursuit. SIAM Review 43(1), 129–159 (2001)
81. Cherry, A., Jones, R.: Fuzzy logic control of an automotive suspension system. IEEE Proceedings Control Theory and Applications 142, 149–160 (1995)
82. Chiang, L., Russell, E., Braatz, R.: Fault Detection and Diagnosis in Industrial Systems. Springer, Heidelberg (2001)
83. Chiu, S.: Fuzzy model identification based on cluster estimation. Journal of Intelligent and Fuzzy Systems 2(3), 267–278 (1994)
84. Chuang, Y., Chen, L.: How to evaluate 100 visual stimuli efficiently. International Journal of Design 2(1), 31–43 (2008)
85. Cleveland, W.: Robust locally weighted regression and smoothing scatterplots. Journal of the American Statistical Association 74(368), 829–836 (1979)
86. Cohen, F., Fan, Z., Attali, S.: Automated inspection of textile fabrics using textural models. IEEE Transactions on Pattern Analysis and Machine Intelligence 13(8), 803–808 (1991)
87. Cohn, D., Atlas, L., Ladner, R.: Improving generalization with active learning. Machine Learning 15(2), 201–221 (1994)
88. Collins, M., Schapire, R., Singer, Y.: Logistic regression, adaboost and bregman distances. Machine Learning 48(1-3), 253–285 (2002)
89. Condurache, A.: A two-stage-classifier for defect classification in optical media inspection. In: Proceedings of the 16th International Conference on Pattern Recognition (ICPR 2002), Quebec City, Canada, vol. 4, pp. 373–376 (2002)
90. Constantinescu, C., Storer, A.J.: Online adaptive vector quantization with variable size codebook entries. Information Processing & Management 30(6), 745–758 (1994)
91. Crawford, S.: Extensions to the CART algorithm. International Journal of Man-Machine Studies 31(2), 197–217 (1989)
92. Crespo, F., Weber, R.: A methodology for dynamic data mining based on fuzzy clustering. Fuzzy Sets and Systems 150(2), 267–284 (2005)
93. Culp, M., Michailidis, G.: An iterative algorithm for extended learners to a semi-supervised setting. Journal of Computational and Graphical Statistics 17(3), 545–571 (2008)
94. Dagan, I., Engelson, S.: Committee-based sampling for training probabilistic classifier. In: Proceedings of 12th International Conference on Machine Learning, pp. 150–157 (1995)
95. Dara, R., Kremer, S., Stacey, D.: Clustering unlabeled data with SOMs improves classification of labeled real-world data. In: Proceedings of the 2002 International Joint Conference on Neural Networks (IJCNN 2002), Honolulu, Hawaii, pp. 2237–2242 (2002)
96. Daubechies, I., Defrise, M., Mol, C.D.: An iterative thresholding algorithm for linear inverse problems with a sparsity constraint. Communications on Pure and Applied Mathematics 57(11), 1413–1457 (2004)

97. Delany, S.J., Cunningham, P., Tsymbal, A., Coyle, L.: A case-based technique for tracking concept drift in spam filtering. Knowledge-Based Systems 18(4-5), 187–195 (2005)
98. Delgado, M.R., Zuben, F.V., Gomide, F.: Hierarchical genetic fuzzy systems: accuracy, interpretability and design autonomy. In: Casillas, J., Cordón, O., Herrera, F., Magdalena, L. (eds.) Interpretability Issues in Fuzzy Modeling, pp. 379–405. Springer, Berlin (2003)
99. Demant, C., Streicher-Abel, B., Waszkewitz, P.: Industrial Image Processing: Visual Quality Control in Manufacturing. Springer, Heidelberg (1999)
100. Demiriz, A., Bennett, K., Embrechts, M.: Semi-supervised clustering using genetic algorithms. In: Proceedings of the Artificial Neural Networks in Engineering (ANNIE 1999), St. Louis, Missouri, pp. 809–814 (1999)
101. Dempster, A., Laird, N., Rubin, D.: Maximum likelihood from incomplete data via the EM algorithm. Journal of the Royal Statistical Society, Series B 39(1), 1–38 (1977)
102. Diehl, C., Cauwenberghs, G.: SVM incremental learning, adaptation and optimization. In: Proceedings of the International Joint Conference on Neural Networks, Boston, vol. 4, pp. 2685–2690 (2003)
103. Dietterich, T.: Ensemble methods in machine learning. In: Kittler, J., Roli, F. (eds.) MCS 2000. LNCS, vol. 1857, pp. 1–15. Springer, Heidelberg (2000)
104. Donoho, D., Johnstone, I.: Minimax estimation via wavelet shrinkage. Annual Statistics 26(3), 879–921 (1998)
105. Dorf, R., Bishop, R.: Modern Control Systems, 11th edn. Prentice Hall, Upper Saddle River (2007)
106. Douglas, S.: Efficient approximate implementations of the fast affine projection algorithm using orthogonal transforms. In: Proceedings of the IEEE International Conference on Acoustic, Speech and Signal Processing, Atlanta, Georgia, pp. 1656–1659 (1996)
107. Dragomir, S.: A survey on Cauchy-Bunyakovsky-Schwarz type discrete inequalities. Journal of Inequalities in Pure and Applied Mathematics 4(3), 142 (2003)
108. Draper, N., Smith, H.: Applied Regression Analysis. Probability and Mathematical Statistics. John Wiley & Sons, New York (1981)
109. Dubois, D., Huellermeier, E., Prade, H.: Towards the representation of implication-based fuzzy rules in terms of crisp rules. In: Proceedings of the 9th Joint IFSA World Congress and 20th NAFIPS International Conference, Vancouver, Canada, vol. 3, pp. 1592–1597 (2001)
110. Dubois, D., Prade, H., Ughetto, L.: Checking the coherence and redundancy of fuzzy knowledge bases. IEEE Transactions on Fuzzy Systems 5(3), 398–417 (1997)
111. Duda, R., Hart, P., Stork, D.: Pattern Classification, 2nd edn. Wiley-Interscience (John Wiley & Sons), Southern Gate, Chichester, West Sussex, England (2000)
112. Dy, J., Brodley, C.: Feature selection for unsupervised learning. Journal of Machine Learning Research 5, 845–889 (2004)
113. Economou, C., Morari, M., Palsson, P.: Internal model control: Extension to nonlinear systems. Industrial & Engineering Chemistry Process Design and Development 25(2), 403–411 (1986)
114. Efendic, H., Re, L.D.: Automatic iterative fault diagnosis approach for complex systems. WSEAS Transactions on Systems 5(2), 360–367 (2006)
115. Efendic, H., Schrempf, A., Re, L.D.: Data based fault isolation in complex measurement systems using models on demand. In: Proceedings of the IFAC-Safeprocess 2003, IFAC, Washington D.C., USA, pp. 1149–1154 (2003)
116. Efron, B., Tibshirani, R.: An Introduction to the Bootstrap. Chapman and Hall/CRC (1993)

117. Eitzinger, C., Gmainer, M., Heidl, W., Lughofer, E.: Increasing classification performance with adaptive features. In: Gasteratos, A., Vincze, M., Tsotsos, J. (eds.) ICVS 2008. LNCS, vol. 5008, pp. 445–453. Springer, Heidelberg (2008)
118. Eitzinger, C., Heidl, W., Lughofer, E., Raiser, S., Smith, J., Tahir, M., Sannen, D., van Brussel, H.: Assessment of the influence of adaptive components in trainable surface inspection systems. Machine Vision and Applications 21(5), 613–626 (2010)
119. Elgammal, A., Duraiswami, R., Harwood, D., Davis, L.: Background and foreground modeling using nonparametric kernel density estimation for visual surveillance. Proceedings of the IEEE 90(7), 1151–1163 (2002)
120. Engl, H., Hanke, M., Neubauer, A.: Regularization of Inverse Problems. Kluwer Academinc Publishers, Dordrecht (1996)
121. Espinosa, J., Vandewalle, J.: Constructing fuzzy models with linguistic intergrity from numerical data - AFRELI algorithm. IEEE Transactions on Fuzzy Systems 8(5), 591–600 (2000)
122. Espinosa, J., Wertz, V., Vandewalle, J.: Fuzzy Logic, Identification and Predictive Control (Advances in Industrial Control). Springer, Berlin (2004)
123. Ester, M., Kriegel, H., Sander, J., Xu, X.: A density-based algorithm for discovering clusters in large spatial databases with noise. In: Proceedings of 2nd International Conference on Knowledge Discovery and Data Mining (KDD 1996), Portland, Oregon, pp. 226–231 (1996)
124. Eykhoff, P.: System Identification: Parameter and State Estimation. John Wiley & Sons, Chichester (1974)
125. Fang, C., Ge, W., Xiao, D.: Fault detection and isolation for linear systems using detection observers. In: Patton, R., Frank, P., Clark, R. (eds.) Issues of Fault Diagnosis for Dynamic Systems, pp. 87–113. Springer, Heidelberg (2000)
126. Fernald, A., Kuhl, P.: Acoustic determinants of infant preference for motherese speech. Infant Behavior and Development 10(3), 279–293 (1987)
127. Fink, A., Fischer, M., Nelles, O., Isermann, R.: Supervision of nonlinear adaptive controllers based on fuzzy models. Journal of Control Engineering Practice 8(10), 1093–1105 (2000)
128. Fiordaliso, A.: A constrained Takagi-Sugeno fuzzy system that allows for better interpretation and analysis. Fuzzy Sets and Systems 118(2), 281–296 (2001)
129. Fortuna, L., Graziani, S., Rizzo, A., Xibilia, M.: Soft Sensor for Monitoring and Control of Industrial Processes. Springer, London (2007)
130. Fu, L.: An expert network for DNA sequence analysis. IEEE Intelligent Systems and Their Application 14(1), 65–71 (1999)
131. Fujino, A., Ueda, N., Saito, K.: A hybrid generative/discriminative approach to semi-supervised classifier design. In: Proceedings of the 20th National Conference on Artificial Intelligence (AAAI 2005), Pittsburgh, Pennsylvania, pp. 764–769 (2005)
132. Fukumizu, K.: Statistical active learning in multilayer perceptrons. IEEE Transactions on Neural Networks 11(1), 17–26 (2000)
133. Fukunaga, K.: Statistical Pattern Recognition, 2nd edn. Academic Press, San Diego (1990)
134. Fuller, R.: Introduction to Neuro-Fuzzy Systems. Physica-Verlag, Heidelberg (1999)
135. Furuhashi, T., Hasekawa, T., Horikawa, S., Uchikawa, Y.: An adaptive fuzzy controller using fuzzy neural networks. In: Proceedings Fifth IFSA World Congress, Seoul, pp. 769–772 (1993)
136. Gama, J., Medas, P., Castillo, G., Rodrigues, P.: Learning with drift detection. In: Bazzan, A.L.C., Labidi, S. (eds.) SBIA 2004. LNCS (LNAI), vol. 3171, pp. 286–295. Springer, Heidelberg (2004)

137. Garcia, C., Prett, D., Morari, M.: Model predictive control: Theory and practice — a survey. Automatica 25(1), 335–348 (1989)
138. Gardiner, D.: Practical Raman spectroscopy. Springer, New York (1989)
139. Gay, S.L.: Dynamically regularized fast recursive least squares with application to echo cancellation. In: Proceedings of the IEEE International Conference on Acoustic, Speech and Signal Processing, Atalanta, Georgia, pp. 957–960 (1996)
140. Gerla, G.: Approximate reasoning to unify norm-based and implication-based fuzzy control (2009), `http://citeseerx.ist.psu.edu/viewdoc/summary?doi=10.1.1.144.4468`
141. Gersho, A.: Asymptotically optimal block quantization. IEEE Transactions on Information Theory 25(4), 373–380 (1979)
142. Gertler, J.: Fault Detection and Diagnosis in Engineering Systems. Marcel Dekker, New York (1998)
143. Giurgiutiu, V.: Structural Health Monitoring: Fundamentals and Applications: With Piezoelectric Wafer Active Sensors. Academic Press, San Diego (2007)
144. Goldman, S., Zhou, Y.: Enhanced supervised learning with unlabelled data. In: Proceedings of the 17th International Conference on Machine Learning, Stanford, California, pp. 327–334 (2000)
145. Golub, G., Loan, C.V.: Matrix Computations, 3rd edn. John Hopkins University Press, Baltimore (1996)
146. Golub, H., Kahan, W.: Calculating the singular values and the pseudo-inverse of a matrix. Journal of the Society for Industrial and Applied Mathematics: Series B, Numerical Analysis 2(2), 205–224 (1965)
147. Gonzlez, J., Rojasa, I., Pomaresa, H., Herrera, L., Guillna, A., Palomares, J., Rojasa, F.: Improving the accuracy while preserving the interpretability of fuzzy function approximators by means of multi-objective evolutionary algorithms. International Journal of Approximate Reasoning 44(1), 32–44 (2007)
148. Govindhasamy, J., McLoone, S., Irwin, G.: Second-order training of adaptive critics for online process control. IEEE Transactions on Systems, Man and Cybernetics, Part B: Cybernetics 35(2), 381–385 (2006)
149. Grahn, H., Geladi, P.: Techniques and Applications of Hyperspectral Image Analysis. John Wiley & Sons, Southern Gate (2007)
150. Gray, R.: Vector quantization. IEEE ASSP Magazine 1(2), 4–29 (1984)
151. Griesse, R., Lorenz, D.: A semismooth Newton method for Tikhonov functionals with sparsity constraints. Inverse Problems 24(3), 035,007 (2008)
152. Groetsch, C.: The theory of Tikhonov regularization for Fredholm equations of the first kind, Pitman, Boston (1984)
153. Groißböck, W., Lughofer, E., Klement, E.: A comparison of variable selection methods with the main focus on orthogonalization. In: Lopéz-Díaz, M., Gil, M., Grzegorzewski, P., Hryniewicz, O., Lawry, J. (eds.) Soft Methodology and Random Information Systems, Advances in Soft Computing, pp. 479–486. Springer, Heidelberg (2004)
154. Groissboeck, W., Lughofer, E., Thumfart, S.: Associating visual textures with human perceptions using genetic algorithms. Information Sciences 180(11), 2065–2084 (2010)
155. Grossberg, S.: Nonlinear neural networks: Principles, mechanisms, and architectures. Neural Networks 1(1), 17–61 (1988)
156. Guillaume, S.: Designing fuzzy inference systems from data: an interpretability-oriented review. IEEE Transactions on Fuzzy Systems 9(3), 426–443 (2001)
157. Guo, Y., Woo, P.: Adaptive fuzzy sliding mode control for robotic manipulators. In: Proceedings of the IEEE CDC Conference 2003, Maui, Hawaii, pp. 2174–2179 (2003)

158. Gustafson, D., Kessel, W.: Fuzzy clustering with a fuzzy covariance matrix. In: Proceedings of the IEEE CDC Conference 1979, San Diego, CA, USA, pp. 761–766 (1979)
159. Guyon, I., Elisseeff, A.: An introduction to variable and feature selection. Journal of Machine Learning Research 3, 1157–1182 (2003)
160. Guyon, I., Gunn, S., Nikravesh, M., Zadeh, L.: Feature Extraction. Foundations and Applications. Springer, Heidelberg (2006)
161. Hadamard, J.: Sur les problmes aux drives partielles et leur signification physique. In: Princeton University Bulletin, pp. 49–52 (1902)
162. Haenlein, M., Kaplan, A.: A beginner's guide to partial least squares (PLS) analysis. Understanding Statistics 3(4), 283–297 (2004)
163. Haffari, G., Sarkar, A.: Analysis of semi-supervised learning with the Yarowsky algorithm. In: Proceedings of the 23rd Conference on Uncertainty in Artificial Intelligence (UAI), Vancouver, Canada (2007)
164. Halkidi, M., Batistakis, Y., Vazirgiannis, M.: On clustering validation techniques. Journal of Intelligent Information Systems 17(2-3), 107–145 (2001)
165. Hall, P., Martin, R.: Incremental eigenanalysis for classification. In: British Machine Vision Conference (BMVC) 1998, Southampton, UK, pp. 286–295 (1998)
166. Hämarik, U., Raus, T.: On the choice of the regularization parameter in ill-posed problems with approximately given noise level of data. Journal of Inverse and Ill-posed Problems 14(3), 251–266 (2006)
167. Hamker, F.: RBF learning in a non-stationary environment: the stability-plasticity dilemma. In: Howlett, R., Jain, L. (eds.) Radial Basis Function Networks 1: Recent Developments in Theory and Applications, pp. 219–251. Physica Verlag, Heidelberg (2001)
168. Hansen, L., Salamon, P.: Neural network ensembles. IEEE Transactions on Patterns Analysis and Machine Intelligence 12(10), 993–1001 (1990)
169. Hansen, P.: The truncated SVD as a method for regularization. BIT 27(4), 534–553 (1987)
170. Harrel, F.: Regression Modeling Strategies. Springer, New York (2001)
171. Harris, C., Hong, X., Gan, Q.: Adaptive Modelling, Estimation and Fusion From Data: A Neurofuzzy Approach. Springer, Berlin (2002)
172. den Hartog, M., Babuska, R., Deketh, H., Grima, M., Verhoef, P., Verbruggen, H.: Knowledge-based fuzzy model for performance prediction of a rock-cutting trencher. International Journal of Approximate Reasoning 16(1), 43–66 (1997)
173. Hassibi, B., Stork, D.: Second-order derivatives for network pruning: optimal brain surgeon. In: Hanson, S., Cowan, J., Giles, C. (eds.) Advances in Neural Information Processing, vol. 5, pp. 164–171. Morgan Kaufman, Los Altos (1993)
174. Hastie, T., Tibshirani, R., Friedman, J.: The Elements of Statistical Learning: Data Mining, Inference and Prediction, 2nd edn. Springer, Heidelberg (2009)
175. Haykin, S.: Neural Networks: A Comprehensive Foundation, 2nd edn. Prentice Hall Inc., Upper Saddle River (1999)
176. Hensel, A., Spittel, T.: Kraft- und Arbeitsbedarf bildsamer Formgebungsverfahren. VEB Deutscher Verlag für Grundstoffindustrie (1978)
177. Hernandez, N., Talavera, I., Biscay, R., Porroa, D., Ferreira, M.: Support vector regression for functional data in multivariate calibration problems. Analytica Chimica Acta 642(1-2), 110–116 (2009)
178. Himmelbauer, J., Drobics, M.: Regularized numerical optimization of fuzzy rule bases. In: Proceedings of FUZZ-IEEE 2004, Budapest, Hungary (2004)
179. Hintermüller, M., Ito, K., Kunisch, K.: The primal-dual active set strategy as a semismooth Newton method. SIAM Journal on Optimization 13(3), 865–888 (2003)

180. Ho, W., Tung, W., Quek, C.: An evolving mamdani-takagi-sugeno based neural-fuzzy inference system with improved interpretability–accuracy. In: Proceedings of the WCCI 2010 IEEE World Congress of Computational Intelligence, Barcelona, pp. 682–689 (2010)
181. Holmblad, L., Ostergaard, J.: Control of a cement kiln by fuzzy logic. Fuzzy Information and Decision Processes, 398–409 (1982)
182. Hopkins, B.: A new method for determining the type of distribution of plant individuals. Annals of Botany 18, 213–226 (1954)
183. Howieson, I., Normand, E., McCulloch, M.: Quantum-cascade lasers smell success. Laser Focus World 41(3) (2005)
184. Huang, G., Saratchandran, P., Sundararajan, N.: An efficient sequential learning algorithm for growing and pruning RBF (GAP-RBF) networks. IEEE Transactions on Systems, Man and Cybernetics Part B: Cybernetics 34(6), 2284–2292 (2004)
185. Hühn, J., Hüllermeier, E.: FR3: A fuzzy rule learner for inducing reliable classifiers. IEEE Transactions on Fuzzy Systems 17(1), 138–149 (2009)
186. Humphrey, M., Cunningham, S., Witten, I.: Knowledge visualization techniques for machine learning. Intelligent Data Analysis 2(1-4), 333–347 (1998)
187. Hunt, K., Haas, R., Murray-Smith, R.: Extending the functional equivalence of radial basis function networks and fuzzy inference systems. IEEE Transactions on Neural Networks 7(3), 776–781 (1996)
188. Iivarinen, J., Visa, A.: An adaptive texture and shape based defect classification. In: Procedings of the International Conference on Pattern Recognition, Brisbane, Australia, vol. 1, pp. 117–123 (1998)
189. Isermann, R., Ball, P.: Trends in the application of model-based fault detection and diagnosis of technical processes. In: Proceedings of the 13th IFAC World Congress, pp. 1–12. IEEE Press, San Francisco (1996)
190. Jacobs, R.H., Haak, K., Thumfart, S., Renken, R., Henson, B., Cornelissen, F.: Judgement space as a stepping stone: Finding the cognitive relationship between texture features and beauty ratings. Plos One (2010) (in revision)
191. Jain, A., Dubes, R.: Algorithms for Clustering Data. Prentice Hall, Upper Saddle River (1988)
192. Jakubek, S., Hametner, C.: Identification of neuro-fuzzy models using GTLS parameter estimation. IEEE Transactions on Systems, Man and Cybernetics Part B: Cybernetics 39(5), 1121–1133 (2009)
193. Jakubek, S., Hametner, C., Keuth, N.: Total least squares in fuzzy system identification: An application to an industrial engine. Engineering Applications of Artificial Intelligence 21(8), 1277–1288 (2008)
194. Jang, J.S.: ANFIS: Adaptive-network-based fuzzy inference systems. IEEE Transactions on Systems, Man and Cybernetics 23(3), 665–685 (1993)
195. Jang, J.S., Sun, C.T.: Functional equivalence between radial basis function networks and fuzzy inference systems. IEEE Transactions on Neural Networks 4(1), 156–159 (1993)
196. Jimenez, F., Gomez-Skarmeta, A.F., Sanchez, G., Roubos, H., Babuska, R.: Accurate, transparent and compact fuzzy models by multi-objective evolutionary algorithms. In: Casillas, J., Cordón, O., Herrera, F., Magdalena, L. (eds.) Interpretability Issues in Fuzzy Modeling, pp. 431–451. Springer, Berlin (2003)
197. Jin, Y.: Fuzzy modelling of high dimensional systems: Complexity reduction and interpretability improvement. IEEE Transactions on Fuzzy Systems 8(2), 212–221 (2000)
198. Jin, Y., Seelen, W., Sendhoff, B.: On generating FC^3 fuzzy rule systems from data using evolution strategies. IEEE Transactions on Systems, Man and Cybernetics, Part B: Cybernetics 29(6), 829–845 (1999)

199. Jin, Y., Wang, L.: Fuzzy Systems in Bioinformatics and Computational Biology. Springer, Berlin (2009)
200. Johansen, T., Babuska, R.: Multiobjective identification of Takagi-Sugeno fuzzy models. IEEE Transactions on Fuzzy Systems 11(6), 847–860 (2003)
201. Jolliffe, I.: Principal Component Analysis. Springer, Heidelberg (2002)
202. Juang, C., Lin, C.: An on-line self-constructing neural fuzzy inference network and its applications. IEEE Transactions on Fuzzy Systems 6(1), 12–32 (1998)
203. Juang, C., Tsao, Y.: A self-evolving interval type-2 fuzzy neural network with on-line structure and parameter learning. IEEE Transactions on Fuzzy Systems 16(6), 1411–1424 (2008)
204. Julesz, B.: Experiments in the visual perception of texture. Scientific American 232(4), 34–43 (1975)
205. Kalman, R.: A new approach to linear filtering and prediction problems. Transaction of the ASME, Journal of Basic Engineering 82, 35–45 (1960)
206. Kang, S., Woo, C., Hwang, H., Woo, K.: Evolutionary design of fuzzy rule base for nonlinear system modelling and control. IEEE Transactions on Fuzzy Systems 8(1), 37–45 (2000)
207. Karnik, N., Mendel, J.: Centroid of a type-2 fuzzy set. Information Sciences 132(1-4), 195–220 (2001)
208. Kasabov, N.: Evolving Connectionist Systems - Methods and Applications in Bioinformatics, Brain Study and Intelligent Machines. Springer, London (2002)
209. Kasabov, N.: Evolving Connectionist Systems: The Knowledge Engineering Approach, 2nd edn. Springer, London (2007)
210. Kasabov, N., Zhang, D., Pang, P.: Incremental learning in autonomous systems: evolving connectionist systems for on-line image and speech recognition. In: Proceedings of IEEE Workshop on Advanced Robotics and its Social Impacts, 2005, Hsinchu, Taiwan, pp. 120–125 (2005)
211. Kasabov, N.K.: Evolving fuzzy neural networks for supervised/unsupervised online knowledge-based learning. IEEE Transactions on Systems, Man and Cybernetics, Part B: Cybernetics 31(6), 902–918 (2001)
212. Kasabov, N.K., Song, Q.: DENFIS: Dynamic evolving neural-fuzzy inference system and its application for time-series prediction. IEEE Transactions on Fuzzy Systems 10(2), 144–154 (2002)
213. Kawamoto, N., Soen, T.: Objective evaluation of color design. Color Research and Application 18(4), 260–266 (1993)
214. Keogh, E., Lin, J., Fu, A.: HOT SAX: Efficiently finding the most unusual time series subsequence. In: Proceedings of the 5th IEEE International Conference on Data Mining (ICDM 2005), Houston, Texas, pp. 226–233 (2005)
215. Keogh, E., Lonardi, S., Chiu, W.: Finding surprising patterns in a time series database in linear time and space. In: Proc. of the Eighth ACM SIGKDD Int. Conf. on Knowledge Discovery and Data Mining, Edmonton, Alberta, Canada, pp. 550–556 (2002)
216. Keogh, E., Lonardi, S., Ratanamahatana, C.: Towards parameter-free data mining. In: Proc. of the Tenth ACM SIGKDD Int. Conf. on Knowlede Discovery and Data Mining, Seattle, Washington, pp. 206–215 (2004)
217. Kim, C., Koivo, A.: Hierarchical classification of surface defects on dusty wood boards. Pattern Recognition Letters 15(7), 712–713 (1994)
218. Kim, S., Kim, E.Y., Jeong, K., Kim, J.: Emotion-based textile indexing using colors, texture and patterns. In: Bebis, G., Boyle, R., Parvin, B., Koracin, D., Remagnino, P., Nefian, A., Meenakshisundaram, G., Pascucci, V., Zara, J., Molineros, J., Theisel, H., Malzbender, T. (eds.) ISVC 2006. LNCS, vol. 4292, pp. 9–18. Springer, Heidelberg (2006)

219. Kindermann, S., Ramlau, R.: Surrogate functionals and thresholding for inverse interface problems. Journal Inverse Ill-Posed Problems 15(4), 387–401 (2007)
220. Klement, E., Mesiar, R., Pap, E.: Triangular Norms. Kluwer Academic Publishers, Dordrecht (2000)
221. Klinkenberg, R.: Learning drifting concepts: example selection vs. example weighting. Intelligent Data Analysis 8(3), 281–300 (2004)
222. Klinkenberg, R., Joachims, T.: Detection concept drift with support vector machines. In: Proc. of the Seventh International Conference on Machine Learning (ICML), San Francisco, CA, U.S.A., pp. 487–494 (2000)
223. Klir, G., Yuan, B.: Fuzzy Sets and Fuzzy Logic: Theory and Applications. Prentice Hall PTR, Upper Saddle River (1995)
224. Koch, K.: Process Analytical Chemistry: Control, Optimization, Quality, Economy. Springer, Berlin (1999)
225. Koczy, L., Tikk, D., Gedeon, T.: On functional equivalence of certain fuzzy controllers and RBF type approximation schemes. International Journal of Fuzzy Systems 2(3), 164–175 (2000)
226. Kohonen, T.: An introduction to neural computing. Neural Networks 1(1), 3–16 (1988)
227. Kohonen, T.: Self-Organizing Maps: second extended edition. Springer, Heidelberg (1995)
228. Kohonen, T., Barna, G., Chrisley, R.: Statistical pattern recognition with neural networks: Benchmarking studies. In: Proceedings of the IEEE International Conference on Neural Networks, San Diego, California, pp. 61–68 (1988)
229. Korbicz, J., Koscielny, J., Kowalczuk, Z., Cholewa, W.: Fault Diagnosis - Models, Artificial Intelligence and Applications. Springer, Heidelberg (2004)
230. Kordon, A., Smits, G.: Soft sensor development using genetic programming. In: Proceedings GECCO 2001, San Francisco, pp. 1346–1351 (2001)
231. Kordon, A., Smits, G., Kalos, A., Jordaan, E.: Robust soft sensor development using genetic programming. In: Leardi, R. (ed.) Nature-Inspired Methods in Chemometrics, pp. 69–108 (2003)
232. Kowalski, B., McLennan, F.: Process Analytical Chemistry. Springer, Netherlands (1995)
233. Krishnapuram, R., Freg, C.: Fitting an unknown number of lines and planes to image data through compatible cluster merging. Pattern Recognition 25(4), 385–400 (1992)
234. Krogh, A., Vedelsby, J.: Neural network ensembles, cross validation, and active learning. In: Tesauro, I.G., Touretzky, D., Leen, T. (eds.) Advances in Neural Information Processing Systems, vol. 7, pp. 231–238 (1995)
235. Kruse, R., Gebhardt, J., Palm, R.: Fuzzy Systems in Computer Science. Verlag Vieweg, Wiesbaden (1994)
236. Kuncheva, L.: Fuzzy Classifier Design. Physica-Verlag, Heidelberg (2000)
237. Kuncheva, L.: Combining pattern classifiers: Methods and algorithms. Wiley-Interscience (John Wiley & Sons), Southern Gate (2004)
238. Kuncheva, L.I., Bezdek, J.C., Duin, R.P.W.: Decision templates for multiple classifier fusion: an experimental comparison. Pattern Recognition 34(2), 299–314 (2001)
239. Kurzhanskiy, A.A., Varaiya, P.: Ellipsoidal toolbox. Tech. rep. (2006)
240. Lam, C.: Emotion modelling using neural network. Universiti Utara, Malaysia (2005)
241. Lee, C.: Fuzzy logic in control systems: fuzzy logic controller - part i and ii. IEEE Transactions on Systems, Man and Cybernetics 20(2), 404–435 (1990)
242. Lemos, A., Caminhas, W., Gomide, F.: Fuzzy multivariable gaussian evolving approach for fault detection and diagnosis. In: Hüllermeier, E., Kruse, R., Hoffmann, F. (eds.) IPMU 2010. LNCS, vol. 6178, pp. 360–369. Springer, Heidelberg (2010)

243. Lendasse, A., Francois, D., Wertz, V., Verleysen, M.: Vector quantization: A weighted version for time-series forecasting. Future Generation Computer Systems 21(7), 1056–1067 (2005)
244. Leng, G., McGinnity, T., Prasad, G.: An approach for on-line extraction of fuzzy rules using a self-organising fuzzy neural network. Fuzzy Sets and Systems 150(2), 211–243 (2005)
245. Leng, G., Prasad, G., McGinnity, T.: An new approach to generate a self-organizing fuzzy neural network model. In: Proceedings of the International Confernence of Systems, Man and Cybernetics, Hammamet, Tunisia (2002)
246. Leng, G., Prasad, G., McGinnity, T.: An on-line algorithm for creating self-organizing fuzzy neural networks. Neural Networks 17(10), 1477–1493 (2004)
247. Leondes, C.: Fuzzy Logic and Expert Systems Applications (Neural Network Systems Techniques and Applications). Academic Press, San Diego (1998)
248. Lepskij, O.: On a problem of adaptive estimation in gaussian white noise. Theory of Probability and its Applications 35(3), 454–466 (1990)
249. Leung, C., Wong, K., Sum, P., Chan, L.: A pruning method for the recursive least squares algorithm. Neural Networks 14(2), 147–174 (2001)
250. Lewis, D., Catlett, J.: Heterogeneous uncertainty sampling for supervised learning. In: Proceedings of the 11th International Conference on Machine Learning, New Brunswick, New Jersey, pp. 148–156 (1994)
251. Li, X., Li, H., Guan, X., Du, R.: Fuzzy estimation of feed-cutting force from current measurement - a case study on tool wear monitoring. IEEE Transactions Systems, Man, and Cybernetics Part C: Applications and Reviews 34(4), 506–512 (2004)
252. Li, X., Tso, S.K.: Drill wear monitoring with current signal. Wear 231(2), 172–178 (1999)
253. Li, X., Wang, L., Sung, E.: Multilabel SVM active learning for image classification. In: Proceedings of the International Conference on Image Processing (ICIP), Singapore, vol. 4, pp. 2207–2010 (2004)
254. Li, Y.: On incremental and robust subspace learning. Pattern Recognition 37(7), 1509–1518 (2004)
255. Liang, Q., Mendel, J.: Interval type-2 fuzzy logic systems: Theory and design. IEEE Transactions on Fuzzy Systems 8(5), 535–550 (2000)
256. Lim, C., Harrison, R.: Online pattern classification with multiple neural network systems: An experimental study. IEEE Transactions on Systems, Man, and Cybernetics, Part C: Applications and Reviews 33(2), 235–247 (2003)
257. Lim, R., Phan, M.: Identification of a multistep-ahead observer and its application to predictive control. Journal of Guidance, Control and Dynamics 20(6), 1200–1206 (1997)
258. Lima, E., Gomide, F., Ballini, R.: Participatory evolving fuzzy modeling. In: 2nd International Symposium on Evolving Fuzzy Systems, Lake District, UK, pp. 36–41 (2006)
259. Lima, E., Hell, M., Ballini, R., Gomide, F.: Evolving fuzzy modeling using participatory learning. In: Angelov, P., Filev, D., Kasabov, N. (eds.) Evolving Intelligent Systems: Methodology and Applications, pp. 67–86. John Wiley & Sons, New York (2010)
260. Lin, C., Lee, C.: Neural-network-based fuzzy logic control and decision system. IEEE Transactions on Computation 40, 1320–1336 (1991)
261. Lin, C., Lee, C.: Reinforcement structure/parameter learning for neual-network-based fuzzy logic control systems. IEEE Transactions on Fuzzy Systems 2(1), 46–63 (1994)
262. Lin, C., Lee, C.: Neuro Fuzzy Systems. Prentice Hall, Englewood Cliffs (1996)
263. Lin, C., Segel, L.: Mathematics Applied to Deterministic Problems in the Natural Sciences. SIAM: Society for Industrial and Applied Mathematics, Philadelphia (1988)

264. Ljung, L.: System Identification: Theory for the User. Prentice Hall PTR, Prentic Hall Inc., Upper Saddle River, New Jersey (1999)
265. Lughofer, E.: Process safety enhancements for data-driven evolving fuzzy models. In: Proceedings of 2nd Symposium on Evolving Fuzzy Systems (EFS 2006), Lake District, UK, pp. 42–48 (2006)
266. Lughofer, E.: Evolving vector quantization for classification of on-line data streams. In: Proc. of the Conference on Computational Intelligence for Modelling, Control and Automation (CIMCA 2008), Vienna, Austria, pp. 780–786 (2008)
267. Lughofer, E.: Extensions of vector quantization for incremental clustering. Pattern Recognition 41(3), 995–1011 (2008)
268. Lughofer, E.: FLEXFIS: A robust incremental learning approach for evolving TS fuzzy models. IEEE Transactions on Fuzzy Systems 16(6), 1393–1410 (2008)
269. Lughofer, E.: On dynamic selection of the most informative samples in classification problems. In: Proc. of the 9th International Conference in Machine Learning and Applications, ICMLA 2010. IEEE, Washington D.C. (to appear, 2010)
270. Lughofer, E.: On dynamic soft dimension reduction in evolving fuzzy classifiers. In: Hüllermeier, E., Kruse, R., Hoffmann, F. (eds.) IPMU 2010. LNCS, vol. 6178, pp. 79–88. Springer, Heidelberg (2010)
271. Lughofer, E.: On-line evolving image classifiers and their application to surface inspection. Image and Vision Computing 28(7), 1063–1172 (2010)
272. Lughofer, E.: On-line feature weighing in evolving fuzzy classifiers. Fuzzy Sets and Systems (in press, 2010), doi:10.1016/j.fss.2010.08.012
273. Lughofer, E.: Towards robust evolving fuzzy systems. In: Angelov, P., Filev, D., Kasabov, N. (eds.) Evolving Intelligent Systems: Methodology and Applications, pp. 87–126. John Wiley & Sons, New York (2010)
274. Lughofer, E., Angelov, P.: Detecting and reacting on drifts and shifts in on-line data streams with evolving fuzzy systems. In: Proceedings of the IFSA/EUSFLAT 2009 Conference, Lisbon, Portugal, pp. 931–937 (2009)
275. Lughofer, E., Angelov, P.: Handling drifts and shifts in on-line data streams with evolving fuzzy systems. Applied Soft Computing (in press, 2010), doi:10.1016/j.asoc.2010.07.003
276. Lughofer, E., Angelov, P., Zhou, X.: Evolving single- and multi-model fuzzy classifiers with FLEXFIS-Class. In: Proceedings of FUZZ-IEEE 2007, London, UK, pp. 363–368 (2007)
277. Lughofer, E., Bodenhofer, U.: Incremental learning of fuzzy basis function networks with a modified version of vector quantization. In: Proceedings of IPMU 2006, Paris, France, vol. 1, pp. 56–63 (2006)
278. Lughofer, E., Efendic, H., Re, L.D., Klement, E.: Filtering of dynamic measurements in intelligent sensors for fault detection based on data-driven models. In: Proceedings of the IEEE CDC Conference, Maui, Hawaii, pp. 463–468 (2003)
279. Lughofer, E., Guardiola, C.: Applying evolving fuzzy models with adaptive local error bars to on-line fault detection. In: Proceedings of Genetic and Evolving Fuzzy Systems 2008, pp. 35–40. Witten-Bommerholz, Germany (2008)
280. Lughofer, E., Guardiola, C.: On-line fault detection with data-driven evolving fuzzy models. Journal of Control and Intelligent Systems 36(4), 307–317 (2008)
281. Lughofer, E., Hüllermeier, E., Klement, E.: Improving the interpretability of data-driven evolving fuzzy systems. In: Proceedings of EUSFLAT 2005, Barcelona, Spain, pp. 28–33 (2005)
282. Lughofer, E., Kindermann, S.: Improving the robustness of data-driven fuzzy systems with regularization. In: Proc. of the IEEE World Congress on Computational Intelligence (WCCI) 2008, Hongkong, pp. 703–709 (2008)

283. Lughofer, E., Kindermann, S.: Rule weight optimization and feature selection in fuzzy systems with sparsity constraints. In: Proceedings of the IFSA/EUSFLAT 2009 Conference, Lisbon, Portugal, pp. 950–956 (2009)
284. Lughofer, E., Kindermann, S.: SparseFIS: Data-driven learning of fuzzy systems with sparsity constraints. IEEE Transactions on Fuzzy Systems 18(2), 396–411 (2010)
285. Lughofer, E., Klement, E.: Online adaptation of Takagi-Sugeno fuzzy inference systems. In: Proceedings of CESA—IMACS MultiConference, Lille, France (2003)
286. Lughofer, E., Klement, E.: Premise parameter estimation and adaptation in fuzzy systems with open-loop clustering methods. In: Proceedings of FUZZ-IEEE 2004, Budapest, Hungary (2004)
287. Lughofer, E., Klement, E., Lujan, J., Guardiola, C.: Model-based fault detection in multi-sensor measurement systems. In: Proceedings of IEEE IS 2004, Varna, Bulgaria, pp. 184–189 (2004)
288. Lughofer, E., Macian, V., Guardiola, C., Klement, E.: Data-driven design of Takagi-Sugeno fuzzy systems for predicting NOx emissions. In: Hüllermeier, E., Kruse, R., Hoffmann, F. (eds.) IPMU 2010. Communications in Computer and Information Science, vol. 81, pp. 1–10. Springer, Heidelberg (2010)
289. Lughofer, E., Smith, J.E., Caleb-Solly, P., Tahir, M., Eitzinger, C., Sannen, D., Nuttin, M.: On human-machine interaction during on-line image classifier training. IEEE Transactions on Systems, Man and Cybernetics, Part A: Systems and Humans 39(5), 960–971 (2009)
290. Lukas, M.: Robust generalized cross-validation for choosing the regularization parameter. Inverse Problems 22(5), 1883–1902 (2006)
291. Luo, A.: Discontinuous Dynamical Systems on Time-varying Domains. Springer, Heidelberg (2009)
292. Macias-Hernandez, J., Angelov, P.: Applications of evolving intelligent systems to the oil and gas industry. In: Angelov, P., Filev, D., Kasabov, N. (eds.) Evolving Intelligent Systems: Methodology and Applications, pp. 401–421. John Wiley & Sons, New York (2010)
293. Mackay, D.: Information-based objective functions for active data selection. Neural Computation 4(4), 305–318 (1992)
294. Mackey, M., Glass, L.: Oscillation and chaos in physiological control systems. Science 197(4300), 287–289 (1977)
295. Mahalanobis, P.C.: On the generalised distance in statistics. Proceedings of the National Institute of Sciences of India 2(1), 49–55 (1936)
296. Mahony, S., Hendrix, D., Golden, A., Smith, T.J., Rokhsar, D.S.: Transcription factor binding site identification using the self-organizing map. Bioinformatics 21(9), 1807–1814 (2005)
297. Mamdani, E.: Application of fuzzy logic to approximate reasoning using linguistic systems. Fuzzy Sets and Systems 26(12), 1182–1191 (1977)
298. Mardia, K., Kent, J., Bibby, J.: Multivariate Analysis. Academic Press, New York (1979)
299. Marquardt, D.: An algorithm for least-squares estimation of nonlinear parameters. SIAM Journal on Applied Mathematics 11(2), 431–441 (1963)
300. Massart, D., Vandeginste, B., Buydens, L., Jong, S.D., Lewi, P., Smeyer-Verbeke, J.: Handbook of Chemometrics and Qualimetrics Part A. Elsevier, Amsterdam (1997)
301. Mastorocostas, P., Theocharis, J., Petridis, V.: A constrained orthogonal least-squares method for generating TSK fuzzy models: application to short-term load forecasting. Fuzzy Sets and Systems 118(2), 215–233 (2001)

302. Mendel, J.: Uncertain Rule-Based Fuzzy Logic Systems: Introduction and New Directions. Prentice Hall, Upper Saddle River (2001)
303. Mendel, J.: Type-2 fuzzy sets and systems: an overview. IEEE Computational Intelligence Magazine 2, 20–29 (2007)
304. Mendel, J., John, R.: Type-2 fuzzy sets made simple. IEEE Transactions on Fuzzy Systems 10(2), 117–127 (2002)
305. Merched, R., Sayed, A.: Fast RLS laguerre adaptive filtering. In: Proceedings of the Allerton Conference on Communication, Control and Computing, Allerton, IL, pp. 338–347 (1999)
306. Michels, K., Klawonn, F., Kruse, R., Nürnberger, A.: Fuzzy-Regelung: Grundlagen, Entwurf, Analyse. Springer, Berlin (2002)
307. Mikenina, L., Zimmermann, H.: Improved feature selection and classification by the 2-additive fuzzy measure. Fuzzy Sets and Systems 107(2), 197–218 (1999)
308. Mikut, R., Mäkel, J., Gröll, L.: Interpretability issues in data-based learning of fuzzy systems. Fuzzy Sets and Systems 150(2), 179–197 (2005)
309. Miller, A.: Subset Selection in Regression, 2nd edn. Chapman and Hall/CRC, Boca Raton, Florida (2002)
310. Miller, G.: The magic number seven plus or minus two: some limits on our capacity for processing information. Psychological Review 63(2), 81–97 (1956)
311. Mitchell, T.M.: Machine Learning. McGraw-Hill International Editions, Singapore (1997)
312. Moos, R.: A brief overview on automotive exhaust gas sensors based on electroceramics. International Journal of Applied Ceramic Technology 2(5), 401–413 (2005)
313. Morant, F., Albertos, P., Martinez, M., Crespo, A., Navarro, J.: RIGAS: An intelligent controller for cement kiln control. In: Proceedings of the IFAC Symposium on Artificial Intelligence in Real Time Control. Delft, Netherlands (1992)
314. Morozov, V.A.: On the solution of functional equations by the methdod of regularization. Soviet Mathematics Doklady 7, 414–417 (1966)
315. Muslea, I.: Active learning with multiple views. Ph.D. thesis, University of Southern California (2000)
316. Myers, R.: Classical and Modern Regression with Applications. PWS-KENT, Boston (1990)
317. Narendra, K., Parthasarthy, K.: Identification and control of dynamic systems using neural networks. IEEE Transactions on Neural Networks 1(1), 4–27 (1990)
318. Nauck, D., Kruse, R.: NEFCLASS-X – a soft computing tool to build readable fuzzy classifiers. BT Technology Journal 16(3), 180–190 (1998)
319. Nauck, D., Nauck, U., Kruse, R.: Generating classification rules with the neuro–fuzzy system NEFCLASS. In: Proceedings of the Biennial Conference of the North American Fuzzy Information Processing Society (NAFIPS), Berkeley, CA, pp. 466–470 (1996)
320. Nelles, O.: Nonlinear System Identification. Springer, Berlin (2001)
321. Ngia, L., Sjöberg, J.: Efficient training of neural nets for nonlinear adaptive filtering using a recursive Levenberg-Marquardt algorithm. IEEE Trans. Signal Processing 48(7), 1915–1926 (2000)
322. Nguyen, H., Sugeno, M., Tong, R., Yager, R.: Theoretical Aspects of Fuzzy Control. John Wiley & Sons, New York (1995)
323. Nigam, K., Ghani, R.: Analyzing the effectiveness and applicability of co-training. In: Proceedings of the 9th International Conference on Information and Knowledge Management, Washington, DC, pp. 86–93 (2000)
324. Nigam, K., McCallum, A., Thrun, S., Mitchell, T.: Text classification from labelled and unlabelled documents using EM. Machine Learning 39(2-3), 103–134 (2000)

325. Norman, D.: Emotional Design: Why We Love (or Hate) Everyday Thing? Basic Books, New York (2003)
326. Nyberg, M.: Model based fault diagnosis, methods, theory, and automotive engine application. Ph.D. thesis, Department of Electrical Engineering Linköping University, SE-581 83 Linköping, Sweden (1999)
327. Oliveira, J.V.D.: A design methodology for fuzzy system interfaces. IEEE Transactions on Fuzzy Systems 3(4), 404–414 (1995)
328. Oliveira, J.V.D.: Semantic constraints for membership function optimization. IEEE Transactions on Systems, Man and Cybernetics, Part A: Systems and Humans 29(1), 128–138 (1999)
329. Oppenheim, A., Schafer, R., Buck, J.: Discrete-Time Signal Processing, 2nd edn. Prentice Hall, Upper Saddle River (1999)
330. Oza, N.: Online ensemble learning. Ph.D. thesis. University of California, USA (2001)
331. Özdemir, S., Baykut, A., Meylani, R., Erçil, A., Ertüzün, A.: Comparative evaluation of texture analysis algorithms for defect inspection of textile products. In: Proceedings of the International Conference on Pattern Recognition, Los Alamitos, CA, pp. 1738–1741 (1998)
332. Pal, N., Chakraborty, D.: Mountain and subtractive clustering method: Improvement and generalizations. International Journal of Intelligent Systems 15(4), 329–341 (2000)
333. Palm, R.: Fuzzy controller for a sensor guided robot. Fuzzy Sets and Systems 31(2), 133–149 (1989)
334. Pang, S., Ozawa, S., Kasabov, N.: Incremental linear discriminant analysis for classification of data streams. IEEE Transaction on Systems, Men and Cybernetics, Part B: Cybernetics 35(5), 905–914 (2005)
335. Papari, G., Petkov, N.: Algorithm that mimics human perceptual grouping of dot patterns. In: Brain, Vision, and Artificial Intelligence, pp. 497–506. Springer, Berlin (2005)
336. Pedreira, C.: Learning vector quantization with training data selection. IEEE Transactions on Pattern Analysis and Machine Intelligence 28(1), 157–162 (2006)
337. Pedrycz, W.: An identification algorithm in fuzzy relational systems. Fuzzy Sets and Systems 13(2), 153–167 (1984)
338. Pedrycz, W.: A dynamic data granulation through adjustable clustering. Pattern Recognition Letters 29(16), 2059–2066 (2008)
339. Pedrycz, W., Gomide, F.: Introduction to Fuzzy Sets. MIT Press, Cambridge (1998)
340. Pedrycz, W., Rai, P.: Collaborative clustering with the use of fuzzy c-means and its quantification. Fuzzy Sets and Systems 159(18), 2399–2427 (2008)
341. Pham, B., Brown, R.: Visualisation of fuzzy systems: requirements, techniques and framework. Future Generation Computer Systems 27(7), 1199–1212 (2005)
342. Piegat, A.: Fuzzy Modeling and Control. Physica Verlag, Springer, Heidelberg, New York (2001)
343. Poirier, F., Ferrieux, A.: DVQ: Dynamic vector quantization - an incremental LVQ. In: Kohonen, T., Mäkisara, K., Simula, O., Kangas, J. (eds.) Artificial Neural Networks, pp. 1333–1336. Elsevier Science Publishers B.V., North-Holland (1991)
344. Polat, K., Günes, S.: A novel hybrid intelligent method based on C4.5 decision tree classifier and one-against-all approach for multi-class classification problems. Expert Systems with Applications 36(2), 1587–1592 (2007)
345. Polikar, R.: Ensemble based systems in decision making. IEEE Circuits and Systems Magazine 6(3), 21–45 (2006)
346. Prasad, G., Leng, G., McGuinnity, T., Coyle, D.: Online identification of self-organizing fuzzy neural networks for modeling time-varying complex systems. In: Angelov, P., Filev, D., Kasabov, N. (eds.) Evolving Intelligent Systems: Methodology and Applications, pp. 201–228. John Wiley & Sons, New York (2010)

347. Preparata, F., Hong, S.: Convex hulls of finite sets of points in two and three dimensions. Communication ACM 20(2), 87–93 (1977)
348. Press, W., Teukolsky, S., Vetterling, W., Flannery, P.: Numerical Recipes in C: The Art of Scientific Computing. Cambridge University Press, Cambridge (1992)
349. Provost, F.: Machine learning from imbalanced data sets. In: Proceedings of the AAAI Workshop, Menlo Park, CA, USA, pp. 1–3 (2000)
350. Qin, S., Li, W., Yue, H.: Recursive PCA for adaptive process monitoring. Journal of Process Control 10(5), 471–486 (2000)
351. Quinlan, J.R.: C4.5: Programs for Machine Learning. Morgan Kaufmann Publishers, San Francisco (1993)
352. Jacobs, R., Jordan, M., Nowlan, S.J., Hinton, G.E.: Adaptive mixtures of local experts. Neural Computation 3, 79–87 (1991)
353. Polikar, R., Upda, L., Upda, S.S., Honavar, V.: Learn++: An incremental learning algorithm for supervised neural networks. EEE Transactions on Systems, Man, and Cybernetics, Part C: Applications and Reviews 31(4), 497–508 (2001)
354. Raghavan, H., Madani, O., Jones, R.: Active learning with feedback on both features and instances. Journal of Machine Learning Research 7, 1655–1686 (2006)
355. Raiser, S., Lughofer, E., Eitzinger, C., Smith, J.: Impact of object extraction methods on classification performance in surface inspection systems. Machine Vision and Applications 21(5), 627–641 (2010)
356. Rallo, R., Ferre-Gine, J., Arena, A., Girault, F.: Neural virtual sensor for the inferential prediction of product quality from process variables. Computers and Chemical Engineering 26(12), 1735–1754 (2004)
357. Ramamurthy, S., Bhatnagar, R.: Tracking recurrent concept drift in streaming data using ensemble classifiers. In: Proceedings of the Sixth International Conference on Machine Learning and Applications (ICMLA), 2007, Cincinnati, Ohio, pp. 404–409 (2007)
358. Ramlau, R., Teschke, G.: A tikhonov-based projection iteration for nonlinear ill-posed problems with sparsity constraints. Numerische Mathematik 104(2), 177–203 (2006)
359. Ramos, J., Dourado, A.: Pruning for interpretability of large spanned eTS. In: Proceedings of the 2006 International Symposium on Evolving Fuzzy Systems (EFS 2006), Lake District, UK, pp. 55–60 (2006)
360. Ramos, J.V., Pereira, C., Dourado, A.: The building of interpretable systems in real-time. In: Angelov, P., Filev, D., Kasabov, N. (eds.) Evolving Intelligent Systems: Methodology and Applications, pp. 127–150. John Wiley & Sons, New York (2010)
361. Rao, Y., Principe, J., Wong, T.: Fast RLS-like algorithm for generalized eigendecomposition and its applications. The Journal of VLSI Signal Processing 37(2-3), 333–344 (2004)
362. Raus, T.: About regularization parameter choice in case of approximately given error bounds of data. In: Vainikko, G. (ed.) Methods for Solution of Integral Equations and Ill-Posed Problems, pp. 77–89. Springer, Berlin (1992)
363. Reed, R.: Pruning algorithms - a survey. IEEE Transactions on Neural Networks 4(5), 740–747 (1993)
364. Reeves, J., Delwiche, S.: Partial least squares regression for analysis of spectroscopic data. Journal of Near Infrared Spectroscopy 11(6), 415–431 (2003)
365. Rehm, F., Klawonn, F., Kruse, R.: Visualization of fuzzy classifiers. International Journal of Uncertainty, Fuzziness and Knowledge-Based Systems 15(5), 615–624 (2007)
366. Reinke, R., Michalski, R.: Incremental learning of concept description: A method and experimental results. In: Hayes, J., Michie, D., Richards, J. (eds.) Machine Intelligence, vol. 11, pp. 263–288. Oxford University Press, Inc., New York (1988)

367. Rhee, F.H., Choi, B.I.: A convex cluster merging algorithm using support vector machines. In: Proceedings of the 12th IEEE International Conference on Fuzzy Systems, St. Louis, Missouri, vol. 2, pp. 892–895 (2003)
368. Robinson, T., Moyeed, R.: Making robust the cross-validatory choice of smoothing parameter in spline smoothing regression. Communications in Statistics — Theory and Methods 18(2), 523–539 (1989)
369. Rong, H.J., Sundararajan, N., Huang, G.B., Saratchandran, P.: Sequential adaptive fuzzy inference system (SAFIS) for nonlinear system identification and prediction. Fuzzy Sets and Systems 157(9), 1260–1275 (2006)
370. Ros, L., Sabater, A., Thomas, F.: An ellipsoidal calculus based on propagation and fusion. IEEE Transactions on Systems, Man and Cybernetics - Part B: Cybernetics 32(4), 430–442 (2002)
371. Rothamsted, V., Lewis, T., Barnett, V.: Outliers in Statistical Data. John Wiley & Sons, Chichester (1998)
372. Roubos, H., Setnes, M.: Compact and transparent fuzzy models and classifiers through iterative complexity reduction. IEEE Transactions on Fuzzy Systems 9(4), 516–524 (2001)
373. Roubos, J., Setnes, M., Abonyi, J.: Learning fuzzy classification rules from data. Information Sciences 150(1-2), 77–93 (2003)
374. Rubio, J.: Stability analysis for an on-line evolving neuro-fuzzy recurrent network. In: Angelov, P., Filev, D., Kasabov, N. (eds.) Evolving Intelligent Systems: Methodology and Applications, pp. 173–199. John Wiley & Sons, New York (2010)
375. Rumelhart, D., Hinton, G., Williams, R.: Learning internal representations by error propagation. In: Rumelhart, D., McClelland, J. (eds.) Parallel Distributed Processing: Explorations in the Microstructure of Cognition, vol. 1, pp. 318–362. MIT Press, Cambridge (1986)
376. Ruspini, E.: A new approach to clustering. Information and Control 15(1), 22–32 (1969)
377. Sadeghi-Tehran, P., Angelov, P., Ramezani, R.: A fast recursive approach to autonomous detection, identification and tracking of multiple objects in video streams under uncertainties. In: Hüllermeier, E., Kruse, R., Hoffmann, F. (eds.) IPMU 2010. Communications in Computer and Information Science, vol. 81, pp. 30–43. Springer, Heidelberg (2010)
378. Samanta, B.: Gear fault detection using artificial neural networks and support vector machines with genetic algorithms. Mechanical Systems and Signal Processing 18(3), 625–644 (2004)
379. Sanchez, L., Suarez, M., Villar, J., Couso, I.: Mutual information-based feature selection and partition design in fuzzy rule-based classifiers from vague data. International Journal of Approximate Reasoning 49(3), 607–622 (2008)
380. Sannen, D., Lughofer, E., Brussel, H.V.: Increasing on-line classification performance using incremental classifier fusion. In: Proc. of International Conference on Adaptive and Intelligent Systems (ICAIS 2009), Klagenfurt, Austria, pp. 101–107 (2009)
381. Sannen, D., Lughofer, E., Brussel, H.V.: Towards incremental classifier fusion. Intelligent Data Analysis 14(1), 3–30 (2010)
382. Sannen, D., Nuttin, M., Smith, J., Tahir, M., Lughofer, E., Eitzinger, C.: An interactive self-adaptive on-line image classification framework. In: Gasteratos, A., Vincze, M., Tsotsos, J.K. (eds.) ICVS 2008. LNCS, vol. 5008, pp. 173–180. Springer, Heidelberg (2008)
383. Sato, A., Yamada, K.: Generalized learning vector quantization. In: Tesauro, G., Touretzky, D., Leon, T. (eds.) Advances in Neural Information Processing Systems, vol. 7, pp. 423–429 (1988)

384. Schael, M.: Texture fault detection using invariant textural features. In: Radig, B., Florczyk, S. (eds.) DAGM 2001. LNCS, vol. 2191, pp. 17–24. Springer, Heidelberg (2001)
385. Schaffer, C.: Overfitting avoidance as bias. Machine Learning 10(2), 153–178 (1993)
386. Scheier, E., Slaney, M.: Construction and evaluation of a robust multifeature speech/music discriminator. In: Proceedings of the International Conference on Acoustics, Speech, and Signal Processing (ICASSP) 1997, Munich, pp. 1331–1334 (1997)
387. Schmidt, B., Klawonn, F.: Construction of fuzzy classification systems with the Lukasiewicz-t-norm. In: Proceedings of the 19th International Conference of the North American Fuzzy Information Processing Society (NAFIPS), Atlanta, Georgia, pp. 109–113 (2000)
388. Schoener, H., Moser, B., Lughofer, E.: On preprocessing multi-channel data for online process monitoring. In: Proc. of the Conference on Computational Intelligence for Modelling, Control and Automation (CIMCA 2008), Vienna, Austria, pp. 414–420 (2008)
389. Schölkopf, B., Smola, A.: Learning with Kernels - Support Vector Machines, Regularization, Optimization and Beyond. MIT Press, London (2002)
390. Sebe, N., Lew, M.: Texture features for content-based retrieval. In: Lew, M. (ed.) Principles of visual information retrieval, pp. 51–85. Springer, London (2001)
391. Senge, R., Hüllermeier, E.: Pattern trees for regression and fuzzy systems modeling. In: Proc. of the IEEE World Congress on Computational Intelligence. IEEE, Barcelona (2010)
392. Serafim, A.: Segmentation of natural images based on multiresolution pyramids linking: Application to leather defects detection. In: Proceedings of the International Conference on Pattern Recognition, Kobe, Japan, pp. 41–44 (1992)
393. Setnes, M.: Simplification and reduction of fuzzy rules. In: Casillas, J., Cordón, O., Herrera, F., Magdalena, L. (eds.) Interpretability Issues in Fuzzy Modeling, pp. 278–302. Springer, Berlin (2003)
394. Setnes, M., Babuska, R., Kaymak, U., Lemke, H.: Similarity measures in fuzzy rule base simplification. IEEE Transactions on Systems, Man and Cybernetics, Part B: Cybernetics 28(3), 376–386 (1998)
395. Sherman, J., Morrison, W.: Adjustment of an inverse matrix corresponding to changes in the elements of a given column or a given row of the original matrix. Annals of Mathematical Statistics 20, 621 (1949)
396. Shi, J., Malik, J.: Normalized cuts and image segementation. IEEE Transactions on Pattern Analysis and Machine Intelligence 22(8), 888–905 (2000)
397. Siler, W., Buckley, J.: Fuzzy Expert Systems and Fuzzy Reasoning: Theory and Applications. John Wiley & Sons, Chichester (2005)
398. Silva, L., Gomide, F., Yager, R.: Participatory learning in fuzzy clustering. In: IEEE International Conference on Fuzzy Systems, FUZZ-IEEE 2005, Reno, Nevada, pp. 857–861 (2005)
399. Simani, S., Fantuzzi, C., Patton, R.: Model-based Fault Diagnosis in Dynamic Systems Using Identification Techniques. Springer, Heidelberg (2002)
400. Simoudis, E., Aha, D.: Special issue on lazy learning. Artificial Intelligence Review 11(1-5) (1997)
401. Smith, J., Tahir, M.: Stop wasting time: On predicting the success or failure of learning for industrial applications. In: Yin, H., Tino, P., Corchado, E., Byrne, W., Yao, X. (eds.) IDEAL 2007. LNCS, vol. 4881, pp. 673–683. Springer, Heidelberg (2007)
402. Smith, L.I.: A tutorial on principal component analysis (2002), `http://www.cs.otago.ac.nz/cosc453/student_tutorials/principal_components.pdf`

403. Smithson, M.: Confidence Intervals. SAGE University Paper (Series: Quantitative Applications in the Social Sciences), Thousand Oaks, California (2003)
404. So, C., Ng, S., Leung, S.: Gradient based variable forgetting factor RLS algorithm. Signal Processing 83(6), 1163–1175 (2003)
405. Sobue, S., Huang, X., Chen, Y.: Mapping functions between image features and KANSEI and its application to KANSAI based clothing fabric image retrieval. In: Proceedings of the 23rd International Technical Conference on Circuits/Systems, Computers and Communications (ITC-CSCC 2008), Yamaguchi, Japan, pp. 705–708 (2008)
406. Song, M., Wang, H.: Incremental estimation of gaussian mixture models for online data stream clustering. In: Proceedings of the International Conference on Bioinformatics and its Applications, Fort Lauderdale, Florida, USA (2004)
407. Song, Q.: Adaptive evolving neuro-fuzzy systems for dynamic system identification. Ph.D. thesis. University of Otago, New Zealand (2002)
408. Stephani, H., Hermann, M., Wiesauer, K., Katletz, S., Heise, B.: Enhancing the interpretability of thz data through unsupervised classification. In: Proc. of the IMEKO XIX World Congress, Fundamental and Applied Metrology, Lisbon, Portugal (2009)
409. Stockman, G., Shapiro, L.: Computer Vision. Prentice Hall, Upper Saddle River (2001)
410. Stone, M.: The generalized weierstrass approximation theorem. Mathematics Magazine 29(4), 167–184 (1948)
411. Stone, M.: Cross-validatory choice and assessment of statistical predictions. Journal of the Royal Statistical Society 36(1), 111–147 (1974)
412. Sugeno, M.: Industrial Applications of Fuzzy Control. Elsevier Science, Amsterdam (1985)
413. Sutton, R., Barto, A.: Reinforcement learning: an introduction. MIT Press, Cambridge (1998)
414. Swanson, D.: Signal Processing for Intelligent Sensors. Marcel Dekker, New York (2000)
415. Swierenga, H., de Weier, A., van Wijk, R., Buydens, L.: Strategy for constructing robust multivariate calibration models. Chemometrics and Intelligent Laboratory Systems 49(1), 1–17 (1999)
416. Tahir, M.A., Smith, J.: Improving nearest neighbor classifier using tabu search and ensemble distance metrics. In: Proceedings of the Eleventh International Conference on Machine Learning
417. Takagi, T., Sugeno, M.: Fuzzy identification of systems and its applications to modeling and control. IEEE Transactions on Systems, Man and Cybernetics 15(1), 116–132 (1985)
418. Thompson, C., Califf, M., Mooney, R.: Active learning for natural language parsing and information extraction. In: Proceedings of 16th International Conference on Machine Learning, Bled, Slovenia, pp. 406–414 (1999)
419. Thumfart, S., Jacobs, R., Lughofer, E., Cornelissen, F., Maak, H., Groissboeck, W., Richter, R.: Modelling human aesthetic perception of visual textures. ACM Transactions on Applied Perception (to appear, 2010)
420. Thumfart, S., Jacobs, R.H., Haak, K.V., Cornelissen, F.W., Scharinger, J., Eitzinger, C.: Feature based prediction of perceived and aesthetic properties of visual textures. In: Proc. Materials & Sensations 2008, PAU, France, pp. 55–58 (2008)
421. Tick, J., Fodor, J.: Fuzzy implications and inference processes. Computing and Informatics 24(6), 591–602 (2005)
422. Tickle, A., Andrews, R., Golea, M., Diederich, J.: The truth will come to light: directions and challenges in extracting the knowledge embedded within the trained artificial neural networks. IEEE Transactions on Neural Networks 9(6), 1057–1068 (1998)

423. Tikhonov, A., Arsenin, V.: Solutions of ill-posed problems. Winston & Sons, Washington D.C. (1977)
424. Tikhonov, A., Glasko, V.: Use of the regularization method in non-linear problems. U.S.S.R. Computational Mathematics and Mathematical Physics 5(3), 93–107 (1965)
425. Tong, R., Beck, M., Latten, A.: Fuzzy control of the activated sludege wastewater treatment process. Automatica 16(6), 695–701 (1980)
426. Tong, S., Koller, D.: Support vector machine active learning with application to text classification. Journal of Machine Learning Research 2, 45–66 (2001)
427. Treado, P., Levin, I., Lewis, E.: Near-infrared acousto-optic filtered spectroscopic microscopy: A solid-state approach to chemical imaging. Applied Spectroscopy 46(4), 553–559 (1992)
428. Tschumitschew, K., Klawonn, F.: Incremental quantile estimation. Evolving Systems (in press, 2010), doi:10.1007/s12530-010-9017-7
429. Tsymbal, A.: The problem of concept drift: definitions and related work. Tech. Rep. TCD-CS-2004-15, Department of Computer Science, Trinity College Dublin, Ireland (2004)
430. Turban, E., Aronson, J., Liang, T.P.: Decision Support Systems and Intelligent Systems, 7th edn. Prentice Hall, Upper Saddle River (2004)
431. Utgoff, P.: Incremental induction of decision trees. Machine Learning 4(2), 161–186 (1989)
432. Vaira, S., Mantovani, V.E., Robles, J., Sanchis, J.C., Goicoechea, H.: Use of chemometrics: Principal component analysis (PCA) and principal component regression (PCR) for the authentication of orange juice. Analytical Letters 32(15), 3131–3141 (1999)
433. Vandeginste, B., Massart, D., Buydens, L., Jong, S.D., Lewi, P., Smeyer-Verbeke, J.: Handbook of Chemometrics and Qualimetrics Part B. Elsevier, Amsterdam (1998)
434. Vapnik, V.: Statistical Learning Theory. Wiley and Sons, New York (1998)
435. Wang, H., Fan, W., Yu, P., Han, J.: Mining concept-drifting data streams using ensemble classifiers. In: Proceedings of the 9th ACM SIGKDD International Conference on Knowledge Discovery and Data Mining, New York, USA, pp. 226–235 (2003)
436. Wang, J., Yu, B., Gasser, L.: Classification visualization with shaded similarity matrices. Tech. rep., Graduate school of Library and Information Science, UIUC (2002)
437. Wang, J.H., Sun, W.D.: Online learning vector quantization: a harmonic competition approach based on conservation network. IEEE Transactions on Systems, Man, and Cybernetics, part B: Cybernetics 29(5), 642–653 (1999)
438. Wang, L.: Fuzzy systems are universal approximators. In: Proc. 1st IEEE Conf. Fuzzy Systems, San Diego, CA, pp. 1163–1169 (1992)
439. Wang, L., Mendel, J.: Fuzzy basis functions, universal approximation and orthogonal least-squares learning. IEEE Transactions on Neural Networks 3(5), 807–814 (1992)
440. Wang, L., Yen, J.: Extracting fuzzy rules for system modelling using a hybrid of genetic algorithm and Kalman filter. Fuzzy Sets and Systems 101(3), 353–362 (1999)
441. Wang, W., Vrbanek, J.: An evolving fuzzy predictor for industrial applications. IEEE Transactions on Fuzzy Systems 16(6), 1439–1449 (2008)
442. Wang, X., Kruger, U., Lennox, B.: Recursive partial least squares algorithms for monitoring complex industrial processes. Control-Engineering-Practice 11(6), 613–632 (2003)
443. Ware, M., Frank, E., Holmes, G., Hall, M., Witten, I.: Interactive machine learning: letting users build classifiers. International Journal of Human-Computer Studies 55(3), 281–292 (2001)
444. Wasserman, P.: Advanced Methods in Neural Computing. Van Nostrand Reinhold, New York (1993)

445. Wei, L.Y., Levoy, M.: Fast texture synthesis using tree-structured vector quantization. In: Proceedings of the 27th Annual Conference on Computer Graphics and Interactive Techniques (SIGGRAPH), New York, pp. 479–488 (2000)
446. Werbos, P.: Beyond regression: New tools for prediction and analysis in the behavioral sciences. Ph.D. thesis, Appl. Math., Harvard University, USA (1974)
447. Widmer, G., Kubat, M.: Learning in the presence of concept drift and hidden contexts. Machine Learning 23(1), 69–101 (1996)
448. Wolpert, D.: No free lunch theorems for optimization. IEEE Transactions on Evolutionary Computation 1(1), 67–82 (1997)
449. Wu, S., Er, M., Gao, Y.: A fast approach for automatic generation of fuzzy rules by generalized dynamic fuzzy neural networks. IEEE Transactions on Fuzzy Systems 9(4), 578–594 (2001)
450. Wu, S., Er, M.J.: Dynamic fuzzy neural networks - a novel approach to function approximation. IEEE Transactions on Systems, Man and Cybernetics, Part B: Cybernetics 30(2), 358–364 (2000)
451. Wu, X., Kumar, V., Quinlan, J., Gosh, J., Yang, Q., Motoda, H., MacLachlan, G., Ng, A., Liu, B., Yu, P., Zhou, Z.H., Steinbach, M., Hand, D., Steinberg, D.: Top 10 algorithms in data mining. Knowledge and Information Systems 14(1), 1–37 (2006)
452. Xiaoan, W., Bellgardt, K.H.: On-line fault detection of flow-injection analysis systems based on recursive parameter estimation. Analytica Chimica Acta 313(3), 161–176 (1995)
453. Xu, L., Schuurmans, D.: Unsupervised and semi-supervised multi-class support vector machines. In: Proceedings of the 20th National Conference on Artificial Intelligence (AAAI 2005), Pittsburgh, Pennsylvania, pp. 904–910 (2005)
454. Yager, R., Filev, D.: Learning of fuzzy rules by mountain clustering. In: Proceedings of the SPIE Conference on Application of Fuzzy Logic Technology, Boston, MA, pp. 246–254 (1993)
455. Yager, R., Filev, D.: Approximate clustering via the mountain method. IEEE Transactions on Systems, Man and Cybernetics 24(8), 1279–1284 (1994)
456. Yager, R.R.: A model of participatory learning. IEEE Transactions on Systems, Man and Cybernetics 20(5), 1229–1234 (1990)
457. Yan, W., Shao, H., Wang, X.: Soft sensing modelling based on support vector machine and baysian model selection. Computers and Chemical Engineering 28(8), 1489–1498 (2004)
458. Yang, M., Wu, K.: A new validity index for fuzzy clustering. In: Proceedings of the IEEE International Conference on Fuzzy Systems, Melbourne, Australia, pp. 89–92 (2001)
459. Ye, J., Li, Q., Xiong, H., Park, H., Janardan, R., Kumar, V.: IDR, QR: An incremental dimension reduction algorithms via QR decomposition. IEEE Transactions on Knowledge and Data Engineering 17(9), 1208–1222 (2005)
460. Yen, J., Wang, L.: Application of statistical information criteria for optimal fuzzy model construction. IEEE Transactions on Fuzzy Systems 6(3), 362–372 (1998)
461. Yen, J., Wang, L.: Simplifying fuzzy rule-based models using orthogonal transformation methods. IEEE Transactions on Systems, Man and Cybernetics, Part B: Cybernetics 29(1), 13–24 (1999)
462. Yen, J., Wang, L., Gillespie, C.: Improving the interpretability of TSK fuzzy models by combining global learning and local learning. IEEE Transactions on Fuzzy Systems 6(4), 530–537 (1998)
463. Yeung, K., Ruzzo, W.: Principal component analysis for clustering gene expression data. Bioinformatics 17(9), 763–774 (2001)

464. Zadeh, L.: Fuzzy sets. Information and Control 8(3), 338–353 (1965)
465. Zadeh, L.: Outline of a new approach to the analysis of complex systems and decision processes. IEEE Transactions Systems, Man and Cybernetics 3, 28–44 (1973)
466. Zadeh, L.: The concept of a linguistic variable and its application to approximate reasoning. Information Sciences 8(3), 199–249 (1975)
467. Zadeh, L.: Fuzzy sets and information granulity. In: Advances in Fuzzy Set Theory and Applications, pp. 3–18 (1979)
468. Zhang, Y.Q.: Constructive granular systems with universal approximation and fast knowledge discovery. IEEE Transactions on Fuzzy Systems 13(1), 48–57 (2005)
469. Zhou, S., Gan, J.: Low-level interpretability and high-level interpretability: a unified view of data-driven interpretable fuzzy systems modelling. Fuzzy Sets and Systems 159(23), 3091–3131 (2008)
470. Zhou, X., Angelov, P.: Real-time joint landmark recognition and classifier generation by an evolving fuzzy system. In: Proceedings of FUZZ-IEEE 2006, Vancouver, Canada, pp. 1205–1212 (2006)
471. Zhou, X., Angelov, P.: Autonomous visual self-localization in completely unknown environment using evolving fuzzy rule-based classifier. In: 2007 IEEE International Conference on Computational Intelligence Application for Defense and Security, Honolulu, Hawaii, USA, pp. 131–138 (2007)
472. Zhu, X.: Semi-supervised learning literature survey. Tech. Rep. TR 1530, Computer Sciences, University of Wisconsin - Madison, Wisconsin, U.S.A. (2008)
473. Zoelzer, U.: Digital Audio Signal Processing. John Wiley & Sons, Chichester (2008)

Index

Printed by Books on Demand, Germany